Fluorescent Probes and Sensors

Special Issue Editor
Sheshanath V. Bhosale

MDPI • Basel • Beijing • Wuhan • Barcelona • Belgrade

MDPI

Special Issue Editor
Sheshanath V. Bhosale
Goa University
India

Editorial Office
MDPI
St. Alban-Anlage 66
Basel, Switzerland

This edition is a reprint of the Special Issue published online in the open access journal *Sensors* (ISSN 1424-8220) from 2017–2018 (available at: http://www.mdpi.com/journal/sensors/special issues/fps).

For citation purposes, cite each article independently as indicated on the article page online and as indicated below:

Lastname, F.M.; Lastname, F.M. Article title. *Journal Name* **Year**, *Article number*, page range.

First Editon 2018

ISBN 978-3-03842-927-2 (Pbk)
ISBN 978-3-03842-928-9 (PDF)

Table of Contents

About the Special Issue Editor

Sheshanath V. Bhosale has worked for several different organizations in various countries that include India, Germany, Switzerland and currently Australia. He has established a very good reputation at the international level, and his exceptional international experience and research excellence have given him the dynamic background needed to make a significant contribution to the scientific community in areas associated with advanced technology. In total he has published 170 peer revived publications, 17 cover page articles and overall citations are > 3750 and 9 book chapters on his credit. His h-index is 29, and i-index 82. This was evaluated by according to Google scholar. Under Dr. Bhosale's supervision supervised 9 PhD's, 1 Master by research student and 4 honors students completed. Currently he is working at Goa University as a UGC-Professor, his research interest is synthesis of small organic molecules with possible applications in nanomaterials, supramolecular chemistry, sensors, artificial photosynthesis and organic solar cells.

sensors

MDPI

Article

The Trace Detection of Nitrite Ions Using Neutral Red Functionalized SH-β-Cyclodextrin @Au Nanoparticles

Xiaoyang Du [1], Xiaoxia Zhang [1], Chunlai Jiang [3], Weilu Zhang [1,*] and Lizhu Yang [2,*]

[1] College of Chemistry & Materials Engineering, Wenzhou University, Wenzhou, Zhejiang 325035, China; 15451283264@stu.wzu.edu.cn (X.D.); 16451283269@stu.wzu.edu.cn (X.Z.)
[2] School of Pharmaceutical Sciences, Wenzhou Medical University, Wenzhou, Zhejiang 325035, China
[3] The Atmospheric Environment Department, Chinese Academy for Environmental Planning, Beijing 100012, China; jiangcl@caep.org.cn
* Correspondence: zwl@wzu.edu.cn (W.Z.); yanglz@wmu.edu.cn (L.Y.); Tel.: +86-138-5771-7511(W.Z.); Tel.: +86-135-8759-9522 (L.Y.)

Received: 28 December 2017; Accepted: 21 February 2018; Published: 25 February 2018

Abstract: A novel fluorescence sensor of NR-β-CD@AuNPs was prepared for the trace detection of nitrite in quantities as low as 4.25×10^{-3} µg·mL^{-1} in an aqueous medium. The fluorescence was due to the host-guest inclusion complexes between neutral red (NR) molecules and gold nanoparticles (AuNPs), which were modified by per-6-mercapto-beta-cyclodextrins (SH-β-CDs) as both a reducing agent and a stabilizer under microwave radiation. The color of the NR-β-CD@AuNPs changed in the presence of nitrite ions. A sensor was applied to the determination of trace nitrites in environmental water samples with satisfactory results.

Keywords: gold nanoparticles; cyclodextrin; neutral red; nitrite ions; fluorescence sensor; trace detection

1. Introduction

Gold nanoparticles (AuNPs) have important applications in the fields of nanoscience and nanotechnology because of their unique optical, electronic, and catalytic properties [1]. First, the distance-dependent surface plasmon resonance (SPR) band of AuNPs makes them vital units for establishing assembly/disassembly modulated colorimetric sensors [2,3]. Second, the high specific surface areas of AuNPs result in their surfaces being modified with multiple ligands [4]. Meanwhile, AuNPs are an ideal energy acceptor in structured fluorescence resonance energy transfer systems (FRET) due to their high extinction coefficient [5–7]. Additionally, the major advantage of AuNPs-based sensors is that the molecular recognition can appear as a color change, which can be easily observed by the naked eye [8]. To date, AuNPs have been applied to the fabrication of assembly/disassembly modulated colorimetric sensors [9], as well as various types of optical [10,11] and electrochemical [12,13] sensors and biosensors [14]. Among them, the interactions of AuNPs with macrocycles such as cyclodextrins, calixarenes, and cucurbiturils [15–17] have received considerable attention for their special and potential properties, for example, the application of resveratrol-stabilized AuNPs in the anticancer field [18].

As a well-known molecular receptor, β-cyclodextrin (β-CD) can form host-guest inclusion complexes with a wide variety of organic, inorganic, and biologic guest molecules in their hydrophobic cavities [19,20]. In parallel, β-CD is water-soluble and environmentally friendly, and is useful in improving the dispersibility of the functional materials [21–24]. On the basis of host-guest interactions, these complexes have been well applied to self-assembly, drug/gene delivery, separation, and sensing applications [15,25]. Considering the unique topological structures that macrocyclic supramolecules possess, several novel3452 properties and corresponding new applications may be presented when β-CD is attached to the surfaces of AuNPs [26]. For example, β-CD-capped AuNPs assembled on ferrocene-functionalized indium tin oxide surfaces were applied to enhance the voltammetric analysis of ascorbic acid [27].

The concentration of NO_2^- is one of the most important parameters in water quality [28]. The maximum allowable amount of nitrite in drinking water is 100 ng·mL^{-1}, according to the regulation of the European Community [29]. The rapid detection of trace concentrations of NO_2^- in water bodies is essential [30,31]. Many analytical methods for the trace detection of nitrite and nitrate have been reported, including colorimetric methods [32,33], fluorometric methods [34], and electrochemical methods [35,36]. However, these methods have limitations such as poor sensitivity, anti-interference, and the use of expensive experimental apparatus. The chemiluminescent methods have proven to be more sensitive and selective in the measurement of nitrite and nitrate [37–42]. Some of the typical methods are summarized in Table S1 (see Supplementary materials). Neutral red (NR) exists in two different prototropic forms in aqueous solutions, namely, the cationic/protonated (NRH$^+$) and neutral (NR) forms, depending on the pH of the solution. NR is a type of dye containing a primary amine structure, which can interact with NO_2^- and lead to fluorescence quenching. Meanwhile, it has been reported that hydroxyls in the cavities of β-CD form inclusion complexes with the nitrogen atoms on heterocyclic molecules of NR [43,44].

In this paper, a sensitive sensor is established for the trace detection of NO_2^- in water because of the observation of a color change. Ultraviolet–visible spectroscopy (UV-Vis), transmission electron microscopy (TEM), and Fourier transform infrared spectroscopy (FT-IR) spectra are explored to understand the quenching interaction and corresponding binding forces. AuNPs modified by SH-β-CD were used as both the reducing agent and stabilizer in this method. Monodispersed β-CD@AuNPs with 10 nm diameters are synthesized in an eco-friendly way, which is different than previous approaches used for the fabrication of β-CD@AuNPs [45,46]. No harsh reagents are used in this method. NR-β-CD@AuNPs were synthesized by host-guest recognition between the β-CD@AuNPs and NR. The host was β-CD@AuNPs, and the guest was NR. The detection of nitrite ions was traced by the diazonium reaction of NO_2^- and the primary amine of NR. The fabrication of the NR-β-CD@AuNP sensor and nitrite detection are shown in Scheme 1.

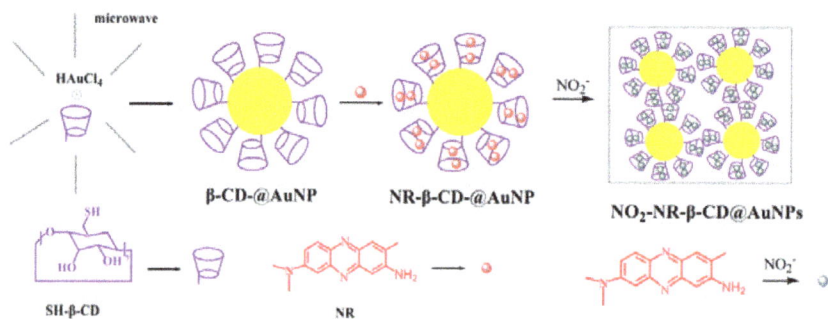

Scheme 1. Schematic representation of the fabrication of the NR-β-CD@AuNP sensor and nitrite detection.

2. Materials and Methods

2.1. Reagents and Materials

Chloroauric acid trihydrate (HAuCl$_4$·3H$_2$O, 99.99%), sodium nitrite (NaNO$_2$, 99.0%), hydrochloric acid (HCl, 36%), borax (Na$_2$B$_4$O$_7$·10H$_2$O, 99.0%), sodium bicarbonate (Na$_2$CO$_3$, 99.0%), neutral red (NR, 4% in water), sodium bicarbonate (NaHCO$_3$, 99.0%), disodium hydrogen phosphate (Na$_2$HPO$_4$, 99.0%), sodium sulfate (Na$_2$SO$_4$, 99.0%), sodium chloride (NaCl, 99.0%), sodium fluoride (NaF, 99.0%), sodium dihydrogen phosphate (NaH$_2$PO$_4$, 99.0%), and sodium nitrate (NaNO$_3$, 99.0%) were purchased from Aladdin Industrial Corporation (Shanghai, China). Per-6-mercapto-beta-cyclodextrin (SH-β-CD, 99.0%) was purchased from Shandong Binzhou Zhiyuan Bio-Technology Co., Ltd (Shandong, China). Other reagents were of analytical grade and directly used without further purification. All solutions were prepared using ultra-pure water (=18.20 MΩ·cm).

2.2. Apparatus

The morphology and the size of products were obtained from TEM, JEM-2100 (JEOL, Tokyo, Japan). The absorption spectra were obtained using a UV-2600 spectrophotometer (SHIMADZU, Tokyo, Japan). The fluorescence spectra were obtained using a FluoroMAX-4-TCSPC detector (HORIBA Jobin Yvon, Paris, France). The AuNPs were prepared with the microwave reactor Discover CEM (CEM, Matthews, NC, USA).

2.3. Preparation of the SH-β-CD Functionalized AuNPs (β-CD@AuNPs)

The β-CD@AuNPs were synthesized by the SH-β-CD reduction of HAuCl$_4$. Briefly, 0.010 g HAuCl$_4$·3H$_2$O and 15.0 mg SH-β-CD were dissolved in 30.0 mL ultra-pure water using an ultrasonication for 5 min. The mixture was stirred for 3 min at 120°C under microwave radiation of 150 W. A suspension of the β-CD@AuNPs characterized by a wine-red color was finally obtained and stored at 4 °C. The reaction was different from previous approaches for the preparation of β-CD@AuNPs because no sodium borohydride was used.

2.4. Preparation of Fluorescence Dye-Incorporated SH-β-CD Functionalized Gold Nanoparticles (NR-β-CD@AuNPs)

In a typical experiment, 5 mL of a NaHCO$_3$-borax buffer solution and 5 mL of NR (5 × 10^{-6} mol·L^{-1}) were added into 5 mL of the β-CD@AuNP solution, and the solution was stirred in a dark environment at room temperature for 80 min. A solution of NR-β-CD@AuNPs was obtained, which became orange-red.

2.5. Detection of Nitrite Ions

NaNO$_2$ (14.99 mg) was dissolved in ultra-pure water to prepare a 100.0 mg·L^{-1} standard solution, which was diluted to the desired concentrations for further use. A NaNO$_2$ standard solution (2.1 mL) and a HCl (1.50 mg·L^{-1}) solution (0.2 mL) were added sequentially into a 5-mL colorimetric tube, followed by the addition of 0.7 mL of the above-prepared NR-β-CD@AuNP solution. Fluorescence spectra were obtained after 5 min.

2.6. Detection of Nitrite Ions in Real Samples

The water samples were obtained from local ponds and Oujiang river (Wenzhou City, China). Then, the samples underwent filtration and centrifugal separation, after which the NR-β-CD@AuNPs were added to the samples, and then the fluorescence spectra were collected.

3. Results

3.1. Characterization of the β-CD@AuNPs

The UV-Vis spectra of the β-CD@AuNPs are shown in Figure 1A. An absorption band at 526 nm indicated the typical feature of the AuNPs and the localized surface plasmon resonance of the dispersed β-CD@AuNPs. The absorption peak was sharper than that prepared from HAuCl$_4$ reduced by sodium citrate (see Supplementary materials, Figure S1A). The color also exhibited a slight variation that can be seen from the inner illustration of Figure 1A. The different colors of the AuNPs obtained using SH-β-CD and sodium citrate may be due to their different sizes and morphologies, as mentioned in Reference [47]. When the surfaces of the AuNPs were decorated with SH-β-CD molecules, they could be employed as scaffolds and energy acceptors for fluorescent sensing by host-guest interactions. The binding of SH-β-CD with AuNPs was verified by comparing the FT-IR spectra between the SH-β-CDs and β-CD@AuNPs, as shown in Figure 1B. The spectrum of SH-β-CD (b) had a band at 1647 cm^{-1} that corresponds to the stretching vibration peak of -C=O. The bands at 1157 and 938 cm^{-1} correspond to the stretching vibration peak of -C-O. The peak at 1590 cm^{-1} of the spectrum of β-CD@AuNPs (a) is the stretching vibration peak of -C=O. The two peaks at approximately 1155 and 1028 cm^{-1} correspond to the stretching vibration peak of -C-O. Moreover, the S-H stretching band at 2576 cm^{-1} of SH-β-CD (b) disappeared in the FT-IR spectrum of β-CD@AuNPs (a), which proved the formation of an Au-S bond, according to References [8] and [26].

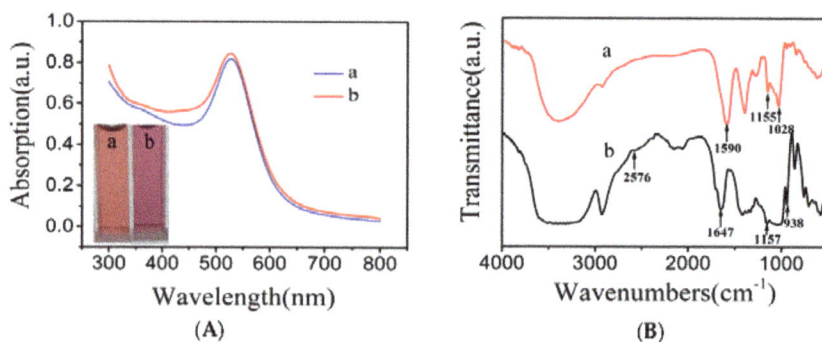

Figure 1. **(A)** UV-Vis spectra of the β-CD@AuNPs (a) and AuNPs (b); **(B)** FT-IR spectra of the β-CD@AuNPs (a) and SH-β-CD (b).

To further confirm their nanostructure and atomic composition, the β-CD@AuNPs were analyzed by transmission electron microscopy (TEM), and the images are shown in Figure 2. The β-CD@AuNPs were nearly spherically shaped with an average size of 10 nm. Energy dispersive spectrometry (EDS) element mappings of β-CD@AuNPs are also shown in Figure 2 using different colors, in which the red and green areas correspond to elemental Au and S, respectively.

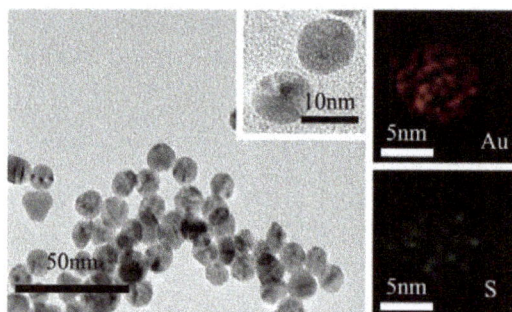

Figure 2. TEM images and energy dispersive spectrometry (EDS) element mappings of the β-CD@AuNPs; the red and green colors correspond to elemental Au and S, respectively.

The preparation of AuNPs from HAuCl$_4$ reduced by sodium citrate was also tested in this paper. Additionally, the morphological characteristics of the TEM images are shown in Figure S1B (see Supplementary materials). Several of the nanoparticles were approximately 10 nm in size and some exhibited an irregular spherical shape.

3.2. Characterization of the NR-β-CD@AuNPs

To demonstrate the potential application of NR-β-CD@AuNPs during the trace detection of NO_2^- in water, the host-guest recognition of β-CD@AuNPs and NR molecules was studied in this paper. Nitrogen heterocyclic molecules reacted with the hydroxyls of β-CD when the guest molecules of NR entered the cavities of the β-CD@AuNPs, used as the host molecules. The AuNPs were designed especially for their signal amplification in this paper. Therefore, a sensor of NR-β-CD@AuNPs exhibits a higher sensitivity than a sensor of both NR and NR-β-CD (see Supplementary materials, Figure S2A). The fluorescence of the NR-β-CD@AuNPs were gradually quenched with the addition of the host molecules of β-CD@AuNPs, and ultimately a stable quenching rate was attained when the volume of the β-CD@AuNPs was 5 mL. The fluorescence spectra are shown in Figure 3A. The relationship between fluorescence intensity and the volume of the β-CD@AuNPs solution is shown in Figure 3B. The fluorescence intensity of the NR-β-CD@AuNPs gradually decreased with an increase in β-CD@AuNPs. The fluorophores entered into the macrocyclic cavities of the β-CD@AuNPs for structure matching by host-guest interactions. As a consequence, the quenching efficiency achieved a constant value when the volume of the β-CD@AuNPs solution reached 5 mL. The average size of the NR-β-CD@AuNPs was 10 nm (see inserted TEM image in Figure 3B) and the dispersion of size was even and comparable to that of the β-CD@AuNPs (shown in Figure 2).

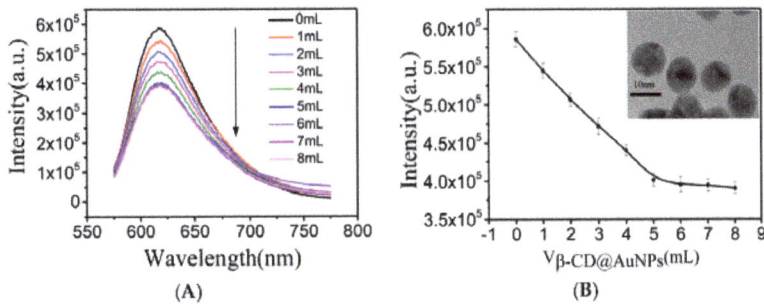

Figure 3. (**A**) Fluorescence spectra of the NR-β-CD@AuNPs when the volume of β-CD@AuNPs was 1, 2, 3, 4, 5, 6, 7, and 8 mL. (**B**) Plots of the NR-β-CD@AuNPs fluorescence intensities versus the volume of the β-CD@AuNPs with error bars. Inserted image: TEM image of the NR-β-CD@AuNPs.

The energy of NR was transferred to the AuNPs through β-CD during the synthesis of NR-β-CD@AuNPs [47]. The quenching constant of K_{sv} was 1.68×10^4 L·mol^{-1}, which was calculated according to the Stern-Volumer equation [48]:

$$F/F_0 = 1 + K_{SV}[Q] \tag{1}$$

where F_0 and F are the fluorescence intensities before and after the addition of β-CD@AuNPs, respectively; K_{sv} is the static quenching constant; and [Q] is the concentration of β-CD@AuNPs.

3.3. Effect of pH on the Fluorescence Property of the NR-β-CD@AuNPs

The prototropic equilibrium shifted from NRH^+ to NR in the cavity of SH-β-CD with a change in the solution pH. NRH^+ was the main form in an acidic aqueous solution with an absorption peak at 530 nm, and NR was the dominant form in a weakly alkaline media with an absorption peak at 450 nm. The fluorescence spectra of the NR-β-CD@AuNPs were changed with various pH values accordingly to the reaction between the NR-β-CD@AuNPs and NO_2^-, which could be clearly monitored by the fluorescence spectra. A diazonium group was formed by the selective reaction between NO_2^- and the primary amine group of NR, which is unstable in weakly acidic and alkaline media, and rapidly converted to another stable form with nitrogen (N_2) released [49]. On the other hand, diazonium salts easily react with surplus aromatic amine groups in NR with a deficiency of NO_2^- during the diazotization reaction. As a result, the acid-base properties of the solution and the concentrations of NO_2^- were the principal factors for the diazo coupling reaction.

Experiments were carried out to explore the fluorescence properties of NR-β-CD@AuNPs at different pH values in the range of 2–9. The effect of pH on the excitation spectra of the NR-β-CD@AuNPs is shown in Figure 4A. The excitation peak appeared at 448 nm when the solution was weakly alkaline, which corresponds to the neutral form of NR. Another excitation peak appeared at 532 nm (pH = 7) due to the increasing amount of the protonated form of NRH^+. Only the excitation peak at 532 nm remained when the aqueous solution was acidic. The corresponding emission peak (red line) shifted from 621 to 627 nm with an increase in the fluorescence intensity in Figure 4B. Because the ground-state pK_a value was 6.8 of NR in water, the critical point of the excitation peak at pH = 7 appeared. Changing from rose-red to purple, the colors of the NR-β-CD@AuNPs were different in weakly alkaline and acidic solutions, as shown in the inserted image of Figure 4B. The TEM image in Figure 4C shows the morphology of the NR-β-CD@AuNPs at pH 5, which is similar to that at pH 9. In the presence of NO_2^-, the fluorescence intensity of the solution clearly decreased compared to that of the NR-β-CD@AuNPs solution, as shown in Figure 4C. The NR-β-CD@AuNPs can detect trace amounts of NO_2^- in acidic to weakly alkaline aqueous solutions. It was demonstrated that NR-β-CD@AuNPs have a broad detection range. The relatively wide detection range of NR-β-CD@AuNPs may have contributed to the structure of β-CD@AuNPs.

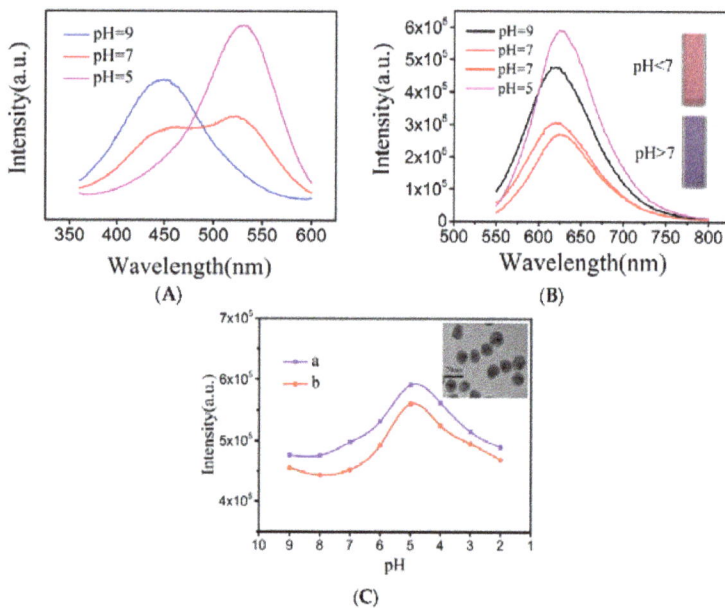

Figure 4. (**A**) Excitation spectra of the NR-β-CD@AuNPs at pH values of 9, 7, and 5; (**B**) Emission spectra of the NR-β-CD@AuNPs at pH values of 9, 7, and 5; (**C**) Fluorescence intensity of the NR-β-CD@AuNPs (a) and the NR-β-CD@AuNPs in the presence of NO_2^- (100 μg·L^{-1}) (b); Inserted image: TEM image of the NR-β-CD@AuNPs at a pH value of 5.

3.4. The Detection of NO_2^- in an Aqueous Solution

To investigate the detection sensitivity of NR-β-CD@AuNPs to NO_2^- in broad ranges, experiments were designed in both weakly alkaline and acidic media. When the solution was weakly alkaline, the fluorescence intensity of the NR-β-CD@AuNPs at 623 nm was clearly gradually quenched with an increasing concentration of NO_2^- ($[NO_2^-]$), as shown in Figure 5A. A linear relationship ($R^2 = 0.998$) was obtained between the fluorescence intensity and $[NO_2^-]$ in the range of 0.0–0.9 $\mu g \cdot mL^{-1}$. When $[NO_2^-]$ exceeded 0.9 $\mu g \cdot mL^{-1}$, the rate of fluorescence quenching reached 100%, and the color changed accordingly. The regression equation was $F = 473893 - 539242C$, where F represents the fluorescence intensity of the solution, and C represents $[NO_2^-]$ (see Supplementary materials, Figure S3A). The detection limit was as low as 5.78×10^{-3} $\mu g \cdot mL^{-1}$, which was calculated as follows: the blank solution was measured 11 times, and its standard deviation was multiplied by 3 and divided by the slope of the linear relationship. The fluorescence quenching was static because the non-fluorescent diazonium groups were produced, and K_{sv} was 9.8×10^4 $L \cdot mol^{-1}$, as calculated by Equation (1). The experiments under acidic conditions were performed in the same way as those performed under weakly alkaline conditions, apart from the employed pH values, and the results are shown in Figure 5B. The regression equation was $F = 574156 - 673222C$ (see Supplementary materials, Figure S3B), and the detection limit was 4.25×10^{-3} $\mu g \cdot mL^{-1}$, which is better than that of the weakly alkaline conditions. The K_{sv} was 2.1×10^5 $L \cdot mol^{-1}$, as determined by Equation (1).

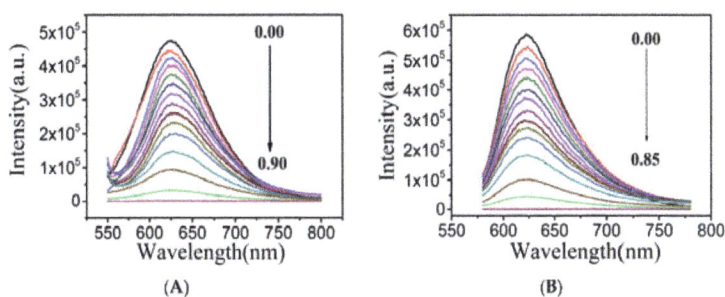

Figure 5. (**A**) Emission spectra of the NR-β-CD@AuNPs in the presence of different $[NO_2^-]$ concentrations, including 0, 0.05, 0.10, 0.15, 0.20, 0.25, 0.30, 0.35, 0.40, 0.45, 0.50, 0.60, 0.70, 0.80, and 0.90 $\mu g \cdot mL^{-1}$ under a weakly alkaline medium; (**B**) Emission spectra of the NR-β-CD@AuNPs in the presence of different $[NO_2^-]$ concentrations including 0, 0.05, 0.10, 0.15, 0.20, 0.25, 0.30, 0.35, 0.40, 0.45, 0.50, 0.60, 0.70, 0.80, and 0.85 $\mu g \cdot mL^{-1}$ under an acid medium.

As shown in Figure 6, the colorimetric response was recorded. It was obvious that the color changed from light purple to light blue, and could be observed by the naked eye, when $[NO_2^-]$ was approximately 0.30 $\mu g \cdot mL^{-1}$.

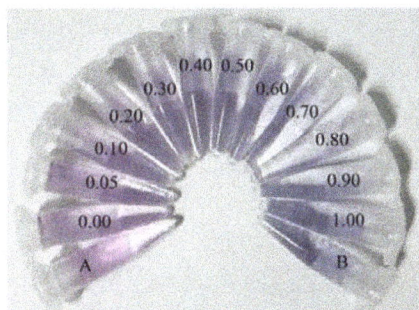

Figure 6. Photograph of the NR-β-CD@AuNPs in the presence of different $[NO_2^-]$ concentrations including 0, 0.05, 0.10, 0.20, 0.30, 0.40, 0.50, 0.60, 0.70, 0.80, 0.90, and 1.00 $\mu g \cdot mL^{-1}$; A: 0 $\mu g \cdot mL^{-1}$ of $[NO_2^-]$ and competing ions; and B: 1.00 $\mu g \cdot mL^{-1}$ of $[NO_2^-]$ and competing ions.

Both NR and NR-β-CD could be used to detect NO_2^- based on our results (see Supplementary materials, Figure S4), and the detection limit was 0.56 µg·mL^{-1} and 5.6×10^{-2} µg·mL^{-1}, respectively. The solutions must be under a strongly acidic condition of pH 1 for higher detection limits. The NR-β-CD@AuNP sensor exhibited a good sensitivity of 5.78×10^{-3} µg·mL^{-1}.

The diazonium group between NO_2^- and the primary amine group of NR was more stable in an acid solution [49]. Compared with other sensors [50,51], this sensor displayed a wide detection range and good sensitivity. Some of the typical methods are summarized in Table S1 (see Supplementary materials).

After reacting with NO_2^-, the product of NO_2-NR-β-CD@AuNPs was analyzed by UV-Vis spectra (Figure 7). The UV-Vis spectrum of the NR-β-CD@AuNPs (a) had two absorption bands at 520 nm and 450 nm, which correspond to the two states of NR. There were two new absorption bands appearing at 583 nm and 349 nm for the NO_2-NR-β-CD@AuNPs. The color was also different when the NR-β-CD@AuNPs reacted with NO_2^-, as shown in Figure 7.

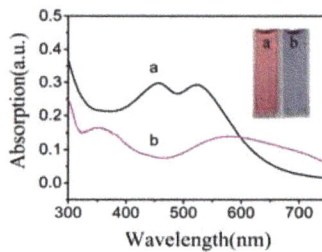

Figure 7. UV-Vis spectra of the NR-β-CD@AuNPs (a) and NO_2-NR-β-CD@AuNPs (b).

3.5. Selectivity

A 100-fold concentration of the other common ions were selected as competing ions in order to demonstrate the selectivity of this sensor, including Cl^-, CO_3^{2-}, HCO_3^-, F^-, SO_4^{2-}, $H_2PO_4^-$, HPO_4^{2-}, and NO_3^-. Meanwhile, $[NO_2^-]$ of 0.35 µg·mL^{-1} was measured as the control group. As shown in Figure 8, the fluorescence quenching only occurred at 623 nm in the presence of NO_2^-, even though it contained other ions. These results emphasized the high selectivity of the fluorescence sensor for the trace detection of NO_2^-.

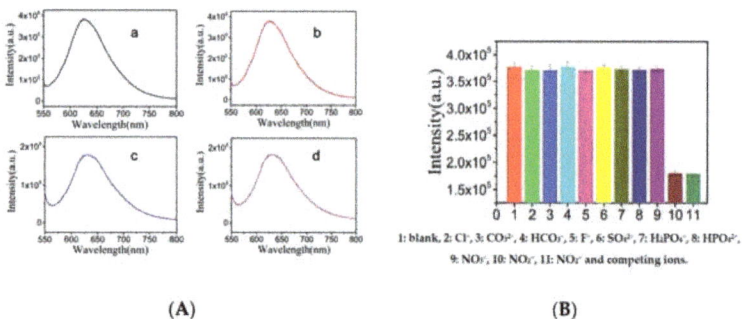

(A)   (B)

Figure 8. (A) Fluorescence spectra of the (a) NR-β-CD@AuNPs; (b) NR-β-CD@AuNPs and competing ions (Cl^-, CO_3^{2-}, HCO_3^-, F^-, SO_4^{2-}, $H_2PO_4^-$, HPO_4^{2-} and NO_3^-); (c) NR-β-CD@AuNPs and NO_2^-; and (d) NR-β-CD@AuNPs, NO_2^- and competing ions. (B) Fluorescence intensity of the NR-β-CD@AuNPs and 1: blank, 2: Cl^-, 3: CO_3^{2-}, 4: HCO_3^-, 5: F^-, 6: SO_4^{2-}, 7: $H_2PO_4^-$, 8: HPO_4^{2-}, 9: NO_3^-, 10: NO_2^-, 11: NO_2^- and competing ions.

3.6. Application of NO_2^- Detection in Real Samples

The sensor was used to detect NO_2^- in the waters of a local river or pond to determine the realistic efficiency of an NR−β−CD@AuNP sensor. Recovery experiments were carried out on samples by adding nitrite ion standards [52]. The data are listed in Table 1, and the method has a recovery of 98.6−102.5% with a relative standard deviation (RSD) of less than 3%. The result indicated that the sensor was reliable for detecting NO_2^- in outdoor waters. It is confirmed that the NR−β−CD@AuNPs could be available for the detection of NO_2^- in real samples.

Table 1. Results of the detection of nitrite ions in river (1) and pond (2) water (n = 6).

Samples	Content (NO_2^-, $\mu g \cdot mL^{-1}$)	Added (NO_2^-, $\mu g \cdot mL^{-1}$)	Found (NO_2^-, $\mu g \cdot mL^{-1}$)	Recovery (%)	RSD (%)
1	0.23±0.01	0.10	0.33±0.01	101.00±1.5	1.20±0.5
2	0.32±0.02	0.10	0.42±0.01	99.80±1.2	2.10±0.7

4. Conclusions

A fluorescence sensor was fabricated by modifying β−CD@AuNPs with NR for the trace detection of NO_2^-. The optional condition for the sensor was an acidic aqueous solution, and the detection limit was as low as 4.25×10^{-3} $\mu g \cdot mL^{-1}$. This sensor can selectively recognize NO_2^- through a visual color change from light purple or pink to light blue when the $[NO_2^-]$ concentration is 0.30 $\mu g \cdot mL^{-1}$. When $[NO_2^-]$ exceeded 0.9 $\mu g \cdot mL^{-1}$, the rate of fluorescence quenching reached 100%, and the color changed. The sensor was applied to the detection of NO_2^- in local waters with a low detection limit, wide linear concentration range, good reproducibility, and anti−interference ability.

Supplementary Materials: The following are available online at www.mdpi.com/1424-8220/18/3/681/s1. Figure S1: (A) UV-Vis spectrum of AuNPs; (B) TEM image of AuNPs, Figure S2: (A) Emission spectra of NR (a) NR-β-CD@AuNPs (b) and NR-β-CD(c); (B) TEM images of NO_2-NR-β-CD., Figure S3: The regression equation in a weakly alkaline (A) and an acidic medium (B)., Figure S4: (A) Emission spectra of NR in the presence of different $[NO_2^-]$, including 0, 0.10, 0.20, 0.30, 0.40, 0.50 and 0.60 $\mu g \cdot mL^{-1}$. (B) Emission spectra of NR-β-CD in the presence of different $[NO_2^-]$, including 0, 0.20, 0.40, 0.60, 0.80, 1.00, 1.20, 1.40, and 1.6 $\mu g \cdot mL^{-1}$. Table S1: Comparison of the fabricated sensor with other reported sensors for nitrite ions.

Acknowledgments: This work was supported by the Projects of Wenzhou Science and Technology Bureau (S20150012) and (W20170006), the Science Foundation of Zhejiang Province (LY13B040002), and the National Natural Science Foundation of China (No. 31300819).

Author Contributions: Weilu Zhang and Lizhu Yang conceived and designed the experiments. Xiaoyang Du performed the experiments and analyzed the data. Xiaoxia Zhang contributed the reagents and materials. Chunlai Jiang validated the experiments. Weilu Zhang and Xiaoyang Du wrote the paper.

Conflicts of Interest: The authors declare no conflict of interest.

References

1. Daniel, M.C.; Astruc, D. Gold nanoparticles: assembly, supramolecular chemistry, quantum−size−related properties, and applications toward biology, catalysis, and nanotechnology. *Chem. Rev.* **2004**, *104*, 293–346. [CrossRef] [PubMed]

2. Saha, K.; Agasti, S.S.; Kim, C.; Li, X.; Rotello, V.M. Gold nanoparticles in chemical and biological sensing. *Chem. Rev.* **2012**, *112*, 2739–2779. [CrossRef] [PubMed]

3. Kelly, K.L.; Coronado, E.; Zhao, L.L.; Schatz, G.C. The optical properties of metal nanoparticles: The influence of size, shape, and dielectric environment. *J. Phys. Chem. B* **2003**, *107*, 668–677. [CrossRef]

4. Rana, S.; Bajaj, A.; Mout, R.; Rotello, V.M. Monolayer coated gold nanoparticles for delivery applications. *Adv. Drug Delivery Rev.* **2012**, *64*, 200–216. [CrossRef] [PubMed]

5. Yang, X.; Yang, M.; Pang, B.; Vara, M.; Xia, Y. Gold nanomaterials at work in biomedicine. *Chem. Rev.* **2015**, *115*, 10410–10488. [CrossRef] [PubMed]

6. Sapsford, K.E.; Algar, W.R.; Berti, L.; Gemmill, K.B.; Casey, B.J.; Oh, E.; Stewart, M.H.; Medintz, I.L. Functionalizing nanoparticles with biological molecules: developing chemistries that facilitate nanotechnology. *Chem. Rev.* **2013**, *113*, 1904–2074. [CrossRef] [PubMed]

7. Ray, P.C.; Fan, Z.; Crouch, R.A.; Sinha, S.S.; Pramanik, A. Nanoscopic optical rulers beyond the FRET distance limit: fundamentals and applications. *Chem. Soc. Rev.* **2014**, *43*, 6370–6404. [CrossRef] [PubMed]

8. Liu, C.W.; Lian, J.Y.; Liu, Q.; Xu, C.L.; Li, B.X. β−Cyclodextrin−modified silver nanoparticles as colorimetric probes for the direct visual enantioselective recognition of aromatic α−amino acids. *Anal. Methods* **2016**, *8*, 5794–5800. [CrossRef]

9. Liu, J.; Lu, Y. Preparation of aptamer−linked gold nanoparticle purple aggregates for colorimetric sensing of analytes. *Nat. Protoc.* **2006**, *1*, 246–252. [CrossRef] [PubMed]

10. Deeb, C.; Zhou, X.A.; Gerard, D.; Bouhelier, A.; Jain, P.K.; Plain, J.; Soppera, O.; Royer, P.; Bachelott, R. Off−resonant optical excitation of gold nanorods: nanoscale imprint of polarization surface charge distribution. *J. Phys. Chem. Lett.* **2011**, *2*, 7–11. [CrossRef] [PubMed]

11. Stender, A.S.; Wang, G.F.; Sun, W.; Fang, N. Influence of gold nanorod geometry on optical response. *ACS Nano* **2010**, *4*, 7667–7675. [CrossRef] [PubMed]

12. Guo, S.J.; Wang, E.K. Synthesis and electrochemical applications of gold nanoparticles. *Anal. Chim. Acta* **2007**, *598*, 181–192. [CrossRef] [PubMed]

13. Serafin, V.; Eguilaz, M.; Agui, L.; Yanez−Sedeno, P.; Pingarron, J.M. An electrochemical immunosensor for testosterone using gold nanoparticles–carbon nanotubes composite electrodes. *Electroanalysis* **2011**, *23*, 169–176. [CrossRef]

14. Pingarron, J.M.; Yanez−Sedeno, P.; Gonzalez−Cortes, A. Gold nanoparticle−based electrochemical biosensors. *Electrochim. Acta* **2008**, *53*, 5848–5866. [CrossRef]

15. Dsouza, R.N.; Pischel, U.; Nau, W.M. Fluorescent dyes and their supramolecular host/guest complexes with macrocycles in aqueous solution. *Chem. Rev.* **2011**, *111*, 7941–7980. [CrossRef] [PubMed]

16. Yang, Y.W. Towards biocompatible nanovalves based on mesoporous silica nanoparticles. *Med. Chem. Commun.* **2011**, *2*, 1033–1049. [CrossRef]

17. Sun, Y.L.; Yang, B.J.; Zhang, X.A.; Yang, Y.W. Cucurbit [7] uril pseudorotaxane−based photo responsive supramolecular nanovalve. *Chem. Eur. J.* **2012**, *18*, 9212–9216. [CrossRef] [PubMed]

18. Mohanty, R.K.; Thennarasu, S.; Mandal, A.B. Resveratrol stabilized gold nanoparticles enable surface loading of doxorubicin and anticancer activity. *Colloids Surf. B* **2014**, *114*, 138–143. [CrossRef] [PubMed]

19. Palanisamy, S.; Sakthinathan, S.; Chen, S.M.; Thirumalraj, B.; Wua, T.H.; Loub, B.S.; Liuc, X.H. Preparation of β−cyclodextrin entrapped graphite composite for sensitive detection of dopamine. *Carbohydr. Polym.* **2016**, *135*, 267–273. [CrossRef] [PubMed]

20. Freeman, R.; Finder, T.; Bahshi, L.; Willner, I. β−cyclodextrin−modified CdSe/ZnS quantum dots for sensing and chiroselective analysis. *Nano Lett.* **2009**, *9*, 2073–2076. [CrossRef] [PubMed]

21. Wayu, M.B.; Schwarzmann, M.A.; Gillespie, S.D.; Leopold, M.C. Enzyme−free uric acid electrochemical sensors using β−cyclodextrin−modified carboxylic acid−functionalized carbon nanotubes. *J. Mater. Sci.* **2017**, *52*, 6050–6062. [CrossRef]

22. Du, D.; Wang, M.H.; Cai, J.; Zhang, A.D. Sensitive acetylcholinesterase biosensor based on assembly of β−cyclodextrins onto multiwall carbon nanotubes for detection of organophosphates pesticide. *Sens. Actuators B Chem.* **2010**, *146*, 337–341. [CrossRef]

23. Pourjavadi, A.; Eskandari, M.; Hosseini, S.H.; Nazari, M. Synthesis of water dispersible reduced graphene oxide via supramolecularcomplexation with modified β−cyclodextrin. *Int. J. Polym. Mater. Polym. Biomater.* **2017**, *66*, 235–242. [CrossRef]

24. Abbaspour, A.; Noori, A. A cyclodextrin host–guest recognition approach to an electrochemical sensor for simultaneous quantification of serotonin and dopamine. *Biosens. Bioelectron.* **2011**, *26*, 4674–4680. [CrossRef] [PubMed]

25. Chen, Y.; Liu, Y. Construction and function of cyclodextrin−based 1D supramolecular strands and their secondary assemblies. *Adv. Mater.* **2015**, *27*, 5403–5409. [CrossRef] [PubMed]

26. Guo, Y.Q.; Zhao, Y.M.; Lu, D.T.; Wu, H.J.; Fan, M.; Wei, Y.L.; Shuang, S.M.; Dong, C. β−Cyclodextrin functionalized gold nanoparticles: Characterization and its analytical application for L−tyrosine. *J. Inclusion Phenom. Macrocyclic Chem.* **2014**, *78*, 275–286. [CrossRef]

27. Luo, C.H.; Zheng, Z.H.; Ding, X.B.; Peng, Y.X. Supramolecular assembly of β−cyclodextrin−capped gold nanoparticles on ferrocene−functionalized ITO surface for enhanced voltammetric analysis of ascorbic acid. *Electroanalysis* **2008**, *20*, 894–899. [CrossRef]

28. Kodamatania, H.; Yamazakib, S.; Saitoc, K.; Tomiyasua, T.; Komatsud, Y. Selective determination method for measurement of nitrite and nitrate in water samples using high−performance liquid chromatography with post−column photochemical reaction and chemiluminescence detection. *J. Chromatogr. A* **2009**, *1216*, 3163–3167. [CrossRef] [PubMed]

29. Beamonte, E.; Bermudez, J.D.; Casino, A. A statistical study of the quality of surface water intended for human consumption near Valencia (Spain). *J. Environ. Manag.* **2007**, *83*, 307–314. [CrossRef] [PubMed]

30. Ito, K.; Takayama, Y.; Makabe, N.; Mitsui, R.; Hirokawa, T. Ion chromatography for determination of nitrite and nitrate in seawater using monolithic columns. *J. Choromatogr. A* **2005**, *1083*, 63–67. [CrossRef]

31. Zuo, Y.G.; Wang, C.J.; Van, T. Simultaneous determination of nitrite and nitrate in dew, rain, snow and lake water samples by ion−pair high−performance liquid chromatography. *Talanta* **2006**, *70*, 281–285. [CrossRef] [PubMed]

32. Zhang, H.; Qi, S.D.; Dong, Y.L. A sensitive colorimetric method for the determination of nitrite in water supplies, meat and dairy products using ionic liquid−modified methyl red as a colour reagent. *Food Chem.* **2014**, *151*, 429–434. [CrossRef] [PubMed]

33. Aydın, A.; Ercan, Ö.; Taşçıoğlu, S. A novel method for the spectrophotometric determination of nitrite in water. *Talanta* **2005**, *66*, 1181–1186. [CrossRef] [PubMed]

34. Wang, L.L.; Li, B.; Zhang, L.M.; Zhang, L.G.; Zhao, H.F. Fabrication and characterization of a fluorescent sensor based on Rh 6G−functionlized silica nanoparticles for nitrite ion detection. *Sens. Actuators B Chem.* **2012**, *171–172*, 946–953. [CrossRef]

35. Ojani, R.; Raoof, J.B.; Zarei, E. Electrocatalytic reduction of nitrite using ferricyanide; Application for its simple and selective determination. *Electrochim. Acta* **2006**, *52*, 753–759. [CrossRef]

36. Paixão, T.R.L.C.; Cardoso, J.L.; Bertotti, M. Determination of nitrate in mineral water and sausage samples by using a renewable in situ copper modified electrode. *Talanta* **2007**, *71*, 186–191. [CrossRef] [PubMed]

37. Lu, C.; Lin, J.M.; Huie, C.W.; Yamada, M. Chemiluminescence study of carbonate and peroxynitrous acid and its application to the direct determination of nitrite based on solid surface enhancement. *Anal. Chim. Acta* **2004**, *510*, 29–34. [CrossRef]

38. Pelletier, M.M.; Kleinbongard, P.; Ringwood, L.; Hito, R.; Hunter, C.J.; Schechter, A.N.; Gladwin, M.T.; Dejam, A. The measurement of blood and plasma nitrite by chemiluminescence: Pitfalls and solutions. *Free Radic. Biol. Med.* **2006**, *41*, 541–548. [CrossRef] [PubMed]

39. He, D.Y.; Zhang, Z.J.; Huang, Y.; Hu, Y.F. Chemiluminescence microflow injection analysis system on a chip for the determination of nitrite in food. *Food Chem.* **2007**, *101*, 667–672. [CrossRef]

40. Mikuška, P.; Večeřa, Z. Chemiluminescent flow−injection analysis of nitrates in water using on−line ultraviolet photolysis. *Anal. Chim. Acta* **2002**, *474*, 99–105. [CrossRef]

41. Mikuška, P.; Večeřa, Z. Simultaneous determination of nitrite and nitrate in water by chemiluminescent flow−injection analysis. *Anal. Chim. Acta* **2003**, *495*, 225–232. [CrossRef]

42. Zhang, T.; Fan, H.L.; Jin, Q.H. Sensitive and selective detection of nitrite ion based on fluorescence super quenching of conjugated polyelectrolyte. *Talanta* **2010**, *81*, 95–99. [CrossRef] [PubMed]

43. Singh, M.K.; Pal, H.; Koti, A.S.R.; Sapre, A.V. Photophysical properties and rotational relaxation dynamics of neutral red bound to β−cyclodextrin. *J. Phys. Chem. A* **2004**, *108*, 1465–1474. [CrossRef]

44. Mohanty, J.; Bhasikuttan, A.C.; Nau, W.M.; Pal, H. Host−guest complexation of neutral red with macrocyclic host molecules: Contrasting pK_a shifts and binding affinities for cucurbit [7] uril and β−cyclodextrin. *J. Phys. Chem. B* **2006**, *110*, 5132–5138. [CrossRef] [PubMed]

45. Li, H.; Chen, D.X.; Sun, Y.L.; Zheng, Y.B.; Tan, L.L.; Weiss, P.S.; Wang, Y.Y. Viologen−mediated assembly of and sensing with carboxylatopillar [5] arene−modified gold nanoparticles. *J. Am. Chem. Soc.* **2013**, *135*, 1570–1576. [CrossRef] [PubMed]

46. Huang, T.; Meng, F.; Qi, L. Facile synthesis and one−dimensional assembly of cyclodextrin−capped gold nanoparticles and their applications in catalysis and surface−enhanced Raman scattering. *J. Phys. Chem. C* **2009**, *113*, 13636–13642. [CrossRef]

47. Zhao, Y.; Huang, Y.C.; Zhu, H.; Zhu, Q.Q.; Xia, Y.S. Three−in−One: Sensing, Self−Assembly, and Cascade Catalysis of Cyclodextrin Modified Gold Nanoparticles. *J. Am. Chem. Soc.* **2016**, *138*, 16645–16654. [CrossRef] [PubMed]

48. Eftink, M.R.; Ghiron, C.A. Anal Biochem, Fluorescence quenching studies with proteins. *Anal. Biochem.* **1981**, *114*, 199–227. [CrossRef]

49.	Liu, Y.L.; Kang, N.; Ke, X.B.; Wang, D.; Ren, L.; Wang, H.J. A fluorescent nanoprobe based on metal−enhanced fluorescence combined with Förster resonance energy transfer for the trace detection of nitrite ions. *RSC Adv.* **2016**, *6*, 27395–27403. [CrossRef]

50.	Adarsh, N.; Shanmugasundaram, M.; Ramaiah, D. Efficient reaction based colorimetric probe for sensitive detection, quantification, and on−site analysis of nitrite ions in natural water resources. *Anal. Chem.* **2013**, *85*, 10008–10012. [CrossRef] [PubMed]

51.	Chen, J.H.; Pang, S.; He, L.L.; Nugen, S.R. Highly sensitive and selective detection of nitrite ions using Fe$_3$O$_4$@SiO$_2$/Au magnetic nanoparticles by surface−enhanced Raman spectroscopy. *Biosens. Bioelectron.* **2016**, *85*, 726–733. [CrossRef] [PubMed]

52.	Huang, X.; Li, Y.X.; Chen, Y.L.; Wang, L. Electrochemical determination of nitrite and iodate by use of gold nanoparticles/poly(3−methylthiophene) composites coated glassy carbon electrode. *Sens. Actuators B Chem.* **2008**, *134*, 780–786. [CrossRef]

sensors

MDPI

Article

A Red-Emitting, Multidimensional Sensor for the Simultaneous Cellular Imaging of Biothiols and Phosphate Ions [†]

Pilar Herrero-Foncubierta [1,2], Jose M. Paredes [1], Maria D. Giron [3], Rafael Salto [3], Juan M. Cuerva [2], Delia Miguel [1] and Angel Orte [1,*]

[1]   Department of Physical Chemistry, Faculty of Pharmacy, University of Granada, Campus Cartuja, 18071 Granada, Spain; pilarhf@ugr.es (P.H.-F.); jmparedes@ugr.es (J.M.P.); dmalvarez@ugr.es (D.M.)
[2]   Department of Organic Chemistry, Faculty of Sciences, University of Granada, C. U. Fuentenueva s/n, 18071 Granada, Spain; jmcuerva@ugr.es
[3]   Department of Biochemistry and Molecular Biology, Faculty of Pharmacy, University of Granada, Campus Cartuja, 18071 Granada, Spain; mgiron@ugr.es (M.D.G.); rsalto@ugr.es (R.S.)
*   Correspondence: angelort@ugr.es; Tel.: +34-958-243-825
†   This paper is dedicated to Prof. Jose M. Alvarez-Pez for his retirement.

Received: 17 November 2017; Accepted: 2 January 2018; Published: 9 January 2018

Abstract: The development of new fluorescent probes for cellular imaging is currently a very active field because of the large potential in understanding cell physiology, especially targeting anomalous behaviours due to disease. In particular, red-emitting dyes are keenly sought, as the light in this spectral region presents lower interferences and a deeper depth of penetration in tissues. In this work, we have synthesized a red-emitting, dual probe for the multiplexed intracellular detection of biothiols and phosphate ions. We have prepared a fluorogenic construct involving a silicon-substituted fluorescein for red emission. The fluorogenic reaction is selectively started by the presence of biothiols. In addition, the released fluorescent moiety undergoes an excited-state proton transfer reaction promoted by the presence of phosphate ions, which modulates its fluorescence lifetime, τ, with the total phosphate concentration. Therefore, in a multidimensional approach, the intracellular levels of biothiols and phosphate can be detected simultaneously using a single fluorophore and with spectral clearing of cell autofluorescence interferences. We have applied this concept to different cell lines, including photoreceptor cells, whose levels of biothiols are importantly altered by light irradiation and other oxidants.

Keywords: dual probes; fluorescent sensors; fluorescence lifetime imaging; FLIM; cellular stress; photoreceptor cells

1. Introduction

Intracellular sensing by using fluorescent probes is a well-established approach to monitor relevant biological processes at the cellular level. Understanding cellular function in terms of metabolism, differentiation, homeostasis, gene expression, or inter-cellular communication is a major aim for many interdisciplinary research branches, since these processes may be strongly affected by pathological states, such as neurodegenerative diseases or cancer [1,2]. Therefore, a thorough understanding of how diseases work at the molecular and cellular levels will provide an invaluable background to set the basis for new therapeutic tools. Nevertheless, the actual quantification of analytes or metabolites of interest has not been a trivial problem since the early days of immunofluorescence experiments [3]. The widespread use of fluorescent protein mutants [4] and advanced microscopy techniques, such as super-resolution nanoscopy [5], has resulted in a substantial boost of fluorescence-based cellular sensing.

An additional stimulating challenge in intracellular sensing is the capability of multiplexing; this is the simultaneous measurement of more than one parameter, which can provide important information on the correlation between cellular events and cause-effect relations. For the quantification of a single analyte, a widely accepted approach is based on ratiometric methods that use spectral separation in two channels for reconstructing ratio pseudo-images. However, intracellular multiplexing using ratiometric methods would require at least four different excitation/emission channels that make it a challenging problem that suffers from numerous complications and the lack of robustness, which limits their use. A powerful alternative to ratiometric methods for intracellular sensing is fluorescence lifetime imaging microscopy (FLIM) [6]. A group of fluorescent molecules that have been promoted to the excited state by a pulse of light will emit fluorescence, usually following exponential decay kinetics, whose decay rate (k) defines the so-called fluorescence lifetime (τ) as $\tau = 1/k$, normally on the order of a few nanoseconds. FLIM microscopy exhibits many unique advantages for quantitative sensing compared to ratiometric fluorescence imaging, especially in terms of removing the contribution of cellular autofluorescence, a problem that may cause systematic errors in ratiometric methods [7]. Several FLIM-based intracellular sensors have been reported recently regarding the quantification of different analytes such as pH [8,9] or calcium [10], as well as other physical parameters such as temperature [11] or microviscosity [12,13]. Our research group has recently concentrated efforts on the development of different FLIM-based intracellular sensors [14,15]. Specifically, our thorough studies on the excited-state proton transfer (ESPT) reactions of xanthene derivatives, mediated by the presence of suitable proton acceptor/donor pairs [16–18], led us to propose a FLIM methodology and a family of sensors for the intracellular quantification of the total phosphate ions concentration [19,20]. The presence of a suitable proton donor/acceptor, such as the pair $H_2PO_4^-$ and HPO_4^{2-}, does promote an inter-molecular proton transfer to the prototropic species of xanthene dyes. When this transfer occurs sufficiently rapid to compete with the fluorescence emission, it results in excited-state dynamics, thus altering the fluorescence emission properties of the dye [21,22]. The main parameters that define the ESPT reaction are: the total concentration of the proton donor/acceptor, the pH, and the excited-state acidity of the dye. Although this is not a specific feature of phosphate and other pairs can promote the reaction, such as acetate [16,23] and certain amino acids [24], not all the buffers are capable of promoting the ESPT reaction [25]. Interestingly, the family of xanthene derivatives so-called Tokyo Green dyes [26], which have the main feature of an off acidic prototropic form, showed fluorescence decay traces that were mostly mono-exponential when undergoing the buffer-mediated ESPT reaction, and the decay time of such fluorescence kinetics was dependent on the total phosphate concentration [17]. Based on these results, we started the development of fluorescent phosphate sensors in which the analytical parameter was the fluorescence lifetime of the dye, with particular usefulness for intracellular sensing using FLIM microscopy [19]. Phosphate ions are ubiquitous in many important processes such as energy storage, signal transduction, a myriad of phosphorylation/dephosphorylation reactions, and other processes such as osteoblast differentiation and bone deposition. Hence, the intracellular detection and quantification of these ions is a relevant matter for understanding cellular physiology [27–29].

Interestingly, the emission of fluorescent photons by electronically excited molecules is a multilayered, or multidimensional phenomenon. This process is defined by many different parameters that are orthogonal to each other, including the excitation and emission energy (colours), the emission intensity, the emission efficiency (quantum yield), the fluorescence lifetime (τ), and the polarization of the emitted light. This multidimensionality makes the fluorescence spectroscopy an extremely versatile tool, offering varied sources of information on the studied systems. Recently, we have taken advantage of the multidimensional character of the fluorescence emission process to develop an actual intracellular multiplexed sensing approach using a single dye sensor [30]. Our methodology consisted of a fluorogenic xanthene-based sensor that is reactive towards biological thiols, widely present in living cells but directly related to the presence of oxidative stress. Due to the high sensitivity of fluorescence techniques, the fluorogenic approach has been widely employed in biothiol sensing [31]. The novelty of our method is that the fluorescent molecule cleaved after the reaction with biothiols was a carefully selected xanthene derivative whose fluorescence lifetime value was dependent on the total phosphate ions concentration, which could be obtained through FLIM imaging. Therefore, by focusing on the fluorescence emission intensity, the thiol levels were accessible, whereas by inspecting the fluorescence lifetime of the release dye, the phosphate concentration was estimated. Other studies have shown the possibility of fluorogenic intracellular sensors for FLIM microscopy [32,33]; however, these sensors respond to a single analyte of interest. In contrast, our work exemplifies a novel and elegant use of multidimensional information to simultaneously report on two different analytes using a single sensor. In fact, the simultaneous estimation of these two parameters represents an invaluable tool to study dysfunctional cellular statuses, such as in obesity and diabetes, that exhibit alterations in bone metabolism with increased oxidative stress [34].

One of the advantages of FLIM microscopy for intracellular imaging is the possibility of discarding all interferences of cellular autofluorescence by applying a time-gated filtering method, by which the short-lived photons coming from cellular species are removed, leaving in the image only those photons arising from the species of interest [6]. However, this approach is more efficient when fluorophores exhibiting a long lifetime are employed. Unfortunately, one of the drawbacks of our xanthene derivatives to date is their spectral overlap with the cellular autofluorescence and their short lifetimes (usually <4 ns). This problem was addressed by including a lifetime contribution representing the cell autofluorescence and a second contribution for the specific dye. Although majorly corrected, some of the autofluorescence photons may be misplaced, causing an apparent decrease in the estimated lifetime, which leads to slightly overestimating the total phosphate concentration.

It is well known that the spectral region in which the cell autofluorescence is practically negligible is in the red and near-infrared (NIR) regions. Hence, the development of NIR fluorescent probes is currently a very active research field [35]. One of the most striking alternatives to achieve red-emitting dyes is the insertion of silicon atoms within the conjugated π moiety of the dye's core. Silicon-substituted rhodamines [36,37], fluoresceins [38,39], and other xanthenes [40] have been reported in the literature as redshifted fluorophores for bioimaging probes. Interestingly, silicon-substituted fluoresceins, so-called Tokyo magenta (TM) dyes, still maintain the ability of undergoing an ESPT reaction mediated by the presence of the phosphate species present at a near-neutral pH [18]. A thorough investigation of the photophysics of 7-hydroxy-5,5- dimethyl-10-(*o*-tolyl)dibenzo[*b*,*e*]silin-3(5*H*)-one (2-Me TM), a dye of the Tokyo magenta family, described the kinetics of the excited-state reaction between the prototropic species of this dye and the $H_2PO_4^-$/HPO_4^{2-} pair as the proton donor/acceptor, reporting the values of all the kinetic rate constants involved in the reaction [18]. This reaction causes the fluorescence lifetimes of 2-Me TM to be dependent on the total phosphate concentration. Although this would be a primary condition to suggest this dye as an intracellular FLIM sensor of phosphate ions, 2-Me TM does not exhibit an on/off behaviour in its prototropic equilibrium, which results in complex bi-exponential decay kinetics of the fluorescence emission. This bi-exponential fluorescence decay dramatically hinders the usefulness of the dye as a FLIM sensor. Therefore, a clear step forward in the development of redshifted FLIM phosphate sensors entails the design of silicon-substituted xanthenes with on/off prototropic schemes.

In this work, we have synthesized a red-emitting, dual probe for multiplexed intracellular detection of biothiols and phosphate ions based on a fluorogenic construct involving silicon-substituted fluorescein, whose acidic form is eminently non-fluorescent and whose basic form is highly fluorescent. These prototropic features lead to virtually mono-exponential fluorescence decay traces, which can be phosphate-sensitive owing to the ESPT reaction. All these characteristics combine to represent a step forward with respect to our previous dual probe [30] because of the spectral clearing of cell autofluorescence interferences, which should provide a better response towards both biothiols and phosphate ions.

2. Materials and Methods

2.1. Synthesis: General Aspects

All reactions were performed in dry glassware and an air atmosphere. All commercially available solvents (dry dichloromethane (CH_2Cl_2), methanol (MeOH)) and reagents (4-bromo-3-methylanisole, *tert*-butyl litium solution, hydrochloric acid (HCl), 4-(dimethylamine)pyridine (DMAP) and 2,4-dinitrobenxenesulfonyl chloride) were used without further purification. For reactions involving organolithium derivatives, freshly distilled tetrahydrofuran (THF) in presence of Na wires and benzophenone was used. Thin-layer chromatography analysis was performed on aluminium-backed plates coated with silica gel 60 (230–240 mesh) with an F254 indicator. The spots were visualized with ultraviolet (UV) light ($\lambda = 254$ nm) and stained with phosphomolybdic acid solution and subsequent heating. NMR spectra were collected at room temperature at 400 MHz for ^{1}H NMR and 101 MHz for ^{13}C-NMR. Carbon multiplicities were assigned by DEPT techniques. HRMS were carried out by atmospheric-pressure chemical ionization (APCI+) or electrospray ionization (ESI).

The details of all the synthetic protocols, spectroscopic data and copies of NMR spectra of new compounds are shown in Appendix A.

2.2. Instrumentation

Absorption spectra were collected on a Lambda 650 UV-visible spectrophotometer (PerkinElmer, Waltham, MA, USA). Fluorescence emission spectra and kinetics were obtained on a Jasco FP-8300 spectrofluorimeter (Jasco, Tokyo, Japan), at the excitation wavelength λ_{ex} of 530 nm. Fluorescence quantum yields were obtained using Rhodamine 101 as a reference, as an average of 12 independent measures, of two different concentrations of the probe, two different concentrations of the reference, and three different excitation wavelengths.

Images of the fluorescence emission intensities and fluorescence lifetimes were recorded on a MicroTime 200, fluorescence-lifetime microscope system (PicoQuant GmbH, Berlin, Germany). The excitation source consisted of a pulsed laser diode head pulsed laser at $\lambda = 530$ nm (LDH-P-FA-530B, PicoQuant, Berlin, Germany), operated by a PDL-800 driver (PicoQuant, Berlin, Germany) at a repetition rate of 20 MHz. The light beam was directed onto a dichroic mirror (Z532RDC, Chroma, Bellows Falls, VT, USA) to the oil immersion objective ($100\times$, 1.4 NA) of an inverted microscope system (IX-71, Olympus, Tokyo, Japan). The fluorescence emission was directed to a 550-nm long-pass filter (AHF analysentechnik AG, Tübingen, Germany) and focused to a 75-µm pinhole. The fluorescence then passed through a bandpass filter (D630/60M, Chroma, Bellows Falls, VT, USA) and focused into a single-photon avalanche diode (SPCM-AQR 14, PerkinElmer). Imaging reconstruction, photon counting, and data acquisition were realized with a TimeHarp 200 TCSPC module (PicoQuant, Berlin, Germany). Raw images were obtained at a 512×512-pixel resolution over an area of 80×80 µm^2.

2.3. Cell Culture and Lysates

The human hepatocellular carcinoma HepG2 (ATCC no. HB-8065™) and MC3T3-E1 preosteoblasts (ECACC 99072810) cell lines were provided by the Cell Culture Facility, University of Granada. The mouse retinal cone-cell line 661W is a transformed cell line derived from mouse retinal tumours and was a gift from Dr. Enrique de la Rosa (CIB, CSIC, Madrid, Spain). HepG2 and 661W cells were grown at 37 °C in Dulbecco's modified Eagle's medium (DMEM) supplemented with 10% (v/v) fetal bovine serum (FBS), 2 mM glutamine plus 100 U/mL penicillin, and 0.1 mg/mL streptomycin. MC3T3-E1 cells were grown in alpha minimum essential medium (α-MEM) with ribonucleosides, deoxyribonucleosides, 2 mM L-glutamine and 1 mM sodium pyruvate, 10% (v/v) FBS and 100 U/mL penicillin, and 0.1 mg/mL streptomycin.

For the FLIM microscopy experiments, the HepG2, MC3T3-E1 preosteoblasts and retinal cone 661W cell lines were seeded onto circular coverslips (diameter of 25 mm) in six-well plates at a density of 2.3×10^5 cells per well.

For cell lysates, MC3T3-E1 cells were seeded at a density of 1×10^6 cells/well in a p100 plate and incubated at 37 °C for 24 h to reach a cell confluence of 80–90%. Cells were washed twice with Hepes buffer 10 mM, pH 7.4, and scrapped in the same buffer. Cells were centrifuged at $800 \times g$ for 10 min and the pellet of cells was resuspended in 100 μL of the Hepes buffer. The cell suspension was sonicated and centrifuged at $12,000 \times g$ for 10 min. The supernatant was diluted 1:200, in the absence or the presence of 6×10^{-4} M of N-methylmaleimide (NMM), and DNBS-2Me-4OMe-TM was added to a final concentration of 6×10^{-7} M for measuring the fluorogenic response to cell lysates.

2.4. FLIM Imaging Experiments

For both HepG2 and preosteoblasts cell lines, the reaction of DNBS-2Me-4OMe-TM with intracellular biothiols was initially followed. First, cells were washed twice with Krebs-Ringer buffer solution (118 mM NaCl, 5 mM KCl, 1.2 mM MgSO$_4$, 1.3 mM CaCl$_2$, 1.2 mM KH$_2$PO$_4$, 30 mM 4-(2-hydroxyethyl)-1-piperazineethanesulfonic acid (HEPES), at pH 7.4). The cover was then placed in the holder, and 1 mL of Krebs-Ringer buffer was added. Once the sample holder was placed in a microscope, the cellular autofluorescence was measured. The buffer solution was then removed and 1 mL of a 6×10^{-7} M solution of DNBS-2Me-4OMe-TM in the Krebs-Ringer buffer was added. Intensity images were collected every 5 min.

In a second step, once the biothiols had been measured, the fluorescence lifetime variation of the released 2Me-4OMe-TM with different amounts of phosphate ions was also analysed in the same cell lines. Since the plasma membrane is impermeable to phosphate ions, 1 μg/mL of α-toxin from *Staphylococcus Aureus* was added to the cells and incubated for 20 min. After this time, solutions of different total phosphate concentrations (10, 20, 30, 50 and 100 mM) were added. For the preparation of the phosphate buffer solutions, the individual phosphate species (NaH$_2$PO$_4$·H$_2$O and Na$_2$HPO$_4$·7H$_2$O; both from Fluka, puriss, p.a.) were mixed, and the pH was adjusted to 7.35. After adding each concentration, a FLIM image of the cell was collected.

The cellular stress in photoreceptor cells was promoted by a preincubation with H$_2$O$_2$ (0-1 mM) for 12 h. Then, a similar protocol was used to measure the reaction of DNBS-2Me-4OMe-TM with biothiols. For the samples incubated with either 0.75 or 1 mM H$_2$O$_2$ cell death was observed before the FLIM measurements; thus, only results from the control (absence), and 0.25 and 0.5 mM concentrations are described.

All the data were analysed using the SymPhotime 32 (PicoQuant) package and the ImageJ distribution, FIJI [41]. Before the analysis, a 5×5 spatial binning was performed, for a final pixel size of 0.78×0.78 μm^2, in order to increase the number of photons per pixel, to gain statistical robustness in the fitting of the fluorescence decay traces. The 2Me-4OMe-TM FLIM imaging was performed by fitting the fluorescence decay traces in each pixel to a bi-exponential decay function, using a reconstructed instrument response function (IRF) for the deconvolution analysis based on the maximum likelihood estimator. A short decay time of 1.5 ns was kept fixed to account for the contribution of the cell autofluorescence. The long decay time was an adjustable parameter, whose value was phosphate-dependent and assigned to the released 2Me-4OMe-TM.

3. Results

To achieve the objectives showed in the introduction, our new probe must be sensitive to both biothiol and phosphate concentrations in an independent but simultaneous way and exhibit emission in the red spectral range. To attain the dual sensing features, we previously used a highly fluorescent ON-OFF xanthene-based dye modified with a 2,4-dinitrobenzenesulfonic group (DNBS), susceptible to suffering a thiolysis reaction by biothiols present in the medium [30]. On the other hand, the substitution of the oxygen in the xanthone by a silicon atom can promote a bathochromic shift in the emission wavelength. The first precedent for this reaction was described by Maeda and colleagues [42]. In a first step, the sulphur nucleophile attacks at the aromatic ring thus releasing the fluorescent probe, sulphur dioxide and the (2,4-dinitrophenyl)biothiol sulphide (Scheme 1). Nevertheless, it is also known that, after the reaction, the (2,4-dinitrophenyl)biothiol sulphide can be hydrolysed in water, recovering a thiol group. This adds certain catalytic activity to the complex model of reaction. Thus, our working hypothesis consists of a weakly fluorescent compound that, in the presence of biothiols, will release a highly fluorescent dye, which in turn will respond to the phosphate ion concentration during its fluorescence lifetime (Scheme 1). In this sense, we prepared DNBS-2Me-4OMe-TM by the reaction of 2,4-dinitrobenzenesulfonyl chloride with the silicon-substituted xanthene 2Me-4OMe-TM, which was in turn prepared by a nucleophilic reaction of in situ prepared 2-methyl-4-methoxy-lithiobenzene to the corresponding silicon-substituted xanthone (See Appendix A for more details).

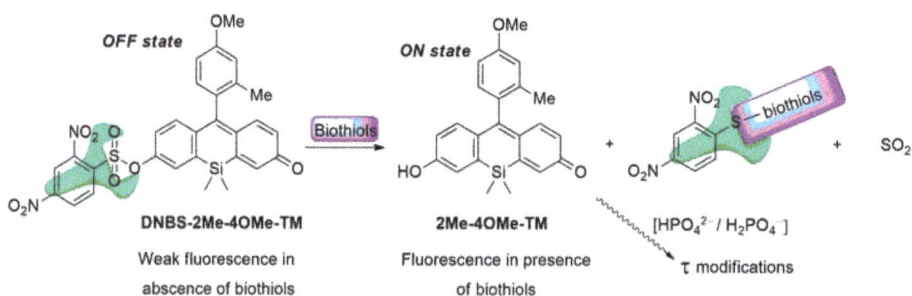

Scheme 1. Working hypothesis of DNBS-2Me-4OMe-TM as a dual probe sensor for biothiols and phosphate ions (at a near neutral pH).

Once we have synthesized DNBS-2Me-4OMe-TM, we tested the response towards glutathione (GSH), the most abundant source of thiols in live cells. Although free cysteine and homo-cysteine may be found in the cell cytoplasm, their relative abundance is usually lower than 10% that of GSH; hence, we focused our experiments in solution on the response of the probe to GSH. Once the DNBS-2Me-4OMe-TM probe is dissolved in the presence of GSH, the emission intensity shows a time-dependent increase (Figure 1a) due to the release of 2Me-4OMe-TM. To quantify the extent of the fluorogenic reaction, we employed the area under the curve (AUC) of the emission intensity at the emission maximum, 597 nm, with λ_{ex} = 530 nm. We also obtained other spectroscopic features of both the DNBS-2Me-4OMe-TM and the 2Me-4OMe-TM to ensure the fluorogenic nature of the reaction. Whereas the DNBS-2Me-4OMe-TM exhibited absorbance and emission values within background levels, at the experimental conditions, the 2Me-4OMe-TM dye showed an absorbance maximum at 583 nm, with molar absorptivity of 165,300 M^{-1} cm^{-1}, and a fluorescence quantum yield of 0.44 ± 0.03. We then tested the response of the DNBS-2Me-4OMe-TM probe to different concentrations of GSH (Figure 1b). Using as a reference the AUC of a molar ratio [GSH]/[probe] = 1, the decrease to a ratio of 0.75 caused a 19% decrease in the signal (representing a signal of 0.81). Likewise, a [GSH]/[probe] molar ratio of 0.5 resulted in a decrease in the reference signal of 51% (representing a signal of 0.49). This means that linearity is well fulfilled in such concentration ranges. However, when the GSH concentration was further decreased, the response was no longer linear, reaching a background limiting value. This suggests that a potential interfering reaction may be causing the hydrolysation of the probe, and the release of 2Me-4OMe-TM.

We then performed a study of the stability and selectivity of the DNBS-2Me-4OMe-TM probe (Figure 1c). Using as a reference the AUC of the reaction with GSH, we first explored the stability of the probe, in the absence of thiols, in aqueous solution at pH 7.0 and 4.5 (the latter being the pH in the cellular lysosomes). Importantly, we found release of 2Me-4OMe-TM to a certain extent (20.6% of the signal with GSH) at pH 7.0, whereas the DNBS-2Me-4OMe-TM was totally stable at pH 4.5, exhibiting no fluorescence increase. Given that the DNBS group is similar to those employed for photo-uncaging, fluorogenic reactions, such as α-carboxy nitroveratryl [43,44] and other *o*-nitrobenzyl protecting groups [45], one could think on the spontaneous photolysis of the DNBS group as the cause for the release of 2Me-4OMe-TM at pH 7.0. However, the photolytic uncaging of such groups requires light in the near-UV spectral region, below 420 nm. Hence, it is not likely that a lower energy radiation, such as that used in our experiments (530 nm), was capable of producing the photo-uncaging effect. Despite this, we also explored the effect of white light irradiation of the DNBS-2Me-4OMe-TM probe, and found an increase in the fluorescence emission of 30.1% (referred to the presence of GSH). Therefore, photolysis of the DNBS group, caused by external light, may have certain weight on the stability of the off probe DNBS-2Me-4OMe-TM. Moreover, we studied other potential interferences that could cause a fluorogenic response of the probe, such as oxidants (H_2O_2) and nucleophile amino acids (alanine, Ala; and serine, Ser). The amino acids showed a response similar to that of the probe alone at pH 7.0, whereas H_2O_2 caused a hydrolysation similar to that of the irradiated sample (30.2%), meaning that these are not causing any additional interference than that already in place. We also found that N-methylmaleimide (NMM) exhibited a protective effect, so that the fluorogenic dye was less hydrolysed at pH 7.0.

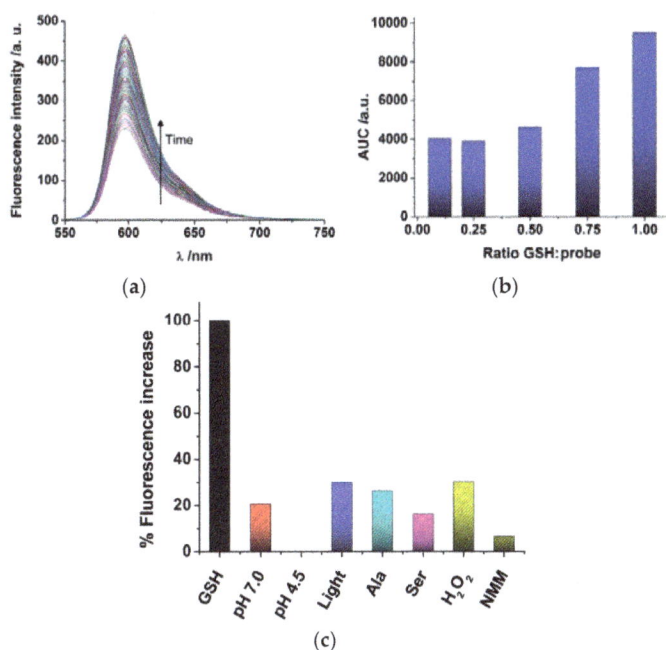

Figure 1. (a) Temporal evolution of the fluorescence emission spectra (λ_{ex} = 530 nm) of DNBS-2Me-4OMe-TM (6×10^{-7} M) in the presence of a stoichiometric amount of glutathione (GSH). (b) Area under the curve (AUC) of the response of DNBS-2Me-4OMe-TM (6×10^{-7} M) with different concentrations of GSH. (c) Stability and selectivity study of the fluorogenic reaction of DNBS-2Me-4OMe-TM (6×10^{-7} M).

Finally, before applying the probe to the intracellular environment, we investigated the behaviour of the DNBS-2Me-4OMe-TM fluorogenic reaction in cell lysate, blocking the GSH and other thiols with NMM. With this experiment, we can identify other sources of interference in the cell cytoplasm. We obtained that when the thiols were blocked with NMM the AUC of the release of 2Me-4OMe-TM was 24% of that response in cell lysate in the absence of NMM. This means that the endogenous thiols in the cell extract were effectively blocked, and that other potential factors of interference have a similar effect to those already described above. Hence, the experiment confirms the selectivity of the probe towards thiols, although not yet totally specific.

To test the intracellular performance of the DNBS-2Me-4OMe-TM dual probe, it was added to the extracellular medium of HepG2 and preosteoblast cells, resulting in the spontaneous and fast incorporation of the probe inside the cells. Immediately, the thiolysis reaction started releasing the fluorescent moiety 2-Me-4OMe-TM, thus increasing the intensity with time, as seen in Figure 2a,c, where images of the two cell lines are represented (additional examples can be found in Figure A1 in Appendix B). As observed, the fluorescence emission exhibits higher values and faster kinetics in HepG2 cells than it does in the preosteoblasts, suggesting higher levels of biothiols in the former. Figure 2b,d represent the kinetics of the average intensity using different images from different experiments of HepG2 and preosteoblast cell lines, respectively.

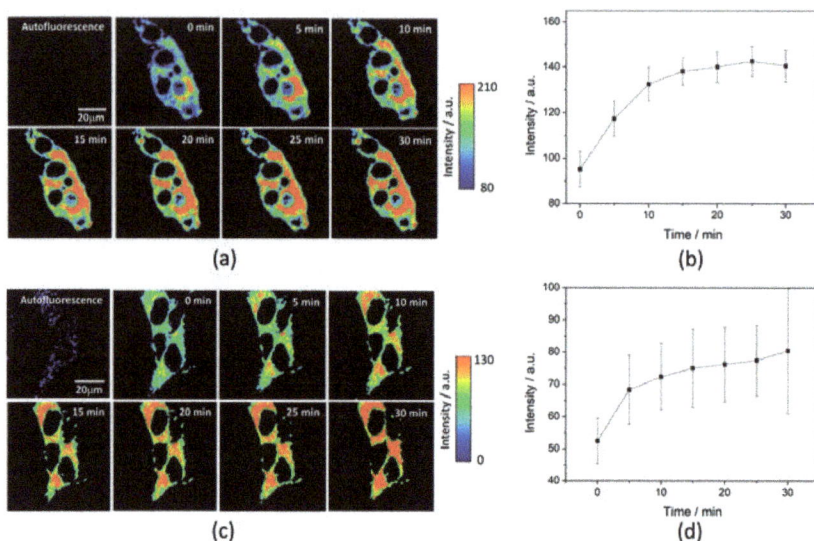

Figure 2. Fluorogenic reaction of DNBS-2Me-4OMe-TM (6×10^{-7} M) in response to intracellular biothiols. (**a**) Representative example of fluorescence intensity imaging with time in HepG2 cells, and (**b**) the emission intensity averaged over the pixels in the cytoplasm of 10 repeated experiments containing several cells. Error bars represent the standard errors. (**c**) Representative example of fluorescence intensity imaging with time in the preosteoblasts, and (**d**) the emission intensity averaged over the pixels in the cytoplasm of five repeated experiments. Error bars represent the standard errors.

Simultaneously, and by focusing on the fluorescence lifetime, τ, of the released 2Me-4OMe-TM, the intracellular phosphate concentration can be estimated. As it is described by the theory of buffer-mediated ESPT reactions [21,22], the fluorescence lifetime is dependent on the presence of an adequate proton/donor acceptor as the phosphate buffer. For this ESPT reaction to occur, the 2Me-4OMe-TM must undergo an acid-base equilibrium with a pK_a value that, optimally, matches that of the proton donor/acceptor buffer. Indeed, we obtained that the 2Me-4OMe-TM has a neutral, non-emissive form, whose molar absorptivity was 12,300 M^{-1} cm^{-1} at 499 nm, and a ground-state pK_a of 7.26 ± 0.03. To test the capability of the released 2Me-4OMe-TM to report on the intracellular phosphate ions concentration, we opened small pores in the cell membrane using α-toxin, allowing small molecules (such as phosphate ions) to enter and exit freely from the medium to the cell without any loss of high molecular weight cellular compounds. Next, we obtained FLIM images using different extracellular buffers with different phosphate concentrations. Figure 3a,c show FLIM images of the released 2Me-4OMe-TM in HepG2 and preosteoblast cell lines at different phosphate concentrations. Using the arbitrary colour scale (from 4.6 to 3.7 ns) to represent the fluorescence lifetime, it is possible to directly notice a decrease in the τ values with the addition of phosphate. We repeated the experiment with 10 (for HepG2) and five (for preosteoblasts) different sets of cells (additional examples can be found in Figure A2 in Appendix B), to obtain the average τ values as a function of the total phosphate concentration (Figure 3b,d). The fluorescence lifetime exhibits a marked decrease with phosphate concentration up to approximately 50 mM, as the ESPT reaction theory predicts. Higher phosphate concentrations did not cause any further decrease in the τ values. These results combined with those obtained in the fluorogenic thiolysis of DNBS-2Me-4OMe-TM support the capability of this probe to simultaneously determine amounts of both biothiols and phosphate anions in cell media.

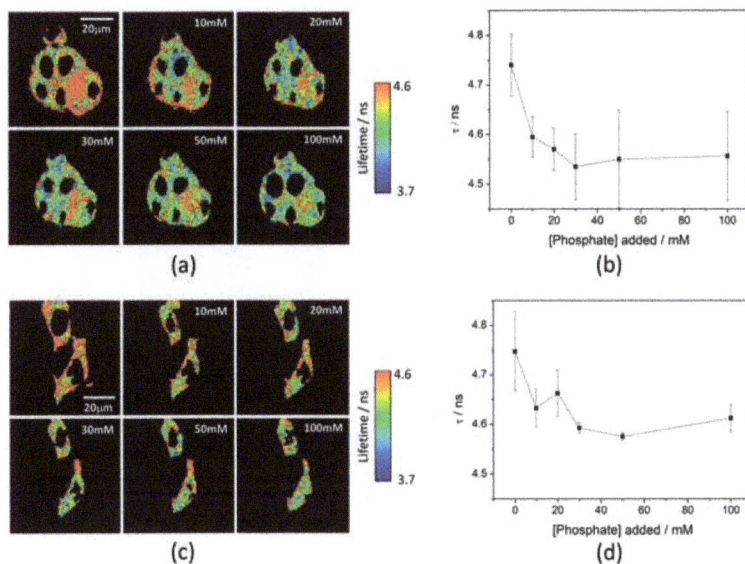

Figure 3. Phosphate response of the released 2Me-4OMe-TM in fluorescence lifetime imaging microscopy (FLIM) imaging. (**a**) Representative FLIM images of released 2Me-4OMe-TM in HepG2 cells, incubated with α-toxin for 20 min at different extracellular phosphate concentrations, and (**b**) the corresponding fluorescence lifetime τ in the cytoplasm pixels averaged over 10 repeated experiments. Error bars represent the standard errors. (**c**) Representative FLIM images of released 2Me-4OMe-TM in preosteoblast cells, incubated with α-toxin for 20 min at different extracellular phosphate concentrations, and (**d**) the corresponding fluorescence lifetime τ in the cytoplasm pixels averaged over four repeated experiments. Error bars represent the standard errors.

Finally, we tested the probe in the 661W cell line, previously used as model of the photoreceptor cells [46]. Multiple diseases related with the degeneration of the retina are caused by high oxidative stress generated in the photoreceptor cells. As an indirect application to detect cellular stress through biothiols, we incubated 661W cells for 12 h with different concentrations of hydrogen peroxide (H_2O_2). To fight against the cellular stress generated by H_2O_2, these cells synthesize higher concentrations of glutathione and other biothiols. This behaviour can be observed in the illustrative examples given in Figure 4a in which the fluorogenic reaction of DNBS-2Me-4OMe-TM is shown in the control cells (without H_2O_2 addition), and in those incubated with 0.25 mM and 0.5 mM of H_2O_2. These images clearly illustrate that not only is the intensity increase faster, but also it achieves a higher value in the presence of H_2O_2. In Figure 4b, the intensity curves at different reaction times (averaged over different experiments) are shown. Finally, Figure 4c represents the AUC calculated from all the measurements from Figure 4b. As is observed, the AUC values can be related to the presence of intracellular biothiols generated by cellular stress.

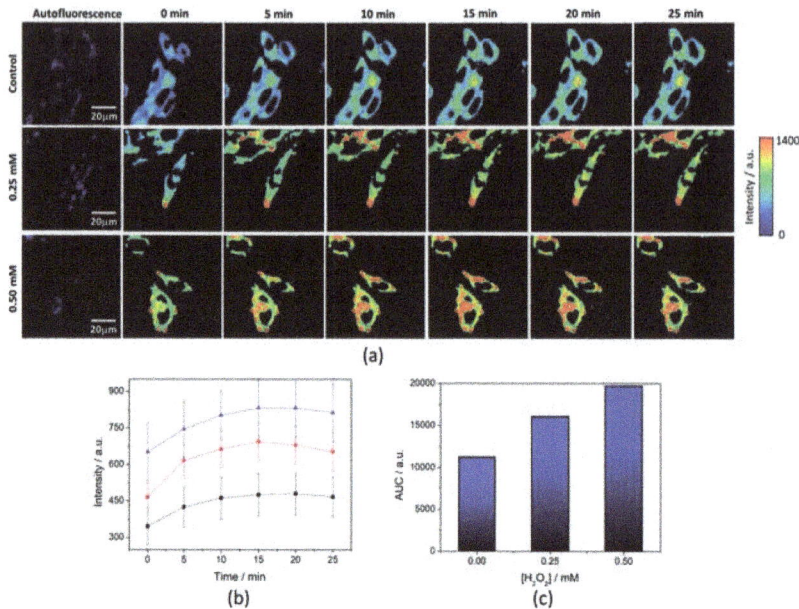

Figure 4. (a) Fluorescence intensity images of DNBS-2Me-4OMe-TM (6×10^{-7} M) in 661W cells after 12 h incubation with 0, 0.25, and 0.50 mM of H_2O_2, and (b) the corresponding emission intensity averaged over the pixels in the cytoplasm of at least four repeated experiments containing several cells (H_2O_2 concentration: 0 mM, black; 0.25 mM, red; 0.50 mM, blue). Error bars represent the standard errors. (c) AUC values for the data are shown in panel (b).

4. Discussion and Conclusions

The simultaneous detection of different analytes is an important concept in the study of live cell physiology and pathology because it allows the direct correlation of homeostatic responses as well as the cause-effect relations. Indeed, multiplex detection in live cells is a key approach to the study of physiological pathways and metabolic mechanisms in which studied analytes are involved. In this work, we have tested the effectiveness as a dual probe for the simultaneous detection of intracellular phosphate and biothiols. Following a recent methodology [30], a new compound was designed to combine some appropriate characteristics for the analysis in live cells. First, the use of red fluorescence (λ_{em} = 595 nm) as an analytical signal has great interest in biological measurements because it allows minimizing the interferences that come from natural green autofluorescence. The redshift is caused by the incorporation of a Si atom instead of O in position 10 of a xanthene moiety [38,40]. Second, the fluorophore was specifically designed to preserve a fast and spontaneous incorporation inside the cells due to its low hydrophilic character [18,19,26]. As biothiol detection involves the incorporation of the dinitrobenzenesulfonate (DNBS) group to the fluorophore, we checked that the increase in size of the molecule and the incorporation of the Si atom do not alter the natural membrane permeability of this dye. The acquired images demonstrate excellent cell incorporation, as shown in Figures 2–4.

The sensor that we describe herein can act as a real-time, dual probe. The simultaneous detection is based on the measurement of two different parameters; for biothiols, we used the increase in intensity due to the thiolysis of the DNBS-2Me-4OMe-TM releasing the high fluorescent moiety 2Me-4OMe-TM [30]. In contrast, the phosphate determination is based on the changes in the fluorescence lifetimes due to an ESPT reaction promoted by the phosphate ions [18,21]. The presence of phosphate in the medium makes the ESPT reaction faster than the fluorescence process and consequently, it produces a decrease in the fluorescence lifetime. This decrease is dependent on the phosphate concentration and allows us to determine the intracellular phosphate flux.

To asseverate that the compound responds correctly to both analytes, we introduced it into two different cell lines (HepG2 and MC3T3-E1 cells). We monitored the fluorescence intensity increase mediated by biothiols through the thiolysis reaction. As seen in Figure 2, DNBS-2Me-4OMe-TM has a sensitive response to the intrinsic biothiols, avoiding potential issues of cellular autofluorescence due to spectral separation. As previously described, the sensor is based in the quenching produced by the DNBS group by photoinduced electron transfer [42]. A well-described thiolysis reaction can release the attached fluorophore, producing an increase in fluorescence. This strategy has been previously employed in other thiol probes [42,47–50]. Similar to other DNBS-based probes, DNBS-2Me-4OMe-TM presents a rapid intracellular response, where biothiol levels can be assessed in approximately 20 min. using a semi-quantitative approach through the AUC method. The AUC method of the fluorescence kinetics has become an adequate parameter to estimate the biothiol levels synthesized in cells due to oxidative stress. Although several studies demonstrated the intracellular use as biothiol sensors of other DNBS-based molecules [47,50,51], these studies did not perform an intracellular semi-quantitative estimation of biothiol levels.

Once the fluorogenic reaction was finished, with the use of α-toxin from *Staphylococcus aureus*, we generated membrane pores that allow the diffusion of low molecular weight molecules, such as phosphates, and we added increasing phosphate concentrations to the medium, thus reaching an equilibrium between the intra- and extracellular phosphate concentrations. Fluorescence lifetime images in Figure 3 were acquired at different phosphate concentrations, and they show that the fluorescence lifetime of the released 2Me-4OMe-TM is sensitive to intracellular phosphate concentrations, as the ESPT reaction theory predicts [21]. The obtained τ values exhibit a slight decrease with the phosphate concentration. Although this decrease is enough to determine the intracellular phosphate concentration, its dependency is less marked than that obtained with other xanthenic derivatives.

This multiplex detection of biothiols jointly with the phosphate ions, based on the multiparametric character of the fluorescence emission, is the distinctive feature of our approach. To the best of our knowledge, only our previous biothiol/phosphate probe is reported [30] in the literature, but as a novelty, in this work, we have focused on the optimization for an actual intracellular application. For this purpose, we have shifted the fluorescence from green to red and reduced the time to measure intracellular biothiols from 60 to 20 min., and the lifetime sensitivity has been improved at a low phosphate concentration (approximately 0.2 ns with the addition of 30 mM of phosphate). However, at a higher phosphate concentration (approximately >50 mM), the probe loses its sensitivity. The main reason behind this lower response may come from the extra hydrophobicity that the silicon atom confers to the molecule. This effect may induce strong interactions with the intracellular membranous compartments, keeping the molecule protected from the phosphate proton donors/acceptors that would promote the ESPT reaction. We are currently investigating the effect of microheterogeneous systems, such as micelles, to the phosphate-mediated ESPT reaction of the 2Me-4OMe-TM dye. These studies will shed some light on the excited-state behaviour of the dye and will allow us to have a better control over the ESPT reaction, thus providing invaluable information for a rational design of further modifications to red-emitting silicon-fluoresceins as phosphate sensors.

To consolidate the multiplexing approach of our method, we have analysed the lifetime images of the experiments shown in Figures 2 and A1. These images show the increase in the fluorescence intensity with time upon release of the 2Me-4OMe-TM dye. When we analysed the associated FLIM images, we obtained average lifetimes of the 2Me-4OMe-TM dye in the images of HepG2 and MC3T3-E1 cells. By interpolating these values into the calibration curves depicted in Figure 3, we quantified the intracellular phosphate levels to be 1.9 ± 1.1 mM in the HepG2 cells and 4.2 ± 2.1 mM in the MC3T3-E1 preosteoblasts. These values are in good agreement with the typical concentration of free phosphate ions in the cell cytoplasm, dynamically kept around 10 mM, in the absence of specific alterations or signalling events [52]. The large associated error may come from the fact that, at least,

20 different cells in homeostatic equilibrium and different physiological stages are probed. Likewise, pH level variations between the different cellular compartments may contribute to the broadening of lifetime distributions, and hence, to increase the width of the distribution of total phosphate concentration values.

Once the performance of the DNBS-2Me-4OMe-TM dye as a dual probe was confirmed, we planned its use in biological applications of interest. In this sense, we paid attention to the oxidative stress consequences over cellular homeostasis. Specifically, we focused on the loss of vision generated by oxidative stress in the photoreceptor cells in age-related macular degeneration (AMD) or diabetic retinopathy [53–56]. For this purpose, we induced oxidative stress in 661W cells (a mouse-derived photoreceptor cell line) by adding different concentrations of H_2O_2 during 12 h of incubation. The oxidative stress generated produced an increase in the intracellular biothiol synthesis as a protective cell response. After the 12-h incubation, we added DNBS-2Me-4OMe-TM and measured the fluorogenic response to the biothiol levels. The results obtained (Figure 3) show a direct dependence between the increase in fluorescence and the oxidative stress generated. The presence of reactive oxygen species (ROS) produces a higher intracellular biothiols synthesis and thus a faster and larger increase in the fluorescence intensity. These findings suggest that this technique is a direct methodology that can be used to develop high-throughput tests for ROS generating molecules and that it may easily be extended to study antioxidant drugs.

Acknowledgments: This work was funded by grants CTQ2014-56370-R, CTQ2014-53598, and CTQ2014-55474-C2-2-R from the Spanish Ministry of Economy and Competitiveness and the European Regional Development Fund (ERDF), and grant FQM2012-790 from the Consejería de Innovación, Ciencia y Empresa (Junta de Andalucía), including costs to publish in open access.

Author Contributions: P.H.F. and D.M. performed the synthesis of compounds I–IV, **2Me-4OMe-TM** and **DNBS-2Me-4OMe-TM**. J.M.C. supervised and designed the synthetic protocol. P.H.F. and J.M.P. performed the microscopy measurements and analysed the results. M.D.G. and R.S. provided and cultured the cell lines and performed the incubations. J.M.P. and A.O. designed the experiments. P.H.F., J.M.P., D.M., and A.O. wrote the paper, collecting contributions from all authors.

Conflicts of Interest: The authors declare no conflict of interest.

Appendix A. Synthesis of DNBS-2Me-4OMe-TM

Synthesis and Spectroscopic Data of Compounds DNBS-2Me-4OMe-TM and 2-Me-4OMe-TM

Compound DNBS-2-Me-4OMe-TM was synthesized by nucleophilic addition of 2Me-4OMe-TM to 2,4-dinitrobenzenesulfonic acid in basic media and CH_2Cl_2 as solvent (See Scheme A1). Thus, 4-(dimethylamine)piridyne (24 mg, 0.196 mmol) was added to a solution of 2Me-4OMe-TM (49 mg, 0.131 mmol) in dry CH_2Cl_2. Then, 2,4-dinitrobenzenesulfonyl chloride (42 mg, 0.157 mmol) was added. After 10 min. stirring at room temperature, the solvent was removed under reduced pressure, and the residue was purified by column chromatography with CH_2Cl_2/MeOH mixtures as eluent, giving the corresponding dinitrobenzenesulfonate derivative as a yellow-orange solid (44.4 mg, 56%).

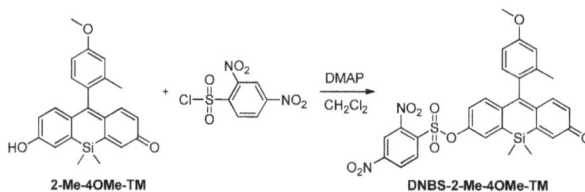

Scheme A1. Synthesis of DNBS-2Me-4-OMe-TM.

The spectroscopic data of compound DNBS-2Me-4OMe-TM are as follows: **^{1}H NMR (400 MHz, Acetone-d$_6$)** δ 8.97 (d, *J* = 2.3 Hz, 1H); 8.73 (dd, *J* = 8.7, 2.3 Hz, 1H); 8.38 (d, *J* = 8.7 Hz, 1H); 7.63 (d, *J* = 2.8 Hz, 1H); 7.19 (dd, *J* = 8.9, 2.8 Hz, 1H); 7.07–6.96 (m, 4H); 6.94 (dd, *J* = 8.3, 2.6 Hz, 1H); 6.81 (d, *J* =

2.2 Hz, 1H); 6.18 (dd, *J* = 10.2, 2.2 Hz, 1H); 3.88 (s, 3H); 2.00 (s, 3H); 0.49 (s, 3H); 0.47 (s, 3H). ^{13}C NMR **(101 MHz, Acetone-d$_6$)** δ 184.2 (C), 160.87 (C), 153.2 (C), 150.1 (C), 146.8 (C), 141.9 (C), 141.8 (C), 141.6 (CH), 138.4 (CH), 135.3 (CH), 134.8 (CH), 133.1 (C), 132.1 (C), 131.7 (C), 131.3 (CH), 129.2 (CH), 128.4 (CH), 128.1 (CH), 124.5 (CH), 121.8 (CH), 116.6 (CH), 112.4 (CH), 55.7 (CH$_3$), 19.9 (CH$_3$), −1.5 (CH$_3$), −1.9 (CH$_3$). **HRMS (ESI):** [M + H]$^+$ calcd. for C$_{29}$H$_{25}$N$_2$O$_9$SSi: 605.1050, obtained: 605.1040.

Compound 2-Me-4-OMe-TM was prepared by nucleophilic addition of the organolithium reagent generated by the halogen-lithium exchange reaction of 4-bromo-3-methylphenol with *t*-BuLi to the silicon-substituted ketone **IV**. Then, the corresponding alcohol obtained underwent a dehydration reaction by acidic hydrolysis, giving Tokyo Magenta derivative (see Scheme A2). In this sense, 4-bromo-3-methylanisole (100 mg, 0.497 mmol) was dissolved in dry THF under an Ar atmosphere in a Schlenck tube, and the solution was cooled to −78°C. Then, *tert*-butyllithium solution (0.58 mL, 0.995 mmol) was added at low temperature, and after stirring 30 min. at −78°C, a solution of ketone **IV** (124 mg, 0.249 mmol) in dry THF (2 mL) was slowly added. The solution was stirred at low temperature for 20 min. and then let warm to room temperature. After one hour at room temperature diluted HCl (1 mL, 10% solution) was added to the reaction. The solvent was then removed under low pressure, and the residue was purified by column chromatography using CH$_2$Cl$_2$ / MeOH mixtures as eluent. 2Me-4OMe-TM was obtained as a pink solid (55 mg, 59.14%), giving the following spectroscopic data: ^{1}H NMR (400 MHz, Methanol-d$_4$) δ 7.02 (d, *J* = 2.5 Hz, 2H); 7.00 (s, 1H), 6.98 (s, 2H); 6.95–6.90 (m, 2H); 6.46 (d, *J* = 2.5 Hz, 1H); 6.44 (d, *J* = 2.5 Hz, 1H); 3.87 (s, 3H); 2.01 (s, 3H); 0.49 (s, 3H); 0.48 (s, 3H). ^{13}C NMR (101 MHz, Methanol-d$_4$) δ 162.0 (C), 161.3 (C), 154.1 (C), 154.0 (C), 141.30 (CH), 141.26 (CH), 138.6 (C), 132.7 (C), 131.4 (CH), 129.7 (C), 122.7 (CH), 116.5 (CH), 112.3 (CH), 55.8 (CH$_3$), 19.9 (CH$_3$), −1.3 (CH$_3$), −1.6 (CH$_3$). **HRMS (APCI+):** [M+H]$^+$ calcd. for C$_{23}$H$_{23}$O$_3$Si: 375.1411, obtained: 375.1424.

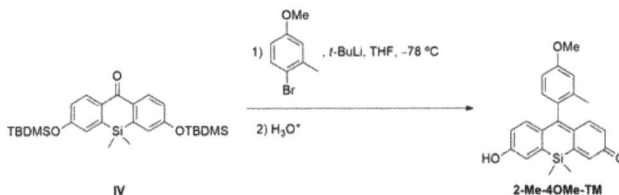

Scheme A2. Synthesis of 2Me-4-OMe-TM.

Finally, compound **IV** was synthesized according to the protocol described by Best et al. [57]. Thus, condensation of 5-bromophenol with formaldehyde gave compound **I**, whose hydroxyl groups were protected as *tert*-butyldimethylsilane (TBDMS) groups. Treatment of **II** with *n*-BuLi at −78°C followed by chlorodimethylsilane addition led to compound **III**. Then, the deprotection of TBDMS groups and later methylene group oxidation gave rise to the corresponding ketone group at the 9 position of the silicon-substituted xanthene. Finally, precursor **IV** was obtained after a double protection of hydroxyl groups as TBDMS (Scheme A3). Spectroscopic data of compounds **I-IV** were identical to those previously described [57].

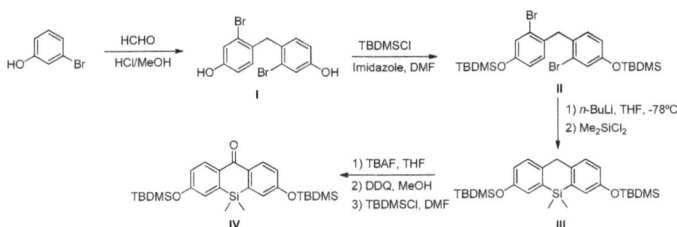

Scheme A3. Synthesis of precursor **IV**.

Appendix B. Additional Figures

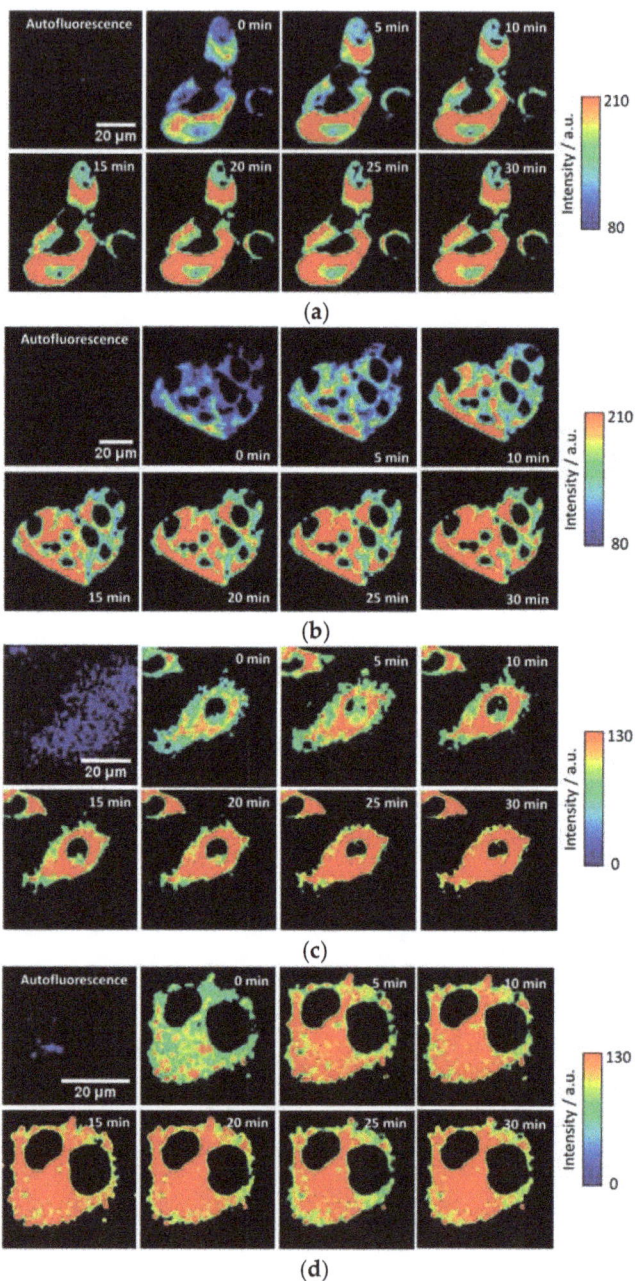

Figure A1. Other representative examples of the fluorogenic reaction of DNBS-2Me-4OMe-TM (6×10^{-7} M) in response to intracellular biothiols in (**a**,**b**) Hep2G and (**c**,**d**) MC3T3-E1 cells. Only pixels in the cytoplasm are shown.

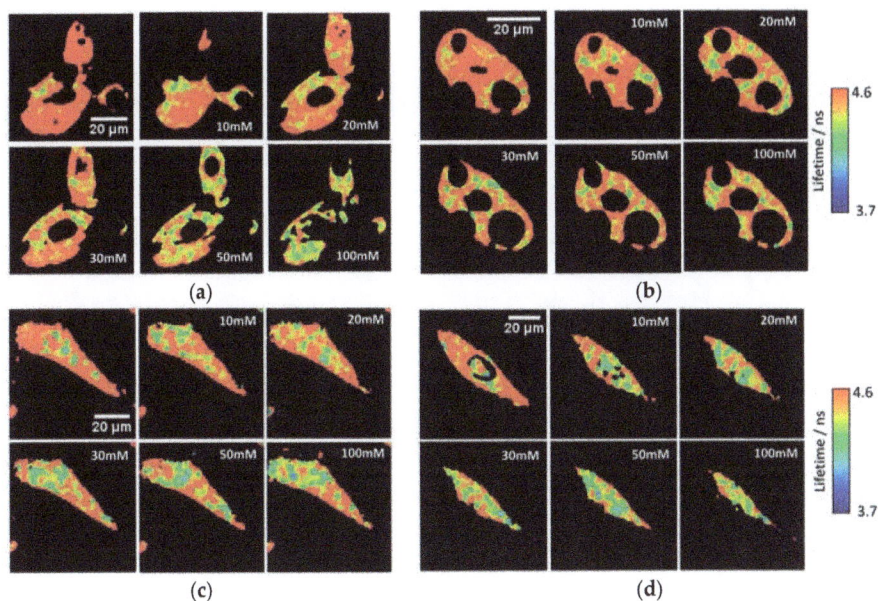

Figure A2. Other representative examples of the variation of the phosphate response of the released 2Me-4OMe-TM in FLIM imaging in (**a**,**b**) Hep2G and (**c**,**d**) MC3T3-E1 cells, permeabilised with α-toxin and different extracellular phosphate concentrations.

References

1. Bayani, U.; Ajay, V.S.; Paolo, Z.; Mahajan, R.T. Oxidative Stress and Neurodegenerative Diseases: A Review of Upstream and Downstream Antioxidant Therapeutic Options. *Curr. Neuropharmacol.* **2009**, *7*, 65–74.
2. Wang, W.-A.; Groenendyk, J.; Michalak, M. Endoplasmic reticulum stress associated responses in cancer. *Biochim. Biophys. Acta* **2014**, *1843*, 2143–2149. [CrossRef] [PubMed]
3. Coons, A.H.; Creech, H.J.; Jones, R.N. Immunological properties of an antibody containing a fluorescent group. *Proc. Soc. Exp. Biol. Med.* **1974**, *47*, 200–202. [CrossRef]
4. Day, R.N.; Davidson, M.W. The fluorescent protein palette: Tools for cellular imaging. *Chem. Soc. Rev.* **2009**, *38*, 2887–2921. [CrossRef] [PubMed]
5. Sydor, A.M.; Czymmek, K.J.; Puchner, E.M.; Mennella, V. Super-Resolution Microscopy: From Single Molecules to Supramolecular Assemblies. *Trends Cell Biol.* **2015**, *25*, 730–748. [CrossRef] [PubMed]
6. Ruedas-Rama, M.; Alvarez-Pez, J.; Crovetto, L.; Paredes, J.; Orte, A. FLIM Strategies for Intracellular Sensing. In *Advanced Photon Counting*; Kapusta, P., Wahl, M., Erdmann, R., Eds.; Springer International Publishing: Cham, Switzerland, 2015; Volume 15, pp. 191–223.
7. Chen, L.-C.; Lloyd, W.R., III; Chang, C.-W.; Sud, D.; Mycek, M.-A. Fluorescence Lifetime Imaging Microscopy for Quantitative Biological Imaging. In *Methods in Cell Biology*; Greenfield, S., David, E.W., Eds.; Academic Press: Cambridge, MA, USA, 2013; Chapter 20; Volume 114, pp. 457–488.
8. Hille, C.; Berg, M.; Bressel, L.; Munzke, D.; Primus, P.; Löhmannsröben, H.G.; Dosche, C. Time-domain fluorescence lifetime imaging for intracellular pH sensing in living tissues. *Anal. Bioanal. Chem.* **2008**, *391*, 1871–1879. [CrossRef] [PubMed]
9. Tantama, M.; Hung, Y.P.; Yellen, G. Imaging Intracellular pH in Live Cells with a Genetically Encoded Red Fluorescent Protein Sensor. *J. Am. Chem. Soc.* **2011**, *133*, 10034–10037. [CrossRef] [PubMed]
10. Sagolla, K.; Löhmannsröben, H.-G.; Hille, C. Time-resolved fluorescence microscopy for quantitative Ca^{2+} imaging in living cells. *Anal. Bioanal. Chem.* **2013**, *405*, 8525–8537. [CrossRef] [PubMed]

11. Okabe, K.; Inada, N.; Gota, C.; Harada, Y.; Funatsu, T.; Uchiyama, S. Intracellular temperature mapping with a fluorescent polymeric thermometer and fluorescence lifetime imaging microscopy. *Nat. Commun.* **2012**, *3*, 705. [CrossRef] [PubMed]

12. Kuimova, M.K.; Yahioglu, G.; Levitt, J.A.; Suhling, K. Molecular Rotor Measures Viscosity of Live Cells via Fluorescence Lifetime Imaging. *J. Am. Chem. Soc.* **2008**, *130*, 6672–6673. [CrossRef] [PubMed]

13. Shimolina, L.E.; Izquierdo, M.A.; López-Duarte, I.; Bull, J.A.; Shirmanova, M.V.; Klapshina, L.G.; Zagaynova, E.V.; Kuimova, M.K. Imaging tumor microscopic viscosity in vivo using molecular rotors. *Sci. Rep.* **2017**, *7*, 41097. [CrossRef] [PubMed]

14. Orte, A.; Alvarez-Pez, J.M.; Ruedas-Rama, M.J. Fluorescence Lifetime Imaging Microscopy for the Detection of Intracellular pH with Quantum Dot Nanosensors. *ACS Nano* **2013**, *7*, 6387–6395. [CrossRef] [PubMed]

15. Ripoll, C.; Martin, M.; Roldan, M.; Talavera, E.M.; Orte, A.; Ruedas-Rama, M.J. Intracellular Zn^{2+} detection with quantum dot-based FLIM nanosensors. *Chem. Commun.* **2015**, *51*, 16964–16967. [CrossRef] [PubMed]

16. Orte, A.; Crovetto, L.; Talavera, E.M.; Boens, N.; Alvarez-Pez, J.M. Absorption and Emission Study of 2′,7′-Difluorofluorescein and Its Excited-State Buffer-Mediated Proton Exchange Reactions. *J. Phys. Chem. A* **2005**, *109*, 734–747. [CrossRef] [PubMed]

17. Paredes, J.M.; Crovetto, L.; Rios, R.; Orte, A.; Alvarez-Pez, J.M.; Talavera, E.M. Tuned lifetime, at the ensemble and single molecule level, of a xanthenic fluorescent dye by means of a buffer-mediated excited-state proton exchange reaction. *Phys. Chem. Chem. Phys.* **2009**, *11*, 5400–5407. [CrossRef] [PubMed]

18. Crovetto, L.; Orte, A.; Paredes, J.M.; Resa, S.; Valverde, J.; Castello, F.; Miguel, D.; Cuerva, J.M.; Talavera, E.M.; Alvarez-Pez, J.M. Photophysics of a Live-Cell-Marker, Red Silicon-Substituted Xanthene Dye. *J. Phys. Chem. A* **2015**, *119*, 10854–10862. [CrossRef] [PubMed]

19. Paredes, J.M.; Giron, M.D.; Ruedas-Rama, M.J.; Orte, A.; Crovetto, L.; Talavera, E.M.; Salto, R.; Alvarez-Pez, J.M. Real-Time Phosphate Sensing in Living Cells using Fluorescence Lifetime Imaging Microscopy (FLIM). *J. Phys. Chem. B* **2013**, *117*, 8143–8149. [CrossRef] [PubMed]

20. Martinez-Peragon, A.; Miguel, D.; Orte, A.; Mota, A.J.; Ruedas-Rama, M.J.; Justicia, J.; Alvarez-Pez, J.M.; Cuerva, J.M.; Crovetto, L. Rational design of a new fluorescent 'ON/OFF' xanthene dye for phosphate detection in live cells. *Org. Biomol. Chem.* **2014**, *12*, 6432–6439. [CrossRef] [PubMed]

21. Alvarez-Pez, J.M.; Ballesteros, L.; Talavera, E.; Yguerabide, J. Fluorescein Excited-State Proton Exchange Reactions: Nanosecond Emission Kinetics and Correlation with Steady-State Fluorescence Intensity. *J. Phys. Chem. A* **2001**, *105*, 6320–6332. [CrossRef]

22. Boens, N.; Basaric, N.; Novikov, E.; Crovetto, L.; Orte, A.; Talavera, E.M.; Alvarez-Pez, J.M. Identifiability of the Model of the Intermolecular Excited-State Proton Exchange Reaction in the Presence of pH Buffer. *J. Phys. Chem. A* **2004**, *108*, 8180–8189. [CrossRef]

23. Orte, A.; Bermejo, R.; Talavera, E.M.; Crovetto, L.; Alvarez-Pez, J.M. 2′,7′-Difluorofluorescein Excited-State Proton Reactions: Correlation between Time-Resolved Emission and Steady-State Fluorescence Intensity. *J. Phys. Chem. A* **2005**, *109*, 2840–2846. [CrossRef] [PubMed]

24. Crovetto, L.; Orte, A.; Talavera, E.M.; Alvarez-Pez, J.M.; Cotlet, M.; Thielemans, J.; De Schryver, F.C.; Boens, N. Global Compartmental Analysis of the Excited-State Reaction between Fluorescein and (±)-N-Acetyl Aspartic Acid. *J. Phys. Chem. B* **2004**, *108*, 6082–6092. [CrossRef]

25. Paredes, J.M.; Orte, A.; Crovetto, L.; Alvarez-Pez, J.M.; Rios, R.; Ruedas-Rama, M.J.; Talavera, E.M. Similarity between the kinetic parameters of the buffer-mediated proton exchange reaction of a xanthenic derivative in its ground- and excited-state. *Phys. Chem. Chem. Phys.* **2010**, *12*, 323–327. [CrossRef] [PubMed]

26. Urano, Y.; Kamiya, M.; Kanda, K.; Ueno, T.; Hirose, K.; Nagano, T. Evolution of Fluorescein as a Platform for Finely Tunable Fluorescence Probes. *J. Am. Chem. Soc.* **2005**, *127*, 4888–4894. [CrossRef] [PubMed]

27. Molony, D.A.; Stephens, B.W. Derangements in Phosphate Metabolism in Chronic Kidney Diseases/Endstage Renal Disease: Therapeutic Considerations. *Adv. Chronic Kidney Dis.* **2011**, *18*, 120–131. [CrossRef] [PubMed]

28. Majed, N.; Li, Y.; Gu, A.Z. Advances in techniques for phosphorus analysis in biological sources. *Curr. Opin. Biotechnol.* **2012**, *23*, 852–859. [CrossRef] [PubMed]

29. Khoshniat, S.; Bourgine, A.; Julien, M.; Weiss, P.; Guicheux, J.; Beck, L. The emergence of phosphate as a specific signaling molecule in bone and other cell types in mammals. *Cell. Mol. Life Sci.* **2011**, *68*, 205–218. [CrossRef] [PubMed]

30. Resa, S.; Orte, A.; Miguel, D.; Paredes, J.M.; Puente-Muñoz, V.; Salto, R.; Giron, M.D.; Ruedas-Rama, M.J.; Cuerva, J.M.; Alvarez-Pez, J.M.; et al. New Dual Fluorescent Probe for Simultaneous Biothiol and Phosphate Bioimaging. *Chem. Eur. J.* **2015**, *21*, 14772–14779. [CrossRef] [PubMed]

31. Chen, X.; Zhou, Y.; Peng, X.; Yoon, J. Fluorescent and colorimetric probes for detection of thiols. *Chem. Soc. Rev.* **2010**, *39*, 2120–2135. [CrossRef] [PubMed]

32. Rood, M.; Raspe, M.; Hove, J.; Jalink, K.; Velders, A.; van Leeuwen, F.W.B. MMP-2/9-Specific Activatable Lifetime Imaging Agent. *Sensors* **2015**, *15*, 11076–11091. [CrossRef] [PubMed]

33. Chen, N.-T.; Cheng, S.-H.; Liu, C.-P.; Souris, J.; Chen, C.-T.; Mou, C.-Y.; Lo, L.-W. Recent Advances in Nanoparticle-Based Förster Resonance Energy Transfer for Biosensing, Molecular Imaging and Drug Release Profiling. *Int. J. Mol. Sci.* **2012**, *13*, 16598–16623. [CrossRef] [PubMed]

34. Yokota, T.; Kinugawa, S.; Yamato, M.; Hirabayashi, K.; Suga, T.; Takada, S.; Harada, K.; Morita, N.; Oyama-Manabe, N.; Kikuchi, Y.; et al. Systemic Oxidative Stress Is Associated With Lower Aerobic Capacity and Impaired Skeletal Muscle Energy Metabolism in Patients With Metabolic Syndrome. *Diabetes Care* **2013**, *36*, 1341–1346. [CrossRef] [PubMed]

35. Hong, G.; Antaris, A.L.; Dai, H. Near-infrared fluorophores for biomedical imaging. *Nat. Biomed. Eng.* **2017**, *1*, 0010. [CrossRef]

36. Fu, M.; Xiao, Y.; Qian, X.; Zhao, D.; Xu, Y. A design concept of long-wavelength fluorescent analogs of rhodamine dyes: Replacement of oxygen with silicon atom. *Chem. Commun.* **2008**, *15*, 1780–1782. [CrossRef] [PubMed]

37. Koide, Y.; Urano, Y.; Hanaoka, K.; Piao, W.; Kusakabe, M.; Saito, N.; Terai, T.; Okabe, T.; Nagano, T. Development of NIR Fluorescent Dyes Based on Si–rhodamine for in Vivo Imaging. *J. Am. Chem. Soc.* **2012**, *134*, 5029–5031. [CrossRef] [PubMed]

38. Egawa, T.; Koide, Y.; Hanaoka, K.; Komatsu, T.; Terai, T.; Nagano, T. Development of a fluorescein analogue, TokyoMagenta, as a novel scaffold for fluorescence probes in red region. *Chem. Commun.* **2011**, *47*, 4162–4164. [CrossRef] [PubMed]

39. Hirabayashi, K.; Hanaoka, K.; Takayanagi, T.; Toki, Y.; Egawa, T.; Kamiya, M.; Komatsu, T.; Ueno, T.; Terai, T.; Yoshida, K.; et al. Analysis of Chemical Equilibrium of Silicon-Substituted Fluorescein and Its Application to Develop a Scaffold for Red Fluorescent Probes. *Anal. Chem.* **2015**, *87*, 9061–9069. [CrossRef] [PubMed]

40. Kushida, Y.; Nagano, T.; Hanaoka, K. Silicon-substituted xanthene dyes and their applications in bioimaging. *Analyst* **2015**, *140*, 685–695. [CrossRef] [PubMed]

41. Schindelin, J.; Arganda-Carreras, I.; Frise, E.; Kaynig, V.; Longair, M.; Pietzsch, T.; Preibisch, S.; Rueden, C.; Saalfeld, S.; Schmid, B.; et al. Fiji: An open-source platform for biological-image analysis. *Nat. Methods* **2012**, *9*, 676–682. [CrossRef] [PubMed]

42. Maeda, H.; Matsuno, H.; Ushida, M.; Katayama, K.; Saeki, K.; Itoh, N. 2,4-Dinitrobenzenesulfonyl Fluoresceins as Fluorescent Alternatives to Ellman's Reagent in Thiol-Quantification Enzyme Assays. *Angew. Chem. Int. Ed.* **2005**, *44*, 2922–2925. [CrossRef] [PubMed]

43. Wysocki, L.M.; Grimm, J.B.; Tkachuk, A.N.; Brown, T.A.; Betzig, E.; Lavis, L.D. Facile and General Synthesis of Photoactivatable Xanthene Dyes. *Angew. Chem. Int. Ed.* **2011**, *50*, 11206–11209. [CrossRef] [PubMed]

44. Grimm, J.B.; Gruber, T.D.; Ortiz, G.; Brown, T.A.; Lavis, L.D. Virginia Orange: A Versatile, Red-Shifted Fluorescein Scaffold for Single- and Dual-Input Fluorogenic Probes. *Bioconjug. Chem.* **2016**, *27*, 474–480. [CrossRef] [PubMed]

45. Yu, H.; Li, J.; Wu, D.; Qiu, Z.; Zhang, Y. Chemistry and biological applications of photo-labile organic molecules. *Chem. Soc. Rev.* **2010**, *39*, 464–473. [CrossRef] [PubMed]

46. Remé, C.E.; Braschler, U.F.; Roberts, J.; Dillon, J. Light Damage in the Rat Retina: Effect of a Radioprotective Agent (WR-77913) on Acute Rod Outer Segment Disk Disruptions. *Photochem. Photobiol.* **1991**, *54*, 137–142. [CrossRef] [PubMed]

47. Bouffard, J.; Kim, Y.; Swager, T.M.; Weissleder, R.; Hilderbrand, S.A. A Highly Selective Fluorescent Probe for Thiol Bioimaging. *Org. Lett.* **2008**, *10*, 37–40. [CrossRef] [PubMed]

48. Shao, J.; Guo, H.; Ji, S.; Zhao, J. Styryl-BODIPY based red-emitting fluorescent OFF–ON molecular probe for specific detection of cysteine. *Biosens. Bioelectron.* **2011**, *26*, 3012–3017. [CrossRef] [PubMed]

49. Peng, H.; Chen, W.; Cheng, Y.; Hakuna, L.; Strongin, R.; Wang, B. Thiol Reactive Probes and Chemosensors. *Sensors* **2012**, *12*, 15907–15946. [CrossRef] [PubMed]

50. Yin, C.; Zhang, W.; Liu, T.; Chao, J.; Huo, F. A near-infrared turn on fluorescent probe for biothiols detection and its application in living cells. *Sens. Actuators B Chem.* **2017**, *246*, 988–993. [CrossRef]
51. Li, X.; Qian, S.; He, Q.; Yang, B.; Li, J.; Hu, Y. Design and synthesis of a highly selective fluorescent turn-on probe for thiol bioimaging in living cells. *Org. Biomol. Chem.* **2010**, *8*, 3627–3630. [CrossRef] [PubMed]
52. Bergwitz, C.; Jüppner, H. Phosphate Sensing. *Adv. Chronic Kidney Dis.* **2011**, *18*, 132–144. [CrossRef] [PubMed]
53. Blackshaw, S.; Fraioli, R.E.; Furukawa, T.; Cepko, C.L. Comprehensive Analysis of Photoreceptor Gene Expression and the Identification of Candidate Retinal Disease Genes. *Cell* **2001**, *107*, 579–589. [CrossRef]
54. Klein, R.; Chou, C.; Klein, B.K.; Zhang, X.; Meuer, S.M.; Saaddine, J.B. Prevalence of age-related macular degeneration in the us population. *Arch. Ophthalmol.* **2011**, *129*, 75–80. [CrossRef] [PubMed]
55. Jarrett, S.G.; Boulton, M.E. Consequences of oxidative stress in age-related macular degeneration. *Mol. Asp. Med.* **2012**, *33*, 399–417. [CrossRef] [PubMed]
56. Shaw, P.X.; Stiles, T.; Douglas, C.; Ho, D.; Fan, W.; Du, H.; Xiao, X. Oxidative stress, innate immunity, and age-related macular degeneration. *AIMS Mol. Sci.* **2016**, *3*, 196–221. [CrossRef] [PubMed]
57. Best, Q.A.; Sattenapally, N.; Dyer, D.J.; Scott, C.N.; McCarroll, M.E. pH-Dependent Si-Fluorescein Hypochlorous Acid Fluorescent Probe: Spirocycle Ring-Opening and Excess Hypochlorous Acid-Induced Chlorination. *J. Am. Chem. Soc.* **2013**, *135*, 13365–13370. [CrossRef] [PubMed]

sensors

MDPI

Article

A Ratiometric Fluorescent Sensor for Cd^{2+} Based on Internal Charge Transfer

Dandan Cheng [1,†], Xingliang Liu [1,†], Yadian Xie [1], Haitang Lv [1], Zhaoqian Wang [1], Hongzhi Yang [1], Aixia Han [1,2,*], Xiaomei Yang [2] and Ling Zang [2,*]

[1] Chemical Engineering College, Qinghai University, Xining 810016, China; 1994990022@qhu.edu.cn (D.C.); liuxingliang@qhu.edu.cn (X.L.); 1991990011@qhu.edu.cn (Y.X.); 1989990029@qhu.edu.cn (H.L.); 1990990009@qhu.edu.cn (Z.W.); yhz17@mails.tsinghua.edu.cn (H.Y.)

[2] Department of Materials Science and Engineering, University of Utah, Salt Lake City, UT 84108, USA; jaimee@eng.utah.edu

[*] Correspondence: hanaixia@tsinghua.org.cn (A.H.); lzang@eng.utah.edu (L.Z.); Tel.: +86-971-5310-427 (A.H.); +1-801-587-1551 (L.Z.)

[†] These authors contributed equally to this work.

Received: 9 October 2017; Accepted: 31 October 2017; Published: 2 November 2017

Abstract: This work reports on a novel fluorescent sensor **1** for Cd^{2+} ion based on the fluorophore of tetramethyl substituted bis(difluoroboron)-1,2-bis[(1*H*-pyrrol-2-yl)methylene]hydrazine (Me$_4$BOPHY), which is modified with an electron donor moiety of *N,N*-bis(pyridin-2-ylmethyl)benzenamine. Sensor **1** has absorption and emission in visible region, at 550 nm and 675 nm, respectively. The long wavelength spectral response makes it easier to fabricate the fluorescence detector. The sensor mechanism is based on the tunable internal charge transfer (ICT) transition of molecule **1**. Binding of Cd^{2+} ion quenches the ICT transition, but turns on the $\pi - \pi$ transition of the fluorophore, thus enabling ratiometric fluorescence sensing. The limit of detection (LOD) was projected down to 0.77 ppb, which is far below the safety value (3 ppb) set for drinking water by World Health Organization. The sensor also demonstrates a high selectivity towards Cd^{2+} in comparison to other interferent metal ions.

Keywords: ratiometric fluorescent sensor; Me$_4$BOPHY; Cd^{2+} ion; ICT

1. Introduction

Cadmium represents a highly toxic industrial and environmental pollutant, and it is classified as a human carcinogen. Exposure to cadmium may cause cancer mutation of some organs, such as lung, endometria, prostate, kidney, etc. [1]. World Health Organization (WHO) underlines drinking water value for cadmium as 3 ppb [2]. So, detection of cadmium at trace level remains an important task, for which cadmium ion (Cd^{2+}) usually remains as the target for chemical sensors to monitor the cadmium pollution in water environment. Current methods for Cd^{2+} detection include UV-Vis spectrometry [3], atomic absorption spectrometry (AAS) [4], inductively coupled plasma atomic emission spectroscopy (ICP-AES) [5], and fluorescent sensors [6–16]. Among these, fluorescent sensors are uniquely compelling due to their high sensitivity, good selectivity [6–16], and capability for ratiometric sensing to further improve the detection sensitivity [17–20]. However, many fluorescence sensors for Cd^{2+} ion reported thus far have some technical drawbacks, for example, a poor limit of detection (LOD) [7,8,18], complicated synthesis of sensor molecules [6], solvent toxicity [7], and a hardly controlled fluorescence change [14]. In order to develop high performance fluorescent sensors, the fluorophore must be designed with both high quantum efficiency and chemical tunability in response to metal binding [21,22]. Borondipyrromethene (BODIPY) has long been studied as an outstanding organoboron fluorophore and been used in the development of fluorescent sensors for many metal ions [23–27], including Cd^{2+} ion [28,29]. In 2014, a novel organoboron compound bis(difluoroboron)-1,2-bis[(1*H*-pyrrol-2-yl)methylene]hydrazine (BOPHY) (Scheme 1) was reported [30–32]. BOPHY has distinctive absorption and emission features that are suited for sensor development, particularly when compared to those with spectral response in high energy blue or UV region. Many BOPHY derivatives have ever since synthesized [33–35], including several from our lab (F-BOPHY1-3) [36], which all showed high efficiency of fluorescence.

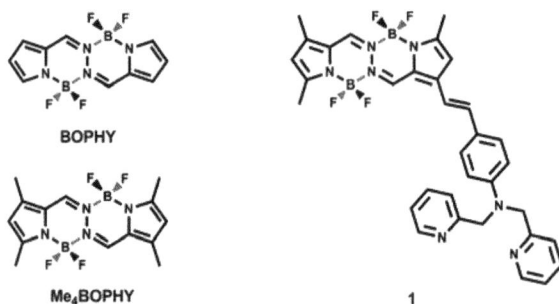

Scheme 1. Molecule structures of bis(difluoroboron)-1,2-bis[(1*H*-pyrrol-2-yl)methylene]hydrazine (BOPHY), tetramethyl substituted BOPHY (Me$_4$BOPHY) and sensor **1**.

We noticed that there have been only two BOPHY fluorescent sensors reported so far, which were used for detecting Cu^{2+} and H^{+}, respectively [37,38]. In this paper, we report on synthesis of a novel fluorescent sensor **1** for Cd^{2+} (Scheme 1) based on a BOPHY fluorophore substituted by tetramethyl group (Me4BOPHY), in conjugation through a vinyl link with an electron donor moiety *N,N*-bis(pyridin-2-ylmethyl)benzenamine (BPA). BPA is also a strong chelator to Cd^{2+} ion, thus affording high sensing sensitivity. Pristine sensor **1** exhibits a significant internal charge transfer (ICT) transition between Me4BOPHY and BPA, with an absorption and fluorescence extending into long wavelength, 550 nm and 675 nm, respectively. When chelated with Cd^{2+} the electron-donating power of BPA will be reduced, thus quenching the ICT transition and turning on the $\pi - \pi$ transition of the fluorophore, which combined the results in blue-shift of the absorption and fluorescence of **1**. Such dramatic spectral change can be used to develop efficient fluorescence sensor for Cd^{2+} detection, particularly through the ratiometric fluorescence modulation [39–43].

2. Experimental Methods

2.1. Materials and Instrumentation

All of the solvents and chemicals were purchased in analytical grade and were used as received. Column chromatography used 300–400 mesh silica gels. Ultrapure water was produced by a Milli-Q Direct 16 system of Millipore. UV-Vis absorption spectra were gained on a Shimadzu UV-2550 spectrophotometer (Shimadzu, Kyoto, Japan). Fluorescence spectra were obtained on a Cary Eclipse fluorescence spectrophotometer from Agilent. ^{1}H- and ^{13}C-NMR spectra were recorded with a Mercury plus instrument at 400 and 100 MHz by using DMSO-d_6 as the solvents. MS spectra were recorded on a MALDI-TOF MS Performance (Shimadzu, Japan).

2.2. Molecular Synthesis

Compound **2** [30] and **3** [29] were synthesized according to literatures, while **1** was synthesized as illustrated in Figure 1. Dry toluene used in synthesis was distilled over sodium and benzophenone. A mixture of **2** (0.50 g, 1.48 mmol), **3** (0.45 g, 1.48 mmol), and *p*-toluenesulfonicacid (1 g, 5.81 mmol) was dissolved in dry toluene (50 mL), followed by the addition of 1 mL piperidine as catalyst. The mixture was refluxed with stirring for 12 h under an atmosphere of nitrogen, during which time the color of the reaction mixture changed from pale yellow to red. After cooling to room temperature, the mixture was poured into H_2O (100 mL) and extracted with CH_2Cl_2. After solvent removal, the crude product was purified by column chromatography (silica gel, CH_2Cl_2/petroleum ether, $v/v = 2/1$), producing a dark purple solid (0.45 g), yield 49%. ^{1}H-NMR (400 MHz, DMSO-d_6) δ = 8.61–8.60 (m, 2H), 7.92 (s, 1H), 7.84 (s, 1H), 7.65–7.62 (m, 2H), 7.38 (d, *J* = 5.6 Hz, 2H), 7.25 (d, *J* = 5.2 Hz, *J* = 5.2 Hz, 2H), 7.20–7.18 (m, 2H), 7.17 (d, *J* = 2.8 Hz, 2H), 6.72 (d, *J* = 6.0 Hz, 2H), 6.68 (s, 1H), 6.16 (s, 1H), 4.87 (s, 4H), 2.48 (s, 3H), 2.32 (s, 3H), 2.31 (s, 3H). (Figure S1, Supplementary Materials). ^{13}C-NMR (100 MHz, DMSO-d_6) δ = 157.67, 150.75, 150.07, 149.40, 148.72, 140.24, 139.84, 136.69, 136.49, 133.39, 132.27, 128.63, 124.96, 124.08, 122.98, 121.81, 120.34, 117.88, 114.17, 113.24, 112.24, 56.83, 13.66, 10.73, 10.65. (Figure S2, Supplementary Materials). MALDI-TOFMS: *m/z* calculated for $C_{33}H_{31}B_2F_4N_7$: 623.28; found: 623.47. (Figure S3, Supplementary Materials).

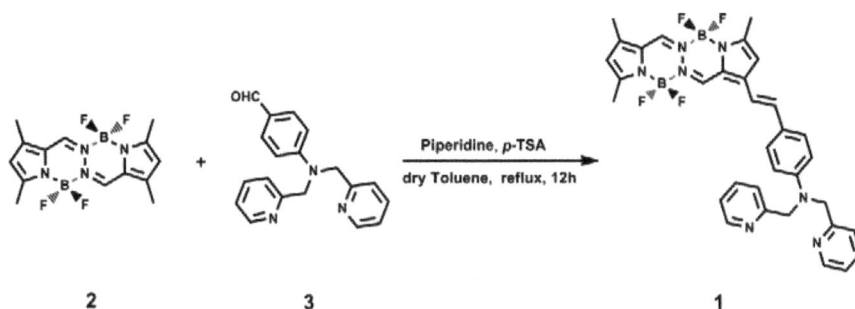

Figure 1. The synthesis route of **1**.

2.3. Sample Preparation and Spectral Measurements

A stock solution (0.5 mM) of sensor **1** was prepared in acetonitrile. Metal ion solutions of Cd^{2+}, Zn^{2+}, Mn^{2+}, Pb^{2+}, Cu^{2+}, Co^{2+}, Mg^{2+}, Ca^{2+}, Ba^{2+}, Fe^{2+}, and Hg^{2+} were prepared by dissolving the corresponding nitrate salts in acetonitrile. These stock solutions were diluted to needed concentrations for sensor testing. UV-Vis and fluorescent spectra were measured under room temperature. Briefly, 2.5 mL solution of **1** (2 μM) was put into a 1 cm quartz cuvette, followed by addition of different concentrations of metal ion. The series of concentrations of metal ions were thus added and were measured for the absorption and fluorescence spectra. Since added volume of the metal ion stock solution was small (up to 8 μL), the concentration of sensor **1** would remain almost unchanged. For fluorescence spectra measurement, the excitation wavelength was set at 410 nm and slit widths at 5 nm/10 nm.

3. Results and Discussion

3.1. Spectral Change of 1 Upon Titration with Cd²⁺

As shown in Figure 2, the absorption spectrum of pristine **1** has two pronounced peaks around 505 nm and 550 nm. These two absorption peaks are significantly red-shifted in comparison with those of Me_4BOPHY, which has the corresponding two peaks at 444 nm and 467 nm. Such spectral redshift is due to the ICT electronic transition, as previously observed in other electron donor-acceptor molecules [29]. In molecule **1** the fluorophore Me_4BOPHY is in full conjugation with the aniline group of BPA through the vinyl bridge (Scheme 1), thus facilitating the ICT transition. Upon titration with Cd^{2+} ion, the absorption at 550 nm gradually decreased, accompanied by a rising blue-shifted absorption peak centered at 475 nm. An isosbestic point was clearly seen around 520 nm, indicating the stoichiometric conversion of molecule **1** from unbound to the Cd^{2+}-bound state. As the concentration of Cd^{2+} increased, the color of the solution turned from red to bright yellow, consistent the absorption spectral change shown in Figure 2. The observed spectral change is due to the binding of Cd^{2+} at the BPA chelator (Scheme 2), which in turn reduces the electron-donating capability of the aniline moiety. As a result, the ICT transition of molecule **1** is diminished. Indeed, as molecule **1** is fully chelated, the absorption spectrum becomes mostly characteristic of the $\pi - \pi$ transition of the Me_4BOPHY part, centered around 475 nm (Figure 2).

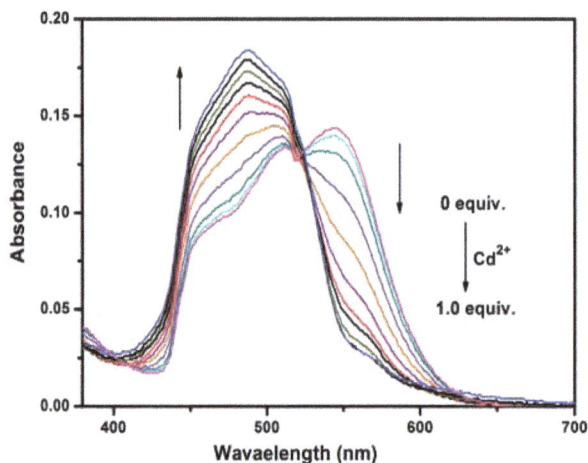

Figure 2. UV-vis absorption spectral change recorded for an acetonitrile solution of sensor **1** (2 μM) upon the titration of Cd^{2+} ion.

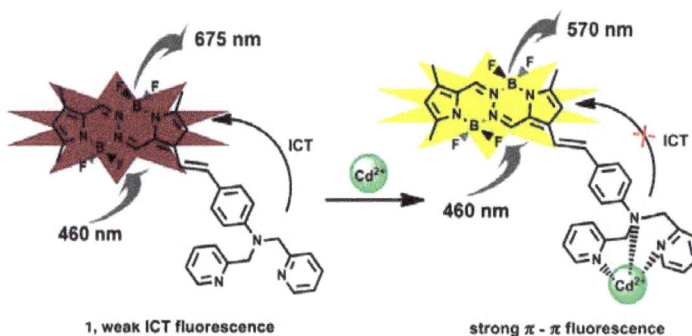

Scheme 2. Sensing mechanism of **1** towards Cd^{2+}.

The same series of titration of Figure 2 was also monitored for fluorescence spectral change, as shown in Figure 3a. The unbound molecule **1** has an emission band centered at 675 nm, which is significantly red-shifted in comparison to the two emission bands (485 nm and 518 nm) that are typically observed for the fluorophore of tetramethyl substituted BOPHY (Me_4BOPHY). The strong redshift is mainly a result of the ICT transition (Scheme 2), which in turn is caused by the BPA substitution. Upon binding with the Cd^{2+} ion, the emission peak was blue-shifted to 570 nm, implying that the ICT transition is diminished, as discussed above. The fluorescence quantum yield of pristine **1** determined as 7.6% by using Rhodamine B in acetonitrile as a standard ($\varphi_F = 0.89$, $\lambda_{ex} = 495$ nm). By comparing the total fluorescence intensity and the absorbance at the same excitation wavelength 495 nm between the unbound and Cd^{2+}-bound state of **1**, the fluorescence quantum yield of Cd^{2+}-bound **1** can be estimated to be 44.2%. The spectral change shown in Figure 3a enables ratiometric sensing by plotting the ratio of fluorescent intensity at 570 nm and 730 nm (*F570/F730*) as a function of the concentration of Cd^{2+} (relative to that of **1**), as shown in Figure 3b. An approximately linear relationship was obtained, allowing for determining the concentration of Cd^{2+} using this linear calibration. The limit of detection (LOD) can be projected by taking three times the standard deviation of measurement as the detectable signal, that is, 0.3 in this study. Using the slope of the linear fitting of Figure 3b, we can determine the LOD to be 6.9 nM, or 0.77 ppb, which is far below the safety value set for drinking water by WHO (3 ppb), indicating a strong feasibility of using sensor **1** for trace level detection of Cd^{2+}. The ratiometric sensing, relying on the fluorescence measurement of both bound and unbound state of **1**, could potentially improve the robustness of signal by canceling the interference from the environment. By comparing with other fluorescence sensors for Cd^{2+} reported in literature (Table 1), sensor **1** developed in this study has many advantages over other Cd^{2+} sensors.

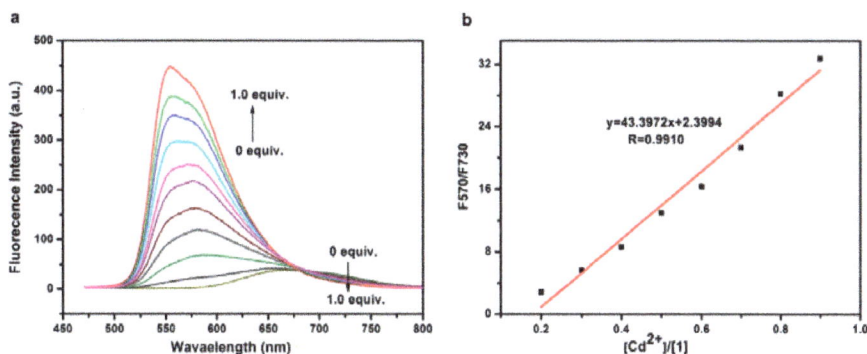

Figure 3. (**a**) Fluorescence spectral change recorded for an acetonitrile solution of sensor **1** (2 μM) upon titration of Cd^{2+} ion; (**b**) The ratio of fluorescence intensity (*F570/F730*) measured for the same solutions at 570 nm and 730 nm as a function of the concentration of Cd^{2+} (relative to that of **1**), showing linear fitting as indicated in the plot.

Table 1. The comparison of **1** with other Cd^{2+} sensors in literature.

Refs.	LOD (mol L^{-1})	Wavelength of Emission	Solvent Used
[7]	1.97×10^{-7}	456 nm	CH_2Cl_2/CH_3CN (1/9)
[8]	2.76×10^{-7}	500 nm	H_2O
[18]	1.76×10^{-7}	495 nm/558 nm	HEPES
[29]	Not available	597 nm	CH_3COCH_3/H_2O (9/1)
This work	6.9×10^{-9}	570 nm/730 nm	CH_3CN

3.2. Sensing Mechanism and Job's Plot

As illustrated in Scheme 2, the sensing mechanism of **1** relies on switching the fluorescence from ICT transition to local $\pi - \pi$ transition at the BOPHY site. The BPA chelator affords strong binding to the Cd^{2+} ion, and this weakens the electron donating power of the aniline moiety, thus diminishing the ICT transition. The tridentate chelation of BPA forms 1:1 complex with Cd^{2+} ion, as also reported in other studies wherein the same chelator was used [29]. The 1:1 chelation stoichiometry was also confirmed in this study through a Job's plot approach [44], as shown in Figure 4. Job's plot is commonly used to determine the stoichiometry of a complex between two species, for which the total molar concentrations of the two species (here molecule **1** and Cd^{2+} ion) are kept constant, while their relative concentrations are varied. A measured variable (here the fluorescence intensity ratio, *F570/F730*) that is dependent on the complex formation can be plotted as a function of the molar fractions of the binding species. The maximum of the plot corresponds to the stoichiometry of the complex formed. In this study, the total concentration of molecule **1** and Cd^{2+} ion was fixed at 2 μM, and the molar ratio of the two species was changed from 1:9 to 9:1, and the fluorescence intensity ratio *F570/F730* was measured under the same conditions. Clearly, as shown in Figure 4, the maximum of the plot corresponds to a 1:1 complex between **1** and Cd^{2+}.

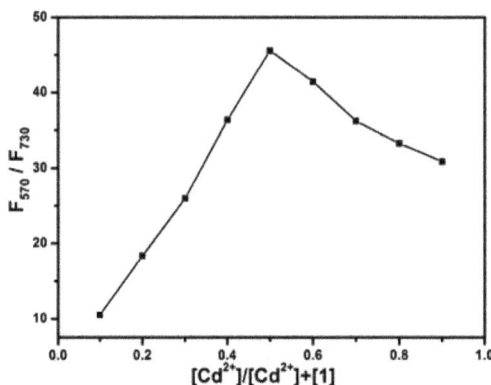

Figure 4. Job's plot of the binding between **1** and Cd^{2+} in acetonitrile, with the total concentration of the two species fixed at 2 μM.

3.3. Sensing Selectivity

The high selectivity of **1** towards Cd^{2+} ion was examined by comparative experiments, which were conducted by repeating the same fluorescence measurements shown in Figure 3 but in the presence of 10 other common metal ions, Mn^{2+}, Pb^{2+}, Cu^{2+}, Co^{2+}, Mg^{2+}, Ca^{2+}, Ba^{2+}, Fe^{2+}, Hg^{2+}, and Zn^{2+}. In contrast to the efficient spectral change observed for Cd^{2+} (far left bar in the figure), all of the other metal ions (except for Zn^{2+}) demonstrated almost no spectral change, as indicated by the low values of *F570/F730* measured under the same experimental conditions (Figure 5a). However, upon the addition of Cd^{2+} ion at the same concentration, all of the 10 solutions containing the different metal ions showed dramatic fluorescence change at the same degree as that observed for the solution of **1** + Cd^{2+}. This observation indicates good sensing selectivity for molecule **1** towards Cd^{2+}, which in turn is largely due to the strong chelation, as illustrated in Scheme 2. The mild fluorescence response observed for Zn^{2+} ion is not surprised considering the similar coordination property between Zn^{2+} and Cd^{2+}. However, due to the weaker electron affinity of Zn^{2+} ion (with standard reduction potential of -0.7 V, as compared to that of Cd^{2+}, -0.4 V), the binding with Zn^{2+} cannot block the ICT transition as effectively as Cd^{2+}. Indeed, as shown in Figure 5b, under the same concentration the solution of **1** + Cd^{2+} exhibited a dramatic fluorescence color change (consistent with the spectral measurement shown Figure 3), whereas the solution of **1** + Zn^{2+} remained about the same color as the solution of **1**. Such dramatic difference in color change provides additional feature to distinguish Cd^{2+} from other metal ions when using **1** as sensor.

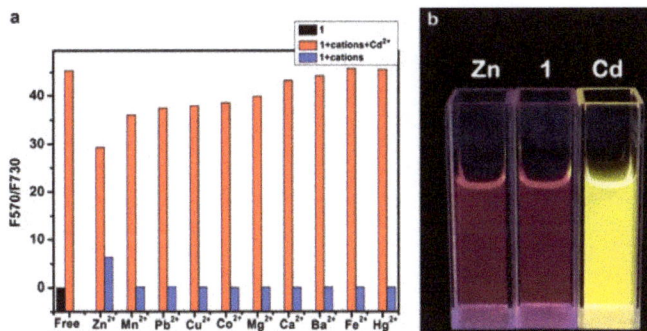

Figure 5. (a) Fluorescence intensity ratio (*F570/F730*) measured for sensor **1** in acetonitrile (2 μM) in the absence of metal ions (black), and in the presence of various metal ions (2 μM), (blue), followed by addition of 2 μM Cd^{2+} into each of the eleven solutions (red); (b) Photographs taken for the 2 μM solution of **1**, in comparison to the ones containing 2 μM of Zn^{2+} and Cd^{2+}.

3.4. Fast Sensor Response

The sensor **1** could rapidly detect Cd^{2+} ion, as shown in Figure 6. When we put 2 μM Cd^{2+} into 2 μM sensor **1** solution, the fluorescence intensity ratio (*F570/F730*) of sensor **1** quickly increased and reached a stable value within 1 min. This is a good trait for fast and real-time determination.

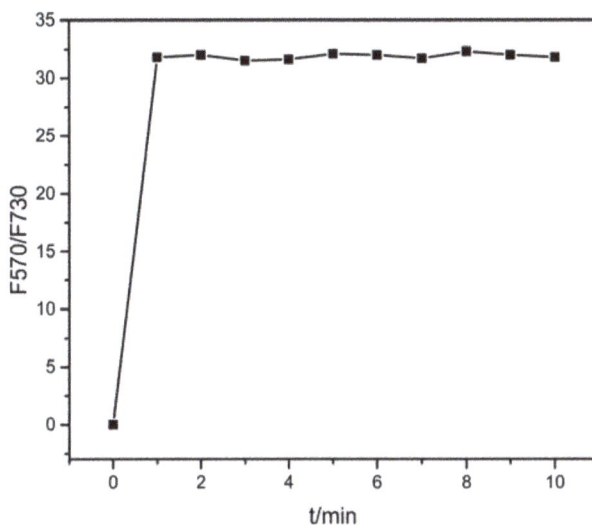

Figure 6. Time course of the fluorescence intensity ratio (*F570/F730*) change measured on an acetonitrile solution of sensor **1** (2 μM) upon addition of Cd^{2+} ion (2 μM).

In addition to the high sensitivity and selectivity observed above, sensor **1** also demonstrated a fast response, consistent with the strong chelation with Cd^{2+}. As shown in Figure 6, the ratiometric fluorescence response of **1** was finished in one min upon addition of 1:1 Cd^{2+} ion. Due to the experimental operation limit, we could not monitor the sensor response in any faster time scale, though the real response time of **1** seems to be in seconds or even faster. This fast fluorescence response makes sensor **1** ideal for real-time monitoring, particularly for in-field detection. Also indicated from

Figure 6 is the high photostability sensor **1**, wherein the fluorescence of **1** was measured ten times after binding with Cd^{2+} ion, but no significant change in the fluorescence intensity was observed.

4. Conclusions

We report on a novel fluorescence sensor **1** for the selective detection of Cd^{2+} ion with LOD down to 0.77 ppb. The sensor molecule is based on a fluorophore of Me_4BOPHY in conjugation with an electron donor group, namely BPA, which also affords strong binding with Cd^{2+}. The electron donor-acceptor conjugation enables ICT fluorescence at long wavelength, desired for sensor development. Upon binding with the Cd^{2+} ion, the fluorescence is switched from ICT transition to be the $\pi - \pi$ transition, which dominated by the Me_4BOPHY fluorophore, which is located in much shorter wavelength region. Such dramatic fluorescence change enables ratiometric sensing by measuring the relative emission intensity at the two wavelengths as a function of the concentration of Cd^{2+} ion, thus allowing for quantitative detection of Cd^{2+}. High selectivity towards Cd^{2+} was also evidenced for the sensor as examined with ten other common metal ions.

Supplementary Materials: The Supplementary Materials are available online at http://www.mdpi.com/1424-8220/17/11/2517/s1.

Acknowledgments: This work was financially supported by the Qinghai Science & Technology Department of China (Grant No. 2016-HZ-806), the National Natural Science Foundation of China (Grant No. 21362027), the China Scholarship Council (CSC) and the Qinghai University (Grant No. 2015-QGY-5).

Author Contributions: Dandan Cheng carried out the majority of experiments and wrote the article; Xingliang Liu designed and synthesized sensor **1**; Yadian Xie did part of the experiments under the help of Dandan Cheng; Haitang Lv measured the fluorescent spectra; Zhaoqian Wang was responsible for UV spectra measuriment; Hongzhi Yang did part of synthesis experiments; Aixia Han was responsible for the whole work; Xiaomei Yang helped on data analysis and manuscript editing; Ling Zang helped supervising the research design.

Conflicts of Interest: The authors declare no conflict of interest.

References

1. Huff, J.; Lunn, R.M.; Waalkes, M.P.; Tomatis, L.; Infante, P.F. Cadmium-induced Cancers in Animals and in Humans. *Int. J. Occup. Environ. Health* **2007**, *13*, 202–212. [CrossRef] [PubMed]
2. World Health Organization. Available online: http://www.who.int/water_sanitation_health/publications/drinking-water-quality-guidelines-4-including-1st-addendum/en/ (accessed on 1 November 2017).
3. Wen, X.D.; Yang, Q.L.; Yan, Z.D.; Deng, Q.W. Determination of cadmium and copper in water and food samples by dispersive liquid–liquid microextraction combined with UV–vis spectrophotometry. *Microchem. J.* **2011**, *97*, 249–254. [CrossRef]
4. Manzoori, J.L.; Bavili-Tabrizi, A. Cloud point preconcentration and flame atomic absorption spectrometric determination of Cd and Pb in human hair. *Anal. Chim. Acta* **2002**, *470*, 215–221. [CrossRef]
5. Rao, K.S.; Balaji, T.; Rao, T.P.; Babu, Y.; Naidu, G.R.K. Determination of iron, cobalt, nickel, manganese, zinc, copper, cadmium and lead in human hair by inductively coupled plasma-atomic emission spectrometry. *Spectrochim. Acta Part B* **2002**, *57*, 1333–1338.
6. Xue, L.; Liu, C.; Jiang, H. Highly Sensitive and Selective Fluorescent Sensor for Distinguishing Cadmium from Zinc Ions in Aqueous Media. *Org. Lett.* **2009**, *11*, 1655–1658. [CrossRef] [PubMed]
7. Zhao, Q.; Li, R.F.; Xing, S.K.; Liu, X.M.; Hu, T.L.; Bu, X.H. A Highly Selective On/Off Fluorescence Sensor for Cadmium(II). *Inorg. Chem.* **2011**, *50*, 10041–10046. [CrossRef] [PubMed]
8. Gunnlaugsson, T.; Lee, T.C.; Parkesh, R. Highly selective fluorescent chemosensors for cadmium in water. *Tetrahedron* **2004**, *60*, 11239–11249. [CrossRef]
9. Zhang, X.X.; Wang, R.J.; Fan, C.B.; Liu, G.; Pu, S.Z. A highly selective fluorescent sensor for Cd^{2+} based on a new diarylethene with a 1,8-naphthyridine unit. *Dyes Pigments* **2017**, *139*, 208–217. [CrossRef]
10. Khani, R.; Ghiamati, E.; Boroujerdi, R.; Rezaeifard, A.; Zaryabi, M.H. A new and highly selective turn-on fluorescent sensor with fast response time for the monitoring of cadmium ions in cosmetic, and health product samples. *Spectrochim. Acta Part A* **2016**, *163*, 120–126. [CrossRef] [PubMed]

11. Chao, D.B. Highly selective detection of Zn^{2+} and Cd^{2+} with a simple amino-terpyridine compound in solution and solid state. *J. Chem. Sci.* **2016**, *128*, 133–139. [CrossRef]

12. Zhou, X.Y.; Li, P.X.; Shi, Z.H.; Tang, X.L.; Chen, C.Y.; Liu, W.S. A Highly Selective Fluorescent Sensor for Distinguishing Cadmium from Zinc Ions Based on a Quinoline Platform. *Inorg. Chem.* **2012**, *51*, 9226–9231. [CrossRef] [PubMed]

13. Goswami, P.; Das, D.K. A New Highly Sensitive and Selective Fluorescent Cadmium Sensor. *J. Fluoresc.* **2012**, *22*, 391–395. [CrossRef] [PubMed]

14. Zhou, Y.; Xiao, Y.; Qian, X.H. A highly selective Cd^{2+} sensor of naphthyridine: fluorescent enhancement and red-shift by the synergistic action of forming binuclear complex. *Tetrahedron Lett.* **2008**, *49*, 3380–3384. [CrossRef]

15. Mameli, M.; Aragoni, M.C.; Arca, M.; Caltagirone, C.; Demartin, F.; Farruggia, G.; De Filippo, G.; Devillanova, F.A.; Garau, A.; Isaia, F.; et al. A Selective, Nontoxic, OFF–ON Fluorescent Molecular Sensor Based on 8-Hydroxyquinoline for Probing Cd^{2+} in Living Cells. *Chem. Eur. J.* **2010**, *16*, 919–930. [CrossRef] [PubMed]

16. Liu, Y.; Qiao, Q.L.; Zhao, M.; Yin, W.T.; Miao, L.; Wang, L.Q.; Xu, Z.C. Cd^{2+}-triggered amide tautomerization produces a highly Cd^{2+}-selective fluorescent sensor across a wide pH range. *Dyes Pigments* **2016**, *133*, 339–344. [CrossRef]

17. Lu, C.L.; Xu, Z.C.; Cui, J.N.; Zhang, R.; Qian, X.H. Ratiometric and Highly Selective Fluorescent Sensor for Cadmium under Physiological pH Range: A New Strategy to Discriminate Cadmium from Zinc. *J. Org. Chem.* **2007**, *72*, 3554–3557. [CrossRef] [PubMed]

18. Xue, L.; Li, G.P.; Liu, Q.; Wang, H.H.; Liu, C.; Ding, X.L.; He, S.G.; Jiang, H. Ratiometric Fluorescent Sensor Based on Inhibition of Resonance for Detection of Cadmium in Aqueous Solution and Living Cells. *Inorg. Chem.* **2011**, *50*, 3680–3690. [CrossRef] [PubMed]

19. Taki, M.; Desaki, M.; Ojida, A.; Iyoshi, S.; Hirayama, T.; Hamachi, I.; Yamamoto, Y. Fluorescence Imaging of Intracellular Cadmium Using a Dual-Excitation Ratiometric Chemosensor. *J. Am. Chem. Soc.* **2008**, *130*, 12564–12565. [CrossRef] [PubMed]

20. Chiu, T.Y.; Chen, P.H.; Chang, C.L.; Yang, D.M. Live-Cell Dynamic Sensing of Cd^{2+} with a FRET-Based Indicator. *PLoS ONE* **2013**, *8*, e65853. [CrossRef] [PubMed]

21. Wang, C.; Huang, H.L.; Bunes, B.R.; Wu, N.; Xu, M.; Yang, X.M.; Yu, L.; Zang, L. Trace Detection of RDX, HMX and PETN Explosives Using a Fluorescence Spot Sensor. *Sci. Rep.* **2016**, *6*, 25015. [CrossRef] [PubMed]

22. Xu, M.; Han, J.M.; Wang, C.; Yang, X.M.; Pei, J.; Zang, L. Fluorescence Ratiometric Sensor for Trace Vapor Detection of Hydrogen Peroxide. *ACS Appl. Mater. Interfaces* **2014**, *6*, 8708–8714. [CrossRef] [PubMed]

23. Kamiya, M.; Johnsson, K. Localizable and Highly Sensitive Calcium Indicator Based on a BODIPY Fluorophore. *Anal. Chem.* **2010**, *82*, 6472–6479. [CrossRef] [PubMed]

24. Atilgan, S.; Kutuk, I.; Ozdemir, T. A near IR distyryl BODIPY-based ratiometric fluorescent chemosensor for Hg(II). *Tetrahedron Lett.* **2010**, *51*, 892–894. [CrossRef]

25. Qi, X.; Jun, E.J.; Xu, L.; Kim, S.J.; Joong Hong, J.S.; Yoon, Y.J.; Yoon, J. New BODIPY Derivatives as OFF–ON Fluorescent Chemosensor and Fluorescent Chemodosimeter for Cu^{2+}: Cooperative Selectivity Enhancement toward Cu^{2+}. *J. Org. Chem.* **2006**, *71*, 2881–2884. [CrossRef] [PubMed]

26. Wu, Y.K.; Peng, X.J.; Guo, B.C.; Fan, J.L.; Zhang, Z.C.; Wang, J.Y.; Cui, A.J.; Gao, Y.L. Boron dipyrromethene fluorophore based fluorescence sensor for the selective imaging of Zn (II) in living cells. *Org. Biomol. Chem.* **2005**, *3*, 1387–1392. [CrossRef] [PubMed]

27. Liu, J.; Wu, K.; Li, S.; Song, T.; Han, Y.F.; Li, X. A highly sensitive and selective fluorescent chemosensor for Pb^{2+} ions in an aqueous solution. *Dalton Trans.* **2013**, *42*, 3854–3859. [CrossRef] [PubMed]

28. Cheng, T.Y.; Xu, Y.F.; Zhang, S.Y.; Zhu, W.P.; Qian, X.H.; Duan, L.P. A Highly Sensitive and Selective OFF-ON Fluorescent Sensor for Cadmium in Aqueous Solution and Living Cell. *J. Am. Chem. Soc.* **2008**, *130*, 16160–16161. [CrossRef] [PubMed]

29. Peng, X.J.; Du, J.J.; Fan, J.L.; Wang, J.Y.; Wu, Y.K.; Zhao, J.Z.; Sun, S.G.; Xu, T. A Selective Fluorescent Sensor for Imaging Cd^{2+} in Living Cells. *J. Am. Chem. Soc.* **2007**, *129*, 1500–1501. [CrossRef] [PubMed]

30. Tamgho, I.S.; Hasheminasab, A.; Engle, J.T.; Nemykin, V.N.; Ziegler, C.J. A new highly fluorescent and symmetric pyrrole–BF2 chromophore: BOPHY. *J. Am. Chem. Soc.* **2014**, *136*, 5623–5626. [CrossRef] [PubMed]

31. Yu, C.J.; Jiao, L.J.; Zhang, P.; Feng, Z.Y.; Cheng, C.; Wei, Y.; Mu, X.L.; Hao, E.H. Highly fluorescent BF2 complexes of hydrazine–Schiff base linked bispyrrole. *Org. Lett.* **2014**, *16*, 3048–3051. [CrossRef] [PubMed]

32. Huaulmé, Q.; Mirloup, A.; Retailleau, P.; Ziessel, R. Synthesis of highly functionalized BOPHY chromophores displaying large stokes shifts. *Org. Lett.* **2015**, *17*, 2246–2249. [CrossRef] [PubMed]

33. Rhoda, H.M.; Chanawanno, K.; King, A.J.; Zatsikha, Y.V.; Ziegler, C.J.; Nemykin, V.N. Unusually Strong Long Distance Metal-Metal Coupling in Bis (ferrocene) Containing BOPHY: An Introduction to Organometallic BOPHYs. *Chem. Eur. J.* **2015**, *21*, 18043–18046. [CrossRef] [PubMed]

34. Wang, L.; Tamgho, I.S.; Crandall, L.; Rack, J.; Ziegler, C. Ultrafast dynamics of a new class of highly fluorescent boron difluoride dyes. *Phys. Chem. Chem. Phys.* **2015**, *17*, 2349–2351. [CrossRef] [PubMed]

35. Sekhar, A.R.; Sariki, S.K.; Reddy, R.V.R.; Bisai, A.; Sahu, P.K.; Tomar, R.S.; Sankar, J. Zwitterionic BODIPYs with large stokes shift: Small molecular biomarkers for live cells. *Chem. Commun.* **2017**, *53*, 1096–1099. [CrossRef] [PubMed]

36. Zhou, L.; Xu, D.F.; Gao, H.Z.; Zhang, C.; Ni, F.F.; Zhao, W.Q.; Cheng, D.D.; Liu, X.L.; Han, A.X. β-Furan-Fused bis (Difluoroboron)-1,2-bis ((1*H*-pyrrol-2-yl) methylene) hydrazine Fluorescent Dyes in the Visible Deep-Red Region. *J. Org. Chem.* **2016**, *81*, 7439–7447. [CrossRef] [PubMed]

37. Li, Y.X.; Zhou, H.P.; Yin, S.H.; Jiang, H.; Niu, N.; Huang, H.; Shahzad, S.A.; Yu, C. A BOPHY probe for the fluorescence turn-on detection of Cu^{2+}. *Sens. Actuators B* **2016**, *235*, 33–38. [CrossRef]

38. Jiang, X.D.; Su, Y.J.; Yue, S.; Li, C.; Yu, H.F.; Zhang, H.; Sun, C.L.; Xiao, L.J. Synthesis of mono-(p-dimethylamino)styryl-containing BOPHY dye for a turn-on pH sensor. *RSC Adv.* **2015**, *5*, 16735–16739. [CrossRef]

39. Xu, Z.C.; Xiao, Y.; Qian, X.H.; Cui, J.N.; Cui, D.W. Ratiometric and selective fluorescent sensor for CuII based on internal charge transfer (ICT). *Org. Lett.* **2005**, *7*, 889–892. [CrossRef] [PubMed]

40. Wang, J.B.; Qian, X.H.; Cui, J.N. Detecting Hg^{2+} ions with an ICT fluorescent sensor molecule: Remarkable emission spectra shift and unique selectivity. *J. Org. Chem.* **2006**, *71*, 4308–4311. [CrossRef] [PubMed]

41. Bozdemir, O.A.; Guliyev, R.; Buyukcakir, O.; Selcuk, S.; Kolemen, S.; Gulseren, G.; Nalbantoglu, T.; Boyaci, H.; Akkaya, E.U. Selective manipulation of ICT and PET processes in styryl-bodipy derivatives: Applications in molecular logic and fluorescence sensing of metal ions. *J. Am. Chem. Soc.* **2010**, *132*, 8029–8036. [CrossRef] [PubMed]

42. Srikun, D.; Miller, E.W.; Domaille, D.W.; Chang, C.J. An ICT-based approach to ratiometric fluorescence imaging of hydrogen peroxide produced in living cells. *J. Am. Chem. Soc.* **2008**, *130*, 4596–4597. [CrossRef] [PubMed]

43. Thiagarajan, V.; Ramamurthy, P.; Thirumalai, D.; Ramakrishnan, V.T. A novel colorimetric and fluorescent chemosensor for anions involving PET and ICT pathways. *Org. Lett.* **2005**, *7*, 657–660. [CrossRef] [PubMed]

44. Zhao, W.Q.; Liu, X.L.; Lv, H.T.; Fu, H.; Yang, Y.; Huang, Z.P.; Han, A.X. A phenothiazine–rhodamine ratiometric fluorescent probe for Hg^{2+} based on FRET and ICT. *Tetrahedron Lett.* **2015**, *56*, 4293–4298. [CrossRef]

sensors

MDPI

Article

Development and Elucidation of a Novel Fluorescent Boron-Sensor for the Analysis of Boronic Acid-Containing Compounds

Yoshihide Hattori * , Takuya Ogaki, Miki Ishimura, Yoichiro Ohta and Mitsunori Kirihata

Research Center of Boron Neutron Capture Therapy, Osaka Prefecture University, 1-1 Gakuen-cho, Nakaku, Sakai, Osaka 599-8531, Japan; takuya.ogaki@riken.jp (T.O.); ishimura@21c.osakafu-u.ac.jp (M.I.); yohta@bioinfo.osakafu-u.ac.jp (Y.O.); kirihata@biochem.osakafu-u.ac.jp (M.K.)
* Correspondence: y0shi_hattori@riast.osakafu-u.ac.jp; Tel.: +81-72-254-6423

Received: 14 September 2017; Accepted: 20 October 2017; Published: 24 October 2017

Abstract: Novel boron-containing drugs have recently been suggested as a new class of pharmaceuticals. However, the majority of current boron-detection techniques require expensive facilities and/or tedious pretreatment methods. Thus, to develop a novel and convenient detection method for boron-based pharmaceuticals, imine-type boron-chelating-ligands were previously synthesized for use in a fluorescent sensor for boronic acid containing compounds. However, the fluorescence quantum yield of the imine-type sensor was particularly low, and the sensor was easily decomposed in aqueous media. Thus, in this paper, we report the development of a novel, convenient, and stable fluorescent boron-sensor based on *O*- and *N*-chelation (i.e., 2-(pyridine-2yl)phenol), and a corresponding method for the quantitative and qualitative detection of boronic acid-containing compounds using this commercially available sensor is presented.

Keywords: fluorescent boron sensor; BNCT; boron pharmaceutical; boron(III) complex

---

1. Introduction

The use of boronic acid-containing compounds has recently received growing interest in a range of fields, such as materials science, analytical chemistry, chemical biology, and pharmacology. In particular, a large number of studies focusing on the interactions between boron compounds and biomolecules such as sugars, proteins and peptides have been reported in recent years [1–6]. For example, in the field of pharmacology, boronic acid-containing compounds have been developed as boron carriers for boron neutron capture therapy (BNCT) [7], and as antibacterial agents, protease inhibitors, sugar sensors, and cell-penetrating peptides. In addition, new boron-containing drugs, including *p*-borono-L-phenylalanine (L-BPA, for BNCT), L-BPA-fructose complex (L-BPA-Fc) [8–10], bortezomib (for the treatment of multiple myeloma) [11,12], and tavaborole (an antifungal drug for the treatment of onychomycosis) [13] have been suggested as a new class of pharmaceuticals [14,15] (Figure 1).

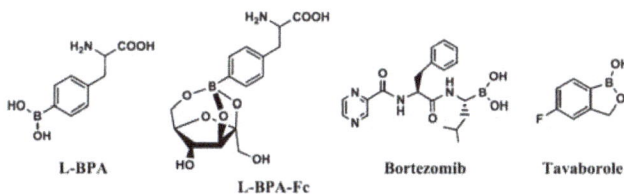

Figure 1. A selection of boron-containing pharmaceuticals.

During the development of novel pharmaceutical compounds, drug distribution analysis is of particular importance. For example, the distribution of boron-based pharmaceuticals in tumor tissue and normal tissue must be determined to elucidate the suitability of pharmaceuticals for BNCT, and monitoring of the boron concentration in bloodstream is of key importance in the BNCT treatment process. In this context, evaluation of the concentration of boron-based pharmaceuticals in biological environments should be possible through the detection of boron, since biological tissues generally contain little or no boron. However, the majority of boron detection methods, such as those based on prompt gamma-ray analysis (PGA) and inductively coupled plasma optical emission spectrometry (ICP-OES) are costly, and often require extensive sample pretreatment.

In this context, we recently designed and synthesized a novel fluorescent probe for boronic acid, namely DAHMI (**1**) (Figure 2) [16]. This probe reacted rapidly with boron-based pharmaceuticals in aqueous media, resulting in the emission of blue fluorescence by the formed DAHMI-boron complexes. Furthermore, DAHMI allowed visualization of the distribution of boron-containing pharmaceuticals such as L-BPA, and tavabolore in live tumor cells without complicated pretreatment [17]. However, the fluorescence quantum yields of the resulting DAHMI-boron complexes were extremely low (i.e., <1%), and DAHMI was easily decomposed by hydrolysis in aqueous media. As such, DAHMI is not particularly suitable for use as a boron sensor in the quantitative analysis of boron compounds in biological samples.

Thus, in this paper, we report the development of a novel, convenient, and stable fluorescent probe for boronic acid-containing compounds, and present a subsequent elucidation of the fluorescence properties of the resulting sensor-boronic acid complexes.

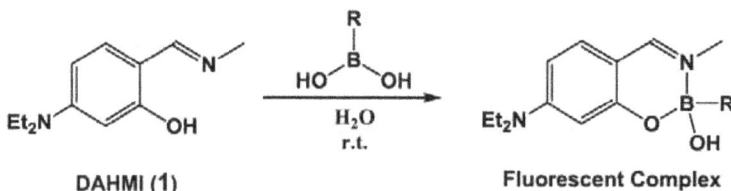

DAHMI (1)　　　　　　　　　　　　　　　　**Fluorescent Complex**

Figure 2. Complex formation using the previously reported fluorescent boron sensor DAHMI (**1**).

2. Results and Discussion

In order for a fluorescent sensor to detect boronic acid derivatives in biological environments, the following criteria must be met: (i) rapid reaction with boronic acid at room temperature; (ii) a change in emission upon complexation with boronic acid; (iii) fluorescence in water; and (iv) stability in water.

Thus, we herein employed a range of commercially available and easily prepared chelating ligands **2–6** as precursors for the boron sensor (Figure 3) [18–21]. These ligands were considered suitable as they were both soluble and stable in a 50 vol % DMSO/H_2O solution, and because they emit very weak fluorescence in aqueous solutions. To confirm the potential of our previously prepared fluorescent probe DAHMI in addition to the five chelating ligands for use as boron sensors, we measured the fluorescence spectra of compounds **1–6** and their corresponding BPA complexes in a 50 vol % DMSO/PBS (phosphate-buffered saline) solution at 25 °C (Table 1). Although compounds **2** and **6** did not react with BPA under these conditions, DAHMI and compounds **3–5** reacted to give fluorescent complexes with BPA. This result suggests that *N*, *O*-type chelating ligands are suitable for use as fluorescent boron-sensors. In addition, we observed that the excitation and fluorescence wavelengths of these complexes differed from those of the corresponding free ligands. In particular, the fluorescence wavelength of the **5**-BPA (**5** = 2-(pyridine-2-yl)phenol) complex was blue shifted by ~20 nm, and the Stokes shift of the complex was extremely large. Furthermore, the maximum fluorescence quantum yield was observed for **5**-BPA, with a 15-fold higher yield being observed than for the DAHMI-BPA complex. Indeed, this **5**-BPA complex showed particular potential for use as a sensor, as it was prepared rapidly by the simple mixing of a solution of **5** with a solution of BPA (1.0 eq.) at room temperature in 50 vol % DMSO/PBS. Upon mixing, the fluorescence intensity of **5**-BPA complex was reached a maximum within a few seconds, and this complex was stable in solution over 24 h (Figure 4). These results indicate that compound **5** is suitable for use as a fluorescent sensor in the detection of boronic acid-containing compounds.

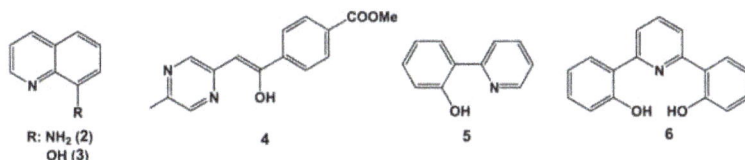

Figure 3. Candidate compounds for the fluorescent boron sensor.

Table 1. Fluorescence properties of the various ligands and ligand-BPA complexes [a].

	Ligand Only				Ligand-BPA Complex			
Ligand	Ex Max (nm)	Em Max (nm)	Stokes Shift (cm^{-1})	ϕ	Ex Max (nm)	Em Max (nm)	Stokes Shift (cm^{-1})	ϕ
1	413	551	6064	—	408	430	1254	0.6%
2	374	531	—	—	374	531	—	—
3	419	477	2902	—	421	516	4373	3.3%
4	340	384	3370	—	352	433	5314	1.2%
5	376	483	5892	—	355	464	6617	9.3%
6	376	—	—	—	376	—	—	—

[a] Measured at a concentration of 1.0 mM in 50 vol % DMSO/PBS at 25 °C.

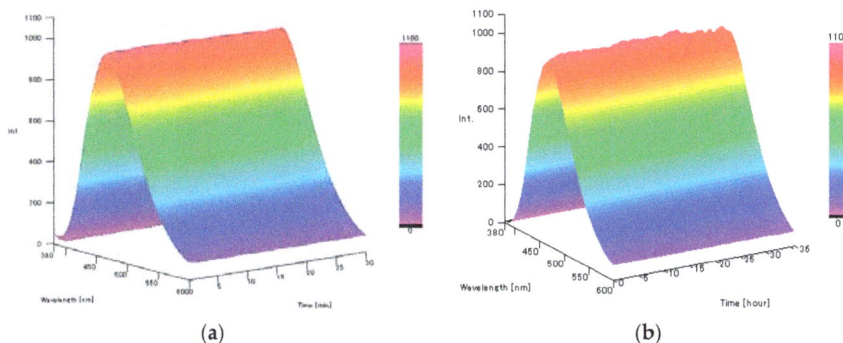

(a)   (b)

Figure 4. Variation in the fluorescence spectra of the 5-BPA complex in 50 vol % DMSO/PBS (0.2 mM, excitation wavelength: 355 nm) at 25 °C. (a) prepared after 0–60 min; (b) prepared after 0–36 h.

Thus, to confirm the potential of boron sensor **5** for use in the qualitative analysis of boronic acid-containing compounds, a number of boron compounds were spotted onto silica gel plates and stained using a 1.0 mM solution of **5** in acetone (Figure 5). Although boron sensor **5** was visible under irradiation using a 254 nm hand-held UV lamp, the spot of **5** did not emit fluorescence under 254 or 365 nm UV lamps. Thus, as shown in Figure 5, all spots corresponding to the different boronic acid-containing compounds, including L-BPA, bortezomib and tavaborole produced bright blue fluorescence using a standard short-wave UV lamp (365 nm) following staining with compound **5**. In particular, boronic ester-type protected derivatives, such as pinacolato, 1,3-dihydroxydimetyl, trifluoroborate and L-BPA-Fc, exhibited strong fluorescence upon complexation with **5**. However, in the case of boronic amide type protected derivatives, such as *N*-methyliminodiacetic acid and 1,8-diaminonaphtnalene, only weak fluorescence was observed. As a control, a number of boron-free compounds were also spotted onto the silica plates, but no fluorescence was observed under UV irradiation at either 254 or 365 nm (Figure 6). These results therefore suggest that boron sensor **5** is a useful tool for the qualitative analysis of boronic acid-containing compounds.

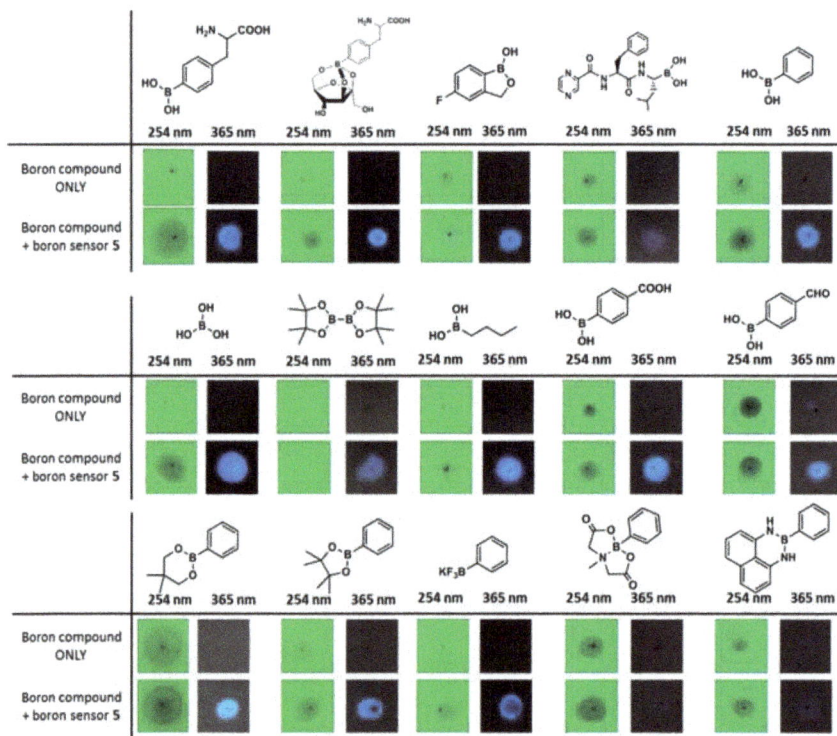

Figure 5. Staining test of various boron-containing compounds spotted onto silica-gel plates followed by the addition of boron sensor **5** (viewed by illumination at 254 or 365 nm using a handheld UV lamp).

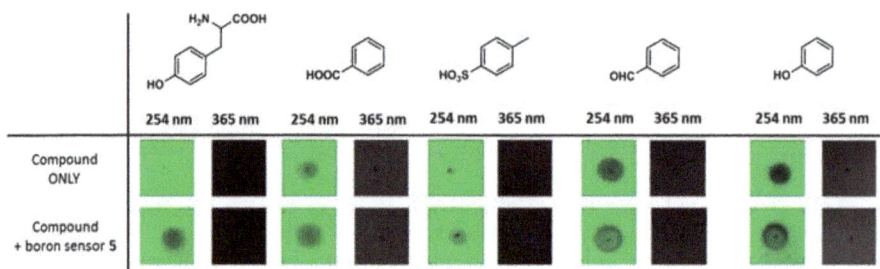

Figure 6. Staining test of boron-free compounds spotted onto silica-gel plates followed by the addition of boron sensor **5** (viewed by illumination at 254 or 365 nm using a handheld UV lamp).

Finally, to determine the potential of boron sensor **5** in the quantitative analysis of boron pharmaceuticals, we examined the relationship between the concentration of L-BPA-Fc and the emission of the complex upon staining with **5** in 50 vol % DMSO/PBS using a standard plate reader. As shown in Figure 7, the emission intensity correlated positively with the concentration of L-BPA-Fc, thereby indicating that boron sensor **5** is suitable for use in the quantitative analysis of L-BPA-Fc (0.5–1000 μM, 0.005–10 Bppm, $R^2 > 0.99$). Furthermore, the detection limit of this method was comparable to ICP-OES.

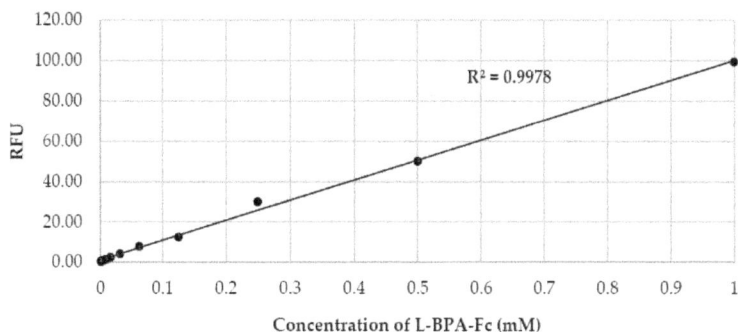

Figure 7. Effect of L-BPA-Fc concentration on the fluorescence intensity following staining with boron-sensor **5** (10 mM in 50 vol % DMSO/PBS, ex λ: 360 nm, em λ: 460 nm).

3. Conclusions

We herein developed and elucidated a novel efficient and commercially available fluorescent sensor based on *O*- and *N*- chelation (i.e., 2-(pyridine-2yl)phenol) for the analysis of boronic acid-containing compounds. We found that this boron sensor reacts rapidly with boronic acid at room temperature, and that it selectively detects boronic acid-containing compounds. Furthermore, the quantitative analyses of boronic acid-containing compounds using this sensor were carried out using a standard plate reader, with the compounds of interest being detected in concentrations of 0.5–1000 μM. These results therefore suggest that this fluorescent boron sensor is suitable for use in the qualitative and quantitative analysis of boronic acid-containing compounds. We therefore expect that the system reported herein will be applicable in the detection of boron-based pharmaceuticals, with the potential to replace current expensive and tedious analytical methods.

4. Experimental Section

4.1. Genaral

L-BPA was provided by the Stella Pharma Corporation (Osaka, Japan), while bortezomib and tavaborole were purchased from Cosmo Bio Co., Ltd. (Tokyo, Japan). Compounds **2**, **3**, and **5** were purchased from Wako Pure Chemical Industries, Ltd. (Osaka, Japan). Compounds **4** and **6** were prepared according to a previous literature method [19,21]. Fluorescence spectra were measured on a FP-8200 spectrometer (JASCO Corporation, Tokyo, Japan). Absolute quantum yields were determined by the Hamamatsu C9920-01 calibrated integrating sphere system (Hamamatsu Photonics K.K., Shizuoka, Japan).

4.2. Preparation of the Sensor-BPA Complexes

A solution of boron sensor (1.0 mM in DMSO, 2.5 mL) was added to a solution of BPA (1.0 mM in PBS buffer, 2.5 mL). After allowing this mixture to stand for 10 min at 37 °C, the resulting solution was employed for fluorescence measurement.

4.3. Silica Gel Plate Staining Using Boron Sensor 5

L-BPA and L-BPA-Fc were prepared as aqueous solutions (10 mM) concentrations while solutions of all other boron-containing compounds were prepared in acetone (10 mM). Each sample solution (5 μL) was applied to a glass-backed silica-gel TLC plate containing the F254 fluorescent indicator (Merck KGaA, Darmstadt, Germany) and allowed to dry under air at room temperature prior to treatment of the test compound spot with a solution of boron sensor 5 (1.0 mM in acetone). After allowing to dry once again under air at room temperature, the plates were visualized by illumination at either 254 or 365 nm using a handheld UV lamp, and photographic images were recorded.

4.4. Effect of BPA Concentration on the Emitted Fluorescence Following Staining with 5

To a 96-well microplate (FluoroNuncTM flat bottom black polystyrene plate, Thermo Fisher Scientific, Waltham, MA) were added of a L-BPA-Fc solution of the desired concentration (50 μL, 0.5–1000 μM in H_2O) and a solution of boron sensor 5 (50 μL, 10 mM in DMSO. After mixing for 5 min at 37 °C, the fluorescence intensity was measured using a Fluoroskan Ascent FL microplate reader (Thermo Fisher Scientific, Waltham, MA, USA).

Supplementary Materials: The following are available online at http://www.mdpi.com/1424-8220/17/10/2436/s1.

Acknowledgments: We would like to thank Yoshihiro Yamaguchi of the Kindai University for the measurement of fluorescence properties. This work was supported by the Nakatani Foundation for Advancement of Measuring Technologies in Biomedical Engineering grant program for biomedical engineering research, and by the Project for Cancer Research and Therapeutic Evolution (PCREATE) from the Japan Agency for Medical Research and Development, AMED.

Author Contributions: Y.H. and M.K. designed the experiments. Y.H., Y.O. and H.T. performed the synthesis of the compounds. Y.H. and M.I. performed the measurement of fluorescence properties. Y.H. wrote the manuscript. All authors discussed the results and implications of the study and approved the final manuscript.

Conflicts of Interest: The authors declare no conflict of interest.

References

1. Yang, W.; Gao, X.; Wang, B. Boronic acid compounds as potential pharmaceutical agents. *Med. Res. Rev.* **2003**, *23*, 346–368. [CrossRef] [PubMed]
2. Chaudhary, P.M.; Murthy, R.V.; Yadav, R.; Kikkeri, R. A rationally designed peptidomimetic biosensor for sialic acid on cell surfaces. *Chem. Commun.* **2015**, *51*, 8112–8115. [CrossRef] [PubMed]
3. Dilek, O.; Lei, Z.; Mukherjee, K.; Bane, S. Rapid formation of a stable boron–nitrogen heterocycle in dilute, neutral aqueous solution for bioorthogonal coupling reactions. *Chem. Commun.* **2015**, *51*, 16992–16995. [CrossRef] [PubMed]
4. Okuro, K.; Sasaki, M.; Aida, T. Boronic acid-appended molecular glues for ATP-responsive activity modulation of enzymes. *J. Am. Chem. Soc.* **2016**, *138*, 5527–5530. [CrossRef] [PubMed]
5. Palombella, V.J.; Conner, E.M.; Fuseler, J.W.; Destree, A.; Davis, J.M.; Larous, F.S.; Wolf, R.E.; Huang, J.; Brand, S.; Elliott, P.J.; et al. Role of the proteasome and NF-kB in streptococcal cell wall-induced polyarthritis. *Proc. Natl. Acad. Sci. USA* **1998**, *95*, 15671–15676. [PubMed]
6. Piest, M.; Ankoné, M.; Engbersen, J.F.J. Carbohydrate-interactive pDNA and siRNA gene vectors based on boronic acid functionalized poly(amido amine)s. *J. Controlled Release* **2013**, *169*, 266–275. [CrossRef] [PubMed]
7. Soloway, A.H.; Tjarks, W.; Barnum, B.A.; Rong, F.G.; Barth, R.F.; Codogni, I.M.; Wilson, J.G. The chemistry of neutron capture therapy. *Chem. Rev.* **1998**, *98*, 1515–1562. [CrossRef] [PubMed]
8. Mishima, Y.; Honda, C.; Ichihashi, M.; Obara, H.; Ichihashi, M.; Obara, H.; Hiratsuka, J.; Fukuda, H.; Karashima, H.; Kobayashi, T.; et al. Treatment of malignant melanoma by single thermal neutron capture therapy with melanoma-seeking [10]B-compound. *Lancet* **1989**, *2*, 388–389. [PubMed]
9. Hattori, Y.; Asano, T.; Kirihata, M.; Yamaguchi, Y.; Wakamiya, T. Development of the first and practical method for enantioselective synthesis of [10]B-enriched p-borono-L-phenylalanine. *Tetrahedron Lett.* **2008**, *49*, 4977–4980. [CrossRef]

10. Andoh, T.; Fujimoto, T.; Sudo, T.; Fujita, I.; Imabori, M.; Moritake, H.; Sugimoto, T.; Sakuma, Y.; Takeuchi, T.; Kawabata, S.; et al. Boron neutron capture therapy for clear cell sarcoma (CCS): Biodistribution study of *p*-borono-L-phenylalanine in CCS-bearing animal models. *Appl. Radiat. Isot.* **2011**, *69*, 1721–1724. [CrossRef] [PubMed]

11. Adams, A. Development of the proteasome inhibitor PS-341. *Oncologist* **2002**, *7*, 9–16. [CrossRef] [PubMed]

12. Girbig, A.K.; Kalesse, M. Synthesis and pharmacology of proteasome inhibitors. *Angew. Chem. Int. Ed.* **2013**, *52*, 5450–5488.

13. Jinna, S.; Finch, J. Spotlight on tavaborole for the treatment of onychomycosis. *Drug Des. Devel. Ther.* **2015**, *9*, 6185–6190. [PubMed]

14. Baker, S.J.; Ding, C.Z.; Akama, T.; Zhang, Y.; Hernandez, V.; Xia, Y. Therapeutic potential of boron-containing compounds. *Future Med. Chem.* **2009**, *1*, 1275–1288. [CrossRef] [PubMed]

15. Smoum, R.; Rubinstein, A.; Dembisky, V.M.; Srebnik, M. Boron containing compounds as protease inhibitors. *Chem. Rev.* **2012**, *112*, 4156–4220. [CrossRef] [PubMed]

16. Hattori, Y.; Ishimura, M.; Ohta, Y.; Takenaka, H.; Watanabe, T.; Tanaka, H.; Ono, K.; Kirihata, M. Detection of boronic acid derivatives in cells using a fluorescent sensor. *Org. Biomol. Chem.* **2015**, *13*, 6927–6930. [CrossRef] [PubMed]

17. Hattori, Y.; Ishimura, M.; Ohta, Y.; Takenaka, H.; Kirihata, M. Visualization of boronic acid-containing pharmaceuticals in live tumor cells using a fluorescent boronic acid sensor. *ACS Sens.* **2016**, *1*, 1394–1397. [CrossRef]

18. Nagata, Y.; Chujo, Y. Synthesis of methyl-substituted main-chain-type organoboron quinolate polymers and their emission color tuning. *Macromolecules* **2008**, *41*, 2809–2813. [CrossRef]

19. Kubota, Y.; Hara, H.; Tanaka, S.; Funabiki, K.; Matsui, M. Synthesis and fluorescence properties of novel pyrazine boron complexes bearing a β-iminoketone ligand. *Org. Lett.* **2011**, *13*, 6544–6547. [CrossRef] [PubMed]

20. Kim, N.G.; Shin, G.H.; Lee, M.H.; Do, Y. Four-coordinate boron compounds derived from 2-(2-pyridyl)phenol ligand as novel hole-blocking materials for phosphorescent OLEDs. *J. Organomet. Chem.* **2009**, *694*, 1922–1928. [CrossRef]

21. Li, Y.; Liu, Y.; Bu, W.; Guo, J.; Wang, Y. A mixed pyridine–phenol boron complex as an organic electroluminescent material. *Chem. Commun.* **2000**, 1551–1552. [CrossRef]

Article

Design and Evaluation of Novel Polymyxin Fluorescent Probes

Bo Yun [1], Kade D. Roberts [1], Philip E. Thompson [2], Roger L. Nation [1], Tony Velkov [1,*] and Jian Li [3,*]

1   Drug Delivery, Disposition and Dynamics, Monash University, Parkville, Victoria 3052, Australia;
    Bo.Yun@petermac.org (B.Y.); Kade.Roberts@monash.edu (K.D.R.); roger.nation@monash.edu (R.L.N.)
2   Medicinal Chemistry, Monash Institute of Pharmaceutical Sciences, Monash University, Parkville,
    Victoria 3052, Australia; Philip.Thompson@monash.edu
3   Monash Biomedicine Discovery Institute, Department of Microbiology, Monash University, Clayton,
    Victoria 3800, Australia
*   Correspondence: tony.velkov@unimelb.edu.au (T.V.); jian.li@monash.edu (J.L.)

Received: 12 October 2017; Accepted: 9 November 2017; Published: 11 November 2017

Abstract: Polymyxins (polymyxin B and colistin) are cyclic lipopeptide antibiotics that serve as a last-line defence against Gram-negative "superbugs". In the present study, two novel fluorescent polymyxin probes were designed through regio-selective modifications of the polymyxin B core structure at the N-terminus and the hydrophobic motif at positions 6 and 7. The resulting probes, FADDI-285 and FADDI-286 demonstrated comparable antibacterial activity (MICs 2–8 mg/L) to polymyxin B and colistin (MICs 0.5–8 mg/L) against a panel of gram-negative clinical isolates of *Acinetobacter baumannii*, *Klebsiella pneumoniae* and *Pseudomonas aeruginosa*. These probes should prove to be of considerable utility for imaging cellular uptake and mechanistic investigations of these important last-line antibiotics.

Keywords: polymyxins; fluorescent; dansyl; probe; gram-negative

1. Introduction

Over the past two decades there has been a pronounced increase in the emergence of multidrug-resistant (MDR) Gram-negative "superbugs", leading to serious infections that are resistant to almost all currently available antibiotics [1]. The dire situation is perpetuated by a lack of novel antibiotics in the developmental pipeline, leaving the world in a vulnerable state against these life-threatening infections [1]. This "perfect storm" has led to the revival of the polymyxin class of antibiotics, polymyxin B and E (the latter also known as colistin), as a last line of defence against MDR Gram-negative "superbugs" [2]. However, despite their excellent antibacterial activity, the use of polymyxins has largely been limited by a high incidence of nephrotoxicity among patients receiving these antibiotics [3–6].

Polymyxins are amphipathic cationic lipopeptides, comprised of hydrophobic and hydrophilic domains that are critical for their antibacterial activity [7]. The general polymyxin structure consists of a cyclic heptapeptide ring with a linear tripeptide segment and an N-terminal fatty acyl tail (Figure 1). Additionally, there are five L-α,γ-diaminobutyric acid (Dab) residues, which contain primary amines that are positively charged at physiological pH (7.4), as well as two hydrophobic residues in positions 6 and 7 of the cyclic ring. The two polymyxins used clinically, polymyxin B and colistin, are differentiated by a single hydrophobic residue at position 6: D-leucine in colistin and D-phenylalanine in polymyxin B [7]. Both polymyxins are products of fermentation and are mixtures, each containing two major components, colistin A and B and polymyxin B_1 and B_2, which differ by one carbon at the fatty acyl tail (Figure 1). The fatty acyl tail is essential for the antibacterial activity of polymyxins, since polymyxin nonapeptide (PMBN) (produced by proteolytic removal of the fatty acyl-Dab1 from the *N*-terminus of the polymyxin) is inactive [8–10]. These structural features of the polymyxin core scaffold are critical for interaction with the initial target, lipid A of the outer membrane.

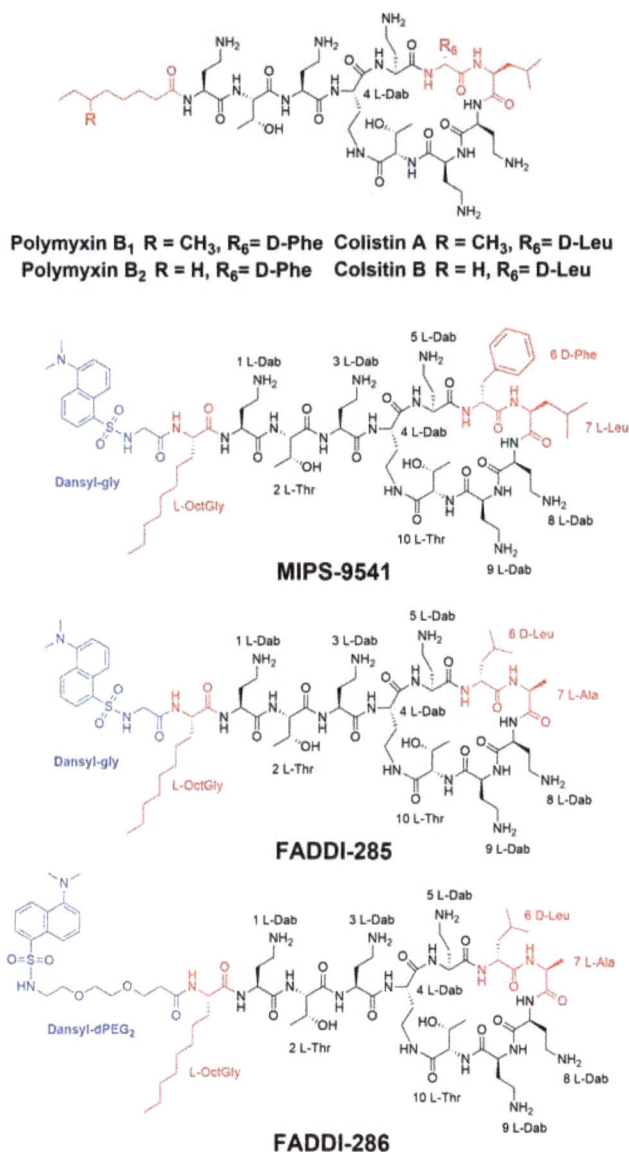

Polymyxin B₁ R = CH₃, R₆= D-Phe Colistin A R = CH₃, R₆= D-Leu
Polymyxin B₂ R = H, R₆= D-Phe Colsitin B R = H, R₆= D-Leu

MIPS-9541

FADDI-285

FADDI-286

Figure 1. Structures of polymyxin B, colistin, MIPS-9541 and the novel fluorescent polymyxin probes FADDI-285 and FADDI-286.

Commercial probes (e.g., dansyl- and BODIPY-polymyxin B) have been utilized in polymyxin mechanistic studies, however, they lack antimicrobial activity due to the blockage of multiple Dab residues (potentially up to all five); therefore, these compounds are not structurally representative of the parent polymyxin [11]. Our group has previously reported the design and synthesis of regio-selectively mono-dansylated polymyxin B probes such as MIPS-9541 (Figure 1) for exploring polymyxin mechanisms of action and imaging of polymyxin interactions with kidney proximal tubular cells [12]. In the current study, we build on our novel design strategy generating the novel fluorescently labelled polymyxin probes, FADDI-285 and FADDI-286 (Figure 1) which are representative of the native polymyxins, and retain antimicrobial activity. These novel fluorescent polymyxin probes should have improved in vivo utility and help facilitate medicinal chemistry strategies to ameliorate unwanted nephrotoxicity and resistance that limit the clinical efficacy of these important last-line lipopeptide antibiotics.

2. Methods

2.1. Chemical Reagents

Diisopropylethylamine (DIPEA) was obtained from Auspep (Melbourne, Australia). Fmoc-L-OctGly-OH and Fmoc-Dab(Boc)-OH were obtained from Try-lead Chem (Hangzhou, China). Fmoc-Dab(ivDde)-OH, Fmoc-D-Leu-OH, 1H-Benzotriazolium-1-[bis(dimethylamino)methylene]-5-chloro hexafluoro- phosphate-(1-),3-oxide (HCTU) and 1,1,1,3,3,3-Hexafluoro-2-propanol (HFIP) were obtained from Chem-Impex International (Wood Dale, IL, USA). Fmoc-Thr(tBu)-OH, Fmoc-Ala-OH and Fmoc-Gly-OH were obtained from Mimotopes (Melbourne, Australia). N-Fmoc-Amido-dPEG$_2$-OH was obtained from Peptides International (Louisville, KY, USA). Dichloromethane (DCM), dimethylformamide (DMF), diethyl ether and acetonitrile were obtained from Merck (Melbourne, Australia). Fmoc-Thr(tBu)-TCP-Resin was obtained from Intavis Bioanyltical Instruments (Köln, Germany). Piperidine, triisopropylsilane (TIPS), trifluoroacetic acid (TFA), dansyl-chloride, ethanedithiol (EDT) and diphenylphosphorylazide (DPPA) were obtained from Sigma-Aldrich (Castle Hill, Australia) Polymyxin B sulfate and colistin sulfate were research grade and obtained from BetaPharma (Shanghai, China).

2.2. HPLC Purification and LC-MS Analysis

Peptides were purified by RP-HPLC on a Waters Prep LC system incorporating a Waters 486 tuneable absorbance detector set at 214 nm and a Phenomenex Luna C8(2) column (250 × 21.2 mm ID, 100 Å, 10 micron). Peptides were eluted with a gradient of 100% Buffer A (0.1% TFA/water) to 60% Buffer B (0.1%TFA/acetonitrile) over 60 min at a flow rate of 15 mL/min. Fractions collected were analysed by LC/MS on a Shimadzu 2020 LCMS system. LC analysis was carried out at 214 nm using a Phenomenex Luna C8(2) column (100 × 2.0 mm ID, 100 Å, 3 micron), eluting with a gradient of 100% Buffer A (0.05% TFA/water) to 60% Buffer B (0.05%TFA/acetonitrile) over 10 min at a flow rate of 0.2 mL/min. Mass spectra were acquired in positive ion mode with a scan range of 200–2000 m/z.

2.3. Synthesis FADDI-285

Synthesis of the protected linear peptide precursor was conducted on a Protein Technologies Prelude automated peptide synthesizer using pre-loaded Fmoc-Thr(tBu)-TCP resin (0.1 mmol scale). Fmoc deprotection was conducted using 20% piperidine in dimethylformamide (1 × 5 min, 1 × 10 min) at room temperature. Coupling of the Fmoc-amino acids for 50 min at room temperature using 3 molar equivalents of the Fmoc-amino acid and HCTU in DMF activated in situ, using 6 molar equivalents of DIPEA. The N-terminal dansyl group was coupled using 3 molar equivalents of dansyl-chloride in DMF in the presence of 6 molar equivalents of DIPEA for 50 min at room temperature. The resin was then treated with 3% Hydrazine/ DMF (4 × 15 min) to remove the ivDde group. The protected linear peptide was then cleaved from the resin by treating the resin with 10% HFIP in DCM (1 × 30 min, 1 × 5 min). This solution was concentrated *in vacuo* to produce the protected linear peptide as a residue. The protected linear peptide was dissolved in DMF (10 mL) to which DPPA 0.3 mmol, 0.65 μL (3 molar equivalents relative to the loading of the resin) and DIPEA 0.6 mmol, 104 μL (6 molar equivalents relative to the loading of the resin) were added. This solution was stirred at room temperature overnight. The reaction solution was then concentrated under vacuum overnight to give the crude protected cyclic peptide as a residue. The resulting residue was taken up in a solution of 2.5% EDT/5% TIPS/TFA and shaken at room temperature for 2 h. To this solution 40 mL of diethyl ether was added. The resulting precipitate was collected by centrifugation and washed twice more with diethyl ether (40 mL) then air-dried in a fume hood to give the crude cyclic peptide as a solid. The resulting solid was taken up in Milli-Q water (5 mL) and de-salted using a Vari-Pure IPE SAX column. The eluent containing the crude cyclic peptide was acidified with TFA (10 L) and subjected to RP-HPLC purification as described above. Fractions collected were analysed by LC-MS as described above. Fractions containing the desired product were freeze-dried to give the **FADDI-285** TFA salt as a pale-yellow solid in a yield of 57.2 mg (>95% purity). Molecular weight was confirmed by ESI-MS analysis; m/z (monoisotopic) calculated: $C_{65}H_{111}N_{19}O_{16}S$ 1446.82, $[M + 2H]^{2+}$ 723.91, $[M + 3H]^{3+}$ 482.93; observed: $[M + 2H]^{2+}$ 724.30. $[M + 3H]^{3+}$ 483.50.

2.4. Synthesis FADDI-286

This peptide was synthesized as described above for **FADDI-285** to give the **FADDI-286** TFA salt as a pale-yellow solid in a yield of 65.2 mg (>95% purity). Molecular weight was confirmed by ESI-MS analysis; m/z (monoisotopic) calculated: $C_{70}H_{121}N_{19}O_{18}S$ $[M + H]^+$ 1548.89, $[M + 2H]^{2+}$ 774.94, $[M + 3H]^{3+}$ 516.96; observed: $[M + 2H]^{2+}$ 775.40, $[M+3H]^{3+}$ 517.60.

2.5. Determination of MICs

MICs against *Pseudomonas aeruginosa*, *Klebsiella pneumoniae*, and *Acinetobacter baumannii* strains were determined by the broth microdilution method (CLSI 2013). Experiments were conducted in 96-well polypropylene microtitre plates with all dilutions using cation-adjusted Mueller-Hinton broth (CaMHB). Bacterial suspension (100 μL, containing ~10^6 colony forming units (CFU) per mL) was added to the wells in the presence of increasing concentrations of polymyxins (0 to 128 mg/L). MICs are defined as the lowest concentration at which visible growth was inhibited after overnight incubation at 37 °C.

3. Results

3.1. Probe Design and Synthesis

Previously we designed the regio-selectively mono-dansylated probe MIPS-9541 in which the *N*-terminal fatty acyl group of polymyxin B was substituted with dansylglycine-octanylglycine (Figure 1) [11]. The hydrophobic dansyl group was utilized as the fluorophore, as its comparatively small size relative to other fluorophores would help to reduce the likelihood of negative steric effects on the polymyxin pharmacophore [7]. The L-octylglycine (C8) residue serves to emulate the eight carbon *N*-terminal fatty acyl chain of the polymyxins and also provides a point for attachment of the dansyl fluorophore [11]. It has been previously demonstrated that a variety of hydrophobic groups are well tolerated at this position and act as mimics of the *N*-terminal saturated alkyl fatty acyl chains of polymyxin B [7]. NMR analysis showed that this probe has a similar mode of binding to lipid A, the initial target of the polymyxins such as the native polymyxin B (Figure 2). One of the concerns with the design of MIPS-9541, was the increase in the overall hydrophobicity of the scaffold resulting from the addition of the dansyl fluorophore which may have a negative effect on the ability of the molecule to closely mimic the native polymyxins. For example, increased hydrophobicity could lead to increased plasma protein binding leading to poor bio-distribution compared to the native polymyxins, which would have a negative impact on in vivo studies utilizing the probe [13]. In the current work, we have made further modification to MIPS-9541 in order to balance the hydrophobicity of the scaffold and improve its in vivo utility. To this end we incorporated a D-leucine residue at position 6 in place of D-Phe as seen in colistin and reduced the hydrophobicity at position 7 by substituting the leucine residue with a less hydrophobic alanine residue. This resulted in the generation of FADDI-285 (Figure 1). Previously it had been shown that the leucine residue at position 7 could be substituted with an alanine residue without loss of antibacterial activity against *P. aeruginosa* [14]. Further to these modifications, the glycine linker between the dansyl fluorophre and octylglycine residue was replaced with a PEG linker to generate FADDI-286. This PEG linker would help to decrease the hydrophobicity of the dansyl-octylglycine *N*-terminal modification.

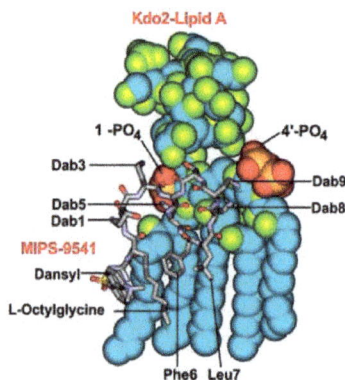

Figure 2. NMR derived model of MIPS-9541 in complex with bacterial Kdo2-lipid.

FADDI-285 and FADDI-286 had to be prepared using a total synthesis approach and were readily synthesized using standard solid-phase peptide synthesis and commercially available chemical reagents. Pleasingly, the modifications made to the polymyxin structure did not have a negative impact on the key cyclisation step to generate the heptapeptide cyclic ring, with final products being obtained in good yield and purity.

3.2. Antibacterial Activity of the Probes

Antibacterial activity of both probes was assessed against the American Type Culture Collection (ATCC) strains and a panel of clinical isolates of polymyxin-susceptible *P. aeruginosa*, *K. pneumoniae* and *A. baumannii* (Table 1). Both probes demonstrated antibacterial activity against the polymyxin-susceptible strains (MICs 2–8 mg/L), comparable to polymyxin B and colistin (MICs 0.5–8 mg/L). Notably, both probes exhibited enhanced activity (MICs ~ 32 mg/L) against the polymyxin-resistant isolates (polymyxin B and colistin MICs ~ 128 mg/L).

Table 1. Minimum inhibitory concentrations (MICs) of each dansylated probe, polymyxin B and colistin against Gram-negative bacteria.

Peptide	*Pseudomonas aeruginosa* ATCC 27853	*Pseudomonas aeruginosa* FADDI-PA022	*Pseudomonas aeruginosa* FADDI-PA025	*Pseudomonas aeruginosa* FADDI-PA070	*Pseudomonas aeruginosa* FADDI-PA060	*Pseudomonas aeruginosa* FADI-PA090	*Acinetobacter baumannii* ATCC 19606	*Acinetobacter baumannii* FADDI-AB034	*Acinetobacter baumannii* ATCC 17978	*Acinetobacter baumannii* ATCC 19606 Col 10	*Acinetobacter baumannii* FADDI-AB156	*Acinetobacter baumannii* FADDI-AB167	*Klebsiella pneumoniae* ATCC 13883	*Klebsiella pneumoniae* FADDI-KP027	*Klebsiella pneumoniae* FADDI-KP003	*Klebsiella pneumoniae* FADDI-KP012
	MIC (mg/L)															
Colistin	1	1	2	>128	>128	8	1	0.5	0.5	128	16	8	1	>128	128	32
PolymyxinB	1	1	1	32	>32	4	1	0.5	1	128	8	16	1	128	>32	16
FADDI-285	2	2	4	>32	2	8	4	4	8	>32	8	8	4	>32	>32	>32
FADDI-286	2	2	4	>32	2	4	4	4	4	>32	8	8	4	>32	>32	>32

4. Discussion and Conclusions

While bacteria have developed resistance to almost all other antibiotics, colistin and polymyxin B remain at the forefront of last-line therapeutics against MDR Gram-negative "superbugs" [15]. The primary focus of this study was to design and synthesize novel polymyxin fluorescent probes with the antibacterial activity representative of the native polymyxins.

Polymyxins elicit their antimicrobial activity by first binding to the lipopolysaccharide in the bacterial outer membrane. The formation of the complex is initiated through the electrostatic interaction of the cationic Dab side-chains with the negatively charged phosphate groups of the lipid a component of LPS. This in turn displaces divalent cations (Ca^{2+} and Mg^{2+}) that bridge adjacent LPS molecules, thereby de-stabilizing the outer membrane [2,7,16]. Subsequently, hydrophobic interactions occur between the *N*-terminal fatty acyl tail and the positions 6 and 7 hydrophobic segment of the polymyxin molecule and the fatty acyl chains of lipid A. The physical integrity of the phospholipid bilayer in the inner membrane appears to be subsequently disrupted [16–20]. This 'self-promoted' uptake mechanism is believed to lead to disruption of the cell envelope and bacterial killing [7,17,18,21]. With these principles in mind, we rationally designed the probes FADDI-285 and FADDI-286 via regio-selective modifications at the hydrophobic *N*-terminus and position 7, without markedly compromising antimicrobial activity of the parent colistin compound. Our group has previously highlighted the pitfalls of directly coupling fluorescent groups such as dansyl directly onto the Dab side chains in semi-synthetic preparations of dansyl-polymyxin B [11]. Furthermore, as polymyxin B and colistin are each comprised of two major components (polymyxin B_1 and B_2; colistin A and B), there is a strong possibility that either of these components will be substituted at any of the five Dab side chains resulting in a highly heterogeneous mixture of dansylated derivatives [11]. Indeed our previously reported mass spectrometry analysis of these semi-synthetic dansyl-polymyxin B preparations indicated the presence of a heterogeneous mixture of *mono-*, *di-*, and *tri-*dansyl Dab-substituted polymyxin B [11]. Accordingly, there is little value in using these semi-synthetic preparations as probes for imaging localization, since they lack the native antibacterial activity of the polymyxin B parent molecule and resultant images would represent the localization of a very complex array of probe molecules. Both FADDI-285 and FADDI-286 possess antibacterial activity similar to colistin and polymyxin B against polymyxin-susceptible isolates and notably, 4-fold enhanced activity against the polymyxin-resistant isolates (Table 1), and as such should prove to be of considerable utility as tools for novel polymyxin lipopeptide discovery programs. The utility of these probes in vivo is being investigated and will be the subject of a future report.

Acknowledgments: J.L. is an Australian NHMRC Senior Research Fellow. T.V. is an Australian NHMRC Industry Career Development Research Fellow.

Author Contributions: B.Y., K.D.R. performed the experiments and helped write the manuscript. R.L.N. and P.T. helped write the manuscript. K.D.R., T.V. and J.L. designed the experiments and wrote the manuscript.

Conflicts of Interest: The authors declare no conflict of interest.

References

1. Boucher, H.W.; Talbot, G.H.; Benjamin, D.K., Jr.; Bradley, J.; Guidos, R.J.; Jones, R.N.; Murray, B.E.; Bonomo, R.A.; Gilbert, D. 10 × '20 Progress—Development of New Drugs Active Against Gram-Negative Bacilli: An Update From the Infectious Diseases Society of America. *Clin. Infect. Dis.* **2013**, *56*, 1685–1694. [CrossRef] [PubMed]

2. Velkov, T.; Roberts, K.D.; Nation, R.L.; Thompson, P.E.; Li, J. Pharmacology of polymyxins: New insights into an "old" class of antibiotics. *Future Microbiol.* **2013**, *8*, 711–724. [CrossRef] [PubMed]

3. Landman, D.; Georgescu, C.; Martin, D.A.; Quale, J. Polymyxins revisited. *Clin. Microbiol. Rev.* **2008**, *21*, 449–465. [CrossRef] [PubMed]

4. Akajagbor, D.S.; Wilson, S.L.; Shere-Wolfe, K.D.; Dakum, P.; Charurat, M.E.; Gilliam, B.L. Higher Incidence of Acute Kidney Injury With Intravenous Colistimethate Sodium Compared with Polymyxin B in Critically Ill Patients at a Tertiary Care Medical Center. *Clin. Infect. Dis.* **2013**, *57*, 1300–1303. [CrossRef] [PubMed]

5. Hartzell, J.D.; Neff, R.; Ake, J.; Howard, R.; Olson, S.; Paolino, K.; Vishnepolsky, M.; Weintrob, A.; Wortmann, G. Nephrotoxicity Associated with Intravenous Colistin (Colistimethate Sodium) Treatment at a Tertiary Care Medical Center. *Clin. Infect. Dis.* **2009**, *48*, 1724–1728. [CrossRef] [PubMed]

6. Kubin, C.J.; Ellman, T.M.; Phadke, V.; Haynes, L.J.; Calfee, D.P.; Yin, M.T. Incidence and predictors of acute kidney injury associated with intravenous polymyxin B therapy. *J. Infect.* **2012**, *65*, 80–87. [CrossRef] [PubMed]

7. Velkov, T.; Thompson, P.E.; Nation, R.L.; Li, J. Structure-activity relationships of polymyxin antibiotics. *J. Med. Microbiol.* **2010**, *53*, 1898–1916. [CrossRef] [PubMed]

8. Ofek, I.; Cohen, S.; Rahmani, R.; Kabha, K.; Tamarkin, D.; Herzig, Y.; Rubinstein, E. Antibacterial synergism of polymyxin B nonapeptide and hydrophobic antibiotics in experimental gram-negative infections in mice. *Antimicrob. Agents Chemother.* **1994**, *38*, 374–377. [CrossRef] [PubMed]

9. Tsubery, H.; Ofek, I.; Cohen, S.; Fridkin, M. Structure-activity relationship study of polymyxin B nonapeptide. *Adv. Exp. Med. Biol.* **2000**, *479*, 219–222. [PubMed]

10. Tsubery, H.; Ofek, I.; Cohen, S.; Fridkin, M. N-terminal modifications of polymyxin B nonapeptide and their effect on antibacterial activity. *Peptides* **2001**, *22*, 1675–1681. [CrossRef]

11. Deris, Z.Z.; Swarbrick, J.D.; Roberts, K.D.; Azad, M.A.K.; Akter, J.; Horne, A.S.; Nation, R.L.; Rogers, K.L.; Thompson, P.E.; Velkov, T.; Li, J. Probing the Penetration of Antimicrobial Polymyxin Lipopeptides into Gram-Negative Bacteria. *Bioconj. Chem.* **2014**, *25*, 750–760. [CrossRef] [PubMed]

12. Yun, B.; Azad, M.A.K.; Nowell, C.J.; Nation, R.L.; Thompson, P.E.; Roberts, K.D.; Velkov, T.; Li, J. Cellular Uptake and Localization of Polymyxins in Renal Tubular Cells Using Rationally Designed Fluorescent Probes. *Antimicrob. Agents Chemother.* **2015**, *59*, 7489–7496. [CrossRef] [PubMed]

13. Velkov, T.; Roberts, K.D.; Nation, R.L.; Wang, J.; Thompson, P.E.; Li, J. Teaching 'Old' Polymyxins New Tricks: New-Generation Lipopeptides Targeting Gram-Negative 'Superbugs'. *ACS Chem. Biol.* **2014**, *9*, 1172–1177. [CrossRef] [PubMed]

14. Kanazawa, K.; Sato, Y.; Ohki, K.; Okimura, K.; Uchida, Y.; Shindo, M.; Sakura, N. Contribution of Each Amino Acid Residue in Polymyxin B3 to Antimicrobial and Lipopolysaccharide Binding Activity. *Chem. Pharm. Bull.* **2009**, *57*, 240–244. [CrossRef] [PubMed]

15. Nation, R.L.; Velkov, T.; Li, J. Colistin and Polymyxin B: Peas in a Pod, or Chalk and Cheese? *Clin. Infect. Dis.* **2014**, *59*, 88–94. [CrossRef] [PubMed]

16. Pristovsek, P.; Kidric, J. The search for molecular determinants of LPS inhibition by proteins and peptides. *Curr. Top. Med. Chem.* **2004**, *4*, 1185–1201. [CrossRef] [PubMed]

17. Hancock, R. The bacterial outer membrane as a drug barrier. *Trends Microbiol.* **1997**, *5*, 37–42. [CrossRef]
18. Hancock, R.E.; Lehrer, R. Cationic peptides: A new source of antibiotics. *Trends Biotechnol.* **1998**, *16*, 82–88. [CrossRef]
19. Clausell, A.; Garcia-Subirats, M.; Pujol, M.; Busquets, M.A.; Rabanal, F.; Cajal, Y. Gram-negative outer and inner membrane models: Insertion of cyclic cationic lipopeptides. *J. Phys. Chem. B* **2007**, *111*, 551–563. [CrossRef] [PubMed]
20. Powers, J.P.; Hancock, R.E. The relationship between peptide structure and antibacterial activity. *Peptides* **2003**, *24*, 1681–1691. [CrossRef] [PubMed]
21. Hancock, R.E. Peptide antibiotics. *Lancet* **1997**, *349*, 418–422. [CrossRef]

sensors

MDPI

Article

Tuning Sensory Properties of Triazole-Conjugated Spiropyrans: Metal-Ion Selectivity and Paper-Based Colorimetric Detection of Cyanide

Juhyen Lee [1], Eun Jung Choi [1], Inwon Kim [1,†], Minhe Lee [1], Chinnadurai Satheeshkumar [2] and Changsik Song [1,*]

1   Department of Chemistry, Sungkyunkwan University, Suwon, Gyeonggi 16419, Korea;
    wngusqq@naver.com (J.L.); cej9658@gmail.com (E.J.C.); kiminwon928@gmail.com (I.K.);
    minhe158@naver.com (M.L.)
2   Graduate School of Nanoscience and Technology, Korea Advanced Institute of Science and
    Technology (KAIST), Daejeon 34141, Korea; vcsatheeshkumar@gmail.com
*   Correspondence: songcs@skku.edu; Tel.: +82-31-299-4567
†   Present address: Department of Chemistry, KAIST and Center for Catalytic Hydrocarbon Functionalizations,
    Institute for Basic Science (IBS), Daejeon 34141, Korea.

Received: 5 July 2017; Accepted: 3 August 2017; Published: 7 August 2017

Abstract: Tuning the sensing properties of spiropyrans (SPs), which are one of the photochromic molecules useful for colorimetric sensing, is important for efficient analysis, but their synthetic modification is not always simple. Herein, we introduce an alkyne-functionalized SP, the modification of which would be easily achieved via Cu-catalyzed azide-alkyne cycloaddition ("click reaction"). The alkyne-SP was conjugated with a bis(triethylene glycol)-benzyl group (EG-BtSP) or a simple benzyl group (BtSP), forming a triazole linkage from the click reaction. The effects of auxiliary groups to SP were tested on metal-ion sensing and cyanide detection. We found that EG-BtSP was more Ca^{2+}-sensitive than BtSP in acetonitrile, which were thoroughly examined by a continuous variation method (Job plot) and UV-VIS titrations, followed by non-linear regression analysis. Although both SPs showed similar, selective responses to cyanide in a water/acetonitrile co-solvent, only EG-BtSP showed a dramatic color change when fabricated on paper, highlighting the important contributions of the auxiliary groups.

Keywords: spiropyran; metal ion; cyanide sensing; side-group effect; click reaction

1. Introduction

A range of stimuli (light, temperature, or metal ions) can induce closed forms of well-known photochromic molecules ie, spiropyrans (SPs), to undergo *cis-trans* isomerization to give rise to open-ring isomers or merocyanines (MCs) with vastly different physicochemical properties [1]. Such transformations of SPs enable their use as colorimetric sensors due to the vivid colors of the MC forms [2]. SPs are usually modified and functionalized for a certain purpose, such as the development of polymer sensors or dynamic materials, giving selectivity and sensitivity to external stimuli, such as metal ions or temperature. Most SPs are synthesized by condensation between indolenine and benzaldehyde, parts of which are modified to form specific functional groups [3,4]. For example, Shiraishi et al. reported that a coumarin-conjugated spiropyran showed blue fluorescence after nucleophilic addition of CN^- under UV light [5]. Stubing et al. compared the absorbance and fluorescence spectra of methyl-1-aza-crown-functionalized SP with different sizes of the aza-crown moiety [6]. These spectra showed the largest changes upon binding of Li^+ (among alkali metal ions). Perry et al. prepared a pyrene-appended SP receptor for Zn^{2+}-selective binding and non-covalent functionalization of carbon surfaces [7]. As described above, evidence from the literature highlights the importance of the addition functional groups to SPs to tailor SP-containing sensors for specific applications. In this respect, the method of easy and high-yielding functionalization to SP needs to be developed. We envisioned that, if an SP has an alkyne moiety [8], the SP can be conjugated with a variety of azide-containing molecules and materials since Cu-catalyzed alkyne-azide cycloaddition (CuAAC) is simple, high-yielding and can tolerate various functional groups [9–11]. Therefore, alkyne-containing SPs should have high utilities for developing novel functional materials.

We can also take advantage of the sensory property of the resulting triazole unit of CuAAC in conjunction with that of SPs. Triazole and its derivatives have been used for selective sensing of metal ions [12–17]. Thakur et al. synthesized triazole-tethered ferrocene-anthracene conjugates for electrochemical and optical sensing of Pb^{2+} [16]. The compound "turned on" its fluorescence up on the binding of Pb^{2+} ions, and also showed a dramatic change from yellow to a greenish-blue color, allowing naked-eye detection. Kim et al. synthesized a rhodamine triazole-based fluorescent probe for Pt^{2+} detection [12]. A triazole moiety aided selectivity and sensitivity of binding to Pt^{2+} rather than other metal ions in aqueous solution. The synthesized probe also showed a change from colorless to a pinkish-red hue upon binding of Pt^{2+}.

In this study, triazole-conjugated SP molecules are tuned by modification of their side groups. Our strategy was to tune their sensing properties by using a click reaction (CuAAC) between propargyl-functionalized SP and azido molecules. Since sensing in an aqueous environment is important, an ethylene glycol moiety was introduced to SP. This modification renders it more hydrophilic and more sensitive to cyanide in a water environment. We demonstrated that the SP's sensing properties for metal ions could be easily tuned by click modification. In addition, ethylene glycol-incorporated EG-BtSP, in constrast to simple BtSP, could be utilized as a paper-based colorimetric sensor.

2. Materials and Methods

2.1. General

All the chemicals were purchased from Sigma-Aldrich (Seoul, Korea), Alfa Aesar (Seoul, Korea), TCI (Tokyo, Japan), Acros Organics (Geel, Belgium), or Samchun Chemical (Seoul, Korea) and were used as received. ^{1}H and ^{13}C NMR spectra were recorded using a Bruker 500 MHz spectrometer. The chemical shifts are reported in ppm (δ) with chloroform-*d* (δ 7.26) as an internal standard, and the coupling constants (*J*) are expressed in Hz. UV-VIS absorption measurements were carried out using a UV-1800 (Shimadzu) spectrophotometer. All the metal ions and the anions used in this research are in the form of a perchlorate salt and tetra-*n*-butyl ammonium salt, respectively. High-resolution mass spectra (HRMS) were obtained on a Bruker Daltonics APEX II 3 T FT-ICR-MS. Mass spectra (ESI) were obtained on an Agilent Model:1100 LC-MS mass spectrometers. Column chromatography was carried out using a 100–200 mesh silica gel. Thin layer chromatograph (TLC) analysis was performed on precoated silica gel 60 F254 slides and visualized by UV irradiation. Deuterated solvents for NMR were purchased from Cambridge Isotope Laboratories (Tewksbury, MA, USA).

2.1.1. Synthesis of 3,5-Bis[2-[2-[2-methoxyethoxy]ethoxy]ethoxy] Benzyl Alcohol (1)

3,5-bis{2-[2-(2-methoxyethoxy)ethoxy]ethoxy}-, methylester [18] (5.68 g, 12.3 mmol) was dissolved in dry tetrahydrofuran (THF) (170 mL) in a 250-mL dropping funnel and this solution was added to LiAlH$_4$ (1.20 g, 24.7 mmol) suspended in a dry THF under N$_2$ atmosphere. The reaction mixture was stirred under reflux for overnight. The mixture was quenched by adding ice at 0 °C, filtered, and the solvent was then removed under vacuum. The residue was then washed with ethyl acetate (EA) and brine. Finally, the organic phases were dried over anhydrous MgSO$_4$, filtered and concentrated to yield **1** (5.36 g, 99%) as a yellow oil. ^{1}H (500 MHz, CDCl$_3$) δ 6.54 (d, *J* = 2.5 Hz, 2H), 6.41 (t, *J* = 2.5 Hz, 1H), 4.61 (d, *J* = 6 Hz, 2H), 4.11 (t, *J* = 5 Hz, 4H), 3.84 (t, *J* = 5 Hz, 4H), 3.74–3.72 (m, 4H), 3.69–3.64 (m,

8H), 3.56–3.54 (m, 4H), 3.38 (s, 6H). ^{13}C (125 MHz, CDCl$_3$) δ 160.1, 143.4, 105.5, 100.9, 71.9, 70.8, 70.7, 70.5, 69.7, 67.5, 65.2, 59.0. MS (HRMS): *m/z* calculated for C$_{21}$H$_{36}$O$_9$ [M]$^+$: 432.2359; found: 432.2361.

2.1.2. Synthesis of 3,5-Bis[2-[2-(2-methoxyethoxy)ethoxy]ethoxy] Benzyl Chloride (2)

A solution of **1** (1.17 g, 2.70 mmol) in dry dichloromethane (100 mL) was added dropwise to the catalytic amounts of dry dimethylformamide and thionyl chloride (0.45 g, 3.79 mmol) at 0 °C. After stirring at room temperature for 10 h, unreacted thionyl chloride and dichloromethane were removed under reduced pressure and extracted with ethyl acetate. The combined organic extracts were dried over Na$_2$SO$_4$, filtered, and evaporated under vacuum. The product was concentrated as a pale-yellow oil **2** (0.94 g, 77%). ^{1}H (500 MHz, CDCl$_3$) δ 6.54 (d, *J* = 2.0 Hz, 2H), 6.44 (d, *J* = 2.0 Hz, 1H), 4.49 (s, 2H), 4.11 (t, *J* = 5.0 Hz, 4H), 3.84 (t, *J* = 5.0 Hz, 4H), 3.74–3.72 (m, 4H), 3.69–3.65 (m, 8H), 3.56–3.54 (m, 4H), 3.38 (s, 6H). ^{13}C (125 MHz, CDCl$_3$) δ 160.0, 139.4, 107.4, 101.6, 71.9, 70.8, 70.7, 70.6, 69.6, 67.6, 59.1, 46.3. MS (HRMS): *m/z* calculated for C$_{21}$H$_{35}$ClO$_8$ [M]$^+$: 450.2020; found: 450.2022.

2.1.3. Synthesis of 3,5-Bis[2-[2-(2-methoxyethoxy)ethoxy]ethoxy] Benzyl Azide (3)

A solution of compound **2** (0.89 g, 2.00 mmol), NaN$_3$ (0.52 g, 8.00 mmol) in dry dimethylformamide (10.0 mL) were stirred at 60 °C for 24 h. The reactant was then cooled to room temperature and quenched by addition of water. The mixture was evaporated under reduced pressure, and the residue was added ethyl acetate and washed with water and brine. The organic layer was dried over Na$_2$SO$_4$, filtered, and concentrated. Compound **3** was obtained as a pale-yellow liquid (0.70 g, 77%). ^{1}H (500 MHz, CDCl$_3$) δ 6.47–6.45 (m, 3H), 4.24 (s, 2H), 4.12–4.10 (m, 4H), 3.86–3.84 (m, 4H), 3.75–3.72 (m, 4H), 3.70–3.67 (m, 4H), 3.67–3.65 (m, 4H), 3.56–3.54 (m, 4H), 3.38 (s, 6H). ^{13}C (125 MHz, CDCl$_3$) δ 160.2, 137.5, 107.0, 101.4, 72.0, 70.9, 70.7, 69.7, 59.1, 31.0. MS (HRMS): *m/z* calculated for C$_{21}$H$_{35}$N$_3$O$_8$ [M]$^+$: 457.2424; found: 457.2422.

2.1.4. Synthesis of 8-Methoxy-3′,3′-dimethyl-6-nitro-1′-(prop-2-yn-1-yl)spiro[chromene-2,2′-indoline] (6)

A mixture of 3,3-dimethyl-2-methylene-1-(prop-2-yn-1-yl)indoline **4** (3.91 g, 125 mmol) was added to 2-hydroxy-3-methoxy-5-nitrobenzaldehyde **5** (2.70 g, 137 mmol) in ethanol (30.0 mL), and sonicated for two hours. The residue was then evaporated and diluted in ethyl acetate. The organic layer was washed with water and brine and dried over Na$_2$SO$_4$. After evaporation, the crude mixture was purified by column chromatography and obtained compound **6** (2.81 g, 60%). ^{1}H (500 MHz, CDCl$_3$) δ 7.70 (d, *J* = 2.5 Hz, 1H), 7.62 (d, *J* = 2.5 Hz, 1H), 7.20–7.23 (td, *J* =7.5 Hz, 1H), 7.09–7.11 (dd, *J* = 7.3 Hz, 1H), 6.90–6.93 (m 2H), 6.81 (d, *J* = 7.5 Hz, 1H), 5.89 (d, *J* = 10 Hz, 1H), 4.04 (dd, *J* = 18 Hz, 2.5 Hz, 1H), 3.86 (dd, *J* = 18.5 Hz, 2.5 Hz, 1H), 3.76 (s, 3H), 2.10 (t, 1H), 1.23 (s, 3H), 1.12 (s, 3H). ^{13}C (125 MHz, CDCl$_3$) δ 149.1, 147.5, 145.8, 140.6, 136.1, 128.7, 127.7, 121.7, 121.3, 120.2, 118.2, 115.4, 108.0, 107.9, 105.8, 79.7, 71.4, 56.3, 52.5, 32.6, 26.0, 20.0. MS (HRMS): *m/z* calculated for C$_{22}$H$_{20}$N$_2$O$_4$ [M]$^+$: 376.1423; found: 376.1418.

2.1.5. Synthesis of EG-BtSP

A solution of compound **6** (0.15 g, 0.39 mmol), compound **3** (0.18 g, 0.39 mmol), CuSO$_4$·H$_2$O (5 mol %), and sodium ascorbate (10 mol %) in a mixture of THF-H$_2$O (1:1 *v/v*) was stirred for 12 h at room temperature. The residue was evaporated under vacuum, dissolved in chloroform and washed with water and brine. The organic layer was dried over Na$_2$SO$_4$ and concentrated to afford the crude product. The product (0.19 g, 56%) was purified by column chromatography (silica gel; eluent, Hexane: EA = 1:10). ^{1}H (500 MHz, CDCl$_3$) δ 7.67 (d, *J* = 2.5 Hz, 1H), 7.54 (d, *J* = 2.5 Hz, 1H), 7.44 (s, 1H), 7.04–7.08 (m, 2H), 6.88 (d, *J* = 10.5 Hz, 1H), 6.82–6.85 (m, 1H), 6.41–6.40 (m, 2H), 6.23 (d, *J* = 2 Hz, 2H), 5.86 (d, *J* = 10 Hz, 1H), 5.36 (d, *J* = 15.5 Hz, 1H), 5.29 (d, *J* = 15 Hz, 1H), 4.71 (d, *J* = 17 Hz, 1H), 4.58 (d, *J* = 17 Hz, 1H), 3.95–3.99 (m, 3H), 3.79 (t, *J* = 5 Hz, 4H), 3.70–3.72 (m, 4H), 3.64–3.68 (m, 12H), 3.53–3.55 (m, 4H), 3.37 (s, 6H), 1.29 (s, 3H), 1.18 (s, 3H). ^{13}C (125 MHz, CDCl$_3$) δ 160.3, 148.7, 147.0,

145.7, 145.6, 140.5, 137.0, 135.8, 128.6, 127.5, 122.7, 121.9, 121.7, 119.6, 118.2, 115.5, 107.8, 107.7, 106.4, 106.2, 101.1, 71.9, 70.8, 70.6, 70.5, 69.5, 67.5, 59.0, 56.1, 53.9, 53.0, 39.6, 26.3, 19.9. MS (ESI): *m/z* calculated for $C_{43}H_{56}N_5O_{12}$ [M + H]$^+$: 834.93; found: 834.47.

2.1.6. Synthesis of BtSP

A solution of compound **6** (0.16 g, 0.43 mmol), benzyl azide (0.06 g, 0.43 mmol), $CuSO_4 \cdot H_2O$ (5 mol %), and sodium ascorbate (10 mol %) in a mixture of THF-H_2O (1:1 *v/v*) was stirred for 12 h at room temperature. The residue was evaporated under vacuum, dissolved in chloroform and washed with water and brine. The organic layer was dried over Na_2SO_4 and concentrated to yield the crude product. The product was purified by column chromatography (silica gel; eluent, Hexane: EA = 3:1; 0.10 g, 48%). ^{1}H (500 MHz, CDCl$_3$) δ 7.66 (d, *J* = 2.5 Hz, 1H), 7.51 (d, *J* = 2.5 Hz, 1H), 7.46 (s, 1H), 7.29–7.30 (m, 3H), 7.04–7.08 (m, 4H), 6.83–6.87 (m, 2H), 6.38 (d, *J* = 7.5 Hz, 1H), 5.85 (d, *J* = 10 Hz, 1H), 5.47 (d, *J* = 15.5 Hz, 1H), 5.39 (d, *J* = 15.5 Hz, 1H), 4.72 (d, *J* = 17 Hz, 1H), 4.58 (d, *J* = 16.5 Hz, 1H), 3.60 (s, 3H), 1.29 (s, 3H), 1.18 (s, 3H). ^{13}C (125 MHz, CDCl$_3$) δ 218.3, 148.7, 147.0, 145.7, 145.6, 140.5, 135.9, 135.1, 129.0, 128.6, 128.5, 127.6, 127.3, 122.7, 121.9, 121.7, 119.6, 118.3, 115.5, 107.8, 107.7, 106.4, 56.1, 53.9, 53.0, 39.6, 26.3, 19.9. MS (HRMS): *m/z* calculated for $C_{29}H_{27}N_5O_4$ [M]$^+$: 509.2063; found: 509.2062.

2.2. UV-VIS Spectrum Measurement for Absorption Spectra

2.2.1. Metal Screening of EG-BtSP and BtSP

Separate EG-BtSP (c = 5 × 10^{-5} M), BtSP (c = 1 × 10^{-4} M) and metal (c = 1 × 10^{-2} M) stock solutions (perchlorate salts in acetonitrile) were prepared for UV-VIS measurements. UV-VIS absorption spectra were obtained with SP solutions (2 mL) 30 min after the addition of 1 equiv. of each metal ion solution.

2.2.2. Dielectric Constant

EG-BtSP and BtSP compounds (c = 5 × 10^{-5} M and 1 × 10^{-4} M, respectively) were prepared in nine solvent mixtures using water/acetonitrile (total 2 mL), varying the water composition up to 90% water (*v/v*) in acetonitrile. After 30 min, UV-VIS absorption analysis was performed on the soluble samples.

2.2.3. Anion Screening of EG-BtSP and BtSP

For UV-VIS anion screening, the stock solutions of EG-BtSP and BtSP (c = 2 × 10^{-5} M) were prepared in a water/CH$_3$CN mixture (9/1 and 1/1 *v/v*, respectively) along with each of anion solution (c = 5 × 10^{-2} M) as a *n*-Bu$_4$N$^+$ salt. UV-VIS spectrophotometric measurements were performed with SP solutions (2 mL) 30 min after addition of each anion solution (40 μL, about 50 equiv.).

2.2.4. Paper-Based Sensor Test of EG-BtSP and BtSP

For the paper sensor test, we prepared a 10^{-1} M SP stock solution in acetonitrile, and filter paper cuts (1 cm × 1 cm) were coated with this solution via a dipping method. After drying for 24 h, the SP-loaded paper was dipped into aqueous cyanide solutions at different concentrations (1 mM, 10 mM, 20 mM, 50 mM, 100 mM, and 500 mM). The paper was then dried in air at room temperature.

3. Results and Discussion

Since azide-containing molecules can be easily conjugated with alkyne-functionalized SP via Cu-catalyzed azide-alkyne cycloaddition (CuAAC), bis(triethylene glycol)-attached benzyl azide **3** was prepared from the corresponding alcohol **1** via reduction [19], chlorination [20], and nucleophilic substitution [21] (Scheme 1). The propargyl-functionalized SP **6** was synthesized by condensation between propargyl indole **4** and 2-hydroxy-3-methoxy-5-nitrobenzaldehyde **5** according to a published method [8]. The CuAAC of propargyl-SP **6** with azide **3** gave rise to a good yield (56%) of a bis(triethylene glycol)-functionalized SP with a triazole linkage EG-BtSP, as illustrated in Scheme 1. BtSP, which does not have ethylene glycol moieties, was also prepared by a similar method. Our hypothesis was that addition of ethylene glycol moieties would render SPs hydrophilic, enhancing their sensing abilities in aqueous environments.

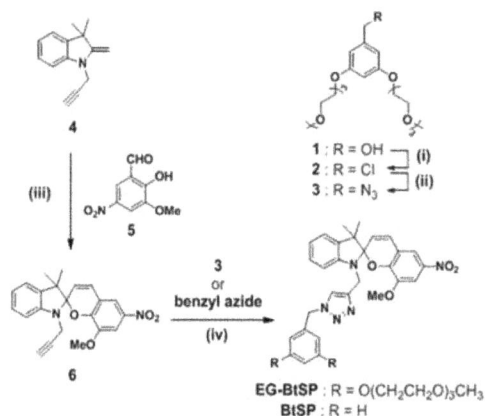

Scheme 1. Synthesis of EG-BtSP and BtSP from an Azide-Functionalized Spiropyran. (i) SOCl$_2$, DMF, DCM, room temperature, 10 h, 77%; (ii) NaN$_3$, DMF, 60 °C, 24 h, 85%; (iii) EtOH, sonication, 2 h, 60%; (iv) CuSO$_4$·5H$_2$O (5 mol %), sodium ascorbate (10 mol %), THF/H$_2$O (1:1 *v/v*), room temperature, 12 h, 56% (EG-BtSP) and 48% (BtSP).

The photochromic behaviors of EG-BtSP and BtSP were investigated in acetonitrile. The initial solutions of EG-BtSP and BtSP in acetonitrile (~0.10 mM) were colorless. When 365-nm UV light was illuminated on the solutions of SP molecules, the color of both solutions changed to blue, showing the same new absorption peak at around 600 nm. UV irradiation caused the C-O bond cleavage of SP molecules, resulting in an open MC form, as also seen in evidence from the literature [22] The spontaneous reverse isomerization from MC to SP occurred under visible light. Understandably, this result showed that the photochromic property of EG-BtSP and BtSP comes from the SP moiety, not from the auxiliary chains.

Interestingly, the modification of the auxiliary chains appeared to exert slight, but important effects on the sensory property of SP, especially for sensing Ca^{2+}. The absorption spectra of EG-BtSP (0.050 mM) and BtSP (0.10 mM) in acetonitrile before and after the addition of 1.0 equiv. of Ca^{2+}, Cd^{2+}, Co^{2+}, Fe^{2+}, Mg^{2+}, Ni^{2+}, Zn^{2+}, and Li$^+$ metal ions are shown in Figure 1a,b. When certain metal ions (especially Zn^{2+}, Mg^{2+}, and Ca^{2+}) were added, SP molecules switched to the colored forms, presumably due to complex formation with metal ions. The absorption maximum of the complex (~500 nm) was blue-shifted from that of the open MC form (~600 nm), which indicated binding of metal ions to the cleaved phenoxide moiety of SP. To compare the reactivities of SPs based on their auxiliary group, the absorbance at 495 nm, which was normalized to the absorbance at 310 nm to correct for the concentration difference and plotted as a function of different metal ions (Figure 1c). Both SPs showed high selectivity toward Zn^{2+}, then to Mg^{2+}, among the metal ions tested. Although the general trends for sensory properties of EG-BtSP and BtSP were very similar, EG-BtSP showed a more sensitive and selective response toward Ca^{2+} than BtSP. The only structural difference between EG-BtSP and BtSP was the presence and absence of a glycol moiety, respectively. However, they showed quite different selectivity and sensitivity to Ca^{2+} metal ions. The above results indicate that metal ions could be selectivity regulated by the choice of the auxiliary group to SP molecules and the click reaction to a propargyl-SP **6** should be useful for introducing various functional groups.

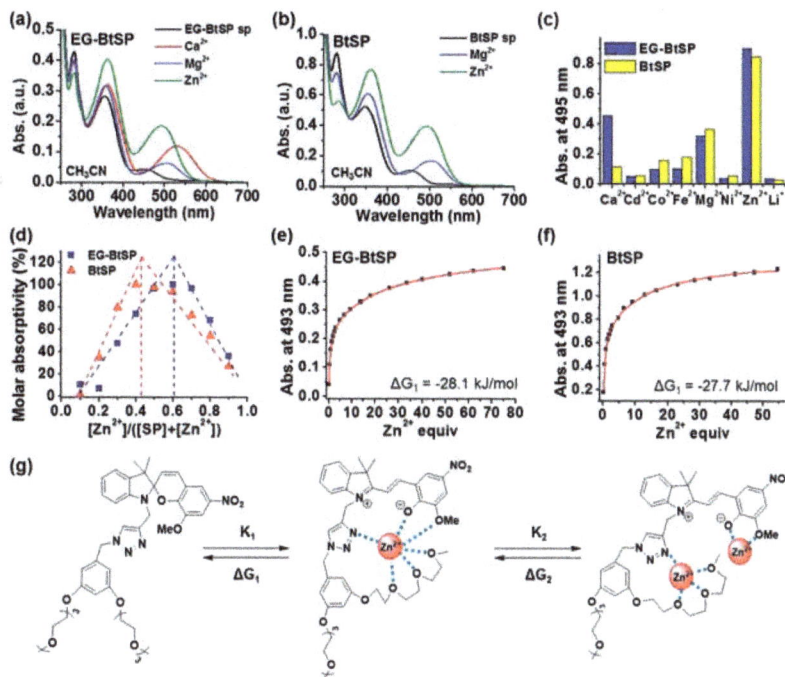

Figure 1. (**a,b**) Absorption spectra of EG-BtSP (a, 5×10^{-5} M) and BtSP (b, 1×10^{-4} M) after addition of 1.0 equiv. of Ca^{2+}, Ma^{2+}, Zn^{2+}, and Ma^{2+}, Zn^{2+} metal perchlorates in CH_3CN at 293 K, respectively; (**c**) Comparison of the absorbance at 495 nm of EG-BtSP and BtSP normalized with the absorbance at 310 nm (the isosbestic point); (**d**) Job's analysis of the EG-BtSP-Zn^{2+} complex ([EG-BtSP] + [Zn^{2+}] = 5×10^{-5} M) and BtSP-Zn^{2+} complex ([BtSP] + [Zn^{2+}] = 1×10^{-4} M) in CH_3CN; (**e,f**) Absorption titration at 493 nm of a solution of EG-BtSP (d, 5×10^{-5} M) and BtSP (e, 1×10^{-4} M) after increasing the concentration of Zn^{2+} in CH_3CN; (**g**) Schematic illustration of the equilibria of the EG-BtSP-Zn^{2+} complex. The first equilibrium constant K_1 is much larger than the second K_2. Although the maximum possible coordination number for Zn^{2+} is 6, the coordination occupied by any solvent molecule(s) were omitted for clarity.

Stoichiometries of binding of metal ions to EG-BtSP and BtSP determined by Job plots, a continuous variation method (Figure 1d, Figures S1 and S2). Figure 1d shows that the Job plots of EG-BtSP and BtSP toward Zn^{2+} deviated from normal triangular shapes and appeared as hyperboles. Since the Job plot is based on the assumption that only one complex H_nG_m (H: host, and G: guest molecule) is formed, it may indicate that SP-M^{2+} complexes with several stoichiometries could be present in the solution, or that the binding constants are relatively small [23] Nevertheless, the Job analyses showed that the simple BtSP appeared to have an absorption maximum at a molar fraction of ~0.5, which indicates 1:1 binding to Zn^{2+}. The triethylene glycol-functionalized EG-BtSP showed a slight shift toward a higher molar fraction of Zn^{2+} (~0.6), which suggests a predominantly 1:1 binding, but it is also possible that more than one Zn^{2+} may bind to EG-BtSP.

For a better understanding of SP-metal ions binding modes, a UV-VIS spectroscopic titration was performed for EG-BtSP-Ca^{2+}, -Mg^{2+}, -Zn^{2+} and BtSP-Mg^{2+}, and -Zn^{2+} (Figure 1e,f and Figures S3–S7). To the solutions of SP host molecules, various equivalents of metal ions were added, and the absorbance at 495 nm was monitored for each SP-metal ion combination. The absorbance datasets were then subjected to non-linear regression analysis, following a procedure developed by Thordarson's group [24,25]. Briefly, the datasets were fitted for 1:1 (SP:M^{2+}) and 1:2 binding systems and the results were qualified by "cov_{fit}" values. The cov_{fit} values are insensitive to the number of parameters used in the fitting process, and showed a numerical representation of experimental data scatter about the fitted lines. Then the "cov_{fit} factor", which is the cov_{fit} value of the 1:1 binding divided by the cov_{fit} value of the 1:2 binding, can be used to determine which binding model can provide the best explanation of the experimental data (Table 1). Based on the analyses, all SP-M^{2+} complexes formed at ratios of 1:2 rather than 1:1, as judged by the cov_{fit} factors, although the second binding constants (K_2) were much smaller than the first ones (K_1). In addition, triethlyene glycol-functionalized EG-BtSP seems to bind to metal ions slightly stronger when compared to BtSP (which lacks the ethylene glycol moiety): for Zn^{2+}, K_1 for EG-BtSP = 85,000 vs. K_1 for BtSP = 72,000, and for Mg^{2+}, K_1 for EG-BtSP = 12,000 vs. K_1 for BtSP = 10,000, which corresponds to 0.4~0.5 kJ/mol difference in ΔG (Tables S1–S5). We attributed this difference to the participation of the triethylene glycol moiety in the binding of metal ions. In our previous research, we have shown that the phenoxide and methoxy groups in the MC form interact with the metal ion, as well as the triazole part of the open isomer. It is reasonable to assume that lone pair electrons of oxygens in the triethylene glycol moiety would favorably interact with metal ions, resulting in a slightly higher binding constant for EG-BtSP. It should be noted here that the triethylene-glycol auxiliary group of EG-BtSP plays an important role in the Ca^{2+} binding, the binding constant of which is larger than that of Mg^{2+}, while the simple BtSP showed little interaction to Ca^{2+}.

Table 1. Plausible binding model, K_1, K_2, and interaction parameter (α), and relative quality of fit for the complexation of EG-BtSP and BtSP toward metal ions obtained from UV-VIS spectroscopy at 298 K in acetonitrile [a].

Spiropyran-Metal		Binding Model	cov_{fit} Factor [b]	K_1 (M^{-1})	K_2 (M^{-1})	α [c]
EG-BtSP	Ca^{2+}	1:2	22.8	4.73×10^4	1803	0.161
	Mg^{2+}	1:2	11.7	1.20×10^4	353	0.118
	Zn^{2+}	1:2	64.7	8.52×10^4	868	0.041
BtSP	Mg^{2+}	1:2	34.8	9.95×10^3	172	0.074
	Zn^{2+}	1:2	48.0	7.20×10^4	636	0.034

[a] The data here are the rounded averages of the triplicate measurements. The fittings with the experimental data are shown in detail in Supporting Information (Tables S1–S5). [b] The relative quality of the fit could be expressed in terms of cov_{fit} factor = cov_{fit} for the 1:1 model divided by the cov_{fit} for the 1:2 model, where cov_{fit} is the (co)variance of the residuals divided by the (co)variances of the raw data. [c] The interaction parameter $\alpha = 4K_2/K_1$ with $\alpha > 1$ indicating positive cooperativity, $\alpha < 1$ negative cooperativity, and $\alpha = 1$ no cooperativity [24,25].

Based on the analyses above, we proposed the binding scheme of EG-BtSP with Zn^{2+} in Figure 1g. When the first Zn^{2+} ion binds to EG-BtSP, the SP form is transformed to the MC form, and the phenoxy, methoxy, and triazole groups all participate in the binding, as well as the triethylene glycol moiety. When the second Zn^{2+} ion binds to the first complex, we assume that the triazole and triethylene glycol moiety are responsible for binding of the second ion, while the MC form retains the first ion. Since the second binding requires the breaking of triazole- and triethylene glycol-Zn^{2+} interactions, although they are weak, K_2 is much smaller than K_1 (negative cooperativity). We do not know the exact conformation of the triethylene glycol moiety in the binding process, but it is certain that the auxiliary group plays an important role.

Modification by an auxiliary group can significantly enhance the utility of SP, especially in an aqueous environment. It was shown that EG-BtSP could be dissolved in the solvent at a water/acetonitrile ratio of up to 9:1. As shown in Figure 2a, EG-BtSP was transformed to the MC form with increasing amounts of water in the co-solvent system. We attributed this transformation to the higher dielectric environment of water, which can enhance the stabilization of strong dipoles of the open MC form (zwitterion). BtSP could also be isomerized to the MC form by increasing the amount of water. However, BtSP was precipitated at water/acetonitrile ratios of 60/40 (60%), indicating poor solubility in water (Figure 2b). As shown in Figure 2c, the dielectric constant of the co-solvent system increased the transformation of both SPs to their MC form. However, the ethylene-glycol auxiliary chain clearly helped the transformation and the stability of EG-BtSP in an aqueous environment.

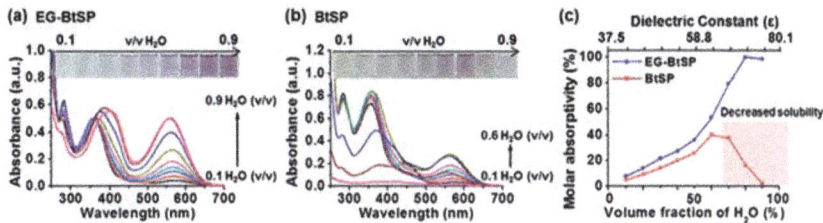

Figure 2. (a,b) UV-absorption spectra of EG-BtSP (a, c = 5 × 10^{-5} M) and BtSP (b, c = 1 × 10^{-4} M) in different water/CH$_3$CN mixtures from 1/1 to 9/1 (*v/v*) at 25 °C. Color changes of the solutions were also presented; **(c)** Relative molar absorptivities of EG-BtSP and BtSP at 564 nm were compared according to the volume fraction of water in CH$_3$CN.

The enhanced stability of EG-BtSP in an aqueous environment by auxiliary-chain modification enabled the fabrication of paper-based colorimetric sensors for a cyanide ion. As reported in the literature, the zwitterionic MC form has an electrophilic site that nucleophiles (e.g., cyanide ion) can attack [26]. However, most SP sensors for the cyanide ion have been tested in organic solvents and only a few examples have been tested in water/acetonitrile co-solvent, as previously reported elsewhere [26–28] Moreover, no paper-based sensor with SP molecules has been developed yet. We investigated the sensory responses of EG-BtSP and BtSP toward various anions in solution. For screening, the absorption spectra of EG-BtSP (0.020 mM) in a 9:1 water/acetonitrile co-solvent system were measured after addition of F$^-$, Cl$^-$, Br$^-$, I$^-$, ClO$_4$$^-$, NO$_3$$^-$, HSO$_4$$^-$, OAc$^-$, and CN$^-$ (50 equiv.). As shown in Figure 3a, only the addition of CN$^-$ induced the blue-shift of λ_{max} in the absorption spectrum of EG-BtSP (from 560 nm to 453 nm); the violet color of the initial EG-BtSP solution turned to a yellowish hue. Due to the poor solubility of BtSP in water, we tested its sensory response at a concentration of 0.02 mM in a 1:1 ratio of water:acetonitrile. As shown in Figure 3b, BtSP also showed a selective response to the cyanide ion. It should be noted here that, unlike BtSP, EG-BtSP showed its cyanide-selective response in a mostly aqueous environment. The effect of auxiliary modification was mostly reflected in the fabrication of SP-utilized paper-based sensors. Pre-cut filter papers were dip-coated with a solution of EG-BtSP or BtSP (100 mM) and then exposed to different concentrations of aqueous cyanide solutions (from 1 mM to 500 mM). As shown in Figure 3c,d, the EG-BtSP-incorporated paper sensors showed apparent color changes (bluish-violet to yellow) after applying aqueous cyanide ions. However, the BtSP-incorporated paper sensors showed very little sensory responses toward aqueous cyanide ions, even at a concentration of 500 mM. In solution, BtSP also operated as a cyanide sensor (Figure S8) similar to EG-BtSP, but on the paper, its sensory response was minimal. When we performed the CN$^-$ sensing test for BtSP in acetonitrile-water 1:1 mixture, a distinct color change was observed over 50 mM of CN$^-$ (Figure S9). This is because the insolubility issue of BtSP was somewhat resolved in the acetonitrile-water mixture. However, it should be noted here that the purpose of our study was to detect CN$^-$ in a purely aqueous environment. Due to its limited solubility in water, BtSP did not respond to CN$^-$ in water only, which highlights the importance of the auxiliary group. The EG-BTSP-incorporated paper sensor showed a dramatic color change toward cyanide ions with the help of the hydrophilic auxiliary chain of triethylene glycols. Interesting CN$^-$-selective paper strip-based sensors with probe molecules based on benzothiazolyl-malononitrile by Hong group [29] and densyl-triazole-glucopyranosyl conjugates by Rao group [30] were separately reported, but they are based on new fluorescence emission or its enhancement, respectively, upon addition of CN$^-$. However, our paper-based SP sensors were colorimetric, which is easy to apply and rapidly detects in a cost-effective manner. Furthermore, our paper-based SP sensor could work in purely aqueous environment. Chow, Tang, and coworkers reported effective colorimetric paper-based sensor with an ethenyl-allylpyridinum derivative, but they utilized an acetonitrile-water (95:5, v/v) mixture [31].

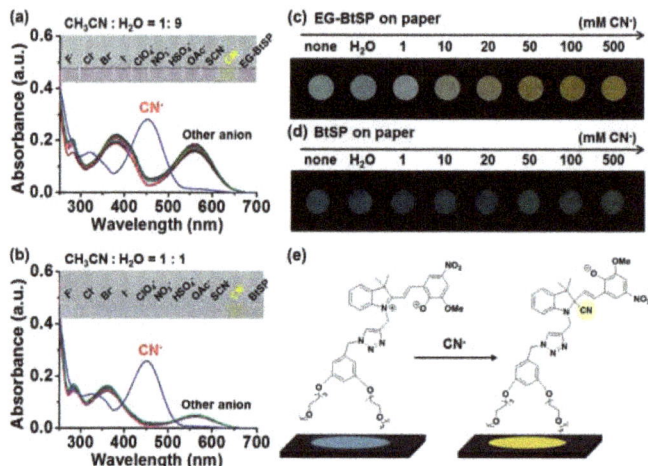

Figure 3. (**a**,**b**) UV-VIS absorption spectra of EG-BtSP (a, 2×10^{-5} M) and BtSP (b, 2×10^{-5} M) measured with 50 equiv. of respective anion (as a n-Bu$_4$N$^+$ salt) in a water/CH$_3$CN mixture (9:1 or 1:1 v/v) at 25 °C. The spectra were obtained 30 min after addition of the anion to the SP solution. Photographs of the solutions were also presented: from left F$^-$, Cl$^-$, Br$^-$, I$^-$, ClO$_4$$^-$, NO$_3$$^-$, HSO$_4$$^-$, OAc$^-$, CN$^-$, without any anion; (**c**,**d**) Colorimetric changes of the papers with SP probes, EG-BtSP (c) and BtSP (d); upon the application of cyanide in H$_2$O. From left to right: probe only, H$_2$O, 1 mM, 10 mM, 20 mM, 50 mM, 100 mM, and 500 mM of cyanide; (**e**) Schematic illustration of paper-based colorimetric detection of cyanide with EG-BtSP.

4. Conclusions

The triethylene glycol-functionalized EG-BtSP and the simple BtSP were synthesized from propargyl-SP **6** via a click reaction, and both SPs were investigated their new sensory properties following the regulation of the SP auxiliary group. Both SPs showed similar sensitivities to Mg^{2+} and Zn^{2+}, but EG-BtSP demonstrated a better sensitivity to Ca^{2+} than BtSP. Higher dielectric constants of the solvent mixtures (water/acetonitrile) were associated with the presence of more zwitterionic forms of both SPs. However, BtSP precipitated at ratios of water/acetonitrile of over 60/40, while EG-BtSP remained stable at ratios of water/acetonitrile of up to 90/10, due to improved hydrophilicity conferred by the presence of a triethylene-glycol auxiliary group. Additionally, in solution, both EG-BtSP and BtSP showed similar selective sensory responses to cyanide. However, when fabricated on paper, only EG-BtSP showed an apparent color change. We showed that the sensory properties of SP molecules could be easily tuned by auxiliary groups, and click chemistry enabled the facile introduction of appropriate auxiliary groups from propargyl-functionalized SPs.

Supplementary Materials: The Supplementary Materials are available online at http://www.mdpi.com/1424-8220/17/8/1816/s1.

Acknowledgments: This work was supported by the Small Grant Exploratory Research (SGER) Program through the National Research Foundation of Korea (NRF), funded by the Ministry of Education, Science and Technology (MEST), Republic of Korea (NRF-2015R1D1A1A02062095). This work was also supported in part by the Nano Material Development Program through the National Research Foundation of Korea (NRF) funded by the Ministry of Education, Science and Technology (MEST), Republic of Korea (2012M3A7B4049644), and in part by the Samsung Advanced Institute of Technology through the Samsung-SKKU Graphene Center.

Author Contributions: J. Lee, C. Satheeshkumar, and C. Song conceived and designed the experiments; J. Lee and E.J. Choi performed the experiments; I. Kim and M. Lee contributed reagents/materials/analysis tools; E.J. Choi and C. Song wrote the paper; and C. Song supervised the research.

Conflicts of Interest: The authors declare no conflict of interest.

References

1. Klajn, R. Spiropyran-based dynamic materials. *Chem. Soc. Rev.* **2014**, *43*, 148–184. [CrossRef] [PubMed]
2. Yagi, S.; Nakamura, S.; Watanabe, D.; Nakazumi, H. Colorimetric sensing of metal ions by bis(spiropyran) podands: Towards naked-eye detection of alkaline earth metal ions. *Dyes Pigment.* **2009**, *80*, 98–105. [CrossRef]
3. Tanaka, M.; Kamada, K.; Ando, H.; Kitagaki, T.; Shibutani, Y.; Yajima, S.; Sakamoto, H.; Kimura, K. Metal-ion stabilization of photoinduced open colored isomer in crowned spirobenzothiapyran. *Chem. Commun.* **1999**, *16*, 1453–1454. [CrossRef]
4. Shao, N.; Wang, H.; Gao, X.D.; Yang, R.H.; Chan, W.H. Spiropyran-Based Fluorescent Anion Probe and Its Application for Urinary Pyrophosphate Detection. *Anal. Chem.* **2010**, *82*, 4628–4636. [CrossRef] [PubMed]
5. Shiraishi, Y.; Sumiya, S.; Hirai, T. Highly sensitive cyanide anion detection with a coumarin-spiropyran conjugate as a fluorescent receptor. *Chem. Commun.* **2011**, *47*, 4953–4955. [CrossRef] [PubMed]
6. Stubing, D.B.; Heng, S.; Abell, A.D. Crowned spiropyran fluoroionophores with a carboxyl moiety for the selective detection of lithium ions. *Org. Biomol. Chem.* **2016**, *14*, 3752–3757. [CrossRef] [PubMed]
7. Perry, A.; Green, S.J.; Horsell, D.W.; Homett, S.M.; Wood, M.E. A pyrene-appended spiropyran for selective photo-switchable binding of Zn(II): UV-visible and fluorescence spectroscopy studies of binding and non-covalent attachment to graphene, graphene oxide and carbon nanotubes. *Tetrahedron* **2015**, *71*, 6776–6783. [CrossRef]
8. Kim, I.; Jeong, D.C.; Lee, M.; Khaleel, Z.H.; Satheeshkumar, C.; Song, C. Triazole-conjugated spiropyran: Synthesis, selectivity toward Cu(II), and binding study. *Tetrahedron Lett.* **2015**, *56*, 6080–6084. [CrossRef]
9. Petrassi, H.M.; Sharpless, K.B.; Kelly, J.W. The copper-mediated cross coupling of phenylboronic acids and N-hydroxyphthalimide at room temperature: Synthesis of aryloxyamines. *Org. Lett.* **2001**, *3*, 139–142. [CrossRef] [PubMed]
10. Bock, V.D.; Hiemstra, H.; van Maarseveen, J.H. Cu-I-catalyzed alkyne-azide "click" cycloadditions from a mechanistic and synthetic perspective. *Eur. J. Org. Chem.* **2006**, *1*, 51–68. [CrossRef]
11. Sokolova, N.V.; Nenajdenko, V.G. Recent advances in the Cu(I)-catalyzed azide-alkyne cycloaddition: Focus on functionally substituted azides and alkynes. *RSC Adv.* **2013**, *3*, 16212–16242. [CrossRef]
12. Kim, H.; Lee, S.; Lee, J.; Tae, J. Rhodamine Triazole-Based Fluorescent Probe for the Detection of Pt^{2+}. *Org. Lett.* **2010**, *12*, 5342–5345. [CrossRef] [PubMed]
13. Chang, K.C.; Su, I.H.; Lee, G.H.; Chung, W.S. Triazole- and azo-coupled calix[4]arene as a highly sensitive chromogenic sensor for Ca^{2+} and Pb^{2+} ions. *Tetrahedron Lett.* **2007**, *48*, 7274–7278. [CrossRef]
14. Hemamalini, A.; Mudedla, S.K.; Subramanian, V.; Das, T.M. Design, synthesis and metal sensing studies of ether-linked bis-triazole derivatives. *New J. Chem.* **2015**, *39*, 3777–3784. [CrossRef]
15. Ornelas, C.; Aranzaes, J.R.; Cloutet, E.; Alves, S.; Astruc, D. Click assembly of 1,2,3-triazole-linked dendrimers, including ferrocenyl dendrimers, which sense both oxo anions and metal cations. *Angew. Chem. Int. Ed.* **2007**, *46*, 872–877. [CrossRef] [PubMed]
16. Thakur, A.; Mandal, D.; Ghosh, S. Sensitive and Selective Redox, Chromogenic, and "Turn-On" Fluorescent Probe for Pb(II) in Aqueous Environment. *Anal. Chem.* **2013**, *85*, 1665–1674. [CrossRef] [PubMed]
17. Ji, X.L.; Xu, H.Y. Preparation of good solubility poly(triazole)s for Hg^{2+} detection via click chemistry. *Mater. Res. Innov.* **2014**, *18*, 37–40. [CrossRef]
18. Chen, H.; Yang, Y.; Wang, Y.; Wu, L. Synthesis, Structural Characterization, and Thermoresponsivity of Hybrid Supramolecular Dendrimers Bearing a Polyoxometalate Core. *Chem. Eur. J.* **2013**, *19*, 11051–11061. [CrossRef] [PubMed]
19. Katritzky, A.R.; Singh, S.K.; Meher, N.K.; Doskocz, J.; Suzuki, K.; Jiang, R.; Sommen, G.L.; Ciaramitaro, D.A.; Steel, P.J. Triazole-oligomers by 1,3-dipolar cycloaddition. *Arkivoc* **2006**, 43–62. [CrossRef]

20. De, P.; Faust, R.; Schimmel, H.; Ofial, A.R.; Mayr, H. Determination of rate constants in the carbocationic polymerization of styrene: Effect of temperature, solvent polarity, and Lewis acid. *Macromolecules* **2004**, *37*, 4422–4433. [CrossRef]

21. Niu, C.; Li, G.; Tuerxuntayi, A.; Aisa, H.A. Synthesis and Bioactivity of New Chalcone Derivatives as Potential Tyrosinase Activator Based on the Click Chemistry. *Chin. J. Chem.* **2015**, *33*, 486–494. [CrossRef]

22. Natali, M.; Soldi, L.; Giordani, S. A photoswitchable Zn(II) selective spiropyran-based sensor. *Tetrahedron* **2010**, *66*, 7612–7617. [CrossRef]

23. Ulatowski, F.; Dabrowa, K.; Balakier, T.; Jurczak, J. Recognizing the Limited Applicability of Job Plots in Studying Host-Guest Interactions in Supramolecular Chemistry. *J. Org. Chem.* **2016**, *81*, 1746–1756. [CrossRef] [PubMed]

24. Thordarson, P. Determining association constants from titration experiments in supramolecular chemistry. *Chem. Soc. Rev.* **2011**, *40*, 1305–1323. [CrossRef] [PubMed]

25. Howe, E.N.; Bhadbhade, M.; Thordarson, P. Cooperativity and complexity in the binding of anions and cations to a tetratopic ion-pair host. *J. Am. Chem. Soc.* **2014**, *136*, 7505–7516. [CrossRef] [PubMed]

26. Shiraishi, Y.; Nakamura, M.; Hayashi, N.; Hirai, T. Coumarin-Spiropyran Dyad with a Hydrogenated Pyran Moiety for Rapid, Selective, and Sensitive Fluorometric Detection of Cyanide Anion. *Anal. Chem.* **2016**, *88*, 6805–6811. [CrossRef] [PubMed]

27. Sumiya, S.; Doi, T.; Shiraishi, Y.; Hirai, T. Colorimetric sensing of cyanide anion in aqueous media with a fluorescein-spiropyran conjugate. *Tetrahedron* **2012**, *68*, 690–696. [CrossRef]

28. Shiraishi, Y.; Itoh, M.; Hirai, T. Rapid colorimetric sensing of cyanide anion in aqueous media with a spiropyran derivative containing a dinitrophenolate moiety. *Tetrahedron Lett.* **2011**, *52*, 1515–1519. [CrossRef]

29. Lee, D.N.; Seo, H.; Shin, I.S.; Hong, J.I. Paper Strip-based Fluorometric Determination of Cyanide with an Internal Reference. *Bull. Korean Chem. Soc.* **2016**, *37*, 1320–1325. [CrossRef]

30. Areti, S.; Bandaru, S.; Yarramala, D.S.; Rao, C.P. Optimizing the Electron-Withdrawing Character on Benzenesulfonyl Moiety Attached to a Glyco-Conjugate to Impart Sensitive and Selective Sensing of Cyanide in HEPES Buffer and on Cellulose Paper and Silica Gel Strips. *Anal. Chem.* **2015**, *87*, 12396–12403. [CrossRef] [PubMed]

31. Ou, X.-X.; Jin, Y.-L.; Chen, X.-Q.; Gong, C.-B.; Ma, X.-B.; Wang, Y.-S.; Chow, C.-F.; Tang, Q. Colorimetric test paper for cyanide ion determination in real-time. *Anal. Methods* **2015**, *7*, 5239–5244. [CrossRef]

sensors

MDPI

Article

Early Identification of Herbicide Stress in Soybean (*Glycine max* (L.) Merr.) Using Chlorophyll Fluorescence Imaging Technology

Hui Li [1,2], Pei Wang [1,2,*], Jonas Felix Weber [2] and Roland Gerhards [2]

[1] College of Engineering and Technology, Southwest University, Chongqing 400716, China; leehui@swu.edu.cn
[2] Institute of Phytomedicine, University of Hohenheim, 70599 Stuttgart, Germany; j.weber@uni-hohenheim.de (J.F.W.); roland.gerhards@uni-hohenhiem.de (R.G.)
* Correspondence: wangpei@live.cn; Tel.: +49-711-459-22940

Received: 27 September 2017; Accepted: 15 December 2017; Published: 22 December 2017

Abstract: Herbicides may damage soybean in conventional production systems. Chlorophyll fluorescence imaging technology has been applied to identify herbicide stress in weed species a few days after application. In this study, greenhouse experiments followed by field experiments at five sites were conducted to investigate if the chlorophyll fluorescence imaging is capable of identifying herbicide stress in soybean shortly after application. Measurements were carried out from emergence until the three-to-four-leaf stage of the soybean plants. Results showed that maximal photosystem II (PS II) quantum yield and shoot dry biomass was significantly reduced in soybean by herbicides compared to the untreated control plants. The stress of PS II inhibiting herbicides occurred on the cotyledons of soybean and plants recovered after one week. The stress induced by DOXP synthase-, microtubule assembly-, or cell division-inhibitors was measured from the two-leaf stage until four-leaf stage of soybean. We could demonstrate that the chlorophyll fluorescence imaging technology is capable for detecting herbicide stress in soybean. The system can be applied under both greenhouse and field conditions. This helps farmers to select weed control strategies with less phytotoxicity in soybean and avoid yield losses due to herbicide stress.

Keywords: herbicide stress; phytotoxicity; soybean; chlorophyll fluorescence imaging

---

1. Introduction

Soybean (*Glycine max* (L.) Merr.) is a worldwide cultivated crop. More than 80% of overall soybeans production originates from the USA, Brazil, and Argentina [1]. Since 1996, the Roundup-Ready (RR) Soybean cultivars have been introduced in the USA, Brazil, and Argentina. Farmers can apply glyphosate as a simple, selective, and effective method for weed control without being concerned about crop injury. In the European Union, weed control in soybean is only performed with conventional herbicides and non-chemical methods. For example, the production of soybean has increased more than 10 times in Germany since 2009 [1]. Pre- and post-emergent herbicide applications are a conventional and effective approach for weed control in soybean cultivations. Occasionally, the herbicides can also damage the crops, delay crop growth, and reduce crop yield when applied under unfavorable soil conditions, weather conditions such as rainfalls and low temperature, or with incorrect timing or mixture [2,3]. Early identification of herbicide stress can contribute to testing the soybean's genotype sensitivity. It can also help to test the management practices, soil, and weather conditions in order to minimize crop damage, and adjust herbicide dose or select proper herbicide for specific conditions.

Conventional estimation of herbicide damage on crops is usually conducted by visual assessments [4]. For instance, the soybean yield loss can be correlated with the injury symptoms of the stressed plants [5,6]. Advances in computer and photography technology enabled a quantitative assessment method by measuring crop ground cover [4]. A linear relationship was presented between the relative soybean yield and percentage of ground cover. The light reflectance is also used to evaluate the herbicide injury to herbicide [7]. However, these methods evaluate the crop healthiness according to the visible features. That usually requires a relatively long period of time so that the phytotoxic symptoms can be identified on the plants or the plants can grow large enough for the distinction of the ground cover rates. Chlorophyll fluorescence imaging technology is a non-destructive method to investigate the physiological reaction of the photosystem II (PS II) of plants. The approach of chlorophyll fluorescence imaging is very sensitive to abiotic and biotic stress detection on plants [8–10]. Some laboratory and greenhouse research demonstrated that, after herbicide application, the chlorophyll fluorescence quantum yield of sensitive weeds was markedly higher than the resistant populations [11–17]. Wang, Peteinatos, Li, and Gerhards [18] successfully practiced this technology in fields for a survey for resistance profiles of 40 *Alopecurus myosuroides* populations. By applying the chlorophyll fluorescence imaging technology, herbicide efficacy on weeds was observed within five days in the above research. However, the studies and measurements were carried out to distinct herbicide injured sensitive weeds from unstressed resistant population. In that case, herbicide would damage the photosystem II of sensitive weeds. Thus, the variation of chlorophyll fluorescence response can be significant. Recovery of herbicide stress in crops has not been investigated.

The objective of this study was to test if the herbicide stress and the respective recovery can be identified on the photosystem II of the soybean plants. The question was whether or not this identification can be performed shortly after the herbicide application, and if so what was the time relation between the induction and the identification of the stress. Furthermore, can this identification be performed both under greenhouse and field conditions using the chlorophyll fluorescence imaging technology.

2. Materials and Methods

2.1. Experimental Design

2.1.1. Greenhouse Experiment

A greenhouse experiment was conducted in the University of Hohenheim (Stuttgart, Germany) from November 2013 until April 2014. Soybeans (Sultana, R.A.G.T. Saaten, Herford, Germany) were sown in pots (15 × 15 × 15 cm) filled with 6.5 kg soil mixture of 50% clay, 25% silt, and 25% sand. The depth of the soil mixture was about 80 mm. The soybeans were sown in a depth of 45 mm with three seeds per pot (equivalent to 96 seeds m^{-2}). Plants were grown in a light cycle of 16 h day and 8 h night. The temperature was kept at 25 °C during the day and 15 °C at night. All pots were placed in a complete randomized block design with four blocks. The following three herbicide combinations with recommended dosages were selected for the treatments:

(i) 0.3 kg ha^{-1} Sencor® WG (700 g a.i. kg^{-1} metribuzin, WG, Bayer CropScience) + 0.25 L ha^{-1} Centium® 36 CS (360 g a.i. L^{-1} clomazone, CS, Cheminova Deutschland GmbH) + 0.8 L ha^{-1} Spectrum® (720 g a.i. L^{-1} dimethenamid-P, EC, BASF);

(ii) 2.0 kg ha^{-1} Artist® (175 g a.i. kg^{-1} metribuzin, 240 g a.i. kg^{-1} flufenacet, WG, Bayer CropScience), Harmony® SX® (500 g a.i. kg^{-1} thifensulfuron, SG, Du Pont);

(iii) Harmony® SX® (500 g a.i. kg^{-1} thifensulfuron, SG, Du Pont) + Basagran® (480 g a.i. L^{-1} bentazon, SL, BASF), Harmony® SX® (500 g a.i. kg^{-1} thifensulfuron, SG, Du Pont) + Fusilade® MAX (125 g a.i. L^{-1} fluazifop-P-butyl, EC, Syngenta).

Table 1. The herbicide application times for the greenhouse experiment (in days after sowing of soybeans). H1, herbicide combination 1; H2, herbicide combination 2; H3, herbicide combination 3; E, early application; L, late application; D$_1$, recommended dosage; D$_{0.5}$, half recommended dosage.

Treatments	Before Emergence		After Emergence				
	4	11	24	31	33	38	45
H1ED$_1$	metribuzin, clomazone, dimethenamid-P						
H1ED$_{0.5}$							
H1LD$_1$		metribuzin, clomazone, dimethenamid-P					
H1LD$_{0.5}$							
H2ED$_1$	metribuzin, flufenacet			thifensulfuron			
H2ED$_{0.5}$							
H2LD$_1$		metribuzin, flufenacet				thifensulfuron	
H2LD$_{0.5}$							
H3ED$_1$			thifensulfuron, bentazon		thifensulfuron, fluazifop-P-butyl		
H3ED$_{0.5}$							
H3LD$_1$					thifensulfuron, bentazon		thifensulfuron, fluazifop-P-butyl
H3LD$_{0.5}$							

Additionally, for the above herbicide combinations, applications with the half recommended dosages were also sprayed as separate treatments. Untreated control pots with and without hand weeding were included respectively in each block. Herbicide treatments were performed pre- and post-emergence depending on the registrations of the products. The application time is given in Table 1. A laboratory track sprayer chamber mounted with a single flat fan nozzle was used for herbicide application (8002 EVS, TeeJet Spraying System Co., Wheaton, IL, USA). The sprayer was calibrated for an applying volume of 200 L ha^{-1}. The applications were performed 500 mm above the soil surface.

2.1.2. Field Experiment

Five field experiments were conducted in 2015. The field trials were located in Southwest Germany at Böblingen, Calw, Nürtingen, Renningen, and Tübingen. All the herbicide combinations were selected according to the local practice of the farmers during the last three years. Seeds of soybeans (Sultana, R.A.G.T. Saaten, Herford, Germany) were sown at a depth of 45 mm between 14 April and 15 May. Approximately 70 seeds m^{-2} were sown with row distance of 170 mm in the fields. The experiments were set up as a randomized complete block design with four blocks and five treatments per block. The size of each plot was 2 × 5 m. The herbicide application was carried out three days after sowing with the following herbicides per treatment,

(i)   2.0 kg ha^{-1} Artist® (175 g a.i. kg^{-1} metribuzin, 240 g a.i. kg^{-1} flufenacet, WG, Bayer CropScience);

(ii)   1.5 kg ha^{-1} Stomp® Aqua (455 g a.i. L^{-1} pendimethalin, CS, BASF) + 2.0 L ha^{-1} Quantum® (600 g a.i. L^{-1} pethoxamid, EC, Cheminova Deutschland GmbH);

(iii) 0.4 L ha^{-1} Sencor® Liquid (600 g a.i. L^{-1} metribuzin, SC, Bayer CropScience) + 0.25 L ha^{-1} Centium® 36 CS (360 g a.i. L^{-1} clomazone, CS, Cheminova Deutschland GmbH);

(iv) 0.4 L ha^{-1} Sencor® Liquid (600 g a.i. L^{-1} metribuzin, SC, Bayer CropScience) + 0.25 L ha^{-1} Centium® 36 CS (360 g a.i. L^{-1} clomazone, CS, Cheminova Deutschland GmbH) + 0.8 L ha^{-1} Spectrum® (720 g a.i. L^{-1} dimethenamid-P, EC, BASF).

An untreated control was included in each block at all sites. Herbicides were sprayed with an electrically motorized plot boom sprayer with Lechler IDK 120-02 nozzles (Metzingen, Germany). The spraying volume was calibrated to 200 L ha^{-1}. No rainfall was recorded within 24 h after treatment.

2.2. Chlorophyll Fluorescence Sensor

The mobile fluorescence sensor, WEED-PAM® system (Figure 1, Heinz Walz GmbH, Effeltrich, Germany), was used to measure the chlorophyll fluorescence in this research. It contains 40 dark adaption cover boxes, a camera head, a tablet computer, and a central control unit. LED lights emitting light at a wavelength of 460 nm were mounted on the camera head to induce chlorophyll fluorescence. The camera detects fluorescence excitation at above 680 nm after an optical red long pass filter. The efficiency of photosystem II (PS II) of soybeans was determined by measuring the maximal PS II quantum yield (Fv/Fm). It is calculated as

$$Fv/Fm = (Fm - F_0)/Fm, \qquad (1)$$

where F_0 is the minimum fluorescence yield, Fm is the maximal fluorescence yield [8]. The WEED-PAM® system was operated by the software "ImagingWin" (Heinz Walz GmbH, Effeltrich, Germany). With this software, the background noise can be removed as described by Kaiser, Menegat, and Gerhards [16].

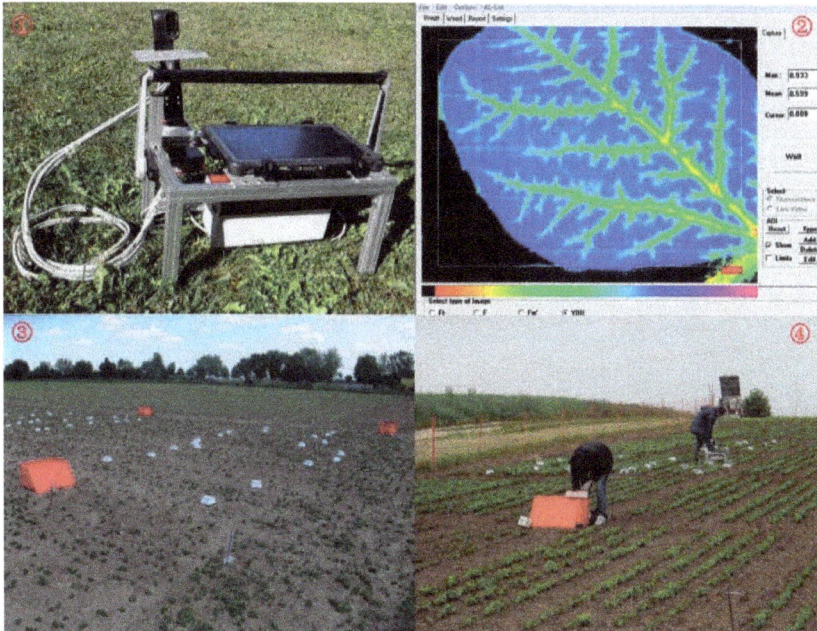

Figure 1. The field chlorophyll fluorescnce sensor WEED-PAM®. ① A picture of the sensor. It consists of the camera control unit and the computer including software; ② The software interface when measuring a herbicide treated leaf of soybean. The purple and blue pixels represent leaf area with higher Fv/Fm values, while the red pixels represent leaf area with lower Fv/Fm values. Blue color represents high Fv/Fm values and healthy tissues while the yellow and red color represents pixels with low Fv/Fm values and plant damage; ③ Dark adaption cover box distribution when conducting the first measurement at the one-leaf stage of soybeans at site Böblingen; ④ Measurement at the two-leaf stage of the soybeans at site Nürtingen.

2.3. Measurements and Data Analysis

For the greenhouse experiment, all measurements with the WEED-PAM® system were conducted 19, 21, 26, 31, 38, and 47 days after sowing (at least one plant had emerged in each pot). One plant per pot was selected for the measurement. All the plants were dark adapted with the dark adaption cover boxes for 30 min before measuring. Whole plants of soybeans were collected and washed 67 days after sowing. The root and aboveground biomass were cut and dried separately. After 48 h drying in a drying chamber under 80 °C, the dry biomass was measured.

For the field trials, three measurements were taken at each site, respectively, when the soybeans were at one-leaf stage (BBCH 10), two-leaf stage (BBCH 11), and three-leaf stage (BBCH 12). Ten soybean plants were measured in each plot. All the plants were dark adapted with the dark adaption cover boxes for 25–30 min before measuring. Values of all 40 plants (four blocks) were averaged. During the measurement, each plant was marked with an orange stick and label, so that the same plants were measured during the experiments. Aboveground biomass was cut on 15 July 2015 (10 to 12 weeks after sowing) at all five sites. Plants were cut in each plot from an area of 0.5 m². The dry aboveground biomass of soybean was weighted after 48 h drying in a drying chamber under 80 °C.

Data were analyzed with R (Version 3.0.2) and the package agricolae and lawstat [19] (R Development Core Team, 2008). The significance of herbicide effect on soybean plants was determined by performing an ANOVA ($p > 0.05$). In order to separate the treatments, a Tukey's HSD test ($p > 0.05$) was used. All the datasets were proved to be normally distributed using Shapiro–Wilk test ($p > 0.05$). Homogeneity of variances was analyzed by Levene's test ($p > 0.05$).

3. Results

3.1. Greenhouse Experiment

In the greenhouse test, at least one plant emerged in each pot at the performance of the measurements (19 days after sowing). As it can also be seen in Table 2, all three herbicide combinations reduced the Fv/Fm value of the soybean plants (several results were ignored because of overexposure during the measurement). The Fv/Fm of soybeans with pre-emergent herbicide treatments were significantly lower than the control plants during the first three weeks after application. However, the Fv/Fm of plants with post-emergent herbicide application dropped to the lower level only for one week after treatment. Meanwhile, soybeans in treatments with half of the recommended dosage mostly presented no significantly different PS II reaction level than the untreated control plants. Both early and late applications of herbicide led to an Fv/Fm reduction of the soybean plants. Dry biomass measurements (Figure 2) demonstrated that the soybean plants in the untreated group without hand-weeding had the lowest weight. The soybean plants in the untreated group with hand-weeding had relatively high biomass. However, the difference from the herbicide treated groups was not significant.

Table 2. The results of chlorophyll fluorescence measurements (Fv/Fm means) of the greenhouse experiment. H1, herbicide combination 1; H2, herbicide combination 2; H3, herbicide combination 3; E, early application; L, late application; D_1, recommended dosage; $D_{0.5}$, half recommended dosage; ConH, control with hand weeding; Con, control without hand weeding; significant differences between mean values are indicated by different letters (Tukey's HSD Test, $p < 0.05$).

Treatments	Days after Sowing									
	19		21		26		31		38		47	
$H1ED_1$	0.264	b	0.241	cd	0.271	cd	0.484	bc	0.717	a	0.724	a
$H1ED_{0.5}$	0.425	ab	0.520	abc	0.483	abc	0.608	abc	0.739	a	0.731	a
$H1LD_1$	0.330	b	0.386	bcd	0.361	bcd	0.605	abc	0.740	a	0.725	a
$H1LD_{0.5}$	0.463	ab	0.466	abcd	0.405	abcd	0.577	abc	0.708	a	0.716	a
$H2ED_1$	0.296	b	0.285	cd	0.295	cd	0.476	bc	0.723	a	-	
$H2ED_{0.5}$	0.420	ab	0.419	abcd	0.336	cd	0.515	abc	0.720	a	0.697	a
$H2LD_1$	0.235	b	0.201	d	0.152	d	0.432	c	0.720	a	0.705	a
$H2LD_{0.5}$	0.306	b	0.267	cd	0.345	cd	0.567	abc	0.727	a	0.714	a
$H3ED_1$	0.655	a	0.695	a	0.425	abcd	0.644	abc	0.737	a	0.724	a
$H3ED_{0.5}$	0.652	a	0.690	a	0.537	abc	0.679	ab	0.746	a	0.729	a
$H3LD_1$	0.641	a	0.691	a	0.667	a	0.668	ab	0.666	a	0.705	a
$H3LD_{0.5}$	0.616	a	0.671	ab	0.650	ab	0.673	ab	0.707	a	0.722	a
ConH	0.641	a	0.674	a	0.694	a	0.720	a	-		0.751	a
Con	0.636	a	0.636	ab	0.643	ab	0.672	ab	-		0.733	a

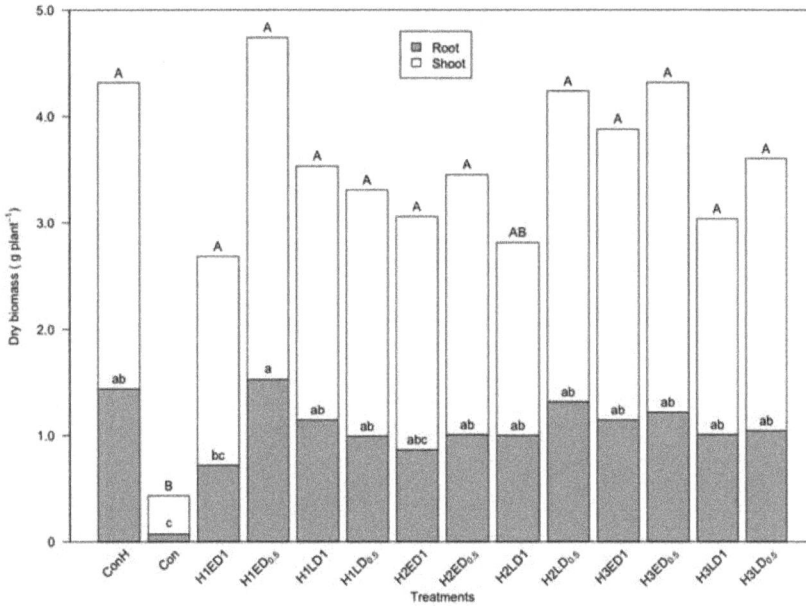

Figure 2. The root and shoot dry biomass per soybean plant on 67 days after sowing. H1, herbicide combination 1; H2, herbicide combination 2; H3, herbicide combination 3; E, early application; L, late application; D1, recommended dosage; $D_{0.5}$, half recommended dosage; significant differences between mean values for the root and the shoot independently are indicated by different letters (Tukey's HSD Test, $p < 0.05$).

3.2. Field Experiment

The results of the Fv/Fm values and the soybeans' biomass are presented in Table 3, and their relative change to plants in control groups after herbicide treatment was shown in Table 4.

Table 3. The results of chlorophyll fluorescence (means of Fv/Fm values) and dry biomass measurements of the field experiment. MoA, Mode of Action; C1, Inhibition of PS II; F4, Inhibition of DOXP synthase; K1, Inhibition of microtubule assembly; K3, Inhibition of cell division (VLCFA); *, stress efficacy indicated by significantly different Fv/Fm values and biomass in both measurements; significant differences between mean values are indicated by different letters (Tukey's HSD Test, $p < 0.05$).

Sites	Treatment	MoA	Fv/Fm			Biomass ($g\ m^2$)	Stress Efficacy
			Date 1	Date 2	Date 3		
Böblingen	Control	-	0.575a	0.587a	0.666a	310b	
	i	C1 K3	0.423b	0.503a	0.681a	394b	*
	ii	K1 K3	0.543a	0.607a	0.639a	476a	
	iii	C1 F4	0.490ab	0.567a	0.674a	450a	
	iv	C1 F4 K3	0.428b	0.524a	0.639a	356b	*
Calw	Control	-	0.584a	0.558ab	0.672a	40b	
	i	C1 K3	0.575a	0.524bc	0.645ab	296a	
	ii	K1 K3	0.585a	0.571ab	0.647ab	226ab	
	iii	C1 F4	0.596a	0.464c	0.563c	130b	*
	iv	C1 F4 K3	0.585a	0.593a	0.627b	248ab	

Table 3. *Cont.*

Sites	Treatment	MoA	Fv/Fm			Biomass (g m²)	Stress Efficacy
			Date 1	Date 2	Date 3		
Nürtingen	Control	-	0.586a	0.602a	0.722a	580a	
	i	C1 K3	0.629a	0.531ab	0.706a	548a	
	ii	K1 K3	0.586a	0.516b	0.644b	490b	*
	iii	C1 F4	0.583a	0.592a	0.714a	558a	
	iv	C1 F4 K3	0.601a	0.577ab	0.709a	526a	
Renningen	Control	-	0.411a	0.472a	0.645ab	102b	
	i	C1 K3	0.440a	0.513a	0.613b	206a	
	ii	K1 K3	-	0.474a	0.666a	242a	
	iii	C1 F4	0.498a	0.490a	0.426c	136b	*
	iv	C1 F4 K3	-	0.514a	0.632ab	216a	
Tübingen	Control	-	0.545a	0.545a	0.662a	85b	
	i	C1 K3	0.529a	0.478a	0.659a	147a	
	ii	K1 K3	0.555a	0.472a	0.663a	125a	
	iii	C1 F4	0.517a	0.518a	0.658a	150a	
	iv	C1 F4 K3	0.545a	0.520a	0.667a	110a	

Table 4. The relative change of the Fv/Fm values and the dry biomass to the untreated control plants of each site and measuring date in the field experiment. The relative Fv/Fm values were calculated on the average Fv/Fm values of the treated plants by the average Fv/Fm values of the relative untreated control plants. MoA, Mode of Action; C1, Inhibition of PS II; F4, Inhibition of DOXP synthase; K1, Inhibition of microtubule assembly; K3, Inhibition of cell division (VLCFA); *, stress efficacy on biomass correlated to significantly different Fv/Fm values in both measurements.

Sites	Treatment	MoA	Relative Fv/Fm			Relative Biomass
			Date 1	Date 2	Date 3	
Böblingen	i	C1 K3	0.736	0.857	1.023	1.271*
	ii	K1 K3	0.944	1.034	0.959	1.535
	iii	C1 F4	0.852	0.966	1.012	1.452
	iv	C1 F4 K3	0.744	0.893	0.959	1.148*
Calw	i	C1 K3	0.985	0.939	0.96	7.4
	ii	K1 K3	1.002	1.023	0.963	5.65
	iii	C1 F4	1.021	0.832	0.838	3.250*
	iv	C1 F4 K3	1.002	1.063	0.933	6.2
Nürtingen	i	C1 K3	1.074	0.882	0.978	0.945
	ii	K1 K3	1	0.857	0.892	0.845*
	iii	C1 F4	0.995	0.983	0.989	0.962
	iv	C1 F4 K3	1.026	0.958	0.982	0.907
Renningen	i	C1 K3	1.071	1.087	0.95	2.02
	ii	K1 K3	-	1.004	1.033	2.373
	iii	C1 F4	1.212	1.038	0.66	1.333*
	iv	C1 F4 K3	-	1.089	0.98	2.118
Tübingen	i	C1 K3	0.971	0.877	0.995	1.729
	ii	K1 K3	1.018	0.866	1.002	1.471
	iii	C1 F4	0.949	0.95	0.994	1.765
	iv	C1 F4 K3	1	0.954	1.008	1.294

At Böblingen, the Fv/Fm of soybean seedlings in the treatment i and iv were significantly lower than in the untreated control plants already at the first measurement. However, the plants recovered until the second measurement. The biomass weight of soybean plants with treatment i and iv were significantly lower than the soybean plants of all the other treatments. The biomass of soybean in the plots without herbicide treatment was lowest probably due to weed competition.

At Calw, the soybean plants presented lower photosystem efficiency in treatment iii. Unlike Böblingen, the herbicide stress on PS II appeared, when plants produced the second leaf.

Moreover, the stress lasted until the end of the measurement. Biomass measurements showed the significantly lower weight of soybean in the control and treatment iii than in the other treatments.

A significant response of PS II was observed in treatment ii at Nürtingen. The Fv/Fm values of the soybeans in treatment ii was reduced from the second measuring date until the end of the measurements similar to the trial at Calw. Weed infestation at this site was very low. Therefore, the biomass of soybeans was not reduced in the untreated plots.

First measurement results of treatment ii and iv at Renningen were lost due to the unexpected power failure when exporting the data from the sensor. At this site, Fv/Fm reduction occurred in treatment iii. However, the difference could only be distinguished until the third leaf of soybeans was produced. The biomass measurements also showed the lower weight of soybeans in the control group and under treatment iii.

At Tübingen, except in the biomass of soybeans in untreated plots, no differences in the PS II quantum yield and the biomass were observed between the treatments.

4. Discussion

The chlorophyll fluorescence measurements showed herbicide induced stress on PS II of young soybeans plants in all treatments in the greenhouse, as well as at four sites out of the five field trials. Herbicides with six modes of action were included in the study, which were: PS II inhibition, DOXP synthase inhibition, microtubule assembly inhibition, cell division inhibition, ALS- and ACCase inhibition. Several authors support our findings, that most herbicides reduce light reactions of photosystems shortly after application. Especially when the herbicide dose absorbed by the plants exceeded a certain critical threshold, the plants' will not be able to metabolize the active ingredients anymore [18,20,21].

Metribuzin rapidly inhibits the PS II after treatment by binding at the QB site of plastoquinone and interrupting the electron transfer flow [22]. Most cultivars of soybean are tolerant to metribuzin. Therefore, metribuzin provides selective weed control in soybean [23,24]. Sultana, which was selected for this research, is a metribuzin tolerant cultivator. According to Falb and Smith [25], tolerant soybean cultivators can detoxify metribuzin within 106 hours after treatment. These finding corresponded to our chlorophyll fluorescence imaging measurements revealing a rapid recovery from metribuzin treatments mainly in the field trial at Böblingen. In treatment iii of the greenhouse test, the stress could also be induced by the PS II inhibitor bentazon, as the separated application of thifensulfuron and fluazifop-P-butyl caused no effect on the Fv/Fm of the soybean plants. Biomass assessment showed that post-emergent ALS- and ACCase-inhibiting herbicides did not cause any stress to soybeans. However, their activity against weed species is limited as well. That is why pre-emergent herbicides in soybean production play a major role in weed management.

In the greenhouse study, early occurrence and long duration of stress effect took place after the treatment of herbicide combinations 1 and 2. Apart from the PS II inhibitor-, DOXP synthase-, and cell division- inhibitors were also included in the herbicide mixtures. Thus, another stress mechanism might take place as well in these groups.

In the field experiments, inhibition of PS II of soybeans at site Calw and Renningen also occurred later and lasted longer than the photosystem regulation at site Böblingen. Besides metribuzin, clomazone (inhibitor of DOXP synthase) was also involved in the stressed treatments. Non-mevalonate 1-deoxy-D-xylulose-5-phosphate (DOXP) pathway is a main biosynthesis approach for plastidic isoprenoids, such as carotenoids, phytol (a side-chain of chlorophylls), plastoquinone-9, isoprene, mono-, and diterpenes [26]. Most of the biosynthesis proceeded inside the chloroplast [27]. Chlorophyll production could be reduced as less phytol was provided due to the DOXP synthase inhibition. Therefore, the photosystem efficiency of DOXP synthase stressed soybeans was lower than the unstressed ones when the plants grew larger. The Fv/Fm reduction of soybean plants in treatment iii at site Calw and Renningen could be attributed to the application of clomazone.

The combined application of pendimethalin (microtubule assembly inhibitor) and pethoxamid (cell division inhibitor) induced stress on PS II at site Nürtingen. „Dinitroaniline herbicides like pendimethalin bind to α-tubulin [28]. Thus, the free tubulin could not group into polymerization as microtubule. Several publications noted that dinitroanilines could interfere with the photosystem II dramatically by oxygen evolution [29,30]. Chloroacetamides inhibits the very-long-chain fatty acids (VLCFA) synthase. The herbicide markedly reduces VLCFA content in the plasma membrane and results in cell death [31]. Some chloroacetamides (e.g., carbetamide) could inhibit electron transport up to 50% as a secondary effect of membrane destabilization [21,32]. Therefore, the chlorophyll fluorescence of plants can be altered. This hypothesis correlated well with the Fv/Fm regulation of soybean under the treatment ii combination of metribuzin and flufenacet in the greenhouse test. However, metribuzin was not to be the only compound causing stress in soybean. As the herbicides inhibiting either cell division or VLCFA synthase might induce the regulation on photosystem, the stress mechanism in the treatment ii at site Nürtingen still could not be clearly explained. Furthermore, considering the long period stress on soybeans under the treatment i in the greenhouse experiment, it could also be induced by the combined effect of DOXP synthase- and cell division-inhibitors after the effect of PS II inhibitor metribuzin.

The biomass assessment on herbicide treated soybean significantly distinct the stressed or non-stressed groups in the field. Apparently, the biomass assessment results were similar to the sensor measurements. A similar relationship was also observed in the greenhouse study.

Beside using the Fv/Fm values, several other parameters of chlorophyll fluorescence measurements, such as ΦPSII (effective quantum yield of photochemical energy conversion in PSII) and NPQ (non-photochemical dissipation of absorbed energy), are also common for stress assessment of plants.

ΦPSII can be measured without dark adaption. However, the measurements require steady-state photosynthesis lighting conditions, which means that plants in the field should be in the full sunlight, and not under any canopy cloudy conditions [8]. As the measurements with Weed-PAM® usually take more than two hours for each site, the weather conditions cannot be ensured for such a long period. Moreover, the system is designed for early herbicide stress detection, the measurements should be conducted within seven days no matter if the weather is sunny, cloudy, or rainy. Thus, spending 20 min for dark adaption and measuring the parameter Fv/Fm should be more appropriate for the sensors field practice.

The other common parameter of chlorophyll fluorescence measurements, NPQ, also can be measured while the plants had a dark adaption period. Yet NPQ is more heavily affected by non-photochemical quenching that reflects heat-dissipation of excitation energy in the antenna system. Thus, it is more often used to indicate the excess radiant energy dissipation to heat in the PSII antenna complexes [33].

WEED-PAM® technology allows quantifying soybean response to herbicide treatments. The variation of plants' chlorophyll fluorescence emission could be detected shortly after treatment. Thus, herbicide damage to soybean can be avoided by proper selection of products. Since soybean cultivars respond differently to herbicides, WEED-PAM® technology can help to select the most tolerant cultivars.

5. Conclusions

Herbicides interfere directly or indirectly with the photosystem of plants and can reduce quantum use efficiency of PSII in soybean plants and result in lower biomass. With the chlorophyll fluorescence imaging technology, we were capable to identify the herbicide stress rapidly in the young growth stages of the soybean plants. This achievement will help farmers to avoid herbicide combinations that reduces crop growth. Besides, this study showed that the Fv/Fm values of the untreated soybean plants were different at each experiment site. A normalized model should be applied in the further development of the sensor system so that a unified assessment of the stress effect on plants can be created and comparisons can be performed.

Acknowledgments: The authors thank Simon Hotz and Moritz Sauter for collecting the data, and our colleagues from the State Plant Protection Services for preparing the field trails. We also thank the Fundamental Research Funds for the Central Universities, the Doctoral Fund of Southwest University (SWU116060), the Chinese Scholarship Council (grant number 201306350053) and the Ministry of Ländlicher Raum Baden-Württemberg for financial support.

Author Contributions: Pei Wang and Roland Gerhards conceived and designed the experiments; Pei Wang and Hui Li performed the experiments; Pei Wang and Hui Li analyzed the data; Jonas Felix Weber, Hui Li, and Roland Gerhards contributed reagents/materials/analysis tools; Hui Li, Pei Wang, and Roland Gerhards wrote the paper.

Conflicts of Interest: The authors declare no conflict of interest.

Abbreviation

ANOVA	Analysis of Variance
BBCH	Biologische Bundesanstalt, Bundessortenamt und Chemische Industrie
C1	Inhibition of Photosystem II
CS	Capsule Suspensions
DOXP	1-Deoxy- D-xylulose 5-phosphate
EC	Emulsifiable Concentrates
F4	Inhibition of DOXP Synthase
Fm	Maximal Fluorescence Yield
Fo	Dark Fluorescence Yield
Fv/Fm	Maximal PS II Quantum Yield
HSD	Honest Significant Difference
K1	Inhibition of Microtubule Assembly
K3	Inhibition of Cell Division
LED	Light-Emitting Diode
MoA	Mode of Action
PAM	Pulse Amplitude Modulation
PS II	Photosystem II
QB	a Protein-bound Plastoquinone
SC	Suspension Concentrates
SG	Soluble Granules
SL	Soluble (liquid) Concentrates
VLCFA	Very Long Chain Fatty Acid
WG	Water-Dispersible Granules

References

1. FAOSTAT (Food and Agriculture Organization of the United Nations: Statistics Division). Soybean Production of Commodity of the World. Available online: http://faostat3.fao.org/browse/Q/QC/E (accessed on 29 October 2016).
2. Salzman, F.P.; Renner, K.A. Response of soybean to combinations of clomazone, metribuzin, linuron, alachlor, and atrazine. *Weed Technol.* **1992**, *6*, 922–929.
3. Johnson, B.F.; Bailey, W.A.; Holshouser, D.L.; Herbert, D.A., Jr.; Hines, T.E. Herbicide effects on visible injury, leaf area, and yield of glyphosate-resistant soybean (*Glycine max*). *Weed Technol.* **2002**, *16*, 554–566. [CrossRef]
4. Donald, W.W. Estimated soybean (*Glycine max*) yield loss from herbicide damage using ground cover or rated stunting. *Weed Sci.* **1998**, *46*, 454–458.
5. Weidenhamer, J.D.; Triplett, G.B.; Sobotka, F.E. Dicamba injury to soybean. *Agron. J.* **1989**, *81*, 637–643. [CrossRef]
6. Bailey, J.A.; Kapusta, G. Soybean (*Glycine max*) tolerance to simulated drift of nicosulfuron and primisulfuron. *Weed Technol.* **1993**, *7*, 740–745.

7. Adcock, T.E.; Nutter, F.W., Jr.; Banks, P.A. Measuring herbicide injury to soybeans (*Glycine max*) using a radiometer. *Weed Sci.* **1990**, *38*, 625–627.
8. Maxwell, K.; Johnson, G.N. Chlorophyll fluorescence—A practical guide. *J. Exp. Bot.* **2000**, *51*, 659–668. [CrossRef] [PubMed]
9. Schreiber, U. Pulse-amplitude-modulation (PAM) fluorometry and saturation pulse method: An overview. In *Chlorophyll a Fluorescence*; Papageorgiou, G.C., Govindjee, G., Eds.; Springer: Dordrecht, The Netherlands, 2004; pp. 279–319.
10. Janka, E.; Körner, O.; Rosenqvist, E.; Ottosen, C.O. Using the quantum yields of photosystem II and the rate of net photosynthesis to monitor high irradiance and temperature stress in chrysanthemum (*Dendranthema grandiflora*). *Plant Phys. Biochem.* **2015**, *90*, 14–22. [CrossRef] [PubMed]
11. Ahrens, W.H.; Arntzen, C.J.; Stoller, E.W. Chlorophyll fluorescence assay for the determination of triazine resistance. *Weed Sci.* **1981**, *29*, 316–322.
12. Ali, A.; Machado, V.S. Rapid detection of 'triazine resistant' weeds using chlorophyll fluorescence. *Weed Res.* **1981**, *21*, 191–197. [CrossRef]
13. Hensley, J.R. A method for identification of triazine resistant and susceptible biotypes of several weeds. *Weed Sci.* **1981**, *29*, 70–73.
14. Vencill, W.K.; Foy, C.L. Distribution of triazine-resistant smooth pigweed (*Amaranthus hybridus*) and common lambsquarters (*Chenopodium album*) in Virginia. *Weed Sci.* **1988**, *36*, 497–499.
15. Van Oorschot, J.L.P.; Van Leeuwen, P.H. Use of fluorescence induction to diagnose resistance of *Alopecurus myosuroides* Huds. (black-grass) to chlorotoluron. *Weed Res.* **1992**, *32*, 473–482. [CrossRef]
16. Kaiser, Y.I.; Menegat, A.; Gerhards, R. Chlorophyll fluorescence imaging: A new method for rapid detection of herbicide resistance in *Alopecurus myosuroides*. *Weed Res.* **2013**, *53*, 399–406. [CrossRef]
17. Zhang, C.J.; Lim, S.H.; Kim, J.W.; Nah, G.; Fischer, A.; Kim, D.S. Leaf chlorophyll fluorescence discriminates herbicide resistance in *Echinochloa* species. *Weed Res.* **2016**, *56*, 424–433. [CrossRef]
18. Wang, P.; Peteinatos, G.; Li, H.; Gerhards, R. Rapid in-season detection of herbicide resistant *Alopecurus myosuroides* using a mobile fluorescence imaging sensor. *Crop Prot.* **2016**, *89*, 170–177. [CrossRef]
19. R Development Core Team. *R: A Language and Environment for Statistical Computing*; R Foundation for Statistical Computing: Vienna, Austria, 2008; ISBN 3-900051-07-0.
20. Dayan, F.E.; Watson, S.B. Plant cell membrane as a marker for light-dependent and light-independent herbicide mechanisms of action. *Pestic. Biochem. Phys.* **2011**, *101*, 182–190. [CrossRef]
21. Dayan, F.E.; Zaccaro, M.L.D.M. Chlorophyll fluorescence as a marker for herbicide mechanisms of action. *Pestic. Biochem. Phys.* **2012**, *102*, 189–197. [CrossRef]
22. Ventrella, A.; Catucci, L.; Agostiano, A. Herbicides affect fluorescence and electron transfer activity of spinach chloroplasts, thylakoid membranes and isolated Photosystem II. *Bioelectrochemistry* **2010**, *79*, 43–49. [CrossRef] [PubMed]
23. Hardcastle, W.S. Differences in the tolerance of metribuzin by varieties of soybeans. *Weed Res.* **1974**, *14*, 181–184. [CrossRef]
24. Barrentine, W.L.; Edwards, C.J.; Hartwig, E.E. Screening soybeans for tolerance to metribuzin. *Agron. J.* **1976**, *68*, 351–353. [CrossRef]
25. Falb, L.N.; Smith, A.E., Jr. Metribuzin metabolism in soybeans. Characterization of the intraspecific differential tolerance. *J. Agric. Food Chem.* **1984**, *32*, 1425–1428. [CrossRef]
26. Lichtenthaler, H.K.; Rohmer, M.; Schwender, J. Two independent biochemical pathways for isopentenyl diphosphate and isoprenoid biosynthesis in higher plants. *Physiol. Plantarum* **1997**, *101*, 643–652. [CrossRef]
27. Lichtenthaler, H.K. The 1-deoxy-D-xylulose-5-phosphate pathway of isoprenoid biosynthesis in plants. *Annu. Rev. Plant Biol.* **1999**, *50*, 47–65. [CrossRef] [PubMed]
28. Morrissette, N.S.; Mitra, A.; Sept, D.; Sibley, L.D. Dinitroanilines bind α-tubulin to disrupt microtubules. *Mol. Biol. Cell* **2004**, *15*, 1960–1968. [CrossRef] [PubMed]
29. Moreland, D.E.; Farmer, F.S.; Hussey, G.G. Inhibition of photosynthesis and respiration by substituted 2, 6-dinitroaniline herbicides: I. Effects on chloroplast and mitochondrial activities. *Pestic. Biochem. Phys.* **1972**, *2*, 342–353. [CrossRef]
30. Moreland, D.E.; Farmer, F.S.; Hussey, G.G. Inhibition of photosynthesis and respiration by substituted 2, 6-dinitroaniline herbicides: II. Effects on responses in excised plant tissues and treated seedlings. *Pestic. Biochem. Phys.* **1972**, *2*, 354–363. [CrossRef]

31. Böger, P. Mode of action for chloroacetamides and functionally related compounds. *J. Pestic. Sci.* **2003**, *28*, 324–329. [CrossRef]

32. Weisshaar, H.; Böger, P. Primary effects of chloroacetamides. *Pestic. Biochem. Phys.* **1987**, *28*, 286–293. [CrossRef]

33. Gilmore, A.M. Mechanistic aspects of xanthophylls cycle-dependent photoprotection in higher plant chloroplasts and leaves. *Physiol Plant* **1997**, *99*, 197–209. [CrossRef]

sensors

MDPI

Article

Methionine-Capped Gold Nanoclusters as a Fluorescence-Enhanced Probe for Cadmium(II) Sensing

Yan Peng [1,2], Maomao Wang [1], Xiaoxia Wu [1], Fu Wang [1,*] and Lang Liu [2,*]

[1] Laboratory of Environmental Sciences and Technology, Xinjiang Technical Institute of Physics & Chemistry, Chinese Academy of Sciences, Urumqi 830011, Xinjiang, China; yanpeng9236@sina.com (Y.P.); wma8899@sina.com (M.W.); wuxx@ms.xjb.ac.cn (X.W.)

[2] College of Chemistry and Chemical Engineering, Xinjiang University, Urumqi 830046, Xinjiang, China

* Correspondence: wangfu@ms.xjb.ac.cn (F.W.); liulang@xju.edu.cn (L.L.); Tel.: +86-991-3835879 (F.W.); +86-991-8588883 (L.L.)

Received: 20 December 2017; Accepted: 5 February 2018; Published: 23 February 2018

Abstract: Gold nanoclusters (Au NCs) have been considered as novel heavy metal ions sensors due to their ultrafine size, photo-stability and excellent fluorescent properties. In this study, a green and facile method was developed for the preparation of fluorescent water-soluble gold nanoclusters with methionine as a stabilizer. The nanoclusters emit orange fluorescence with excitation/emission peaks at 420/565 nm and a quantum yield of about 1.46%. The fluorescence of the Au NCs is selectively and sensitively enhanced by addition of Cd(II) ions attributed to the Cd(II) ion-induced aggregation of nanoclusters. This finding was further used to design a fluorometric method for the determination of Cd(II) ions, which had a linear response in the concentration range from 50 nM to 35 μM and a detection limit of 12.25 nM. The practicality of the nanoprobe was validated in various environmental water samples and milk powder samples, with a fairly satisfactory recovery percent.

Keywords: methionine; gold nanoclusters; fluorescence; enhancing; cadmium ion; water; milk

1. Introduction

Heavy metal ions are prevalent in agriculture, industry and drinking water, causing serious environmental problems. Specifically, cadmium ions (Cd^{2+}), which exist widely in air, soil, and water [1], are extremely toxic, not only causing serious environmental and health problems, but also being listed by the U.S. Environmental Protection Agency (EPA) as one of 126 priority pollutants [2]. Several methods have therefore been established to detect Cd^{2+} ion, such as atomic absorption spectrometry (AAS) [3], atomic fluorescence spectrometry (AFS) [4] and inductively coupled plasma–atomic emission spectroscopy (ICP–AES) [5]. Though they have acceptable sensitivity and selectivity, the operating conditions are normally cumbersome and costly, which inherently limits their wider application. Therefore, it is critical to explore a low-cost and simple method for the determination of Cd^{2+} in real samples.

In recent years, fluorescent sensors have attracted much attention in the detection of Cd^{2+} due to their excellent properties of easy operation, high sensitivity, selectivity, and real-time monitoring [6–9]. In the past few decades, a great variety of fluorescence probes have been reported for the determination of Cd^{2+}, including organic dyes [10–13] and quantum dots (QDs) [14–17]. However, those QDs are limited by the potential leakage of heavy metal elements, where the organic dyes suffer from small Stokes shifts and poor photo-stability [18,19]. Consequently, developing alternative and environmentally friendly materials is of great significance.

Noble metal nanoclusters (NCs), as a star in the family of metal nanomaterials, are gradually coming into view [20]. Owing to their ultrafine size (usually less than 2 nm) [21], which is equivalent to the electronic Fermi wavelength, NCs possess the nature of molecules, including discrete energy level, strong light luminescence, good light stability, biocompatibility and other unique physical and chemical properties, thus exhibiting great potential in the field of sensing and imaging [22–24]. The use of NCs to detect Cd^{2+} has been studied by several groups. In 2016, Niu et al., developed a dumbbell-shaped CQDs/Au NCs nanohybrid as a ratiometric fluorescent sensor for Cd^{2+} through a "turn-off" method [25]. In 2017, Naaz and Chowdhury synthesized a photoluminescent Ag NCs through fine tuning of sunlight and ultrasound to detect thiophilic metal ions (including Cd^{2+} ion), which is also based on fluorescence quenching [26]. However, such "turn-off" modes inevitably produce false results, which is not preferable in practice because other quenchers or environmental stimulus may also cause fluorescence quenching, and thus affect the sensitivity and authenticity of the test [27]. Although considerable progress has been made, finding new rapid and efficient nanoprobes to selectively recognize Cd^{2+} is still of great importance.

Herein, we demonstrated a green and facile strategy to prepare water-soluble, stable orange-emitting gold nanoclusters by using methionine as a stabilizer. The presence of Cd^{2+} ions leads to the aggregation of nanoclusters with enhancement of fluorescence intensity (Figure 1). Moreover, the nanoprobe was successfully applied to detect Cd^{2+} in various real samples with impressive efficiency and satisfactory recovery, showing great potential in practical application.

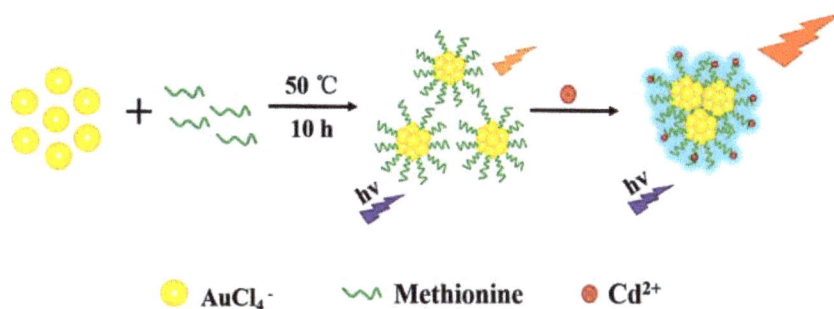

$AuCl_4^-$ Methionine Cd^{2+}

Figure 1. Schematic illustration of the Gold nanoclusters' (Au NCs) formation and the Cd^{2+} induced fluorescence enhancing of Au NCs.

2. Materials and Methods

2.1. Chemicals and Reagents

Chloroauric Acid ($HAuCl_4 \cdot 3H_2O$, 99.9%), DL-methionine ($C_5H_{11}NO_2S$, 99.9%) were obtained from Adamas-beta and Ascorbic Acid ($C_6H_8O_6$, 99.9%) was obtained from Sigma-Aldrich (Shanghai, China). All competitive metal ions and anions used in selectivity testing were acquired from Sinopharm Chemical Reagent Co. Ltd. (Shanghai, China). All the reagents were used as received without further purification. Ultrapure water (resistivity: 18.2 MΩ·cm) was obtained from a Millipore purification system (Shanghai, China).

2.2. Instruments

UV-vis absorption spectra were recorded on a Shimadzu UV-1800 spectrophotometer (Kyoto, Japan). Fluorescence spectra were performed on a Hitachi F-7000 fluorescence spectrometer (Tokyo, Japan). X-ray photoelectron spectroscopy (XPS) measurements were carried out using an ESCALAB 250Xi spectrometer (Thermo Fisher Scientific, Waltham, MA, USA). High-resolution transmission electron microscopy (HR-TEM) and energy-dispersive X-ray spectroscopy (EDX) data were obtained on a FEI Tecnai G2 F20 S-TWIN transmission electron microscopy instrument operating at 200 kV (Hillsboro, OR, USA). Electrospray ionization mass spectrometry (ESI-MS) measurements were conducted on a QSTAR elite liquid chromatography-mass spectrometry (LCMS, SCIEX, Toronto, ON, Canada), equipped with a common ESI source. Time-resolved luminescence intensity decay was recorded on a Horiba JY Fluorolog-3 molecule fluorometer (Paris, France), and samples were excited by a 375 nm laser light source. Inductively coupled plasma-optical emission spectrometry (ICP-OES) data were obtained on VISTA-PRO CCD Simultaneous ICP-OES (Varian, Palo Alto, CA, USA). Dynamic light scattering (DLS) was determined using a Nano-ZS90 (Malvern Instruments, Malvern, UK).

2.3. Synthesis of Au NCs

All glassware used in the following procedures was thoroughly cleaned with freshly prepared aqua regia (3:1 conc. HCl/HNO_3 v/v) and rinsed with ultrapure water prior to use. In a typical procedure, 3.88 mL of $HAuCl_4$ aqueous solution (2.5 mM) was mixed with 8 mL of methionine solution (80 mM) and 1.2 mL of NaOH (0.5 M) for 30 min, and then 3 mL of L-ascorbic acid (20 mM) was added into the solution at 50 °C within 9.5 h, obtaining yellow solutions. Afterwards, the reaction solution was centrifuged at 8000 rpm for 10 min to discard large particles and dialyzed with water via a dialysis membrane (1000 Da) for 48 h to remove the free ions and ligand. The resulting solution was stored in the dark at 4 °C for use. At the same time, the solid powder can be obtained by freeze-drying.

2.4. Fluorescent Detection of Cd^{2+}

$CdCl_2 \cdot 5/2H_2O$ was used for the study of Cd^{2+} detection. A 10 mM stock solution of $CdCl_2 \cdot 5/2H_2O$ was prepared, from which various Cd^{2+} concentration were prepared by serial dilution. The fluorescence spectra were recorded to observe the fluorescence intensity change in the presence of Cd^{2+} at different concentrations. To evaluate the sensitivity toward Cd^{2+}, 100 μL of Cd^{2+} solution of various concentrations was added into 1.9 mL of the prepared Au NCs solutions, and the mixtures were incubated at room temperature for the specified time before spectral measurements. The fluorescence spectra were recorded at room temperature with the maximum excitation wavelength at 420 nm; both the excitation and emission slit widths were 5 nm.

2.5. Selectivity Measurement

To check the selectivity of Au NCs, a series of competitive metal ions and anions with the identical concentration of 1 mM (100 μL), including Na^+, K^+, Mg^{2+}, Ba^{2+}, Pb^{2+}, Ca^{2+}, Al^{3+}, Cr^{3+}, Fe^{3+}, Sn^{2+}, Cl^-, $H_2PO_4^-$, Br^-, SCN^-, NO_3^-, CO_3^-, IO_3^-, SO_4^{2-}, NO_2^-, I^- and ClO_4^-, were respectively introduced to a group of Au NCs solutions (1.9 mL) to measure the change of fluorescence intensity. The resulting solutions were studied by fluorescence spectra at room temperature with excitation wavelength at 420 nm, both the excitation and emission slit widths were 5 nm. For enhancing efficiency measurement, the change of fluorescence (F/F_0) was determined by comparing the intensity of the fluorescence emission of different solutions. F_0 represents the fluorescence intensity of the Au NCs in the absence of metal ions, and F is the fluorescence intensity of the Au NCs in the presence of different concentrations of Cd^{2+} or different interfering ions.

2.6. Analysis of Real Samples

To evaluate the applicability of the Au NCs to real samples, tap water, lake water, milk powders and camel milk powders were spiked with Cd^{2+} and tested in the assay. Tap water was collected from our laboratory (Urumqi, China) and river water from Hong Lake (Urumqi, China). All water samples were freshly collected and filtered with 0.45 μm micropore membrane prior to analysis. Milk powders and camel milk powders were purchased from different food manufacturers and pretreated according to the previous Huang's report [28]. The obtained sample filtrates were adjusted the pH to 8, then analyzed with the proposed strategy and ICP-OES method. A recovery test was performed by spiking the pre-treated samples with the standard solutions of Cd^{2+} ions (3, 5 and 10 μM) and subsequently analyzing them by the aforesaid procedure.

3. Results and Discussion

3.1. Characterization of Au NCs

In the present work, we chose methionine as a stabilizer and chloroauric acid ($HAuCl_4 \cdot 3H_2O$) as metal precursor to prepare Au NCs. The synthetic conditions were systematically optimized, as depicted in Figure S1; the fluorescence intensity of the Au NCs reached its maximum when the concentration of methionine was 80 mM, NaOH was 0.5 M, and ascorbic acid was 20 mM, reaction at 50 °C within 10 h. The spectral characteristics of Au NCs were studied by UV–Vis and luminescence spectrometer. As shown in Figure 2A, the spectrum of as-prepared Au NCs displays insignificant absorption peaks, and does not have the characteristic surface plasma resonance (SPR) peak of larger Au NPs (normally around 520 nm). As previously reported, the ultra-small sizes of methionine-stabilized Au NCs demonstrated molecular-like properties, which were different from relatively large-sized metal nanoparticles [29]. From Figure 2B, it can clearly be observed that the as-prepared Au NCs exhibit strong fluorescence emission at 565 nm with maximum excitation at 420 nm. The Stokes shift of the Au NCs was calculated to be 145 nm. A Stokes shift of less than 150 nm can be attributed to interband transitions of electrons in the Au NCs [30]. The fluorescence emission spectra of Au NCs upon excitation in the range of 360–460 nm showed constant peaks (Figure S2), indicating that the optical signal was actual luminescence from the Au NCs, rather than light scattering [31]. The as-prepared methionine-stabilized Au NCs were highly dispersed in aqueous solution, and exhibited obvious yellow suspension in the room light (inset a of Figure 2B) and emitted intense orange fluorescence (inset b of Figure 2B) under UV light irradiation with the excitation at 365 nm. In addition, a solid powder could be obtained by freeze-drying, which also appeared bright orange in room light (inset c of Figure 2B) and showed an intense yellow fluorescence (inset d of Figure 2B) under UV light. Furthermore, the quantum yield of as-prepared Au NCs was calculated to be about 1.46% (rhodamine B in ethanol was used as reference), which is comparable to other NCs prepared using amino acid as stabilizer [32,33]. Moreover, there was no obvious change in the position of the emission peak of Au NCs after storing in the dark at 4 °C for three months and quenching of the fluorescence intensity (Figure S3), indicating that the Au NCs are suitable for long-term use.

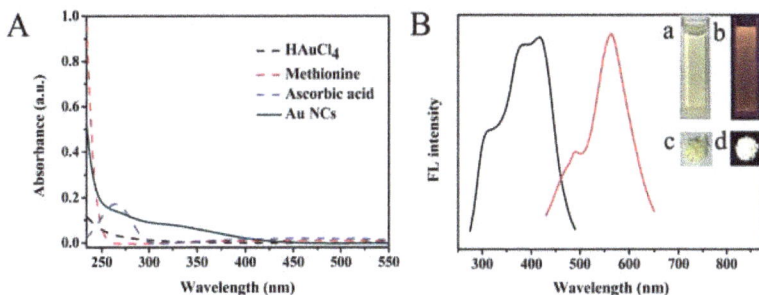

Figure 2. UV–vis absorption spectra (**A**) and fluorescence excitation (black) and emission (red) spectra (**B**) of the as-synthesized Au NCs. The insets of B show photographs of the Au NC aqueous solution in room light (a) and UV light (b), and powder in room light (c) and UV light (d).

The morphology and size distribution of the Au NCs were observed by HR-TEM. Figure 3A,B shows the typical TEM images of the Au NCs and their size ranges, it can be seen that the size of samples is less than 2 nm, and no aggregation appears. The hydrodynamic diameter of Au NCs measured using DLS was approximately 4.24 nm (Figure S7), which further verified the successful synthesis of ultrasmall Au NCs. In addition, the elemental composition of the Au NCs can be confirmed by EDX. From Figure S4 and Table S1, the results indicate that the atomic content of Au in the sample is of 47.10%.

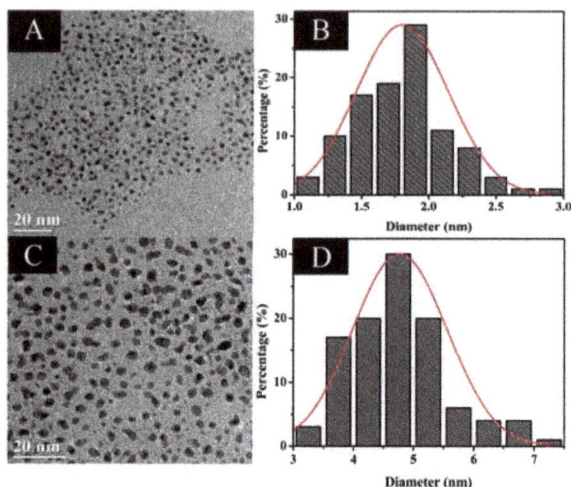

Figure 3. Representative high resolution transmission electron microscopy (HRTEM) images of the as-synthesized luminescent Au NCs (**A**); the particle-size distribution histogram of Au NCs (**B**); Au NCs in the presence of 50 μM Cd^{2+} (**C**); and particle-size distribution histogram (**D**).

Furthermore, the fluorescence decay response of the prepared Au NCs showed three components at 246.92 ns (87.76%), 13.19 ns (7.48%) and 1.39 ns (4.76%); the average fluorescence lifetime of Au NCs was calculated to be 217.74 ns (Figure 4A). As reported in previous research, the long fluorescence lifetime might result from the Au(I)-S complex structure [34], which might have the possibility for fluorescence lifetime imaging in future [35]. Moreover, XPS measurement was performed to verify the valence state of the metal elements in the prepared Au NCs. As exhibited in Figure 4B, it can be seen that the binding energy of Au 4f appeared at 84.3 and 88.1 eV, respectively, which indicated that Au(0) and Au(I) coexist in Au NCs. The Au(I) on the surface of nanoclusters plays a vital role in stabilizing the nanoclusters [36]. In order to further determine the composition of the metal core from as-synthesized Au NCs, ESI-MS was applied. The mass spectrum of Au NCs (negative mode) is displayed in Figure S5, the main charges appearing at 1003.4, 1019.4 Da are assigned to $[Au_6L_6\text{-}2Cl]^{2-}$ and $[Au_6L_6\text{-}Cl\text{-}H]^{2-}$ ("L" refers to "methionine"), suggesting the Au NCs are mainly composed of Au_6 clusters.

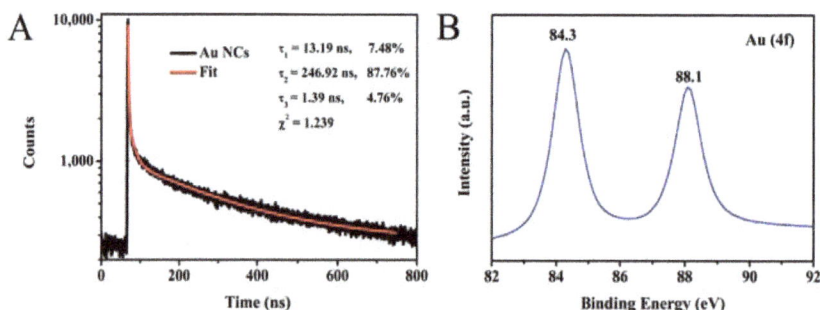

Figure 4. Fluorescence lifetimes of Au NCs in aqueous solution (**A**); data were collected at 565 nm with excitation at 375 nm. X-ray photoelectron spectroscopy (XPS) spectra of Au 4f of Au NCs (**B**).

3.2. Optimization of Sensing Conditions

It is interesting that the fluorescence intensity of the as-prepared Au NCs was significantly enhanced in the presence of Cd^{2+} ions. In order to obtain a highly sensitive response for the detection of Cd^{2+}, the optimization of pH values, ionic strength and incubation time was carried out systematically. The pH value of the reaction solution could greatly influence the interaction between Au NCs and Cd^{2+} ions. The fluctuating pH values in the range of 4.0–11.0 were investigated. As displayed in Figure 5A, the fluorescence intensity increased significantly with the increasing solution pH, and reached its maximum when pH reached 8.0. A possible explanation for this is that the pH influences the Cd^{2+} ion speciation in solution [37]. Therefore, pH 8.0 was selected as the optimum pH value for Cd^{2+} ion detection. Meanwhile, to explore the potential application of the sensing system under high-ionic-strength environments, the stability of the Au NCs was investigated in the presence of various concentrations of NaCl (0.01 mM to 1.0 M). As shown in Figure 5B, the fluorescence intensity of Au NCs in the different salt-containing solutions remained almost unchanged compared with that in the absence of NaCl, indicating the high stability of Au NCs. In addition, the effect of reaction time on the fluorescence intensity was also studied. It can be seen from Figure 5C that the fluorescence intensity increased rapidly, and reached its maximum at around 1 min, after which it remained relatively stable. Therefore, a reaction time of 1 min was chosen in this experiment.

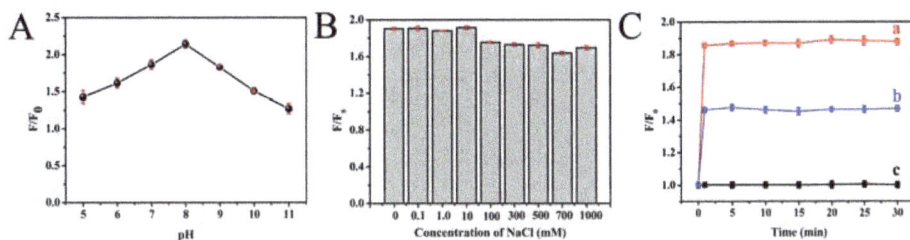

Figure 5. The effect of pH (**A**), concentration of NaCl (**B**) and incubation time (**C**) on fluorescence intensity of Au NCs upon addition of Cd^{2+} ions at different concentrations (35 µM (a); 10 µM (b); and 0 µM (c)), respectively.

3.3. Au NCs Fluorescent Sensing of Cd^{2+}

Under the optimized reaction conditions, the sensing properties were examined in the presence of Cd^{2+}. As shown in Figure 6A, the fluorescence intensity of Au NCs gradually increased with increasing concentration of Cd^{2+}, accompanied with the color changes of the solution from faint orange to bright orange under UV irradiation (Figure S6). Moreover, Figure 6B shows the relationship between fluorescence intensity and Cd^{2+} concentration, and the inset picture represents the linear response over the Cd^{2+} concentration range from 50.0 nM to 35.0 µM; the correlation coefficient R^2 was calculated as 0.9923, and the limit of detection (LOD) was determined to be 12.25 nM, which was lower than the maximum safety level of Cd^{2+} (ca. 44 nM) in drinking water permitted by the U.S. Environmental Protection Agency (EPA) [38]. In comparison with other previously reported methods, the proposed nanoprobe exhibits comparable or more efficient properties for Cd^{2+} detecting (Table S2).

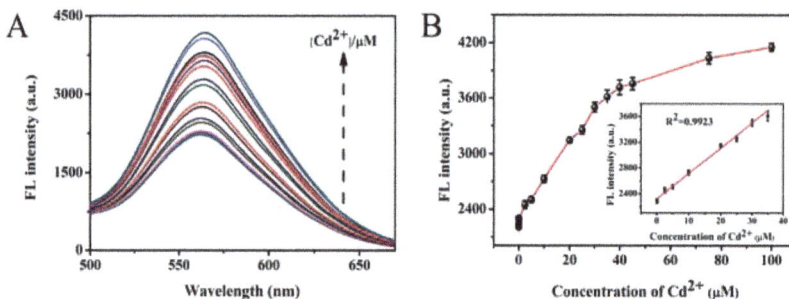

Figure 6. (**A**) The fluorescence response of the Au NCs in the presence of various concentrations of Cd^{2+} (0.0025, 0.005, 0.025, 0.05, 2.5, 5, 10, 15, 20, 25, 30, 35, 40, 45, 50, 75, 100 μM); (**B**) Relationship between fluorescence intensity and Cd^{2+} concentration. The inset picture shows the linear detection range for 0.05–35 μM of Cd^{2+}.

3.4. Selectivity of the Detection System

In order to investigate whether the synthesized probe is specific for Cd^{2+}, the fluorescence response of this sensing system was also tested by other competitive metal cations and anions (Na^+, K^+, Mg^{2+}, Ba^{2+}, Pb^{2+}, Ca^{2+}, Al^{3+}, Cr^{3+}, Fe^{3+}, Sn^{2+}, Cl^-, $H_2PO_4^-$, Br^-, SCN^-, NO_3^-, CO_3^-, IO_3^-, SO_4^{2-}, NO_2^-, I^- and ClO_4^-) under the same conditions. As shown in Figure 7A, no significant increase was observed by adding other ions into the Au NCs solution, only Cd^{2+} showed a fluorescence enhancement phenomenon—about a twofold photoluminescence (PL) increment. It can be clearly seen that only the Cd^{2+}-containing Au NCs solution exhibited greatly enhanced luminescence under UV light irradiation with excitation at 365 nm (Figure 7B). All these results confirm that the present fluorescent probe exhibits the high selectivity required for Cd^{2+} ion assays in real samples.

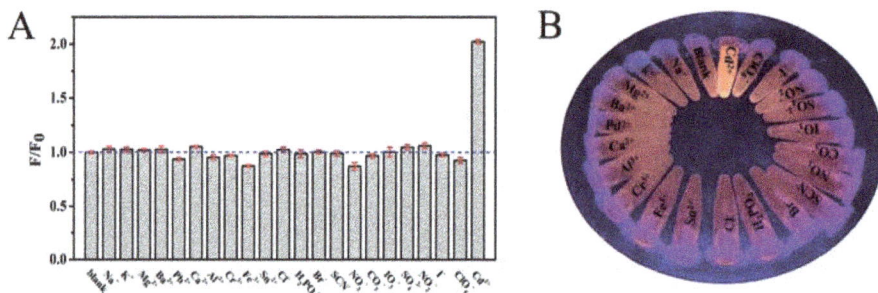

Figure 7. (**A**) Selective experiments of Au NCs for other competitive metal ions and anions; (**B**) Photographs of Au NCs under UV light after being incubated with various ions.

3.5. Mechanism of the Sensing System

The corresponding mechanism of the as-prepared Au NCs for fluorescence-enhanced Cd^{2+} sensing was also explored. In general, both the metal core and the ligand shell of the NCs can be used as recognition components, as they are able to specifically interact with analytes [22]. High resolution transmission electron microscopy (HRTEM) was used to evaluate the differences of as-prepared Au NCs in the absence and presence of Cd^{2+} ions. As displayed in Figure 3A, the Au NCs are well-dispersed irregular spherical shapes with an average size of 1.80 ± 0.34 nm (Figure 3B). After incubation with 50 μM Cd^{2+}, the average size of Au NCs showed an obvious enlargement (Figure 3C) and was about 4.76 ± 0.28 nm (Figure 3D). In addition, as indicated in Figure S7, the hydrodynamic diameter of the Au NC solution obtained from DLS was 4.24 nm, and then increased to 27.63 nm after addition of 50 µM Cd^{2+} into the solution. Therefore, we speculate that the Cd^{2+} ions might link Au NCs via chelating bonds with a carboxyl group or an amino group in methionine adsorbed on the Au nanoclusters, as the ligand shell contributes many surface-related properties to the NCs [22]. However, the detailed mechanisms require further study.

3.6. Real Sample Analysis

To evaluate whether the fluorescent Au NCs probe is applicable to natural systems; tap water, lake water, milk power and camel milk powders were investigated. Analytical results showed that the fluorescence tests were affected in the real samples due to the complex matrix. However, the fluorescence Au NC probe exhibited excellent performance. The average recoveries of four spiked samples ranged from 95.33% to 106.21%, with RSDs of 0.51–3.56% (*n* = 3) at three spiked levels (Table 1), and these results were in good agreement with those obtained by ICP-OES, which indicated the practicality and reliability of the Au NC probe for the detection of Cd^{2+} in various samples.

Table 1. Analytical results for Cd^{2+} sensing in real samples.

Sample	Detected (µM)	Spiked (µM)	Found (µM)	Recovery (%)	RSD (%) (*n* = 3)	ICP-OES (µM)
Tap water	ND [1]	3.00	2.89	96.33	2.56	2.97
		5.00	5.10	101.95	3.56	5.07
		10.00	10.62	106.21	1.45	10.08
Lake water	ND [1]	3.00	2.86	95.33	2.37	3.02
		5.00	4.84	96.74	2.15	5.06
		10.00	10.41	104.14	0.51	10.08
Milk powders	ND [1]	3.00	2.96	98.66	2.56	2.92
		5.00	5.06	101.29	1.68	5.03
		10.00	10.43	104.32	1.56	10.05
Camel milk powders	ND [1]	3.00	2.87	95.67	2.79	2.95
		5.00	4.96	99.23	2.31	5.04
		10.00	10.26	102.62	1.87	10.09

[1] Not detectable.

4. Conclusions

In summary, we successfully prepared water-soluble orange-emitting Au NCs using methionine as stabilizer. The synthesis process is simple, green and environmentally friendly, without using any toxic organic reagents. The resulting Au NCs exhibited impressive properties, such as ultrafine size, long fluorescence lifetime and excellent stability. Au NCs were further used as fluorescence nanoprobe to selectively, sensitively and efficiently recognize Cd^{2+}, with response times of as low as 1 min. Moreover, the sensing system is verified by detecting Cd^{2+} ions in water and milk samples, the average recoveries are in the range of 95.33% to 106.21%, suggesting this strategy may be extended to the efficient detection of Cd^{2+} in various conditions.

Supplementary Materials: The following are available online at http://www.mdpi.com/1424-8220/18/2/658/s1, Figure S1: Optimization of synthesized conditions of Au NCs, Figure S2: Emission spectra of fluorescent Au NCs prepared in a typical synthesis with different excitation wavelengths, Figure S3: The fluorescence spectra of Met-Au NCs storing in dark at 4 °C before and after three months, Figure S4: EDX spectrum for as-prepared Au NCs, Figure S5: ESI mass spectra (negative mode) of as-synthesized Au NCs, Figure S6: The color change of Au NCs under UV irradiation when exposed to different concentration of Cd^{2+} ions, Figure S7: Hydrodynamic diameter measured using DLS of Au NCs (**A**) and Au NCs in the presence of 50 µM Cd^{2+} (**B**) at neutral pH, Table S1: Quantification Results of EDX from Au NCs, Table S2: Comparison of this method with other reported approaches for the detection of Cd^{2+} using different optical systems.

Acknowledgments: Financial support by the National Nature Science Foundation of China (Grant Nos. 21503271, 21362037), the '1000 Talent Program' (The Recruitment Program of Global Experts) and the Joint Funds of NSFC-Xinjiang of China (U1303391) are gratefully acknowledged.

Author Contributions: Y.P. and F.W. conceived and designed the experiments; Y.P. and M.W. performed the experiments; X.W. and L.L. analyzed the data; Y.P. and F.W. wrote the paper; and all authors discussed the results and commented on the manuscript.

Conflicts of Interest: The authors declare no conflict of interest.

References

1. Shi, Z.; Han, Q.; Yang, L.; Yang, H.; Tang, X.; Dou, W.; Li, Z.; Zhang, Y.; Shao, Y.; Guan, L.; et al. A highly selective two-photon fluorescent probe for detection of cadmium(II) based on intramolecular electron transfer and its imaging in living cells. *Chemistry* **2015**, *21*, 290–297. [CrossRef] [PubMed]
2. Miao, Q.; Wu, Z.; Hai, Z.; Tao, C.; Yuan, Q.; Gong, Y.; Guan, Y.; Jiang, J.; Liang, G. Bipyridine hydrogel for selective and visible detection and absorption of Cd²⁺. *Nanoscale* **2015**, *7*, 2797–2804. [CrossRef] [PubMed]
3. Anthemidis, A.N. Flow injection on-line hydrophobic sorbent extraction for flame atomic absorption spectrometric determination of cadmium in water samples. *Microchim. Acta* **2008**, *160*, 455–460. [CrossRef]
4. Li, Y.; Zhu, Z.; Zheng, H.; Jin, L.; Hu, S.H. Significant signal enhancement of dielectric barrier discharge plasma induced vapor generation by using non-ionic surfactants for determination of mercury and cadmium by atomic fluorescence spectrometry. *J. Anal. At. Spectrom.* **2015**, *31*, 383–389. [CrossRef]
5. Davis, A.C.; Calloway, C.P., Jr.; Jones, B.T. Direct determination of cadmium in urine by tungsten-coil inductively coupled plasma atomic emission spectrometry using palladium as a permanent modifier. *Talanta* **2007**, *71*, 1144–1149. [CrossRef] [PubMed]
6. Tan, N.D.; Yin, J.H.; Pu, G.; Yuan, Y.; Meng, L.; Xu, N. A simple polyethylenimine-salicylaldehyde fluorescence probe: Sensitive and selective detection of Zn²⁺ and Cd²⁺ in aqueous solution by adding S²⁻ ion. *Chem. Phys. Lett.* **2016**, *666*, 68–72. [CrossRef]
7. Dai, Y.; Yao, K.; Fu, J.; Xue, K.; Yang, L.; Xu, K. A novel 2-(hydroxymethyl)quinolin-8-ol-based selective and sensitive fluorescence probe for Cd²⁺ ion in water and living cells. *Sens. Actuators B* **2017**, *251*, 877–884. [CrossRef]
8. Liu, Z.; Zhang, C.; He, W.; Yang, Z.; Gao, X.; Guo, Z. A highly sensitive ratiometric fluorescent probe for Cd²⁺ detection in aqueous solution and living cells. *Chem. Commun.* **2010**, *46*, 6138–6140. [CrossRef] [PubMed]
9. Roy, S.B.; Mondal, J.; Khudabukhsh, A.R.; Rajak, K.K. A novel fluorene based "turn on" fluorescent sensor for the determination of zinc and cadmium: Experimental and theoretical studies along with live cell imaging. *New J. Chem.* **2016**, *40*, 9593–9608. [CrossRef]
10. Zhang, Y.; Guo, X.; Zheng, M.; Yang, R.; Yang, H.; Jia, L.; Yang, M. A 4,5-quinolimide-based fluorescent sensor for the turn-on detection of Cd²⁺ with live-cell imaging. *Org. Biomol. Chem.* **2017**, *15*, 2211–2216. [CrossRef] [PubMed]
11. Zhao, Q.; Li, R.F.; Xing, S.K.; Liu, X.M.; Hu, T.L.; Bu, X.H. A highly selective on/off fluorescence sensor for cadmium(II). *Inorg. Chem.* **2011**, *50*, 10041–10046. [CrossRef] [PubMed]
12. Xu, Z.; Li, G.; Ren, Y.Y.; Huang, H.; Wen, X.; Xu, Q.; Fan, X.; Huang, Z.; Huang, J.; Xu, L. A selective fluorescent probe for the detection of Cd²⁺ in different buffer solutions and water. *Dalton Trans.* **2016**, *45*, 12087–12093. [CrossRef] [PubMed]
13. Huang, W.-B.; Gu, W.; Huang, H.-X.; Wang, J.-B.; Shen, W.-X.; Lv, Y.-Y.; Shen, J. A porphyrin-based fluorescent probe for optical detection of toxic Cd²⁺ ion in aqueous solution and living cells. *Dyes Pigments* **2017**, *143*, 427–435. [CrossRef]
14. Brahim, N.B.; Mohamed, N.B.H.; Echabaane, M.; Haouari, M.; Chaâbane, R.B.; Negrerie, M.; Ouada, H.B. Thioglycerol-functionalized CdSe quantum dots detecting cadmium ions. *Sens. Actuators B* **2015**, *220*, 1346–1353. [CrossRef]
15. Banerjee, S.; Kar, S.; Santra, S. A simple strategy for quantum dot assisted selective detection of cadmium ions. *Chem. Commun.* **2008**, *26*, 3037–3039. [CrossRef] [PubMed]

16. Luo, X.; Wu, W.; Deng, F.; Chen, D.; Luo, S.; Au, C. Quantum dot-based turn-on fluorescent probe for imaging intracellular zinc(II) and cadmium(II) ions. *Microchim. Acta* **2014**, *181*, 1361–1367. [CrossRef]

17. Wu, Q.; Zhou, M.; Shi, J.; Li, Q.; Yang, M.; Zhang, Z. Synthesis of water-soluble Ag_2S quantum dots with fluorescence in the second near-infrared window for turn-on detection of Zn(II) and Cd(II). *Anal. Chem.* **2017**, *89*, 6616–6623. [CrossRef] [PubMed]

18. Zhou, T.Y.; Lin, L.P.; Rong, M.C.; Jiang, Y.Q.; Chen, X. Silver-gold alloy nanoclusters as a fluorescence-enhanced probe for aluminum ion sensing. *Anal. Chem.* **2013**, *85*, 9839–9844. [CrossRef] [PubMed]

19. Shang, L.; Dong, S.; Nienhaus, G.U. Ultra-small fluorescent metal nanoclusters: Synthesis and biological applications. *Nano Today* **2011**, *6*, 401–418. [CrossRef]

20. Xu, H.; Suslick, K.S. Water-soluble fluorescent silver nanoclusters. *Adv. Mater.* **2010**, *22*, 1078–1082. [CrossRef] [PubMed]

21. Lu, Y.; Chen, W. Sub-nanometre sized metal clusters: From synthetic challenges to the unique property discoveries. *Chem. Soc. Rev.* **2012**, *41*, 3594–3623. [CrossRef] [PubMed]

22. Yuan, X.; Luo, Z.; Yu, Y.; Yao, Q.; Xie, J. Luminescent noble metal nanoclusters as an emerging optical probe for sensor development. *Chem. Asian J.* **2013**, *8*, 858–871. [CrossRef] [PubMed]

23. Venkatesh, V.; Shukla, A.; Sivakumar, S.; Verma, S. Purine-stabilized green fluorescent gold nanoclusters for cell nuclei imaging applications. *ACS Appl. Mater. Interfaces* **2014**, *6*, 2185–2191. [CrossRef] [PubMed]

24. Knoblauch, C.; Griep, M.; Friedrich, C. Recent advances in the field of bionanotechnology: An insight into optoelectric bacteriorhodopsin, quantum dots, and noble metal nanoclusters. *Sensors* **2014**, *14*, 19731–19766. [CrossRef] [PubMed]

25. Niu, W.J.; Shan, D.; Zhu, R.H.; Deng, S.Y.; Cosnier, S.; Zhang, X.J. Dumbbell-shaped carbon quantum dots/auncs nanohybrid as an efficient ratiometric fluorescent probe for sensing cadmium(II) ions and L-ascorbic acid. *Carbon* **2016**, *96*, 1034–1042. [CrossRef]

26. Naaz, S.; Chowdhury, P. Sunlight and ultrasound-assisted synthesis of photoluminescent silver nanoclusters: A unique 'knock out' sensor for thiophilic metal ions. *Sens. Actuators B* **2017**, *241*, 840–848. [CrossRef]

27. Li, J.; Zhong, X.; Zhang, H.; Le, X.C.; Zhu, J.J. Binding-induced fluorescence turn-on assay using aptamer-functionalized silver nanocluster DNA probes. *Anal. Chem.* **2012**, *84*, 5170–5174. [CrossRef] [PubMed]

28. Huang, P.; Liu, B.; Jin, W.; Wu, F.; Wan, Y. Colorimetric detection of Cd^{2+} using 1-amino-2-naphthol-4-sulfonic acid functionalized silver nanoparticles. *J. Nanopart. Res.* **2016**, *18*, 327–336. [CrossRef]

29. Maruyama, T.; Fujimoto, Y.; Maekawa, T. Synthesis of gold nanoparticles using various amino acids. *J. Colloid Interface Sci.* **2015**, *447*, 254–257. [CrossRef] [PubMed]

30. Zheng, J.; Zhou, C.; Yu, M.; Liu, J. Different sized luminescent gold nanoparticles. *Nanoscale* **2012**, *4*, 4073–4083. [CrossRef] [PubMed]

31. Shang, L.; Dorlich, R.M.; Brandholt, S.; Schneider, R.; Trouillet, V.; Bruns, M.; Gerthsen, D.; Nienhaus, G.U. Facile preparation of water-soluble fluorescent gold nanoclusters for cellular imaging applications. *Nanoscale* **2011**, *3*, 2009–2014. [CrossRef] [PubMed]

32. Yang, X.; Luo, Y.; Zhuo, Y.; Feng, Y.; Zhu, S. Novel synthesis of gold nanoclusters templated with L-tyrosine for selective analyzing tyrosinase. *Anal. Chim. Acta* **2014**, *840*, 87–92. [CrossRef] [PubMed]

33. Mu, X.; Qi, L.; Qiao, J.; Ma, H. One-pot synthesis of tyrosine-stabilized fluorescent gold nanoclusters and their application as turn-on sensors for Al^{3+} ions and turn-off sensors for Fe^{3+} ions. *Anal. Methods* **2014**, *6*, 6445–6451. [CrossRef]

34. Liu, C.L.; Wu, H.T.; Hsiao, Y.H.; Lai, C.W.; Shih, C.W.; Peng, Y.K.; Tang, K.C.; Chang, H.W.; Chien, Y.C.; Hsiao, J.K. Insulin-directed synthesis of fluorescent gold nanoclusters: Preservation of insulin bioactivity and versatility in cell imaging. *Angew. Chem.* **2011**, *50*, 7056–7060. [CrossRef] [PubMed]

35. Shang, L.; Azadfar, N.; Stockmar, F.; Send, W.; Trouillet, V.; Bruns, M.; Gerthsen, D.; Nienhaus, G.U. One-pot synthesis of near-infrared fluorescent gold clusters for cellular fluorescence lifetime imaging. *Small* **2011**, *7*, 2614–2620. [CrossRef] [PubMed]

36. Whetten, R.L.; Price, R.C. Nano-golden order. *Science* **2007**, *318*, 407–408. [CrossRef] [PubMed]

37. Zhu, Y.F.; Wang, Y.S.; Zhou, B.; Yu, J.H.; Peng, L.L.; Huang, Y.Q.; Li, X.J.; Chen, S.H.; Tang, X.; Wang, X.F. A multifunctional fluorescent aptamer probe for highly sensitive and selective detection of cadmium(II). *Anal. Bioanal. Chem.* **2017**, *409*, 4951–4958. [CrossRef] [PubMed]
38. Aragay, G.; Pons, J.; Merkoci, A. Recent trends in macro-, micro-, and nanomaterial-based tools and strategies for heavy-metal detection. *Chem. Rev.* **2011**, *111*, 3433–3458. [CrossRef] [PubMed]

![sensors logo] sensors

MDPI

Article

A Light-Up Probe for Detection of Adenosine in Urine Samples by a Combination of an AIE Molecule and an Aptamer

Yingying Hu [1,†], Jingjing Liu [1,†], Xiangyu You [1], Can Wang [2], Zhen Li [2] and Weihong Xie [1,*]

[1] Department of Food and Pharmaceutical Engineering, Key Laboratory of Fermentation Engineering (Ministry of Education), Hubei University of Technology, Wuhan 430068, China; 15518581008@163.com (Y.H.); jliu3971@gmail.com (J.L.); limnamil@gmail.com (X.Y.)

[2] Department of Chemistry, Wuhan University, Wuhan 430072, China; canwang-scola@whu.edu.cn (C.W.); lizhen@whu.edu.cn (Z.L.)

* Correspondence: weihong.xie@mail.hbut.edu.cn; Tel.: +86-139-7113-5198

† Yingying Hu and Jingjing Liu contributed equally to this work.

Received: 17 August 2017; Accepted: 18 September 2017; Published: 29 September 2017

Abstract: A light-up fluorescent probe for the detection of adenosine was constructed with an AIE (aggregation-induced emission) molecule and a DNA aptamer. The AIE molecule was used as a signal generator, and the DNA aptamer was used as a recognition element for adenosine. The emission of the AIE molecule was due to its intramolecular rotation restriction induced by the aptamer upon binding of adenosine. The optimal component ratio of the probe was AIE molecule/DNA aptamer = 100 (μM/μM). The calibration curve of adenosine detection showed a linear range of 10 pM to 0.5 μM with an R^2 of 0.996, and the detection limit of the probe was 10 pM. The probe exhibited a good selectivity to adenosine against its analogs (uridine, guanosine, and cytidine). The probe was used to detect adenosine in urine samples, a recovery from 86.8% to 90.0% for the spiked concentrations of adenosine (0.01, 0.05, 0.1 μM). The relative standard deviation from 1.2% to 2.0% was obtained. The intra-day and inter-day tests also showed good precisions, with measurement RSD values of 2.3% and 2.1%, respectively.

Keywords: aggregation-induced emission; aptamer; adenosine; 1,2-bis[4-(triethylammoniomethyl)phenyl]-1,2-diphenylethenedibromide (TPE-2N+)

1. Introduction

Aggregation-induced emission (AIE) [1] is a phenomenon that a luminescent molecule is non-emissive when it is dissolved in a solution. However, it becomes emissive when it is in an aggregation state. The unique photo physical phenomenon was first discovered by Tang and his colleagues in 2001. They later explained that the unique phenomenon was caused by a restriction of intramolecular rotation (RIR) of AIE molecules. Light-up probes are more preferred than light-off probes because they give less false-positive responses, so that interests are drawn to design and synthesis of AIE probes for bio-detection purposes [2–7]. In principle, if an analysis target can induce the restriction of intramolecular rotation of an AIE molecule, it will light up the fluorogen: the AIE molecule will give a fluorescent response to the analysis target. To restrict the intramolecular rotation of an AIE molecule, an interaction between the AIE molecule and the target molecule is usually required. Therefore, elements with high affinity for the analysis target must be provided to AIE probes. However, having a high affinity is not enough. AIE probes must also have specificity for their targets. In order to improve the specificity of AIE probes, AIE molecules of different structures have been designed and synthesized to conjugate recognition functional groups or associate with recognition elements to improve the affinity and selectivity of the probes. In this work, we used an adenosine-specific aptamer as the recognition element for the target molecule, adenosine.

Aptamers [8] are single-stranded oligonucleotides that have specific recognition function for peptides, proteins, and small organic molecules. The specific recognition function of aptamers for target molecules is based on their unique sequences and three-dimensional folded structures. Aptamers can undergo significant conformational changes into hairpins, stem-loops, or G-quadruplexs after they bind to their targets. The Aptamer that can specifically bind adenosine has been found. It is a G-rich oligonucleotide and can form G-quartets structures in the presence of adenosine [9,10].

Adenosine (A) [11] is an endogenous nucleoside that plays important roles in many biochemical processes, such as energy transfer by forming molecules like adenosine triphosphate (ATP) and adenosine diphosphate (ADP) and signal transduction by forming signally molecules like cyclic adenosine monophosphate (cAMP). It is also a neuromodulator that plays roles of promotion of sleep and suppression of arousal. In addition, adenosine regulates the blood flow to various organs through vasodilation. In recent years, it was found that adenosine's concentration in the extracellular tissue surrounding tumors was higher than that under healthy conditions. This phenomenon was due to the hypoxic microenvironment of tumors that trigger a strong inflammatory response. Therefore, adenosine is a possible biomarker for cancer [12] and may be used for monitoring progress of diseases.

In this work, a fluoresce probe by the combination of an AIE molecule: 1,2-bis[4-(triethylammoniomethyl)phenyl]-1,2-diphenylethenedibromide (TPE-2N+) and an adenosine-specific aptamer (ABA) for the determination of adenosine in urine samples was described. When adenosine was present, the aptamer bound to the target molecule and formed a G-quadruplex that could bind and aggregate the TPE-2N+. The AIE molecule thus lighted up and gave response to adenosine.

2. Materials and Methods

2.1. Materials

Adenosine (A), cytidine (C), uridine (U), and guanosine (G) were purchased from Shanghai source leaf Biological Technology Co., Ltd. (Shanghai, China). 1,2-bis[4-(triethylammoniomethyl)phenyl]-1,2-diphenylethenedibromide (TPE-2N+) were obtained from the chemistry department, Wuhan University. Three hydroxymethyl aminomethane (Tris) and HCl were bought from Sinopharm Chemical Reagent Co., Ltd. (Shanghai, China). The oligonucleotides were synthesized by Sangon Biotechnology Co. Ltd. (Shanghai, China) with the following sequences:

Adenosine aptamer (ABA): 5′-ACCTGGGGGAGTATTGCGGAGGAAGGT-3′;

All chemicals used were analytical grade, and the ultrapure water was deionized to 18.25 MΩ·cm in a water purification system from Angel Electric Appliance Co., Ltd. (Wuxi, China).

2.2. Measurements

An LS-55 fluorescence spectrometer of PerkinElmer (Shanghai, China) was used to record the fluorescence spectra. The emission spectra were recorded in the range from 350 to 600 nm with both excitation and emission slits of 10 nm. All fluorescence detections were carried out under room temperature.

2.3. The Emission Behavior of (TPE-2N+) in Tris Buffer

The emission of (TPE-2N+) in Tris buffer was measured under an excitation wavelength of 400 nm, the dependence of the fluorescence intensity at the maximum emission wavelength on the concentration of the fluorogen was investigated within a concentration range of 5–100 μM. The variation of TPE-2N+ fluorescence intensity over a time period of 30 min was detected under two concentrations of 5 and 50 μM. The recipe of the (TPE-2N+) solutions is provided in Table S1.

2.4. The Viability of (TPE-2N+) + ABA Probe

To verify the feasibility of the proposed probe, the fluorescence intensity of the following systems was measured:

(1)   10 mM Tris-HCl;
(2)   10 mM Tris-HCl + 10 μM (TPE-2N+);
(3)   10 mM Tris-HCl + 10 μM (TPE-2N+) + 0.1 μM ABA;
(4)   10 mM Tris-HCl + 10 μM (TPE-2N+) + 0.1 μM A;
(5)   10 mM Tris-HCl + 10 μM (TPE-2N+) + 0.1 μM ABA + 0.1 μM A.

The samples were prepared according to the recipe in Table S2. The stock solutions were 100 μM (TPE-2N+), 100 μM ABA, and 1 μM adenosine and 10 mM pH 7.4 Tris-HCl up to 200 μL.

2.5. Optimization of the Concentration and the Concentration Ratio of the (TPE-2N+) and the ABA

The component of the probe was optimized by varying the concentrations of (TPE-2N+) and ABA, and their ratio. (TPE-2N+) solutions of 5 μM, 10 μM, and 20 μM and ABA of 0.1 μM, and 0.3 μM, 0.5 μM were prepared by diluting the stock 100 μM (TPE-2N+) and 100 μM ABA with Tris-HCl buffer, respectively. The concentration ratio of (TPE-2N+): ABA (μM:μM) was adjusted to 5:0.5, 10:0.1, 10:0.3, 10:0.5, and 20:0.5. Adenosine solutions of 0, 0.01, 0.05, 0.1, 0.2, and 0.5 μM were prepared by diluting a 1μM adenosine solution into the detection systems. The fluorescence intensity of the systems was then measured under the same conditions as described in the above sections.

2.6. The Calibration Curve of Adenosine Detection

Solutions of adenosine were prepared with a final concentration of $0, 1 \times 10^{-5}, 5 \times 10^{-5}, 1 \times 10^{-4}$, $5 \times 10^{-4}, 1 \times 10^{-3}, 5 \times 10^{-3}, 1 \times 10^{-2}, 5 \times 10^{-2}$, and 1×10^{-1} μM. (TPE-2N+) and ABA were added into the adenosine solutions with the optimal ratio with a concentration of 10 μM and 0.1 μM, respectively. The fluorescence intensity of the systems was then measured.

2.7. Specificity of (TPE-2N+)-ABA Detection System

The specificity of the probe for adenosine was examined using cytidine, uridine, guanosine, and the mixture of the nucleotides as the controls. The nucleotide samples were prepared according to the recipe in Table S3. The stock solutions were 100 μM of TPE, 100 μM of ABA, 1 μM of nucleotides, and 10 mM pH7.4 Tris-HCl.

2.8. Analytical Application to Urine Samples

To investigate the practical application of the probe in more complicated conditions, the detection of adenosine was carried out in human urine samples. A healthy urine was used as the blank sample; adenosine was added to prepare the spiked samples. The spiking concentrations were 0.01, 0.05, and 0.1 μM. The sample solutions were diluted 100 times before the detection. A working curve was also made using the healthy urine, and a linear regression equation for adenosine was obtained from a calibration curve using concentrations from 0 to 0.1 μM. A number of detections were carried out to investigate the precision and recovery of the probe.

3. Results and Discussions

3.1. The Detection Principle

(TPE-2N+) is an amino functionalized derivative of tetraphenylethene (TPE). With the amino functional groups, the solubility of (TPE-2N+) in an aqueous solution is improved so that it can be applied as a probe for bio-samples. Being an AIE molecule, (TPE-2N+) shares the common (AIE) characteristics of aggregation-induced emission. Therefore, any process or molecules that cause the restriction of intramolecular rotation of (TPE-2N+) will light it up. It has been reported that 1,1,2,2-tetrakis[4-(2-triethylammonioethoxy)phenyl]ethene tetrabromide (TPE-4N+), one of (TPE-2N+)'s structure similar, could bind to a guanine-rich DNA strand and was highly affinitive to the G-quadruplex structure of the DNA strand. The high affinity of the TPE-4N+ to G-quadruplex was associated with a geometric fit aided by an electrostatic attraction. Upon the electrostatic attraction, the TPE's intramolecular rotation was restricted and its emission was turned on. A K+ biosensor was developed using TPE-4N+ because it was specific to the K+-induced and -stabilized G-quadruplex [13]. Similarly, the adenosine-specific aptamer (ABA) used in this work is also a G-rich repeat sequences. It was evidenced that in the presence of adenosine, ABA bound adenosine and performed a conformational change to a G-quadruplex [9,10]. Several aptasensors based on this conformational change had been seen reported [14–18]. However, the combination of ABA with an AIE fluorogen has not been reported. The sensing principle is shown in Scheme 1. In the absence of adenosine, ABA takes a random conformation, and its negatively charged phosphate groups may attract the positive TPE-2N+ via an electrostatic interaction. Upon the electrostatic attraction, at least one rotation of the benzene groups of TPE-2N+ is restricted and emission of the TPE-2N+ is turned on. However, the conformation of ABA is random in the absence of adenosine, so that the ABA-TPE interaction is random and the emission is random. In the presence of adenosine, the aptamer will undergo a conformational change to G-quadruplex. As was reported in previous works, TPE amino functionalized derivatives had strong affinity for G-quadruplex structures. The strong affinity was due to the geometric fit between the TPE molecule and the DNA G-quadruplex, which was aided by the electrostatic attraction between amino groups of the TPE molecule and the DNA phosphate groups and a hydrophobic interaction between the aromatic core of the TPE and the deoxyribose regions of the DNA [19]. The intramolecular rotation of the TPE molecule was strongly restricted and a large enhance in emission was observed. Since the formation of ABA G-quadruplex is specifically induced by adenosine, TPE-2N+ can be used as a bio-probe to detect adenosine.

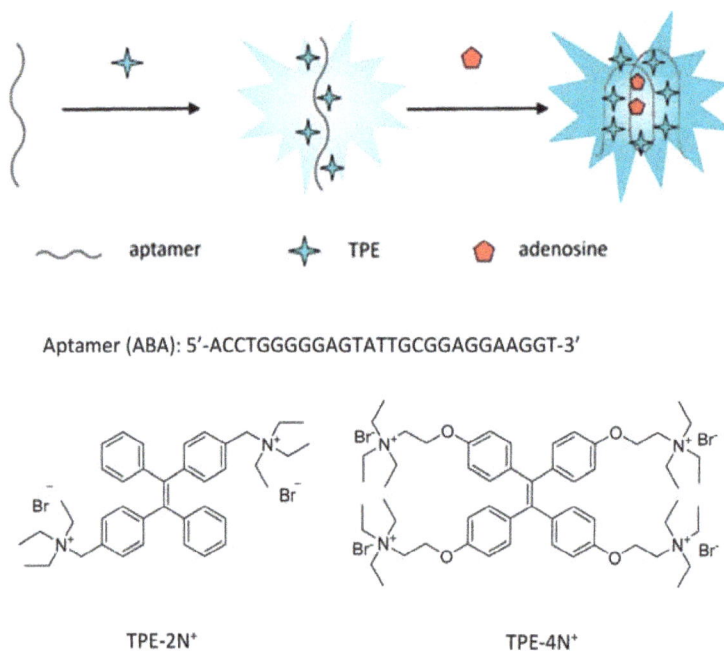

aptamer   TPE   adenosine

Aptamer (ABA): 5'-ACCTGGGGGAGTATTGCGGAGGAAGGT-3'

TPE-2N+   TPE-4N+

Scheme 1. Analytical principle of the tetraphenylethene (TPE)-aptamer probe for adenosine detection.

3.2. The Emission Behavior of TPE Fluorogen in Tris Buffer

The fluorescence intensity of TPE-2N+ in Tris-HCl buffer was measured, the result indicated that under an excitation of 400 nm, the fluorogen gave an emission at 470 nm. The emission intensity was increased gradually with the increase of the fluorogen's concentration (Figure 1A), and a linear relationship was obtained from 0 μM to 100 μM (Figure 1B). The stability of TPE fluorescence in Tris-HCl buffer was also examined, and a variation of TPE fluorescence intensity was measured over a time-period of 30 min. As shown in Figure 2, there is no obvious change in fluorescence intensity for the selected concentrations 5 μM and 50 μM of TPE-2N+ which suggested the TPE fluorogen had a favorable stability in Tris-HCl buffer.

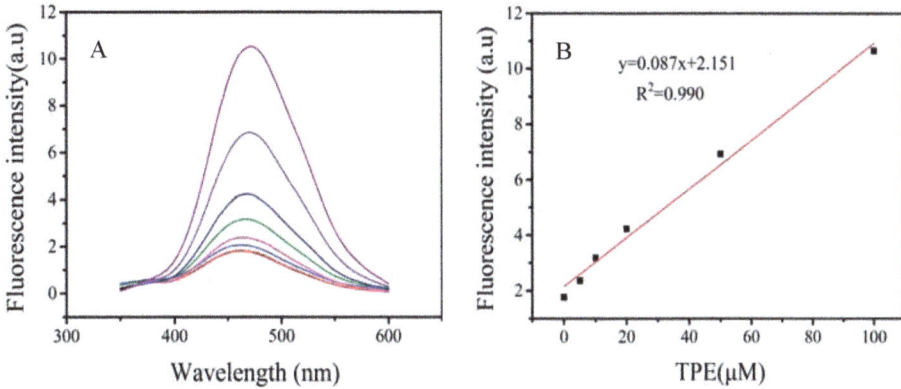

Figure 1. (**A**) Fluorescence emission spectra of different concentrations of TPE-2N+ in Tris-HCL buffer. From bottom to top: 0, 5, 10, 20, 50, 100 μM. (**B**) The linear relationship between the fluorescence intensity and the concentration of TPE-2N+ in Tris-HCL buffer.

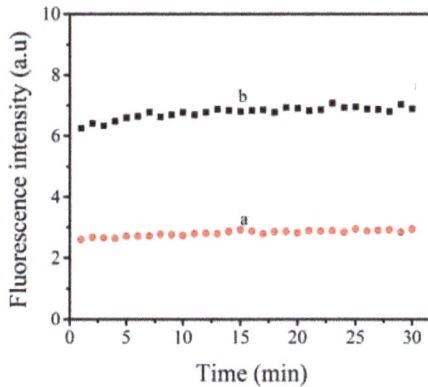

Figure 2. Fluorescence intensity of a: 5 μM and b: 50 μM of TPE-2N+ in Tris-HCL buffer over a time period of 30 min.

3.3. The Viability of (TPE-2N+) + ABA Probe

To verify the feasibility of the present approach, fluorescence intensity of solutions with different components was measured. The fluorescence spectrum of these solutions are presented in Figure 3. Tris buffer exhibited a very low fluorescence intensity (Figure 3, spectrum a). Upon addition of TPE-2N+, fluorescence intensity of the system increased (Figure 3, spectrum b). Further addition of the aptamer

ABA, fluorescent intensity of the system is seen to be enhanced. We attributed the enhancement to the restriction of intramolecular rotation of TPE-2N+ caused by the electrostatic attraction of the negative phosphate groups on the DNA strand (Figur 3, spectrum c). In the absence of the aptamer, TPE-2N+ gave only a small fluorescence response to adenosine that only a small increase in fluorescence intensity upon the addition of the target molecule is observed (Figure 3, spectrum d). As there was no specific interaction between (TPE-2N+) and adenosine, the presence of the target molecule did not cause strong restriction of intramolecular rotation of the fluorogen thus did not cause strong fluorescence response. When the adenosine-specific ABA was added, the aptamer would bind to the target and undergo a secondary conformational change. The G-quadruplex structure of ABA induced by adenosine could attract and restrict the intramolecular rotation of the fluorogen so that the fluorogen become highly emissive (Figure 3, spectrum e). Since the fluorescence response was generated based on the specific recognition between the aptamer and adenosine, the system of TPE-2N+/ABA could be developed as a biopeobe for adenosine sensing.

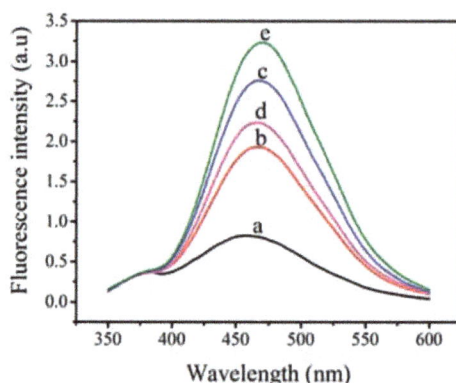

Figure 3. Fluorescence response of the system under different conditions. a: 10 mM Tris-HCL; b: 10 mM Tris-HCL + 10 μM TPE-2N+; c: 10 mM Tris-HCL + 10 μM TPE-2N+ + 0.1 μM ABA; d: 10 mM Tris-HCL + 10 μM TPE-2N+ + 0.1 μM Ade; e: 10 mM Tris-HCL + 10 μM TPE-2N+ + 0.1 μM ABA + 0.1 μM Ade.

3.4. Optimization of the Optimal Ratio of TPE-2N+ and ABA Concentration

In order to get the best response signals, concentrations of TPE-2N+ and ABA and their ratio in the detect system were optimized. A series of concentrations of 5 μM, 10 μM, and 20 μM and 0.1 μM, 0.3 μM, and 0.5 μM, respectively, for TPE-2N+ and ABA were tested. Plots of Figure 4a,d,e show the fluorescent response to adenosine under fixed concentration of ABA (0.5 μM) and vary concentrations (5 μM, 10 μM, and 20 μM) of TPE-2N+, respectively. Under condition a, the detection system does not show an adenosine-dependent fluorescence response; in cases of d and e, the fluorescence signal show an adenosine concentration dependence within 0–0.2 μM. Although the fluorescence response is higher with a higher TPE-2N+ concentration, the background is also higher, so that concentration of 10 μM was chosen for TPE-2N+ for the later experiments. At fixed concentration of TPE-2N+ (10 μM), the fluorescence responses under 0.1, 0.3, and 0.5 μM of ABA were measured and the results are shown respectively in Figure 3b–d. As can be seen, the fluorescence response of the detection system increase with the increase of ABA's concentration. However, the background value also increase, considering the best signal-to-noise level, condition of b is chosen.

The concentration ratio (TPE-2N+): ABA (μM:μM) of the detection system was also optimized. The ratio of (TPE-2N+): ABA as 5:0.5, 10:0.1, 10:0.3, 10:0.5, and 20:0.5 were investigated. As shown in Figure 2, in cases of 5:0.5 and 10:0.5, the fluorescence response shows little dependence on the concentrations of adenosine (Figure 4). The adenosine concentration dependence of fluorescence response was observed under (TPE-2N+): ABA ratio of 10:0.1, 10:0.3 and 20:0.5 (Figure 4, curve b, c and e), the fluorescence intensity of the systems increase greatly within the adenosine concentration range of 0–0.2 µM. Although the fluorescence intensity is higher under (TPE-2N+): ABA ratio of 10:0.3 and 20:0.5 (Figure 4, curve c and e), the background value was much greater than b. Hence 10:0.1 was chosen as the optimal condition for the probe.

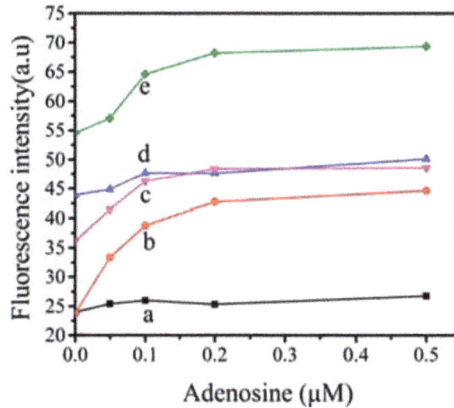

Figure 4. The fluorescent response toward the concentration of adenosine (0, 0.01, 0.05, 0.1, 0.2, and 0.5 µM) under different ratio of TPE-2N+ and ABA concentration, a: 5:0.5; b: 10:0.1; c: 10:0.3; d: 10:0.5, and e: 20:0.5.

3.5. The Sensitivity of the Probe

The ability of this fluorescence aptasensor for the quantitative detection of adenosine was studied under the optimized conditions. Figure 5A shows the different fluorescence responses to adenosine of various concentrations. As can be seen from Figure 5B, the fluorescence response increased dramatically when the concentration of adenosine increases from 0 to 0.1 µM. A linear relationship between the fluorescence response and the concentration of adenosine is obtained according to the equation of $Y = 42.32 + 2.189X$, where Y is the relative fluorescence intensity, and X is the logarithm of adenosine concentration Figure 5C. The linear concentration range is 10 pM to 0.5 µM with an R^2 of 0.996. So that the detection limit of the aptasensor for adenosine can be 10 pM.

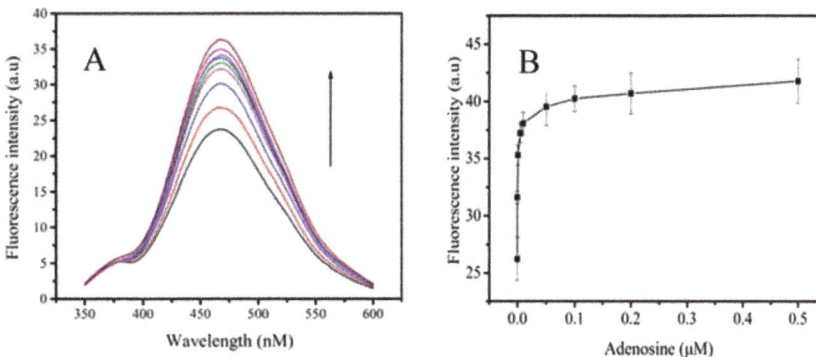

Figure 5. *Cont.*

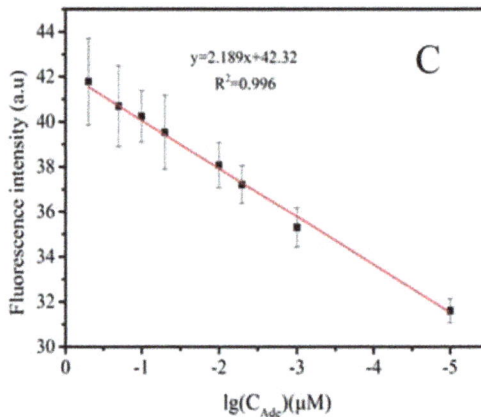

Figure 5. (**A**) Fluorescence emission spectra of the sensing system in the presence of different concentrations of adenosine. From bottom to top: 0, 0.00001, 0.00005, 0.0001, 0.001, 0.005, 0.01, 0.05, and 0.1 μM. (**B**) The relationship between fluorescence intensity and the concentration of adenosine. (**C**) A linearity curve (R2 = 0.996) was confirmed as the logarithm of adenosine concentration changed from 10 pM to 0.5 μM. The error bars are standard deviations of three repetitive measurements.

3.6. Specificity of TPE-ABA Detection System

The selectivity of the probe toward adenosine was examined by measuring the fluorescence responses of several adenosine analogs (uridine, guanosine, and cytidine) under the same experimental conditions. As shown in Figure 6, the TPE-2N+/ABA probe only gives a small fluorescence response to the adenosine analogs and does not show a concentration dependence of the adenosine analogs. However, it gives distinct fluorescence response to adenosine, and the fluorescence response increases with the increase of the concentration of adenosine. This result suggested that uridine, guanosine, and cytidine did not induce a conformational change of the ABA strand, so that the intramolecular rotation of TPE-2N+ was not restricted, and the emission was not turned on. The fluorescence response of TPE-2N+/ABA to adenosine in the presence of cytidine, guanosine and uridine does not show much affect by the adenosine analogs. This result indicated that the TPE-2N+/ABA probe had a sufficient specificity to adenosine against other adenosine analogs.

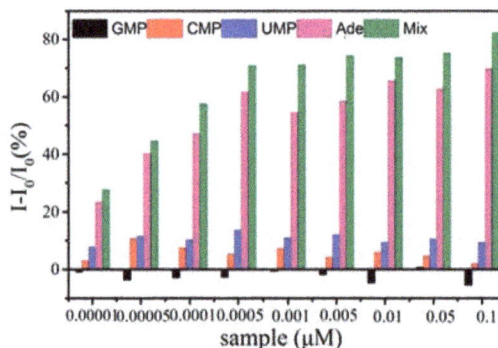

Figure 6. The fluorescent response of the proposed aptasensor towards adenosine and the three analogues: guanosine, uridine and cytidine, and the mixture of adenosine with the analogues. The concentration of all samples were 0.00001, 0.00005, 0.0001, 0.0005, 0.001, 0.005, 0.01, 0.05, and 0.1 μM.

3.7. Analytical Application to Urine Samples

To demonstrate the applicability of the proposed probe in the detection of adenosine in real samples, the recoveries of adenosine in human urine samples by the probe was examined. A working curve was firstly plotted choosing healthy urine as a blank. Adenosine was then spiked to the blank urine and was 100-fold diluted with Tris-HCl buffer to get final concentrations of 0.005, 0.01, 0.03, 0.05, 0.08, and 0.1 μM. A linear relationship between fluorescence intensity and the concentration of adenosine (0.05–0.1 μM) was obtained (Figure 7). Urine samples with three spiked concentrations (0.01, 0.05, and 0.1 μM) of adenosine were measured, and the measurements were quantified by using the working curve. As listed in Table 1, the obtained recoveries of the adenosine concentrations were from 86.8% to 90.0% with the relative standard deviation from 1.2% to 2.0%. The intra-day precision and inter-day precision tests indicated that the RSD values were within 2.3% and 2.1% respectively (Tables S4 and S5). The precision and recovery of this proposed method applied in complex biological samples are satisfactory. Although K+, a physiological component, was well known to stabilize G-quadruplex structure, it has been demonstrated by previous works [9,16] that K+ had little effect on the recognition and binding of the adenosine-specific aptamer to adenosine, in another word, it is not competitive to adenosine to bind the aptamer, so that the sensing based on the aptasensor was less affected by this cation even at a much higher concentration.

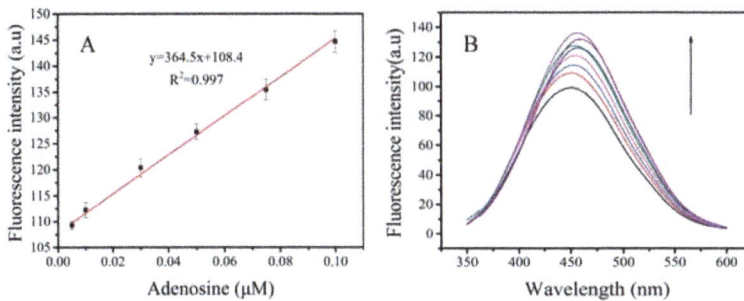

Figure 7. (**A**) The linear relationship between the change of fluorescence intensity and the concentration of adenosine in the urine. (**B**) shows the fluorescence spectra of ABA+TPE in urine diluted with Tris-HCL buffer after put in 0, 0.005, 0.01, 0.03, 0.05, 0.075, 0.1, and 0.12 μM of adenosine with an excitation wavelength of 470 nm.

Table 1. Determination of adenosine in urine samples using the proposed probe.

Samples	Adenosine Spiked (μM)	Adenosine Found (μM)	Recovery (%)	RSD (%, n = 6)
1	0.01	0.0090	90.0	1.9
2	0.05	0.0434	86.8	1.2
3	0.1	0.0892	89.2	2.0

4. Conclusions

A light-up fluorescent probe was developed by the combination of an AIE molecule and a DNA aptamer for the detection of adenosine. The probe device was simple compared with other aptamer-based probes in that the detection did not involve an enzyme strategy. Besides, the DNA strand was label free, and no fluorescent compound was required to conjugate on the DNA strand. This reduced the complications of the work and the expense on the DNA strand. On the other hand, the AIE molecule was also not required to function with a recognition group. The probe showed a lower detection limit when compared with other reported data. The good recovery results and precision of the probe for the urine samples suggested that the device might be further developed into test kit for clinic applications.

Supplementary Materials: The following are available online at www.mdpi.com/1424-8220/17/10/2246/s1, Table S1: The recipe of different concentrations of the TPE-2N+ solutions for the examination of the emission behavior of TPE fluorogen in Tris buffer; Table S2: The recipe for investigation of the viability of (TPE-2N+) + ABA probe; Table S3: The recipe of different concentrations of nucleotide samples for the specificity experiment; Table S4: The intra-day precision of the detection by the probe for urine samples; Table S5: The inter-day precision of the detection by the probe for urine samples.

Acknowledgments: The work was funded by the National Natural Science Foundation of China (No. 20875024) and the Key Laboratory of Fermentation Engineering (Ministry of Education).

Author Contributions: Y.H. performed the experiments and wrote the manuscript; J.L. did the supplementary experiments and data analysis; X.Y. supervised parts of the work; C.W. synthesized TPE-2N+; Z.L. supervised the synthesis work; W.X. conceived, supervised the work and helped with the necessary corrections, data analysis, and publication process.

Conflicts of Interest: The authors declare no conflict of interest.

References

1. Mei, J.; Leung, N.L.C.; Kwok, R.T.K.; Lam, J.W.Y.; Tang, B.Z. Aggregation-induced emission: Together we shine, united we soar! *Chem. Rev.* **2015**, *115*, 11718–11940. [CrossRef] [PubMed]
2. Chen, S.; Wang, H.; Hong, Y.; Tang, B.Z. Fabrication of fluorescent nanoparticles based on AIE luminogens (AIE dots) and their applications in bioimaging. *Mater. Horiz.* **2016**, *3*, 283–293. [CrossRef]
3. Kwok, R.T.K.; Leung, C.W.T.; Lam, J.W.Y.; Tang, B.Z. Biosensing by Luminogens with Aggregation-induced Emission Characteristics. *Chem. Soc. Rev.* **2015**, *44*, 4228–4238. [CrossRef] [PubMed]
4. Ding, D.; Li, K.; Liu, B.; Tang, B.Z. Bioprobes based on AIE fluorogens. *Acc. Chem. Res.* **2013**, *46*, 2441–2453. [CrossRef] [PubMed]
5. Rananaware, A.; Bhosale, R.S.; Patil, H.; Al Kobaisi, M.; Abraham, A.; Shukla, R.; Bhosale, S.V.; Bhosale, S.V. Precise aggregation-induced emission enhancement via H+ sensing and its use in ratiometric detection of intracellular pH values. *RSC Adv.* **2014**, *4*, 59078–59082. [CrossRef]
6. Rananaware, A.; Bhosale, R.S.; Ohkubo, K.; Patil, H.; Jones, L.A.; Jackson, S.L.; Fukuzumi, S.; Bhosale, S.V.; Bhosale, S.V. Tetraphenylethene-based star shaped porphyrins: Synthesis, self-assembly, and optical and photophysical study. *J. Org. Chem.* **2015**, *80*, 3832–3840. [CrossRef] [PubMed]
7. Rananaware, A.; Abraham, A.N.; La, D.D.; Mistry, V.; Shukla, R.; Bhosale, S.V. Synthesis of a tetraphenylethene-substituted tetrapyridinium salt with multifunctionality: Mechanochromism, cancer cell imaging, and DNA marking. *Aust. J. Chem.* **2017**, *70*, 652–659. [CrossRef]
8. McGown, L.B.; Joseph, M.J.; Pitner, J.B.; vonk, G.P.; Linn, C.P. The nucleic acid ligand. A new tool for molecular recognition. *Anal. Chem.* **1995**, *67*, 663A–668A. [CrossRef] [PubMed]
9. Wang, J.; Zhang, P.; Li, J.Y.; Chen, L.Q.; Huang, C.Z.; Li, Y.F. Adenosine-aptamer recognition-induced assembly of gold nanorods and a highly sensitive plasmon resonance coupling assay of adenosine in the brain of model sd rat. *Analyst* **2010**, *135*, 2826–2831. [CrossRef] [PubMed]
10. Hashemian, Z.; Khayamian, T.; Saraji, M.; Shirani, M.P. Aptasensor based on fluorescence resonance energy transfer for the analysis of adenosine in urine samples of lung cancer patients. *Biosens. Bioelectron.* **2016**, *79*, 334–340. [CrossRef] [PubMed]
11. Sachdeva, S.; Gupta, M. Adenosine and its receptors as therapeutic targets: An overview. *Saudi Pharm. J.* **2013**, *21*, 245–253. [CrossRef] [PubMed]
12. Wittmann, B.M.; Stirdivant, S.M.; Mitchell, M.W.; Wulff, J.E.; McDunn, J.E.; Li, Z.; Dennis-Barrie, A.; Neri, B.P.; Milburn, M.V.; Lotan, Y.; Wolfert, R.L. Bladder cancer biomarker discovery using global metabolomic profiling of urine. *PLoS ONE* **2014**, *9*, e115870. [CrossRef] [PubMed]
13. Hong, Y.; Häußler, M.; Lam, J.W.Y.; Li, Z.; Sin, K.K.; Dong, Y.; Tong, H.; Liu, J.; Qin, A.; Renneberg, R.; et al. Label-free fluorescent probing of g-quadruplex formation and real-time monitoring of DNA folding by a quaternized tetraphenylethene salt with aggregation-induced emission characteristics. *Chem. A Eur. J.* **2008**, *14*, 6428–6437. [CrossRef] [PubMed]

14. Yan, X.; Cao, Z.; Lau, C.; Lu, J. DNA aptamer folding on magnetic beads for sequential detection of adenosine and cocaine by substrate-resolved chemiluminescence technology. *Analyst* **2010**, *135*, 2400–2407. [CrossRef] [PubMed]

15. Huang, D.-W.; Niu, C.-G.; Zeng, G.-M.; Ruan, M. Time-resolved fluorescence biosensor for adenosine detection based on home-made europium complexes. *Biosens. Bioelectron.* **2011**, *29*, 178–183. [CrossRef] [PubMed]

16. Zhang, J.Q.; Wang, Y.S.; Xue, J.H.; He, Y.; Yang, H.X.; Liang, J.; Shi, L.F.; Xiao, X.L. A gold nanoparticles-modified aptamer beacon for urinary adenosine detection based on structure-switching/fluorescence-"turning on" mechanism. *J. Pharm. Biomed. Anal.* **2012**, *70*, 362–368. [CrossRef] [PubMed]

17. Sun, J.; Jiang, W.; Zhu, J.; Li, W.; Wang, L. Label-free fluorescence dual-amplified detection of adenosine based on exonuclease iii-assisted DNA cycling and hybridization chain reaction. *Biosens. Bioelectron.* **2015**, *70*, 15–20. [CrossRef] [PubMed]

18. Zhao, H.; Wang, Y.S.; Tang, X.; Zhou, B.; Xue, J.H.; Liu, H.; Liu, S.D.; Cao, J.X.; Li, M.H.; Chen, S.H. An enzyme-free strategy for ultrasensitive detection of adenosine using a multipurpose aptamer probe and malachite green. *Anal. Chim. Acta* **2015**, *887*, 179–185. [CrossRef] [PubMed]

19. Hong, Y.; Xiong, H.; Lam, J.W.Y.; Häußler, M.; Liu, J.; Yu, Y.; Zhong, Y.; Sung, H.H.Y.; Williams, I.D.; Wong, K.S.; et al. Fluorescent bioprobes: Structural matching in the docking processes of aggregation-induced emission fluorogens on DNA surfaces. *Chem. A Eur. J.* **2010**, *16*, 1232–1245. [CrossRef] [PubMed]

sensors

Article

Development of Ratiometric Fluorescent Biosensors for the Determination of Creatine and Creatinine in Urine

Hong Dinh Duong and Jong Il Rhee *

School of Chemical Engineering, Research Center for Biophotonics, Chonnam National University,
Yong-Bong Ro77, 61186 Gwangju, Korea; zink1735@gmail.com
* Correspondence: jirhee@jnu.ac.kr

Received: 21 September 2017; Accepted: 6 November 2017; Published: 8 November 2017

Abstract: In this study, the oxazine 170 perchlorate (O17)-ethylcellulose (EC) membrane was successfully exploited for the fabrication of creatine- and creatinine-sensing membranes. The sensing membrane exhibited a double layer of O17-EC membrane and a layer of enzyme(s) entrapped in the EC and polyurethane hydrogel (PU) matrix. The sensing principle of the membranes was based on the hydrolytic catalysis of urea, creatine, and creatinine by the enzymes. The reaction end product, ammonia, reacted with O17-EC membrane, resulting in the change in fluorescence intensities at two emission wavelengths (λ_{em} = 565 and 625 nm). Data collected from the ratio of fluorescence intensities at λ_{em} = 565 and 625 nm were proportional to the concentrations of creatine or creatinine. Creatine- and creatinine-sensing membranes were very sensitive to creatine and creatinine at the concentration range of 0.1–1.0 mM, with a limit of detection (LOD) of 0.015 and 0.0325 mM, respectively. Furthermore, these sensing membranes showed good features in terms of response time, reversibility, and long-term stability. The interference study demonstrated that some components such as amino acids and salts had some negative effects on the analytical performance of the membranes. Thus, the simple and sensitive ratiometric fluorescent sensors provide a simple and comprehensive method for the determination of creatine and creatinine concentrations in urine.

Keywords: creatine; creatinine; fluorescent sensing membrane; O17-EC membrane; ratiometric calculation

1. Introduction

Creatine, a non-essential nutrient, plays an important role by supplying energy to muscle cells [1]. In the early 1990s, there was a great interest among consumers and researchers concerning therapeutic applications of creatine and benefits of using creatine as a dietary supplement [2]. In healthy adults, creatine levels in biological fluids such as serum or plasma are typically around 0.04–0.15 mM which may rise to above 1 mM under certain pathological conditions such as muscle disorders. Creatine is taken as an ergogenic supplement at a daily dose of 7–30 mM (or up to 140 mM) by athletes to increase their body mass [3]. During oral supplementation, some part of the consumed creatine may be excreted through urine. The difference between the amount of urinary creatine and ingested creatine dose indicates the amount of creatine absorbed by the muscles. Therefore, monitoring creatine levels is important in clinical diagnosis [4].

To our knowledge, few studies have been directed toward the development of creatine sensors/biosensors and the commercially available creatine kits are very expensive. Creatine biosensors have been studied for several decades and are based on the hydrolysis of creatine by creatinase and urease to produce ammonia. Ammonia is usually detected by any pH or ion-selective electrodes [5,6]. Alternatively, a second enzyme-catalyzed hydrolysis of creatine mediated by sarcosine oxidase results in the production of hydrogen peroxide (H_2O_2) as the end product, which can be detected with a chronoamperometric technique [7–9]. Creatine detection has been also performed with flow-injection analysis (FIA) systems combined with immobilized enzyme reactors [10] and more expensive analytical tools such as high-performance liquid chromatography [10,11], capillary electrophoresis [12], and mass spectrometry [13]. However, the use of creatine sensors and other analytical tools is associated with practical issues related to sensitivity, reproducibility, stability, and interferences.

Creatinine is the end product of creatine metabolism in mammalian cells [1]. Muscular creatine is converted to creatinine, which diffuses out of the cells and is excreted by the kidney into the urine. Therefore, creatinine levels in biological fluids are clinically important for the diagnosis of renal, thyroid, and kidney dysfunctions. The normal concentration of creatinine in human blood ranges from 0.04 to 0.15 mM and these values may rise to 1 mM or more in chronic kidney disease patients. A typical urine sample of an adult contains creatinine at 2.5–23 mM concentration [14]. The ratio of creatinine in human serum and urine sample is an important indicator of the kidney function. Therefore, it is important to monitor creatinine concentrations in biological fluids, including urine, for clinical diagnosis [15].

There has been a tremendous improvement in the quality of creatinine biosensors. Some review articles on creatinine biosensors have been published in recent years [16–20]. Creatine and creatinine may interact in a reversible pathway depending on the type of enzymes involved. Therefore, creatine and creatinine biosensors are based on the same principles. Ammonium ion or H_2O_2, the final product of the hydrolysis reaction of creatinine catalyzed by one or three enzymes, is often measured to determine creatinine concentrations. Transducers such as amperometry [21,22], potentiometry [23,24], ion-sensitive field effect transistor [25–27], or spectrophotometry [28] may be used based on the type of the final reaction products. Among these biosensors, the amperometric method was the first system to be successfully commercialized [17].

The use of other biosensors for the determination of creatine and creatinine is still limited. So far, traditional photometric methods based on Jaffe's reaction have been widely used [29], however, colorimetric methods suffer from interferences during analysis. A diamond paste-based biosensor for creatine and creatinine detection is a rapid and simple sensor, but its measurement ranges are too low, 1–500 pM for creatine and 0.01–100 nM for creatinine [30]. The ratiometric fluorescence method offers some advantages over photometric and amperometric methods in terms of high sensitivity, good selectivity, and reproducibility, while enzyme immobilization allows high specificity, stability, and wide detection range to biosensors. Thus, we have developed ratiometric fluorescence biosensors using immobilized enzymes for the detection of creatine and creatinine.

Creatine detection is based on the hydrolysis of urea and creatine catalyzed by two enzymes, urease and creatinase, to produce ammonia in the following reactions:

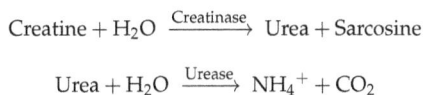

$$\text{Creatine} + H_2O \xrightarrow{\text{Creatinase}} \text{Urea} + \text{Sarcosine}$$

$$\text{Urea} + H_2O \xrightarrow{\text{Urease}} NH_4{}^+ + CO_2$$

The enzyme creatinine deiminase catalyzes the conversion of creatinine to ammonia and *N*-methylhydantoin, as depicted in the following reaction:

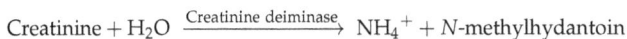

$$\text{Creatinine} + H_2O \xrightarrow{\text{Creatinine deiminase}} NH_4{}^+ + N\text{-methylhydantoin}$$

Ratiometric fluorescence biosensor is based on a fluorescent sensing membrane, which is a double-layer membrane of oxazine 170 perchlorate (O17) and ethylcellulose (EC) as an optical transducer and a layer of enzymes entrapped in the matrix of ethylcellulose (EC) and polyurethane hydrogel (PU) [31,32].

In this study creatine- and creatinine-sensing membranes were prepared (Scheme 1) and their properties were investigated. The sensing membranes were also characterized in terms of their response to creatine and creatinine using ratiometric calculation. These creatine and creatinine biosensors were evaluated for their ability to determine the concentration of creatine and creatinine dissolved in artificial urine solution (AUS).

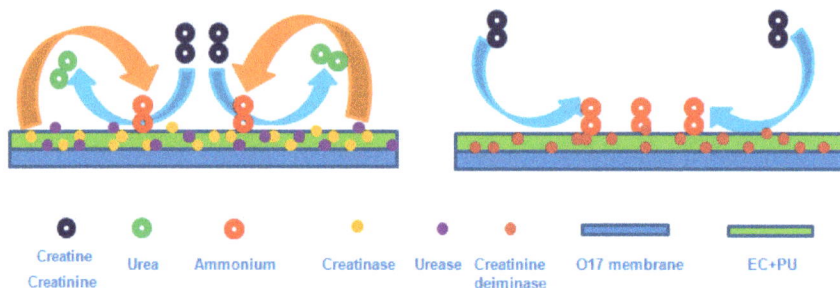

| Creatine Creatinine | Urea | Ammonium | Creatinase | Urease | Creatinine deiminase | O17 membrane | EC+PU |

Scheme 1. Structure of the creatine sensor (**left**) and creatinine sensor (**right**).

2. Materials and Methods

2.1. Materials

Oxazine 170 perchlorate (O17), ethylcellulose (EC), urease (59,400 U/g solid, from *Canavalia ensijormis* (Jack bean)), creatinase (15 U/mg solid, from *Actinobacillus* sp.), creatinine deiminase (20 U/mg solid, *microbial*), creatine, creatinine, glycine, histidine, phenylalanine, tryptophan, and urea were purchased from Sigma-Aldrich Chemical Co. (Seoul, Korea). Tris buffer was obtained from USB Co. (Cleveland, OH, USA) and PU from AdvanceSource Biomaterials Co. (Willmington, MA, USA). Other analytical-grade chemicals such as sodium phosphate, potassium phosphate, sodium chloride, potassium chloride, sodium hydroxide, hydrochloric acid, sodium bicarbonate, magnesium sulfate, sodium sulfate, and calcium chloride were used without further purification.

2.2. Preparation of Creatine- and Creatinine-Sensing Membranes

We prepared O17-EC membranes as previously described [31] by mixing O17 stock (15 μL, 2 mg/mL) with EC (300 μL, 10 wt %) in ethanol. The mixture was incubated for 4 h at room temperature and coated on the bottom of each well of a 96-well microtiter plate (NUNC Co. Copenhagen, Denmark). O17-EC membrane was dried at 60 °C for 12 h. In the next step, 20 μL of the mixture of 10 wt % EC and 10 wt % PU in ethanol and water (9:1 in v/v%) was coated onto O17-EC membrane, followed by the addition of the solution of a given amount of urease and creatinase dissolved in 10 mM phosphate buffer (pH 7.4). The enzymes were entrapped in the polymer matrix of EC and PU and immobilized over O17-EC membrane. The creatine-sensing membrane immobilized with enzymes was incubated for 24 h at 4 °C. Surface morphologies of O17-EC and creatine-sensing membrane were identified by atomic force microscopy (AFM). The preparation of the creatinine-sensing membrane was similar to that of the creatine-sensing membrane. Only one enzyme, creatinine deiminase, dissolved in 10 mM phosphate buffer (pH 7.4) was immobilized on O17-EC membrane.

2.3. Measurements of Creatine and Creatinine

Concentrations of creatine and creatinine for measurements were in the range of 0.1 to 10 mM. Data were collected from the fluorescence intensity of the sensing membranes at two emission wavelengths (λ_{em} = 565 and 625 nm) with an excitation wavelength of 460 nm (λ_{ex} = 460 nm). The fluorescence spectra for the detection of creatine and creatinine were measured using a multifunctional fluorescence microtiter plate reader (Safire[2], Tecan Austria GmbH, Wien, Austria).

2.3.1. Creatine Measurements

The optimization of creatinase amount for immobilization was performed with 3.5, 7, 14, 28, and 42 unit (U) of creatinase and a fixed amount of urease (10 U). The immobilization efficiency of urease and creatinase in the creatine-sensing membrane was calculated by dividing the amount of immobilized enzymes with the total amount of enzymes used for immobilization. The amount of the immobilized urease and creatinase was determined by subtracting the amount of the un-immobilized urease and creatinase from the total amount of the enzymes used. The un-immobilized urease and creatinase were separated from the immobilized urease and creatinase in one well by washing several times with 10 mM phosphate buffer (pH 7.4). The protein values of the washed, un-immobilized urease and creatinase were determined by Bradford method. The kinetic parameters, maximal reaction rate (V_{max}) and Michaelis-Menten constant (K_m), of the enzymes co-immobilized on the supporting material of EC and PU were determined from the Hanes plot using the ratio of fluorescence intensities at λ_{em} = 565 and 625 nm. The sensitivity of creatine-sensing membranes with different amounts of creatinase was evaluated through the slope value (SI), i.e., the ratio of the fluorescence intensities at two emission wavelengths (λ_{em} = 565 and 625 nm) with respect to creatine. The reversibility of the creatine-sensing membrane was tested in distilled water and 1.0 mM creatine prepared in Tris buffer. The creatine-sensing membrane was first exposed to distilled water and then to 1.0 mM creatine solution and the fluorescence measurements were performed on the microtiter plate reader at an interval of 30 s. In addition, the effects of pH and temperature on the creatine-sensing membrane were investigated. The creatine-sensing membrane was exposed to 1.0 mM creatine solutions at a pH range of 5.0–9.0. The membrane was also tested at different temperatures (25, 30, 33, 35, 37, and 40 °C) at creatine concentrations of 0.1–10 mM. The long-term stability of the creatine-sensing membrane in the presence of creatine at various concentrations was evaluated through the determination of its repeatability by measuring the fluorescence intensity obtained initially and after 3 months. The interfering effects of some components in urine samples on the creatine-sensing membrane were investigated using 1.0 mM creatine and 0.2 mM of glycine, histidine, phenylalanine, tryptophan, and creatinine as well as their mixture.

2.3.2. Creatinine Measurements

The optimal amount of creatinine deiminase for immobilization on O17-EC membrane was tested with 2.5, 5, and 7.5 U of creatinine deiminase. The immobilization efficiency of creatinine deiminase in the creatinine-sensing membrane was evaluated with Bradford method, as described to measure the immobilization efficiency of enzymes in part Section 2.3.1. The sensitivity of the creatinine-sensing membrane was also evaluated through SI. The kinetic parameters (V_{max} and K_m) of the immobilized creatinine deiminase were determined from the Lineweaver-Burk plot based on the ratio of the fluorescence intensities at λ_{em} = 565 and 625 nm. The reversibility of the creatinine-sensing membrane was examined with 0.1 and 1.0 mM creatinine dissolved in Tris buffer and the fluorescence measurements were performed on a microtiter plate at an interval of 30 s. The effects of pH and temperature on the creatinine-sensing membrane were investigated with 0.4 mM creatinine solution at a pH range of 5.0 to 10.0 and temperature of 30, 33, 35, 37, and 40 °C using creatinine concentrations from 0.1 to 10 mM. The long-term stability of the creatinine-sensing membrane at various creatinine concentrations was evaluated by measuring the fluorescence intensity obtained initially and after 1.5 months. The interference effects of some components on the creatinine-sensing membrane were also investigated using 1.0 mM creatinine and 0.2 mM of glycine, histidine, phenylalanine, tryptophan, urea, and creatinine as well as their mixture.

2.3.3. Artificial Urine Solution (AUS)

Artificial urine solution containing various concentrations of creatine and creatinine was prepared, and the concentrations of creatine and creatinine were determined with the creatine-

and creatinine-sensing membranes, respectively. The AUS comprised 2.5 mM $CaCl_2$, 45 mM NaCl, 3.5 mM KH_2PO_4, 3.5 mM K_2HPO_4, 2.5 mM $NaHCO_3$, 1 mM $MgSO_4$, 2.5 mM Na_2SO_4, and creatine or creatinine at concentration range of 0.1–10 mM and pH 7.2.

2.4. Ratiometric Method

The ratiometric method for creatine and creatinine biosensors was based on the ratio of the fluorescence intensities of the creatine- and creatinine-sensing membranes at an excitation wavelength of 460 nm (λ_{ex} = 460 nm) and two emission wavelengths (λ_{em} = 565 nm [FI_{565}] and 625 nm [FI_{625}]) as follows:

$$R = FI_{565}/FI_{625}$$

2.5. Data Analysis

The differences in fluorescence intensity of the sensing membranes at different pH, and temperature values were assessed by one-way analysis of variance (ANOVA). A value of $p < 0.05$ was considered as a statistically significant. Statistical tests were performed using the software InStat (v.3.01, GraphPad Software Inc., San Diego, CA, USA).

3. Results and Discussion

3.1. Enzyme Immobilization for Creatine-Sensing Membrane

The response of the creatine-sensing membrane in the presence of different amounts of immobilized creatinase is shown in Figure 1a. The creatine-sensing membrane immobilized with 14 U creatinase showed the highest SI (SI_{14U} = 0.415) at creatine concentrations from 0.1 to 1.0 mM. We failed to see any increase in the sensitivity of the membrane at higher amounts (28 and 42 U) of creatinase, as evident from the lower SI at 28 and 42 U (SI_{28U} = 0.104 and SI_{42U} = 0.127) as compared with 14, 7, and 3.5 U creatinase. Results of Bradford protein assay showed that the immobilization efficiency of urease and creatinase on the second layer of the creatine-sensing membrane was very high at all creatinase amounts. The immobilization efficiency for the two enzymes was 83.7%, 75.8%, 75.5%, 85.1%, and 91.9% with 3.5, 7, 14, 28, and 42 U creatinase, respectively. The high immobilization efficiency resulted in the good enzyme-capturing capacity of EC and PU matrix. The immobilization efficiency of urease (10 U) into the matrix of EC and PU was higher than 90% in preliminary studies. Based on the high immobilization efficiency of urease (10 U), the immobilization efficiency for creatinase into EC and PU matrix may be estimated to be higher than 70%. However, the excess of creatinase immobilization into the matrix of EC and PU may obstruct NH_4^+ transport in O17-EC membrane. The high density of the enzyme-supporting layer obstructed the passage of NH_4^+ in O17-EC membrane below the enzyme layer. Based on these observations, we chose 14 U creatinase to fabricate the creatine-sensing membrane for further work.

The creatine-sensing membrane included two layers. The first layer was O17-EC membrane, while the second layer was the membrane comprising urease and creatinase co-immobilized in the matrix of EC and PU. As seen in AFM images (Figure 1b), O17-EC membrane displayed a smooth surface with a surface mean roughness (Ra) and root mean square roughness (Rq) of 1.07 and 1.343 nm, respectively, whereas the surface of the second layer immobilized with urease and creatinase displayed higher values of Ra and Rq (9.514 and 12.304 nm, respectively). The increase in the roughness of the sensing membrane resulted from the successful immobilization of the enzymes into EC-PU matrix and the subsequent formation of PU hydrogel during the reaction with the aqueous enzyme solution.

(a)

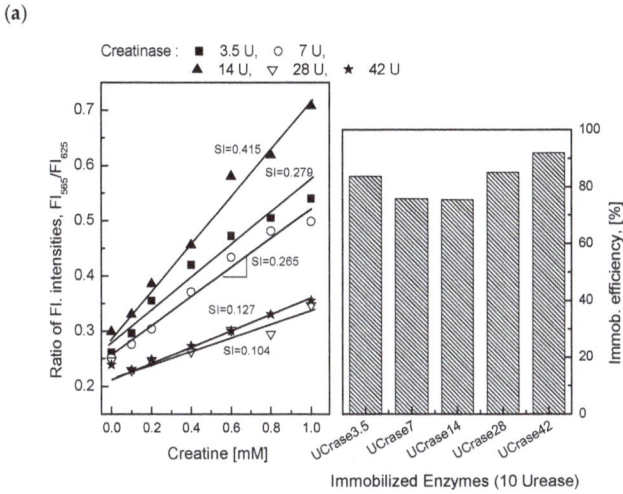

(b)

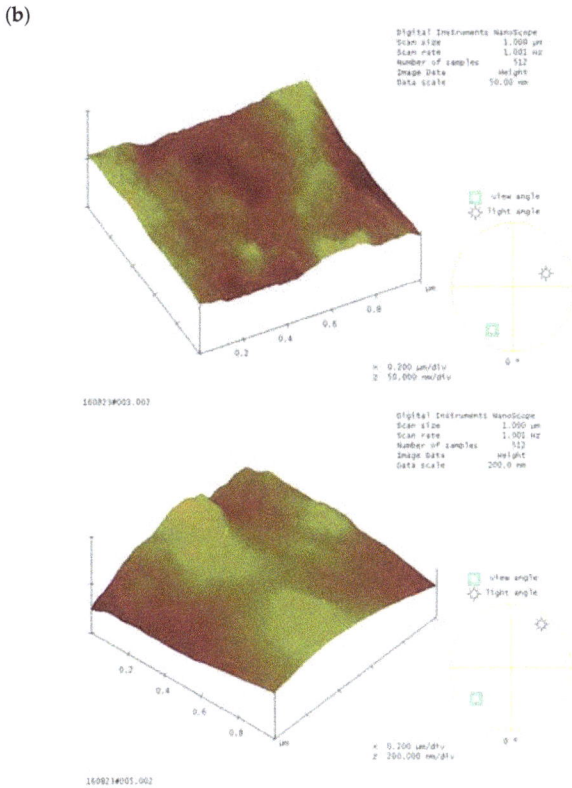

Figure 1. (**a**) The response of the creatine-sensing membrane immobilized with 3.5, 7, 14, 28, and 42 U creatinase and fixed amount (10 U) of urease in the presence of 0.1 to 1.0 mM creatine (**top**). SI presents slope value in the linear detection range of 0.1 to 1.0 mM. The immobilization efficiency for various amounts of creatinase and 10 U of urease (**top**); (**b**) AFM images of O17-EC membrane (**top**) and the second layer containing entrapped enzyme(s) (**below**).

3.2. Characterization of the Creatine-Sensing Membrane

As shown in Figure 2, the creatine-sensing membrane was very sensitive to ammonia produced during the hydrolysis reactions of urea and creatine in the presence of urease and creatinase, respectively. The detection range of creatine based on the ratiometric calculation method could be divided into two linear ranges of 0.1–1.0 mM and 1.0–10 mM with high regression coefficient values of $r^2_{0.1-1.0\,mM}$ = 0.92 and $r^2_{1.0-10\,mM}$ = 0.96, respectively. The relative standard deviation for 1.0 mM creatine was 2.4% (n = 7) based on five repetitive measurements, while the detection limit (LOD, S/N = 3) was 0.015 mM. According to the response of O17-EC membrane to different ammonia concentrations [31], the fluorescence intensity of the creatine-sensing membrane would increase at λ_{em} = 565 nm and decrease at λ_{em} = 625 nm in the presence of the increasing concentrations of creatine. However, a small increase in the fluorescence intensity was observed at λ_{em} = 625 nm at low creatine concentrations. This observation may be attributed to the low amount of ammonia, which reacted with O17 dye as per the following equation

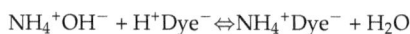

$$NH_4{}^+OH^- + H^+Dye^- \Leftrightarrow NH_4{}^+Dye^- + H_2O$$

A large amount of the dye was in its free state. The reaction mechanism of O17 dye and ammonia produced from the catalytic breakdown of creatine was explained in our previous research [32]. Some modifications in the original sensor may result in unexpected observations, as evident from the behavior of the peak at λ_{em} = 625 nm, which was different from that observed for the original O17 dye. However, the ratiometric method may normalize the change in the fluorescence intensities that are unrelated to the change in the target concentration. Therefore, the values for the ratio of fluorescence intensities at λ_{em} = 565 and 625 nm corresponded well with creatine concentrations. For fluorescence-based sensors, measurements of fluorescence intensity at a single band edge are known to be problematic for practical applications [33]. The ratiometric fluorescence method is useful for the correction of a variety of analyte-independent factors in fluorescent sensors, wherein the temporal and spatial distribution of the measured fluorescence intensity may typically fluctuate owing to the unequal distribution of fluorophores within the sensor, variation in dynamics of fluorophores in different media, and noise in the measurement system (e.g., variations in the illumination intensity). The self-calibration property of the ratiometric method has led to the development of a wide range of ratiometric fluorescent sensors that provide precise quantitative analysis. Nakata et al. described real-time monitoring of saccharide conversion pathway using a seminaphthorhodafluor-conjugated lectin-based ratiometric fluorescent biosensor [34], whereas Xie et al. exploited the enzymatic reaction of hyaluronidase and hyaluronan bound to two fluorescent dyes for fluorescence quenching and dequenching of these dyes before and after enzyme reaction to obtain the proportion between the ratiometric fluorescence intensity and hyaluronidase level [35]. Other researchers have used ratiometric fluorescence biosensors for the detection of DNA [36] and nitric oxide [37].

Kinetic parameters of creatinase (14 U) and urease (10 U) co-immobilized on the supporting material of EC and PU were evaluated with Michaelis-Menten kinetics using the ratio of fluorescence intensities at λ_{em} = 565 and 625 nm. An apparent maximal reaction rate ($V_{max}{}^{app}$) of 0.0406 1/min and apparent Michaelis-Menten constant ($K_m{}^{app}$) of 3.441 mM were obtained using Hanes plot. The value of $K_m{}^{app}$ was greater than that obtained for creatinase immobilized with chitosan-SiO$_2$-multiwall carbon nanotubes nanocomposite ($K_m{}^{app}$ = 0.58 mM, [38]). $K_m{}^{app}$ is usually dependent upon the supporting material and immobilization method. The large value of $K_m{}^{app}$ in this work indicates the low affinity of creatinase to EC and PU matrix over EC-O17 membrane and may be associated with the conformation and arrangement of the enzyme during immobilization. The slow reaction rate of the creatine-sensing membrane may be attributed to the retardation of the reaction product ($NH_4{}^+OH$) toward O17-EC layer through the thick enzyme-immobilized layer.

The response time of the creatine-sensing membrane was approximately $t_{95(0.1-0.4\,mM)}$ = 1 to 2.5 min, $t_{95(0.6-2\,mM)}$ = 2.5 to 3.5 min, and $t_{95(4-10\,mM)}$ = 0.5 to 1.5 min. Moreover, the addition of PU to the supporting material for enzyme immobilization offered a convenient environment to shorten the transport time of ammonia passing through the enzyme-immobilized membrane to contact O17-EC membrane. The use of PU as a supporting material for enzyme immobilization may improve the response time of the enzyme-immobilized membrane.

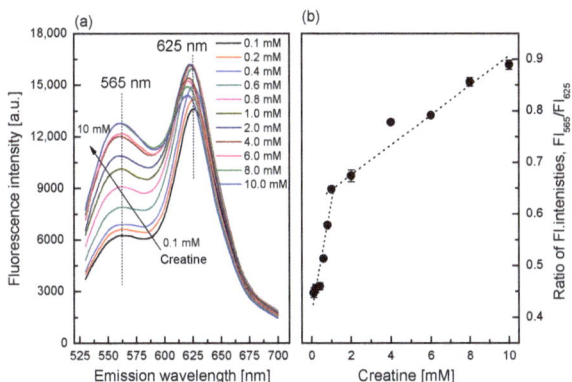

Figure 2. (**a**) Fluorescence emission spectra of the creatine-sensing membrane in the presence of creatine at 0.1 to 10 mM concentration monitored at an excitation wavelength of 460 nm. (**b**) Calibration curve for creatine calculated by the ratio of two fluorescence intensities measured at emission wavelengths of 565 and 625 nm.

As shown in our previous study [32], O17-EC membrane showed high reversibility despite being coated with a second layer of polymer (e.g., EC) and urease. The fluorescent O17 dye could offer a proton in the formation of $NH_4^+OH^-$ by the protonation of ammonia in water and react reversibly upon reduction of $NH_4^+OH^-$. Herein, a sequence of hydrolysis reactions of creatine and urea resulted in the production of ammonia, which reacted with O17-EC membrane and changed the fluorescence intensity of O17 dye. This change was easy to recognize upon repeated exposure of the creatine-sensing membrane to 1.0 mM creatine and distilled water (DW). Figure 3 shows the reversibility of the creatine-sensing membrane as the ratio of the fluorescence intensities at $\lambda_{em} = 565$ and 625 nm. The sensing membrane exhibited a very low relative standard deviation of 0.78% and 2.69% in DW and 1.0 mM creatine, respectively.

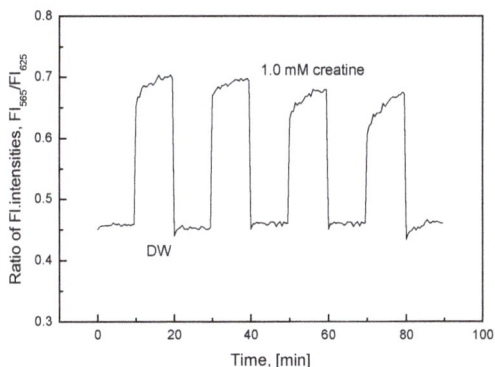

Figure 3. Reversibility of the creatine-sensing membrane in the presence of 1.0 mM creatine and distilled water (DW).

The performance of a biosensor with an enzyme is usually influenced by pH and temperature. pH determines the activity of the enzyme and the consequent efficiency of the catalytic reaction in the biosensor. Similar to pH is the influence of temperature, which may increase or decrease the catalytic reaction rate of the enzyme. The creatine-sensing membrane preferred alkaline (pH range of 7.5 to pH 9.0) to weakly acidic (pH 5.0 to pH 7.0) medium. The ratio of the fluorescence intensities at $\lambda_{em} = 565$ and 625 nm obtained for the creatine-sensing membrane in the presence of 1.0 mM creatine failed to change significantly in the acidic pH range, but considerably increased with an increase in pH from 7.5 to 9.0 (data not shown). Creatine was difficult to dissolve in common aqueous solutions and, hence, was prepared in 10 mM Tris buffer at pH range of 7.5–8.5. Therefore, the creatine-sensing membrane worked well under alkaline conditions induced via ammonia produced from the hydrolysis of creatine and urea or interferences from other alkaline factors. The response of the creatine-sensing membrane to different temperatures was investigated at various creatine concentrations. The temperature range of 25 to 35 °C had no effect on the creatine-sensing membrane in the presence of 0.1 to 10 mM creatine (data not shown). However, the sensitivity of the membrane decreased at high temperatures (37–40 °C) and creatine concentration (1.0–10 mM). However, the sensitivity was unaffected by temperature at low concentrations of creatine (0.1–1.0 mM). Therefore, a temperature of approximately 33 °C was used for creatine measurements to extend the performance of the enzymes as well as the sensitivity of the creatine-sensing membrane for long-term use.

The creatine-sensing membrane was tested after 3 months of use and storage; its sensitivity was found to be quite good (Figure 4), as evident from SI of the linear curve with creatine concentration of 0.1 to 1.0 mM ($SI_{0.1–1.0\,mM}$) (0.22 at initial use and 0.242 after 3 months of use). Thus, the creatine-sensing membrane including O17-EC layer as a transducer and EC and PU matrix layer immobilized with two enzymes (urease and creatinase) showed excellent stability after long-time use. EC and PU were good supporting materials for the fluorescent dye O17 and enzyme immobilization. The layer of EC and PU matrix prevented enzyme leakage [39], but at the same time created an open environment, which improved the sensitivity of the membrane following ammonia production. The presence of PU in the enzyme-immobilized layer induced softening of the supporting layer for enzyme immobilization, thereby allowing higher amount of enzyme immobilization and offering a wider location for the passing ammonia to react with O7-EC membrane. In addition, we suggest that the covalent binding between the isocyanate groups of PU and amine groups of the enzymes increased the lifetime of the sensor. The increase in the background signal was associated with the increased opacity of the sensing membrane. The change probably occurred owing to the long-term soaking in 10 mM phosphate buffer solution that increased the reflection of the incident light and fluorescence emission during creatine measurements.

The recovery capability of a biosensor may suffer from some interferences existing in the sample. Urine samples may contain several types of carbohydrates, metabolites, and electrolytes, which may interfere with the fluorescent creatine biosensor developed here. According to the results of one-way ANOVA with Tukey-Kramer multiple comparison post-test, the measurement for 1.0 mM creatine in the presence of 0.2 mM histidine showed a significant difference from the sample without histidine ($p < 0.001$). The presence of other components such as glycine, phenylalanine, tryptophan, and creatinine or the mixture of these components containing histidine had no significant effect on the measurement ability of the creatine-sensing membrane for 1.0 mM creatine ($p > 0.05$) (Figure 5). In addition, no significant interference was observed with the mixture of cations such as Ca^{2+}, Na^+, K^+, and Mg^{2+} (data not shown).

For biological samples, the ratiometric fluorescence creatine biosensor showed good analytical performance as compared with other creatine biosensors (Table 1). The linear detection range was wide and extended up to 10 mM creatine. The wide linear detection range of the sensing membrane allows quantification of the amounts of creatine in urine samples. The response time of the sensor was relatively fast and its stability was good, attributable to the structure of the supporting materials used for the immobilization of the two enzymes. We speculate the covalent binding between the isocyanate groups in the hydrogel PU and the amine groups of the enzymes, leading to the extension of the shelf-life of the biosensor for several months without any evident loss in the enzyme activity. The ratiometric fluorescent sensing membrane with good sensitivity and long-term stability may be applied to the high-throughput analysis of creatine.

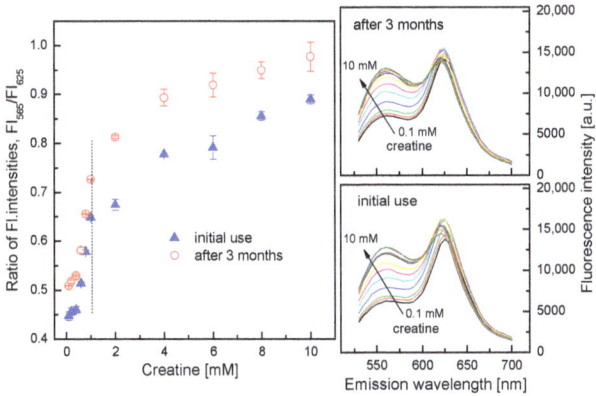

Figure 4. Long-term stability of the creatine-sensing membrane in the presence of creatine 0.1 to 10 mM creatine upon initial use and after 3 months of use (**left**), and two fluorescence emission spectra of the creatine-sensing membrane with respect to creatine concentrations at λ_{ex} = 460 nm.

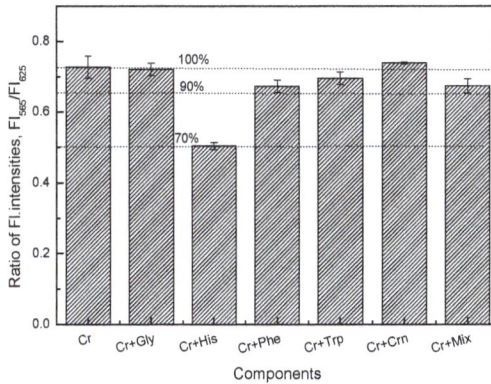

Figure 5. Response of the creatine-sensing membrane to aqueous solutions of 1.0 mM creatine (Cr) and 0.2 mM of each of the following components: Gly; glycine, His; histidine, Phe; phenylalanine, Trp; tryptophan, Crn; creatinine, Mix: mixture of Gly, His, Phe, Trp, and Crn.

Table 1. Summary of the analytical performance of some creatine biosensors.

Detection Method (Enzymes Used)	Linear Detection Range	Limit of Detection	Response Time	Long-Term Stability	Ref.
Potentiometry (CTN + URS)	0.1–30 mM	10 μM	1.5–4 min	14 days	[5]
Potentiometry (CTN + URS)	0.01–1 mM	10 μM	1–2 min	2 months	[6]
Amperometry (CTN + SOD)	0–2 mM	3.4 μM	3 min	3 months	[7]
Amperometry (CTN + SOD)	0–3.5 mM	n.a.	1 min	8 days	[8]
Amperometry (CTN + SOD)	0.2–3.8 μM and 9–120 μM	0.2 μM	10 s	30 days	[9]
Ratiometric fluorescence (CTN + URS)	0.1–1 Mm and 1–10 mM	0.015 mM	3.5 min	>3 months	This work

CTN: creatinase, SOD: sarcosine oxidase, URS: urease, n.a.: not available.

3.3. Enzyme Immobilization for the Creatinine-Sensing Membrane

The optimal amounts of creatinine deiminase for immobilization into the matrix of EC and PU are tested and shown in Figure 6. The sensing membrane immobilized with 2.5 U creatinine deiminase showed the smallest SI (SI_{5U} = 0.109) in the presence of 0.1–1.0 mM creatinine, whereas SI of the membrane immobilized with 5 or 7.5 U creatinine deiminase (SI_{10U} = 0.833 and SI_{15U} = 0.881) were much higher than those obtained for the membrane immobilized with 2.5 U creatinine deiminase. In addition, data collected from Bradford protein assay indicated that the immobilization efficiency of creatinine deiminase on the second layer of the creatinine-sensing membrane was high at all amounts of creatinine deiminase used. The percentage immobilization was 47.1%, 52.9%, and 41.3% for 2.5, 5, and 7.5 U creatinine deiminase. Based on these observations, 5 U creatinine deiminase was used to catalyze the hydrolysis of creatinine at a concentration range of 0.1 to 1.0 mM to produce ammonia.

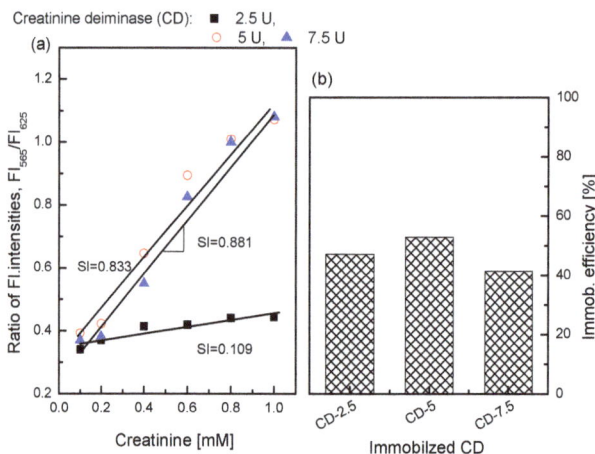

Figure 6. (**a**) The response of the creatinine-sensing membrane immobilized with 2.5, 5, and 7.5 U creatinine deiminase (CD) in the presence of 0.1 to 1.0 mM creatinine. SI represents slope value. (**b**) Immobilization efficiency of various amounts of creatinine deiminase.

3.4. Characterization of the Creatinine-Sensing Membrane

As shown in Figure 7, the linear detection range of creatinine was 0.1–1.0 mM with a high regression coefficient value of r^2 = 0.96, while LOD (S/N = 3) was 0.0325 mM. Given the use of a single enzyme, creatinine deiminase, the response of the creatinine-sensing membrane was similar to that of O17-EC membrane to different ammonia concentrations reported in our previous study [32]. The fluorescence intensity of the creatinine-sensing membrane increased at λ_{em} = 565 nm and decreased at λ_{em} = 625 nm in response to an increase in creatinine concentrations from 0.1 to 10 mM. Thus, the sensing membrane with only one enzyme involved in the direct hydrolysis of the analytes to produce ammonia seems to show higher sensitivity and faster response than that with two enzymes.

The activity of creatinine deiminase immobilized in the matrix of EC and PU was evaluated via Michaelis-Menten kinetics. Kinetic parameters were calculated from the ratio of two emission fluorescence intensities at λ_{em} = 565 and 625 nm. Lineweaver-Burk plot revealed the maximal reaction rate (V_{max}) and Michaelis-Menten constant (K_m) to be 1.1862 1/min and 22.05 mM, respectively. Studies have shown K_m values for free creatinine deiminase to be 1.27 [40] and 0.15 mM [41], while the K_m for the creatinine deiminase immobilized on the polyaniline-copper-nanocomposite was calculated to be 0.163 mM [42]. The higher K_m value observed in our study may be associated with the very low affinitiy of creatinine deiminase to creatinine in EC and PU matrix as compared with EC-O17 membrane. The low affinity may result from the supporting matrix of EC and PU that was partially bound at the active site of the enzyme, owing to its conformational change. The maximal reaction rate (V_{max} = 0.263 1/min) for the immobilized urease reported in our previous study [32] indicates the strong hydrolysis reaction of creatinine by creatinine deiminase, leading to a short response time of the creatinine-sensing membrane of about t_{95} = 1–3 min at all creatinine concentrations.

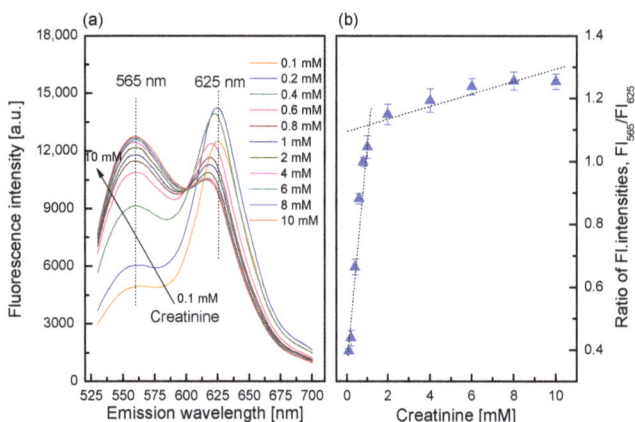

Figure 7. (a) Fluorescence emission spectra of the creatinine-sensing membrane in the presence of creatinine at 0.1 to 10 mM concentration monitored at an excitation wavelength of 460 nm. (b) Calibration curve for creatinine calculated by the ratio of two fluorescence intensities at emission wavelengths of 565 and 625 nm.

Ammonia was probably produced directly through the hydrolysis of creatinine. Therefore, the reproducibility of the creatinine-sensing membrane was rapid, as evident from the repeated exposure of the membrane to 0.1 and 1.0 mM creatinine. Figure 8 shows the reversibility of the creatinine-sensing membrane based on the ratio of the fluorescence intensities at λ_{em} = 565 and 625 nm. The sensing membrane exhibited a very low relative standard deviation of 1.1% and 1.9% for 0.1 and 1.0 mM creatinine, respectively.

The creatinine-sensing membrane also preferred an alkaline (pH range of 8.0 to 9.0) to acidic (pH 5.0 to pH 7.0) medium (data not shown). At 0.4 mM creatinine concentration, the ratio of the fluorescence intensities of the creatinine-sensing membrane at λ_{em} = 565 and 625 nm increased with an increase in the pH to the basic range. The response of the creatinine-sensing membrane at different temperatures was also studied with various creatinine concentrations. The temperature range of 30 to 40 °C failed to exert any significant effect on the sensitivity of the membrane at creatinine concentration of 0.1 to 1.0 mM (p = 0.901) (data not shown).

The long-term stability of the creatinine-sensing membrane was tested after 1.5 months of use and storage. The membrane maintained its high sensitivity to various creatinine concentrations (Figure 9), as evident from the increase in SI of the linear curve at creatinine concentration of 0.1 to 1.0 mM ($SI_{0.1–1.0 mM}$) from 0.832 (initial use) to 0.936 (after 1.5 months). The increase in SI after 1.5 months of use may be related to the good maintenance of the enzyme activity in the polymers EC and PU [39], which created a versatile environment for the enzyme catalysis and response of O17-EC membrane to ammonia produced. In addition, the increase in the background signal of the creatinine-sensing membrane could be recognized after testing the sensing membrane with different pH solutions. This is attributed to the slight leakage of O17 dye upon its exposure to strong pH solutions, thereby increasing the ratio of the fluorescence intensities at λ_{em} = 565 and 625 nm. The long-term soaking of the polymer in the aqueous solution may have contributed to the increase in the reflection of the incident light and fluorescence emission during creatinine measurements.

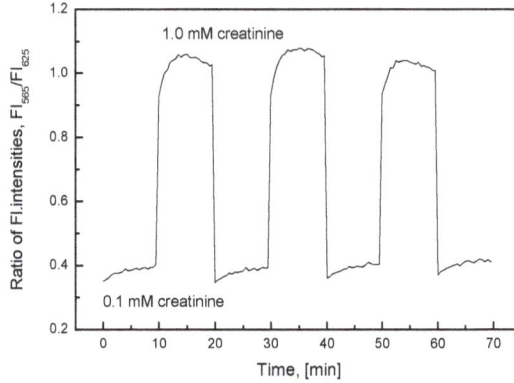

Figure 8. Reversibility of the creatinine-sensing membrane at 0.1 and 1.0 mM creatinine.

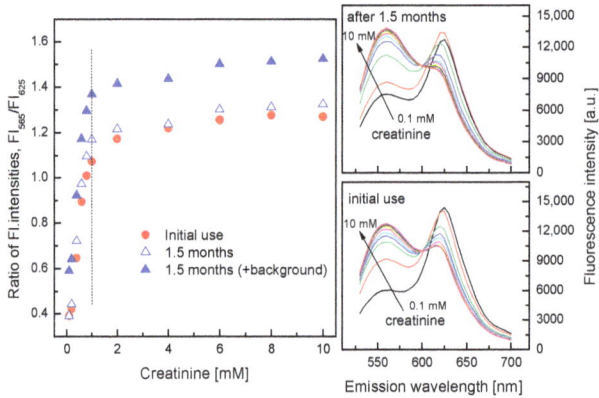

Figure 9. Long-term stability of the creatinine-sensing membrane at 0.1 to 10 mM creatinine concentration upon initial use and after 1.5 months of use (**left**). Two fluorescence emission spectra of the creatinine-sensing membrane with respect to creatinine concentrations at $\lambda_{ex} = 460$ nm.

The limitation of the spectrometric method is the interference from a few factors in the samples [17,18] that have similar absorption and emission wavelengths or the change in the refractive medium. The influence of some components in urine samples on the creatinine-sensing membrane is shown in Figure 10. According to the results of one-way ANOVA, the presence of 0.2 mM histidine in 1.0 mM creatinine sample resulted in a significant difference in the measurement as compared to samples without histidine ($p = 0.003$). However, the presence of other components such as glycine, phenylalanine, tryptophan, urea, and creatine failed to exert any effect on the measurement of 1.0 mM creatinine ($p > 0.05$); however, the mixture of these components containing histidine also had a significant effect on the creatinine measurement using the creatinine-sensing membrane (t-test with $p = 0.005$). No significant interference was observed with a mixture of cations such as Ca^{2+}, Na^+, K^+, and Mg^{2+} (data not shown). Aside from their influence on the pH of the solution, these cations seemed to have less effect on the response of the creatinine-sensing membrane. Maintaining an alkaline medium during creatinine measurements may guarantee the high sensitivity of the creatinine-sensing membrane.

A few studies have systematically reviewed the techniques of electrochemical enzymic/non-enzymic and immuno-sensors for creatinine detection [17]. Aside from electrochemical transducers, some optical methods based on Jaffe's reaction or enzyme-catalyzed reactions were also used for the measurement of creatinine. In Table 2, the analytical performance of some optical methods for creatinine detection is compared with that of the ratiometric fluorescence creatinine biosensor developed in this study. The ratiometric fluorescent creatinine biosensor exhibited good analytical performance in terms of high sensitivity, good selectivity, low-cost, and rapidity. In particular, many samples can be analyzed simultaneously using a 96-well microtiter plate with immobilized creatinine-sensing membranes.

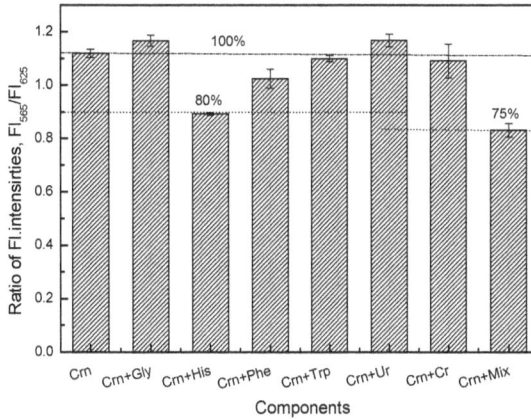

Figure 10. Response of the creatine-sensing membrane to aqueous solutions of 1.0 mM creatinine (Crn) and 0.2 mM of each of the following components: Gly: glycine, His: histidine, Phe: phenylalanine, Trp: tryptophan, Ur: urea, Cr: creatine, and Mix: mixture of Gly, His, Phe, Trp, Ur, and Cr.

Table 2. Summary of the analytical performance of some optical methods for creatinine detection.

Detection Method (Enzyme or Assay Used)	Linear Detection Range	Limit of Detection	Response Time	Ref.
Light diffraction (CD)	0.01–0.7 mM	6 μM	14 min	[43]
Chemiluminescence (H_2O_2-Co^{2+})	0.1–30 μM	0.07 μM	1 min	[44]
Colorimetry (uric acid-Hg^{2+}-AuNPs)	1–12 μM	1.6 nM	5 min	[28]
Colorimetry (citrate-AuNPs)	0.1–20 mM	0.72 mM	24 min	[45]
Colorimetry-enzPAD (CNN + CTN + SOD)	0.22–2.2 mM	0.17 mM	11 min	[46]
Fluorescence (QDs-CD conj.)	0.3–5 mM	n.a.	3–4 min	[47]
Ratiometric fluorescence (CD)	0.1–1.0 mM	0.0325 mM	1–3 min	This work

CD: creatinine deiminase, AuNP: gold nanoparticles, CNN: creatininase, CTN: creatinase, SOD: sarcosine oxidase, QD: quantum dots, enz-PAD: enzyme paper-based analytical device.

3.5. Determination of Creatine and Creatinine in Artificial Urine Solution (AUS)

The creatine- and creatinine-sensing biosensors developed for the detection of creatine and creatinine were used to evaluate their recovery capabilities. The measurement results of creatine and creatinine in the standard solution prepared in 10 mM phosphate buffer saline (pH 7.2) and AUS are shown in Figure 11. The difference plot in Figure 11 also shows the recovery percentage of the concentrations of creatine and creatinine in the AUS. The recovery percentage of two sensing membranes was quite high (85–115%). The large difference in the two measurement results at low concentrations of creatine and creatinine (0–0.2 mM) may be associated with the presence of anions such as SO_4^{2-} in the AUS.

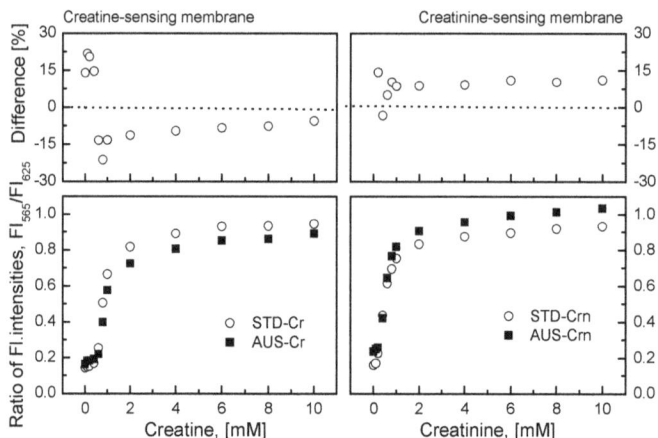

Figure 11. Comparison of creatine (Cr) and creatinine (Crn) concentrations in standard solution (STD) and artificial urine solution (AUS) using creatine- and creatinine-sensing membranes, and their difference plots. The difference represents $100 \times$ (creatine or creatinine in AUS − standard creatine or creatinine)/(standard creatine or creatinine). The dashed lines indicate the recovery percentage of 85–115% for creatine or creatinine in AUS samples.

4. Conclusions

Enzymatic fluorescence assay techniques are highly specific and sensitive, but their use is restricted due to enzyme instability and assay complexity. In this study, ratiometric fluorescent biosensors for creatine and creatinine detection were successfully fabricated. These biosensors showed good sensitivity in the linear concentration range of 0.1–1.0 mM for creatine and 0.1–1.0 mM for creatinine, while their LOD was 0.015 and 0.0325 mM for creatine and creatinine, respectively. The sensing membranes displayed negligible interference from amino acids and salts, with the exception of 0.2 mM histidine. The reproducibility of the sensing membranes for creatine and creatinine was excellent, with a very low relative standard deviation (<2.67%), and their sensitivity to creatine and creatinine was retained for at least 2 months. The high recovery percentage of two sensing membranes in artificial urine samples highlights their potential application for the determination of creatine and creatinine concentrations in clinical chemistry. In addition, the successful development of the ratiometric fluorescent biosensors would be of a great significance in high-throughput screening techniques in analytical biochemistry.

Acknowledgments: This work was partly supported by the National Research Foundation (NRF), Republic of Korea (Grant Number: 2013R1A1A2058628), and by Korea Ministry of Environment as "Global Top Project" (Project No.: 2016002210007).

Author Contributions: H.D. and J.R. conceived and designed the experiments; H.D. performed the experiments and analyzed the data.

Conflicts of Interest: The authors declare no conflict of interest.

References

1. Wyes, M.; Kaddurah-Daouk, R. Creatine and creatinine metabolism. *Physiol. Rev.* **2000**, *80*, 1107–1213.
2. Greenhaff, P.L.; Casey, A.; Short, A.H.; Harris, R.; Soderlund, K.; Hultman, E. Influence of oral creatine supplementation of muscle torque during repeated bouts-of maximal voluntary exercise in man. *Clin. Sci.* **1993**, *84*, 565–571. [CrossRef] [PubMed]
3. Williams, M.H.; Branch, J.D. Creatine supplementation and exercise performance: An update. *J. Am. Coll. Nutr.* **1998**, *17*, 216. [CrossRef] [PubMed]

4. Mora, L.; Sentandreu, M.A.; Toldra, F. Contents of creatine, creatinine and carnosine in porcine muscles of different metabolic types. *Meat Sci.* **2008**, *79*, 709–715. [CrossRef] [PubMed]

5. Koncki, R.; Walcerz, I.; Ruckruh, F.; Glab, S. Bienzymatic potentiometric electrodes for creatine and L-arginine Determination. *Anal. Chim. Acta* **1996**, *333*, 215–222. [CrossRef]

6. Karakus, E.; Pekyardımcı, S.; Kılıc, E. Potentiometric bienzymatic biosensor based on PVC membrane containing palmitic acid for determination of creatine. *Proc. Biochem.* **2006**, *41*, 1371–1377. [CrossRef]

7. Madarag, M.B.; Popescu, I.C.; Ufer, S.; Buck, R.P. Microfabricated amperometric creatine and creatinine biosensors. *Anal. Chim. Acta* **1996**, *319*, 335–345. [CrossRef]

8. Ramanavicius, A. Amperometric biosensor for the determination of creatine. *Anal. Bioanal. Chem.* **2007**, *387*, 1899–1906. [CrossRef] [PubMed]

9. Kacar, C.; Erden, P.E.; Pekyardimci, S.; Kilic, E. An Fe_3O_4-nanoparticles-based amperometric biosensor for creatine determination. *Artif. Cells Nanomed. Biotechnol.* **2013**, *41*, 2–7. [CrossRef] [PubMed]

10. Yao, T.; Kotegawa, K. Simultaneous flow-injection assay of creatinine and creatine in serum by the combined use of a 16-way switching valve, some specific enzyme reactors and a highly selective hydrogen peroxide electrode. *Anal. Chim. Acta* **2002**, *462*, 283–291. [CrossRef]

11. Dash, A.K.; Sawhney, A. A simple LC method with UV detection for the analysis of creatine and creatinine and its application to several creatine Formulations. *J. Pharm. Biomed. Anal.* **2002**, *29*, 939–945. [CrossRef]

12. Smith-Palmer, T. Separation methods applicable to urinary creatine and creatinine. *J. Chromatogr. B Analyt. Technol. Biomed. Life Sci.* **2002**, *781*, 93–106. [CrossRef]

13. Fernandez-Fernanadez, M.; Rodríguez-Gonzalez, P.; Alvarez, M.E.A.; Rodríguez, F.; Menendez, F.V.A.; Alonso, J.I.G. Simultaneous determination of creatinine and creatine in human serum by double-spike isotope dilution liquid chromatography–tandem mass spectrometry (LC-MS/MS) and gas chromatography–mass spectrometry (GC-MS). *Anal. Chem.* **2015**, *87*, 3735–3763. [CrossRef] [PubMed]

14. Liotta, E.; Gottardo, R.; Bonizzato, L.; Pascali, J.P.; Bertaso, A.; Tagliaro, F. Rapid and direct determination of creatinine in urine using capillary zone electrophoresis. *Clin. Chim. Acta* **2009**, *409*, 52–55. [CrossRef] [PubMed]

15. Miura, C.; Funaya, N.; Matsunaga, H.; Haginaka, J. Monodisperse, molecularly imprinted polymers for creatinine by modified precipitation polymerization and their applications to creatinine assays for human serum and urine. *J. Pharm. Biomed. Anal.* **2013**, *85*, 288–294. [CrossRef] [PubMed]

16. Mohabbati-Kalejahi, E.; Azimirad, V.; Bahrami, M.; Ganbari, A. A review on creatinine measurement techniques. *Talanta* **2012**, *97*, 1–8. [CrossRef] [PubMed]

17. Pundir, C.S.; Yadav, S.; Kumar, A. Creatinine sensors. *Trends Anal. Chem.* **2013**, *50*, 42–52. [CrossRef]

18. Randvir, E.P.; Banks, C.E. Analytical methods for quantifying creatinine within biological media. *Sens. Actuators B* **2013**, *183*, 239–252. [CrossRef]

19. Lad, U.; Khokhar, S.; Kale, G.M. Electrochemical creatinine biosensors. *Anal. Chem.* **2008**, *80*, 7910–7917. [CrossRef] [PubMed]

20. Killard, A.J.; Smyth, M.R. Creatinine biosensors: Principles and designs. *Trends Biotechnol.* **2000**, *18*, 433–437. [CrossRef]

21. Shih, Y.T.; Huang, H.J. A creatinine deiminase modified polyaniline electrode for creatinine analysis. *Anal. Chim. Acta* **1999**, *392*, 143–150. [CrossRef]

22. Yadav, S.; Kumar, A.; Pundir, C.S. Amperometric creatinine biosensor based on covalently coimmobilized enzymes onto carboxylated multiwalled carbon nanotubes/polyaniline composite film. *Anal. Biochem.* **2011**, *419*, 277–283. [CrossRef] [PubMed]

23. Radomska, A.; Bodenszac, E.; Glab, S.; Koncki, R. Creatinine biosensor based on ammonium ion selective electrode and its application in flow-injection analysis. *Talanta* **2004**, *64*, 603–608. [CrossRef] [PubMed]

24. Zinchenkoa, O.A.; Marchenkoa, S.V.; Sergeyevaa, T.A.; Kuklab, A.L.; Pavlyuchenkob, A.S.; Krasyukc, E.K.; Soldatkina, A.P.; El'skayaa, A.V. Application of creatinine-sensitive biosensor for hemodialysis control. *Biosens. Bioelectron.* **2012**, *35*, 466–469. [CrossRef] [PubMed]

25. Soldatkin, A.P.; Montoriol, J.; Sant, W.; Martelet, C.; Jaffrezic-Renault, N. Creatinine sensitive biosensor based on ISFETs and creatinine deiminase immobilised in BSA membrane. *Talanta* **2002**, *58*, 351–357. [CrossRef]

26. Sant, W.; Pourciel-Gouzy, M.L.; Launay, J.; Conto, T.D.; Colin, R.; Martinez, A.; Temple-Boyer, P. Development of a creatinine-sensitive sensor for medical analysis. *Sens. Actuators B* **2004**, *103*, 260–264. [CrossRef]

27. Marchenko, S.V.; Soldatkin, O.O.; Kasap, B.O.; Kurc, B.A.; Soldatkin, A.P.; Dzyadevych, S.V. Creatinine deiminase adsorption onto silicalite-modified pH-FET for creation of new creatinine-sensitive biosensor. *Nanoscale Res. Lett.* **2016**, *11*, 173–180. [CrossRef] [PubMed]

28. Du, J.; Zhu, B.; Leow, W.R.; Chen, S.; Sum, T.C.; Peng, X.; Chen, X. Colorimetric detection of creatinine based on plasmonic nanoparticles via synergistic coordination chemistry. *Small* **2015**, *11*, 4104–4110. [CrossRef] [PubMed]

29. Taussky, H.H. A procedure increasing the specificity of the Jaffe reaction for the determination of creatine and creatinine in urine and plasma. *Clin. Chim. Acta* **1956**, *1*, 210–224. [CrossRef]

30. Stefan, R.-L.; Bokretsion, R.G. Determination of creatine and creatinine using a diamond paste based electrode. *Instum. Sci. Technol.* **2003**, *31*, 183–188. [CrossRef]

31. Duong, H.D.; Rhee, J.I. A ratiometric fluorescence sensor for the detection of ammonia. *Sens. Actuators B* **2014**, *190*, 768–774. [CrossRef]

32. Duong, H.D.; Rhee, J.I. Development of a ratiometric fluorescent urea biosensor based on the urease immobilized onto the oxazine 170 perchlorate-ethyl cellulose membrane. *Talanta* **2015**, *134*, 333–339. [CrossRef] [PubMed]

33. Lakowicz, J.R. *Topics in Fluorescence Spectroscopy*; Plenum Press: New York, NY, USA, 1994; Volume 4, pp. 3–6.

34. Nakata, E.; Wang, H.; Hamachi, I. Ratiometric fluorescent biosensor for real-time and label-free monitoring of fine saccharide metabolic pathways. *ChemBioChem* **2008**, *9*, 25–28. [CrossRef] [PubMed]

35. Xie, H.; Zeng, F.; Wu, S. Ratiometric fluorescent biosensor for hyaluronidase with hyaluronan as both nanoparticle scaffold and substrate for enzymatic reaction. *Biomacromolecules* **2014**, *15*, 3383–3389. [CrossRef] [PubMed]

36. Liang, S.S.; Qi, L.; Zhang, R.L.; Jin, M.; Zhang, Z.Q. Ratiometric fluorescence biosensor based on CdTe quantum and carbon dots for double strand DNA detection. *Sens. Actuators B* **2017**, *244*, 585–590. [CrossRef]

37. Barker, S.L.R.; Clark, H.A.; Swallen, S.F.; Kopelman, R.; Tsang, A.W.; Swanson, J.A. Ratiometric and fluorescence-lifetime-based biosensors incorporating cytochrome c and the detection of extra- and intracellular macrophage nitric oxide. *Anal. Chem.* **1999**, *71*, 1767–1772. [CrossRef] [PubMed]

38. Tiwari, A.; Dhakate, S.R. Chitosan–SiO₂–multiwall carbon nanotubes nanocomposite: A novel matrix for the immobilization of creatine amidinohydrolase. *Int. J. Biol. Macromol.* **2009**, *44*, 408–412. [CrossRef] [PubMed]

39. Weidgans, B.M.; Krause, C.; Klimant, I.; Wolfbeis, O.S. Fluorescent pH sensors with negligible sensitivity to ionic strength. *Analyst* **2004**, *129*, 645–650. [CrossRef] [PubMed]

40. Uwajima, T.; Terada, O. Properties of crystalline creatinine deiminase from *Corynebacterium lilium*. *Agric. Biol. Chem.* **1980**, *44*, 1787–1792. [CrossRef]

41. Gottschalk, E.M.; Hippe, H.; Patzke, F. Creatinine deiminase (EC 3.5.4.21) from bacterium BN11: Purification, properties and applicability in a serum/urine creatinine assay. *Clin. Chim. Acta* **1991**, *204*, 223–238. [CrossRef]

42. Zhybak, M.; Beni, V.; Vagin, M.Y.; Dempsey, E.; Turner, A.P.F.; Korpan, Y. Creatinine and urea biosensors based on a novel ammonium ion-selective copper-polyaniline nano-composite. *Biosens. Bioelectron.* **2016**, *77*, 505–511. [CrossRef] [PubMed]

43. Sharma, A.C.; Jana, T.; Kesavamoorthy, R.; Shi, L.; Virji, M.A.; Finegold, D.N.; Asher, S.A. A general photonic crystal sensing motif: Creatinine in bodily fluids. *J. Am. Chem. Soc.* **2004**, *126*, 2971–2977. [CrossRef] [PubMed]

44. Hanif, S.; John, P.; Gao, W.; Saqib, M.; Qi, L.; Xu, G. Chemiluminescence of creatinine/H_2O_2/Co^{2+} and its application for selective creatinine detection. *Biosens. Bioelect.* **2016**, *75*, 347–351. [CrossRef] [PubMed]

45. He, Y.; Zhang, X.; Yu, H. Gold nanoparticles-based colorimetric and visual creatinine-assay. *Microchim. Acta* **2015**, *182*, 2037–2043. [CrossRef]

46. Talalak, K.; Noiphung, J.; Songjaroen, T.; Chailapakul, O.; Laiwattanapaisal, W. A facile low-cost enzymatic paper-based assay for the determination of urine creatinine. *Talanta* **2015**, *144*, 915–921. [CrossRef] [PubMed]

47. Ruedas-Rama, M.J.; Hall, E.A.H. Analytical nanosphere sensors using quantum dot-enzyme conjugates for urea and creatinine. *Anal. Chem.* **2010**, *82*, 9043–9049. [CrossRef] [PubMed]

sensors

MDPI

Article

Theoretical Studies on Two-Photon Fluorescent Hg^{2+} Probes Based on the Coumarin-Rhodamine System

Yujin Zhangand Jiancai Leng *

School of Science, Qilu University of Technology, Jinan 250353, China;
zhangyujin312@163.com or zhangyujin@qlu.edu.cn
* Correspondence: jiancaileng@qlu.edu.cn; Tel.: +86-531-8963-1268

Received: 16 June 2017; Accepted: 16 June 2017; Published: 20 July 2017

Abstract: The development of fluorescent sensors for Hg^{2+} has attracted much attention due to the well-known adverse effects of mercury on biological health. In the present work, the optical properties of two newly-synthesized Hg^{2+} chemosensors based on the coumarin-rhodamine system (named Pro1 and Pro2) were systematically investigated using time-dependent density functional theory. It is shown that Pro1 and Pro2 are effective ratiometric fluorescent Hg^{2+} probes, which recognize Hg^{2+} by Förster resonance energy transfer and through bond energy transfer mechanisms, respectively. To further understand the mechanisms of the two probes, we have developed an approach to predict the energy transfer rate between the donor and acceptor. Using this approach, it can be inferred that Pro1 has a six times higher energy transfer rate than Pro2. Thus the influence of spacer group between the donor and acceptor on the sensing performance of the probe is demonstrated. Specifically, two-photon absorption properties of these two probes are calculated. We have found that both probes show significant two-photon responses in the near-infrared light region. However, only the maximum two-photon absorption cross section of Pro1 is greatly enhanced with the presence of Hg^{2+}, indicating that Pro1 can act as a potential two-photon excited fluorescent probe for Hg^{2+}. The theoretical investigations would be helpful to build a relationship between the structure and the optical properties of the probes, providing information on the design of efficient two-photon fluorescent sensors that can be used for biological imaging of Hg^{2+} in vivo.

Keywords: fluorescent Hg^{2+} probe; time-dependent density functional theory; two-photon absorption

---

1. Introduction

Mercury is a caustic and carcinogenic element with high cellular toxicity which can pass through biological membranes easily and cause serious damage to the neurological and endocrine systems [1–3], which makes the detection of mercuric ion (Hg^{2+}) of great importance in the fields of biology, chemistry and medicine [4–6]. In the past few years, many analytical methods have been developed to monitor the concentration of Hg^{2+}. Thereinto, the fluorescence microscopy technique has attracted much attention due to its high sensitivity and selectivity, a low cost [7,8]. The design of effective fluorescent probes consequently has become a focus of attention in fluorescence microscopy [9–11].

Until now, several recognition mechanisms have been employed in probe design [4,12,13]. At the very beginning, the intramolecular charge transfer (ICT) mechanism was employed. For instance, Srivastava et al. synthesized a fluorescent probe by bridging a benzhydryl moiety and a dansyl fluorophore through a piperazine unit to detect Hg^{2+} [14]. Razi et al. designed a fluorescence turn-on ratiometic probe for Hg^{2+} by bridging imidazole and benzothiazole moieties through a thiophene ring [15]. Even though much success has been achieved in the development of ICT-based fluorescent probes, the shortage of alternatives is evident. It is widely accepted that the detection using ICT-based fluorescent probes depends highly on the intensity of the single characteristic fluorescent band, which is usually affected by the environment and measurement conditions [16–18]. Other than the ICT mechanism, energy transfer-based mechanisms, including Förster resonance energy transfer (FRET) and through bond energy transfer (TBET) can eliminate the mentioned interferences by using the built-in correction provided by two-emission bands. For a probe based on FRET, the donor moiety and acceptor moiety are linked by a non-conjugated spacer. Consequently, the energy transfers from the donor to the acceptor rely on the spectral overlap between the donor emission and acceptor absorption [19,20]. On the other hand, a probe based on TBET is one whose donor is connected to the acceptor via an electronically conjugated linker. As a result, the energy transfer process occurs through the bond without the need for spectral overlap [21,22].

To date, considerable efforts have been devoted to developing energy transfer-based Hg^{2+} fluorescent probes [23–25]. Very recently, Gong et al. designed a coumarin-rhodamine TBET system (named hereafter as Pro1). They have demonstrated that Pro1 was particularly useful for ratiometric Hg^{2+} sensing and bioimaging applications [26]. Adopting the same donor (coumarin) and acceptor (rhodamine), Wang et al. reported another Hg^{2+} fluorescent probe in which a *m*-phenylenediamine spacer was used as the linker (named hereafter as Pro2) [27]. Although the experimental measurements show that both Pro1 and Pro2 are promising fluorescent probes for Hg^{2+}, the underlying mechanism of the probes is insufficiently understood, in particular the role of the spacer in the sensing performance. Meanwhile, energy transfer rate is a very important parameter for evaluating the efficiency of a probe. However, there is no standard approach to evaluate the energy transfer efficiency in different experiments. Thus, comparison between probes' energy transfer rates on the same theoretical level basis is needed. More importantly, in the experiments, Pro1 and Pro2 have been excited by short wavelength one-photon irradiation, which easily results in photobleaching, photodamage and interference from auto-fluorescence. An efficient method that can overcome the shortcomings of a one-photon fluorescent probe is to utilize a two-photon fluorescent probe [28–31], therefore, the potential of the probes Pro1 and Pro2 as two-photon fluorescent Hg^{2+} sensors should be investigated.

In this paper, theoretical studies on the optical properties, including one-photon absorption (OPA), one-photon emission (OPE) and two-photon absorption (TPA) of Pro1 and Pro2 in the absence and presence of Hg^{2+} were carried out. Special attention has been paid to the analysis of probes' recognition mechanisms by illustrating the molecular orbital distributions involved in the photoabsorption and photoemission processes. Importantly, we report a feasible approach to predict the energy transfer rate of the probes. The present research should be helpful to understand the response mechanisms of these fluorescent chemosensors. Most of all, the role of spacer between the donor and acceptor of the probes is demonstrated, providing guidelines for the design of more efficient two-photon fluorescent probes.

2. Theoretical Method and Computational Details

2.1. Theoretical Method

The detailed theory has been reported in [32]. Here, only the main formulations are briefly discussed. One-photon absorption and emission strength between the states i and j can be described by the oscillator strength:

$$\delta_{OPA(OPE)} = \frac{2\omega_{ij}}{3} \sum_{\alpha}^{x,y,z} |\langle i|\mu_\alpha|j\rangle|^2, \tag{1}$$

where ω_{ij} denotes the energy difference between the states i and j, μ_α is the electric dipole moment operator.

The macroscopic TPA cross-section that can directly compare with the experimental value is defined as:

$$\sigma_{TPA} = \frac{4\pi^2 a_0{}^5 \alpha \omega^2 g(\omega)}{15c\Gamma} \delta_{TPA}, \tag{2}$$

where a_0 is the Bohr radius, α the fine structure constant, c is the speed of the light, and $\hbar\omega$ is the incident photon energy. $g(\omega)$ provides the spectral line profile, and the lifetime broadening of the final state Γ is assumed to be a typical value of 0.1 eV [33]. δ_{TPA} is the microscopic TPA cross-section which is given by the orientational averaging over the two-photon transition probability.

2.2. Computational Detail

The geometrical structures of all the studied molecules were fully optimized using the time-dependent density function theory (TD-DFT)/B3LYP exchange functional level with 6-31G(d,p) basis set. In order to verify the stability of the optimized structures, frequency calculations were performed and no imaginary frequency was obtained. On the basis of the optimized structures, OPA properties of the molecules are calculated using the functional of TD-DFT/BLYP and the 6-31G(d,p) basis set. It should be noted that other functionals, such as B3LYP, M06-2X, PBE0 and wB97X are also used for the numerical simulations. The results of BLYP functional agree however better with the experimental measurements. Thus, the BLYP functional was chosen for the property calculations. Meanwhile, simulations on the OPE properties were carried out by optimizing the first excited state of the compounds. All the above mentioned calculations are performed utilizing the Gaussian09 program (Gaussian Inc., Wallingford, CT, USA) [34]. Apart from the OPA and OPE properties, TPA properties of the studied molecules are also investigated based on the optimized ground state structures using the quadratic response theory implemented in the Dalton2013 package [35]. Considering that the experimental measurements are carried out in an aqueous environment, the effect of solvent is taken into account within the polarizable continuum model (PCM) in all calculations.

3. Result and Discussion

3.1. Molecular Geometry

Figure 1 shows the structures of the studied molecules. Both Pro1 and Pro2 employ rhodamine moiety as the acceptor, and a coumarin moiety with excellent biocompatibility as the donor. In Pro1, the acceptor and donor are connected directly, whereas in Pro2, a rigid *m*-phenylenediamine was selected as the spacer group.

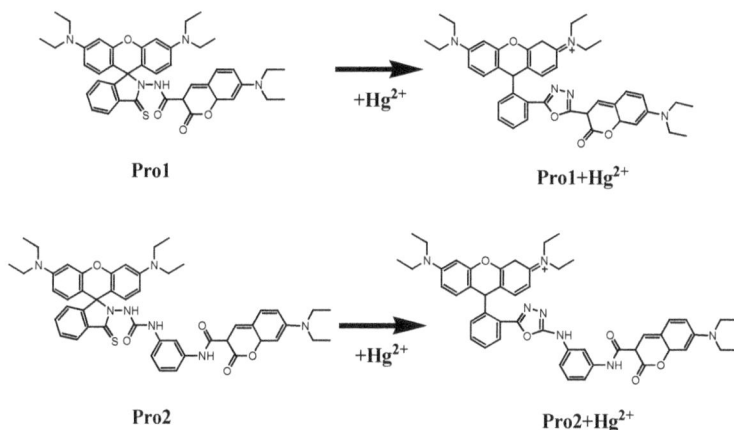

Figure 1. Molecular structures of Pro1, Pro1 + Hg^{2+}, Pro2 and Pro2 + Hg^{2+}.

In the absence of Hg^{2+}, the rhodamine moiety adopts a closed, non-fluorescent spirolactam form, as shown in Pro1 and Pro2. In the presence of Hg^{2+}, a Hg^{2+}-promoted reaction will induce opening of the rhodamine moiety as shown in Pro1 + Hg^{2+} and Pro2 + Hg^{2+}. Optimized ground state molecular geometries are given in Figure 2. In all molecular structures, the donors are planar, and in the acceptors both for closed-ring form and open-ring form, the benzene and the xanthene are vertical. Notably, there is a large torsion between the donor and acceptor of the molecules. This non-coplanar characteristic of the probes prevents the molecule from behaving as a conjugated dye, so that the energy can transfer between the two parts.

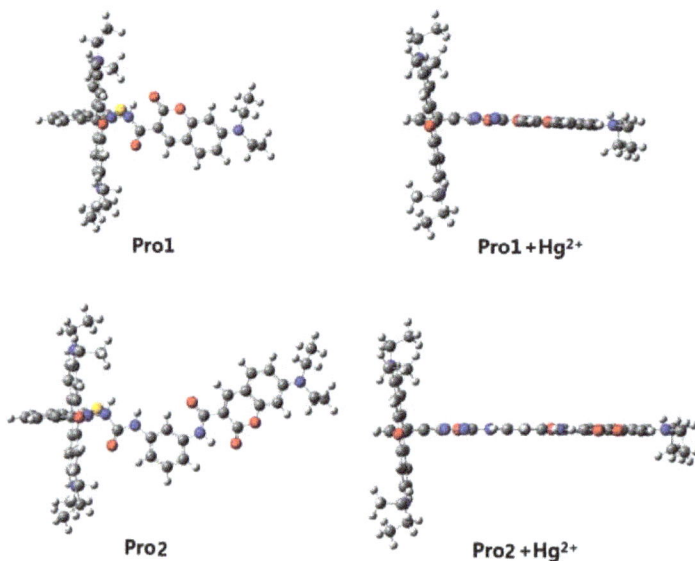

Figure 2. Optimized ground state geometries of Pro1, Pro1 + Hg^{2+}, Pro2 and Pro2 + Hg^{2+} with PCM simulating the dielectric of water.

3.2. One-Photon Absorption

On the basis of the optimized geometries, OPA properties of the studied molecules in H$_2$O were calculated. Figure 3 shows the absorption spectra of Pro1, Pro1 + Hg^{2+}, Pro2 and Pro2 + Hg^{2+}. The details of the OPA peaks, including the one-photon absorption energy, the corresponding wavelength, the oscillator strength and the transition nature are presented in Table 1. It can be seen from Figure 3 and Table 1 that Pro1 and Pro2 have one absorption peak at 425 nm and 434 nm, which agrees well with the experimental data of 420 nm and 440 nm, respectively. When reacts with Hg^{2+}, both Pro1 + Hg^{2+} and Pro2 + Hg^{2+} simultaneously show two absorption peaks, i.e., 486 nm and 520 nm for Pro1 + Hg^{2+}, 447 nm and 521 nm for Pro2 + Hg^{2+}. In comparison with the maximum OPA for the free probes, absorption peaks for the compounds upon combing with Hg^{2+} show redshifts. Predictably, the absorption peak at about 520 nm for Pro1 + Hg^{2+} and Pro2 + Hg^{2+} is dominated by the open-ring rhodamine. The shift of the absorption peak in the short wavelength range is much larger for Pro1 + Hg^{2+} than Pro2 + Hg^{2+}, which is consistent with the trend if the experimental measurements.

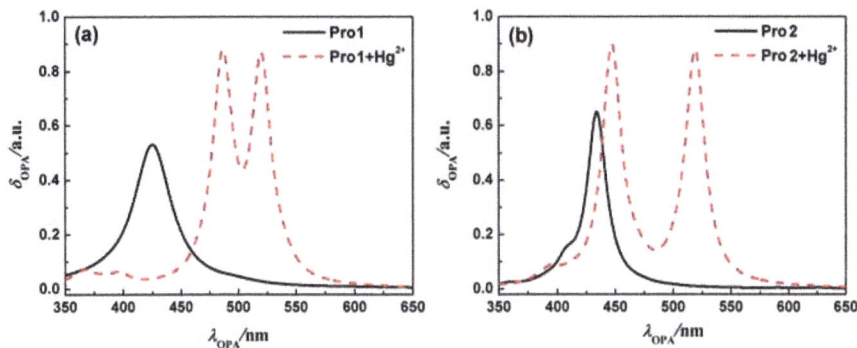

Figure 3. The OPA spectra of (**a**) Pro1 and Pro1 + Hg^{2+}, (**b**) Pro2 and Pro2 + Hg^{2+} with PCM simulating the dielectric of water.

Table 1. The one-photon absorption energy E_{OPA} (eV), the corresponding wavelength λ_{OPA} (nm), the oscillator strength δ_{OPA} (a.u.), and the transition nature of the OPA peaks for Pro1, Pro1 + Hg^{2+}, Pro2 and Pro2 + Hg^{2+} with PCM simulating the dielectric of water.

Molecule	E_{OPA}	λ_{OPA}	λ_{OPA}	Transition Nature	λ_{OPA}^{Exp}
Pro1	2.92	425	0.53	HOMO-2 → LUMO 70%	420 [a]
Pro1 + Hg^{2+}	2.38	520	0.88	HOMO-1 → LUMO 92%	567 [a]
	2.55	486	0.89	HOMO → LUMO + 1 60%, HOMO-3 → LUMO 12%	450 [a]
Pro2	2.86	434	0.64	HOMO-2 → LUMO 98%	440 [b]
Pro2 + Hg^{2+}	2.37	521	0.87	HOMO-2 → LUMO 94%	568 [b]
	2.77	447	0.90	HOMO → LUMO + 1 55%, HOMO-3 → LUMO + 1 30%	440 [b]

[a] Measured in 50:50 (v/v) THF-H_2O [26] and [b] measured in 50:50 (v/v) EtOH-H_2O [27].

To better explain the spectral phenomena, the molecular orbitals contribute to the transitions corresponding to each OPA peak (see Table 1) are shown in Figure 4. It can be observed from Figure 4a that the maximum absorption of Pro1, which originates from the HOMO-2 to the LUMO (HOMO and LUMO represent the highest occupied molecular orbital and the lowest unoccupied molecular orbital, respectively) transition, is distributed on the donor moiety. In the presence of Hg^{2+}, the long wavelength absorption of Pro1 + Hg^{2+} at 520 nm is attributed to the HOMO-1 to LUMO transition and localized on the acceptor part. The short wavelength absorption of Pro1 + Hg^{2+} at 486 nm results from both the HOMO to LUMO + 1 transition and the HOMO-3 to LUMO transition. It should be noted that the HOMO and LUMO + 1 of Pro1 + Hg^{2+} are localized on the donor part, whereas the HOMO-3 and LUMO are localized on the donor and acceptor moiety, respectively. Thus the short wavelength absorption peak for Pro1 + Hg^{2+} at 486 nm relates to the charge transfer process between the donor and the acceptor, which leads to a larger redshift compared with the absorption peak of Pro1. From Figure 4b, it can be seen that similar change trends are shown for the absorption of Pro2 and the long wavelength absorption of Pro2 + Hg^{2+}. However, the short wavelength absorption of Pro2 + Hg^{2+} at 447 nm is contributed by the donor itself, revealing that there is no electronic interaction between the donor and acceptor for Pro2 + Hg^{2+} upon excitation due to the *m*-phenylenediamine spacer. Analyses on the OPA of the molecules suggest that the donor and acceptor of Pro1 + Hg^{2+} and Pro2 + Hg^{2+} can be individually excited at their characteristic absorption peaks, which is conducive to the energy transfer process.

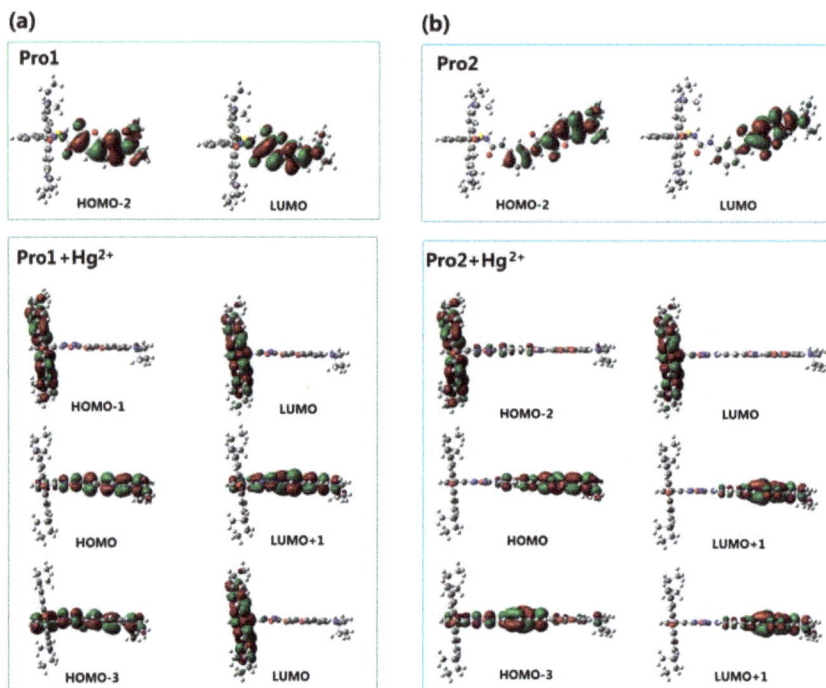

Figure 4. Molecular orbitals involved in the transition of the OPA peaks for (**a**) Pro1 and Pro1 + Hg^{2+}; (**b**) Pro2 and Pro2 + Hg^{2+} with PCM simulating the dielectric of water.

3.3. One-Photon Emission

For a ratiometric fluorescent probe, great changes on the fluorescent signal should be exhibited when it reacts with the target analyte. Calculations on the OPE properties of the studied molecules in H$_2$O are performed on basis of the optimized first excited state geometries. The optimized geometries of Pro1, Pro1 + Hg^{2+}, Pro2 and Pro2 + Hg^{2+} in the first excited state are shown in Figure 5. Compared with the ground state geometries, the first excited state geometries show little change.

At the same time, the fluorescence spectra of the molecules in H$_2$O are presented in Figure 6, and the detailed OPE energy, the corresponding wavelength, the fluorescent intensity and the transition nature are listed in Table 2. Values in Table 2 show that upon the addition of Hg^{2+}, the fluorescent peaks of Pro1 and Pro2 locating at 485 nm and 503 nm are greatly redshifted to 605 nm and 608 nm, showing 120 nm and 105 nm shifts, respectively. The calculated results agree well with the experimental data in trend, despite the quantitative discrepancy which can be attributed to the vibrational contribution, and the short-range interaction that have not been considered in the calculations, etc. By analyzing the OPE properties of the molecules, one can predict that these probes can act as excellent ratiometric chemosensors for Hg^{2+}, and Pro1 is preferable due to its more apparent changes in the fluorescent wavelength. Considering the difference of the molecular structure, the influence of the connection between the donor and acceptor on the molecular fluorescence is revealed.

In order to understand the fluorescence spectra phenomenon more clearly, the frontier molecular orbitals related to the main transitions in the emission process of the molecules are drawn and illustrated in Figure 7. It can be seen that the transitions of the emission for Pro1 and Pro2 are contributed by the orbitals localized on the donor moiety, whereas, the emission of Pro1 + Hg^{2+} and Pro2 + Hg^{2+} are dominated by the orbitals distributing on the acceptor part, resulting in the appearance of a characteristic emission of the acceptor.

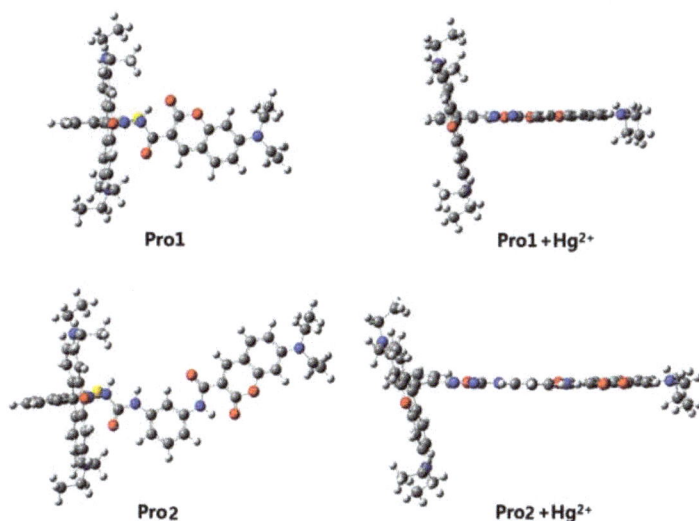

Figure 5. Optimized first excited state geometries of Pro1, Pro1 + Hg^{2+}, Pro2 and Pro2 + Hg^{2+} with PCM simulating the dielectric of water.

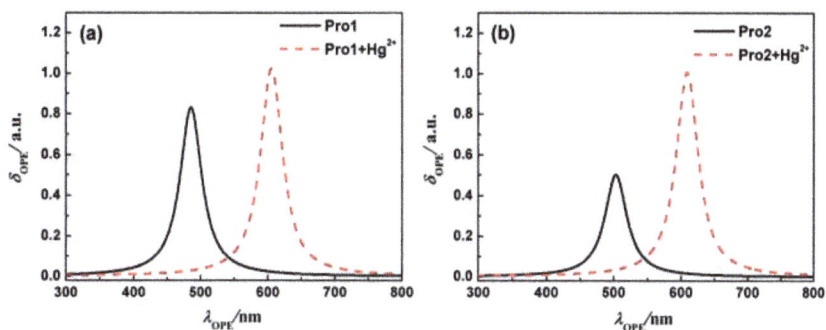

Figure 6. The OPE spectra of (**a**) Pro1 and Pro1 + Hg^{2+}, (**b**) Pro2 and Pro2 + Hg^{2+} with PCM simulating the dielectric of water.

Table 2. The emission energy E_{OPE} (eV), the corresponding wavelength λ_{OPE} (nm), the fluorescent intensity δ_{OPE} (a.u.) and the transition nature of the OPE peaks for Pro1, Pro1 + Hg^{2+}, Pro2 and Pro2 + Hg^{2+} with PCM simulating the dielectric of water.

Molecule	E_{OPE}	λ_{OPE}	δ_{OPE}	Transition Nature	λ_{OPA}^{Exp}
Pro1	2.55	485	0.83	HOMO-3 → LUMO 85%	470 [a]
Pro1 + Hg^{2+}	2.05	605	1.02	HOMO-1 → LUMO 98%	580 [a]
Pro2	2.46	503	0.50	HOMO-2 → LUMO 92%	478 [b]
Pro2 + Hg^{2+}	2.04	608	1.01	HOMO-2 → LUMO 94%	587 [b]

[a] Measured in 50:50 (*v*/*v*) THF-H$_2$O [26] and [b] measured in 50:50 (*v*/*v*) EtOH-H$_2$O [27].

For the probes before and after reacting with Hg^{2+}, the frontier molecular orbitals involved in the fluorescent emission process are entirely localized on the donor or acceptor moiety. This makes the generation of separate characteristic emission peaks possible and is in favor of the recognition of Hg^{2+}.

(a)

(b)

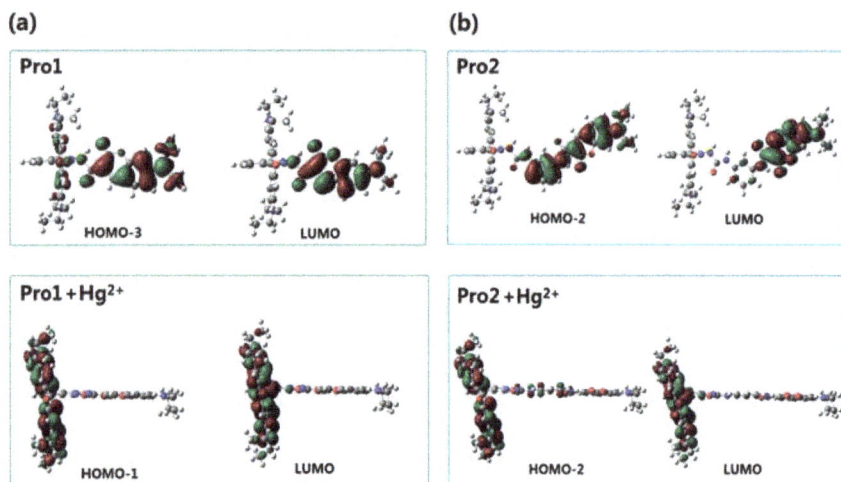

Figure 7. Molecular orbitals involved in the transition of the OPE peaks for (**a**) Pro1 and Pro1 + Hg^{2+}; (**b**) Pro2 and Pro2 + Hg^{2+} with PCM simulating the dielectric of water.

3.4. Recognition Mechanism and Energy Transfer Rate

As reported by Gong and Wang, the recognition of Pro1 and Pro2 for Hg^{2+} pertains to the TBET and FRET mechanism, respectively. However, theoretical analyses on this issue have not been carried out yet. Thus, to further rationalize the recognition mechanisms of Pro1 and Pro2, the orbital distribution diagrams included in the absorptive and emissive processes of the molecules are used to investigate the responsive process of a molecule upon optical excitation. Careful inspection of the absorption and emission of Pro1 + Hg^{2+} (see Figures 4a and 7a) reveals that when excited by the incident light with the wavelength of 486 nm, Pro1 + Hg^{2+} can be excited to a higher excited state, which corresponds to an electronic transition mainly localized on the donor part, whereas, the emission of the molecule originating from the decay from the first excited state to the ground state is distributed on the acceptor moiety with a fluorescence wavelength of 605 nm. To summarize, the fluorescence of the molecular acceptor can be stimulated by exciting the donor moiety, herein a non-radiative energy transfer process occurs between the molecular donor and acceptor. Similarly, the energy transfer process occurs for Pro2 + Hg^{2+} as shown in Figures 4b and 7b. Namely, the emission from the acceptor moiety of Pro2 + Hg^{2+} could be obtained by exciting the donor unit. Further taking the connection and the spectral overlap between the donor and acceptor into consideration, the recognition mechanisms of Pro1 and Pro2 are confirmed as TBET and FRET, respectively.

For an energy transfer-based system, the energy transfer rate between the donor and acceptor is an important parameter to evaluate the efficiency of the probe. According to the Fermi's golden rule, the energy transfer rate from the donor to the acceptor, which represents the probability of the energy transfer per unit time, can be described by [36]:

$$K_{DA} = \sum_{i,j} \frac{2\pi}{\hbar} \int \left| W_{D_i A_j} \right|^2 \delta_{D_i}^2 \delta_{A_j}^2 \frac{\Gamma^2}{\pi^2 \left[(\varepsilon - \varepsilon_{D_i})^2 + \Gamma^2 \right] \left[(\varepsilon - \varepsilon_{A_j})^2 + \Gamma^2 \right]} g(\varepsilon) d\varepsilon, \tag{3}$$

where $\varepsilon_{D_i}(\varepsilon_{A_j})$ is the transition energy of the donor(acceptor), $\delta_{D_i}(\delta_{A_j})$ is the corresponding strength. $g(\varepsilon)$ represents the energy distribution of the incident laser pulse which is assumed to be a Gaussian function. W_{DiAj} denotes the electronic dipole-dipole interaction matrix element. Under the dipole approximation, W_{DiAj} can be defined as:

$$W_{D_iA_j} = \frac{\mu_m \cdot \mu_n}{|R_{mn}|^3} - 3\frac{(\mu_m \cdot R_{mn})(\mu_n \cdot R_{mn})}{|R_{mn}|^5}, \tag{4}$$

here $\mu_m(\mu_n)$ is the transition dipole moment of the molecular fragment $m(n)$, R_{mn} is the distance vector from m to n.

Ideally, the transition dipole moments of donor emission and the acceptor absorption should have large vectors in the same direction, which is beneficial to the long-range dipole-dipole interaction. Here as shown in Figure 8, the X-axis is set to be perpendicular to the plane of the xanthene in rhodamine, the Y-axis on the long axis direction of the xanthene, and the Z-axis on the short axis of the xanthene. After fixing the coordinate direction, the transition wavelength, the corresponding strength and dipole moments for both the emission of the donors and the absorption of the acceptor (open-ring rhodamine) are calculated and the results are listed in Table 3. One can see that the emission wavelength of the donor for Pro2 is much closer to the absorption wavelength of the acceptor part than that for Pro1. Thus, spectral overlap between the donor emission and the acceptor absorption for Pro2 + Hg^{2+} is large while that of Pro1 + Hg^{2+} is small, proving the recognition mechanisms of Pro2 + Hg^{2+} to be FRET. In addition, there are large transition dipole moment vectors in the Z direction for the donors of Pro1 and Pro2, whereas the maximum transition dipole moment of the acceptor part is on the Y direction.

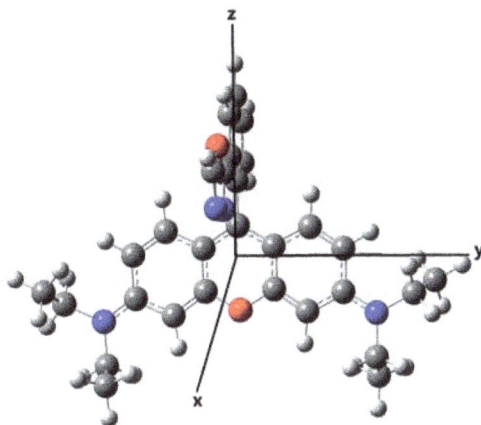

Figure 8. Schematic of coordinate direction.

Table 3. The transition wavelength λ (nm), transition strength δ (a.u.) and the corresponding transition electric dipole moments $\mu_{x,y,z}$ (a.u.) of donor emission and acceptor absorption for Pro1 + Hg^{2+} and Pro2 + Hg^{2+} with PCM simulating the dielectric of water.

Molecule	λ	δ	Transition Electric Dipole Moment		
			μ_x	μ_y	μ_z
Pro1-Donor-emission	396	0.69	−0.18	−0.20	−2.98
Pro2-Donor-emission	441	0.98	−0.06	−0.19	3.78
Acceptor-absorption	474	1.00	−0.14	−3.94	−0.24

According to Equation (4), the energy transfer rate from the donor to the acceptor depends crucially on the distance between the two parts. Table 4 lists the distance vectors between the donor and the acceptor for Pro1 + Hg^{2+} and Pro2 + Hg^{2+}. Obviously, the *m*-phenylenediamine spacer in Pro2

+ Hg^{2+} leads to a long distance between the molecular donor and acceptor. As a result, the energy transfer rate of Pro1 + Hg^{2+} is six times larger than that of Pro2 + Hg^{2+}, suggesting Pro1 + Hg^{2+} to be a more promising energy transfer-based probe. It can be concluded from these results that the spacer group between the donor and acceptor plays an important role on the efficiency of the energy transfer-based fluorescent probes.

Table 4. The distance vector $R_{x,y,z}$ (Å), total distance R_{DA} (Å) and energy transfer rate K_{DA} (10^6) between the donor and acceptor of Pro1 + Hg^{2+} and Pro2 + Hg^{2+} with PCM simulating the dielectric of water.

Molecule	R_x	R_y	R_z	R_{DA}	K_{DA}
Pro1 + Hg^{2+}	8.28	0.55	2.07	8.55	3.7
Pro2 + Hg^{2+}	15.74	0.31	1.98	15.87	0.6

3.5. Two-Photon Absorption

The analyses on the one-photon absorption and emission of the molecules demonstrate that Pro1 and Pro2 can efficiently identify Hg^{2+} through the energy transfer process. This inspired us to further investigate the utility of the probes as two-photon fluorescent chemosensors. The TPA properties of the molecules, including the two-photon absorption energy, the corresponding TPA wavelength and the TPA cross-section are listed in Table 5. In the case of free probes, both Pro1 and Pro2 have one TPA peak at about 750 nm. The maximum TPA cross section of Pro2 is 573.12 GM, which is six times larger than that of Pro1. One can expect the different TPA performances of Pro1 and Pro2 are due to their different spacers. Herein, we analyze the TPA cross-sections of the two probes by a two-level model with the ground state and the maximum charge transfer state [37] according to the transition natures of the probes as shown in Table 1. In this model, the peak TPA cross-section is directly proportional to the square of the transition dipole moment μ_{01}, while inversely proportional to the square of the energy gap E_{01}, namely, $\sigma_{TPA} \propto \mu_{01}{}^2/E_{01}{}^2$. The numerical results are given in Table 6. From Table 6, one can see that Pro1 has a smaller transition dipole moment and larger excitation energy compared with Pro2, thus the TPA cross section of Pro1 is much smaller than that of Pro2.

Table 5. The two-photon absorption energy E_{TPA} (eV), the corresponding TPA wavelength λ_{TPA} (nm) and the TPA cross section σ_{TPA} (GM = 10^{-5} cm^4·s/photon) of the lowest nine excited states for Pro1, Pro1 + Hg^{2+}, Pro2 and Pro2 + Hg^{2+} with PCM simulating the dielectric of water.

Molecule	E_{TPA}	λ_{TPA}	σ_{TPA}	Molecule	E_{TPA}	λ_{TPA}	σ_{TPA}
	2.72	909	3.14		2.38	1040	21.35
	2.87	861	3.45		2.55	972	24.34
	2.92	850	2.47		2.87	861	135.22
	2.98	829	1.99		3.08	802	264.19
Pro1	3.18	777	94.85	Pro1 + Hg^{2+}	3.19	775	11.26
	3.33	742	0.09		3.24	763	33.23
	3.42	723	2.24		3.6	686	1059.1
	3.56	694	11.7		3.67	673	233.94
	3.63	681	0.42		3.79	652	382.14
	2.86	868	0.03		2.37	1042	1.83
	2.88	858	4.71		2.45	1009	0.47
	2.92	846	1.87		2.53	977	23.45
	3	824	0.41		2.77	894	0.36
Pro2	3.06	808	8.44	Pro2 + Hg^{2+}	3.09	800	263.31
	3.18	777	1.17		3.28	754	275.96
	3.34	740	573.12		3.49	708	4.71
	3.5	706	0.02		3.54	698	40.68
	3.55	696	1.55		3.58	690	85.94

Table 6. The transition dipole moment μ (Debye) and excitation energy E_{01} (eV) between the states in two-level model for Pro1 and Pro2.

Molecule	μ_x	μ_y	μ_z	μ_{tot}	E_{01}
Pro1	0.50	0.03	2.66	2.71	2.92
Pro2	0.51	0.83	2.85	3.02	2.86

With the addition of Hg^{2+}, it can be seen that the TPA peak positions of Pro1 + Hg^{2+} and Pro2 + Hg^{2+} are blue shifted and red shifted, respectively. Importantly, the maximum TPA cross sections of Pro1 + Hg^{2+} is largely increased to 1059.10 GM, whereas that of Pro2 + Hg^{2+} is decreased to 275.96 GM. This demonstrates that the spacer group between the donor and acceptor of the probe induces a variation in its TPA property. Considering the molecules have no symmetry, the transition probability between two molecular orbitals is closely related to the overlap of these molecular orbitals [38]. Thus, molecular mainly orbitals involved in the TPA process of the studied molecules are drawn in Figure 9. In comparison with Pro1, the orbitals of Pro1 + Hg^{2+} contributing to the TPA process locate at the same moiety and possess similar distribution. Nevertheless for Pro2 + Hg^{2+}, the overlapped parts take an opposite distribution, namely, color distribution of the overlapped part is opposite, resulting in destructive interference of the corresponding wavefunctions. As a consequence, the TPA cross-section of Pro1 + Hg^{2+} is enhanced and that of Pro2 + Hg^{2+} is reduced compared with Pro1 and Pro2, respectively.

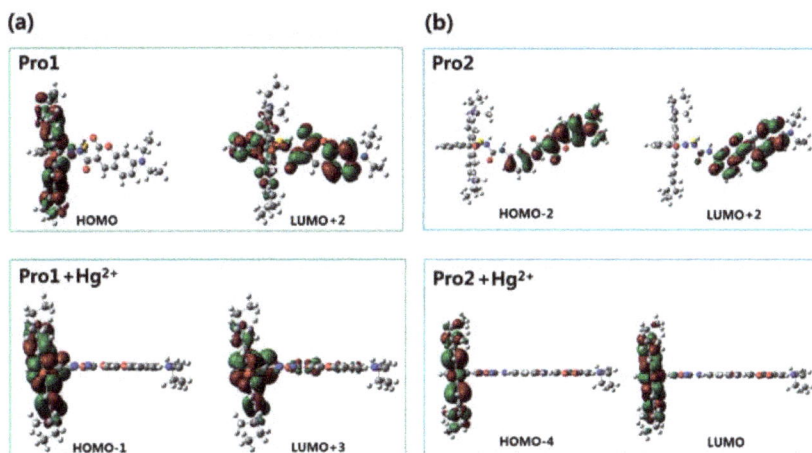

Figure 9. Molecular orbitals involved in the transition of the TPA peaks for (**a**) Pro1 and Pro1 + Hg^{2+}, (**b**) Pro2 and Pro2 + Hg^{2+} with PCM simulating the dielectric of water.

Undoubtedly, an excellent two-photon fluorescent probe should have a significant enhancement in the TPA cross-section when identifying the target. Thus, Pro1 is predicted to be a better two-photon fluorescent probe. For Pro1, using a 777 nm laser as the two-photon excitation light, the probe is excited and the fluorescence emitted from the donor can be observed because no energy transfer process occurs. When Hg^{2+} is present and excited by two-photon irradiation, a TBET process occurs in Pro1 + Hg^{2+}, and the fluorescence of the acceptor can be observed. In a word, Pro1 can be used as an active TBET-based two-photon fluorescent Hg^{2+} probe for live cell imaging.

4. Conclusions

In this work, the photoabsorption and photoemission properties, recognition mechanisms and energy transfer rates of two newly-synthesized fluorescent Hg^{2+} chemosensors were studied by theoretical calculations. The results demonstrate that obvious changes in the absorption and fluorescence signal of the probes are observed upon reaction with Hg^{2+}, which is conductive to the recognition of this cation. The analyses on the energy transfer rate illustrate that Pro1 has a higher efficiency, therefore, Pro1 is proved to be a more effective energy transfer-based ratiometric fluorescent probe for detecting Hg^{2+}. Then the effect of spacer group between the donor and acceptor on the sensing performance and efficiency of the probe is discussed. In particular, the probes show a significant two-photon response in the near-infrared light region, and the largest TPA cross-section of Pro1 is greatly enhanced with the presence of Hg^{2+}. As a result, it is deduced that Pro1 can act as a potential two-photon excited TBET-based ratiometric fluorescent Hg^{2+} probe. The theoretical investigations have explained the experimental results and revealed the underlying response mechanism of the probes. Meanwhile, a new strategy to predict the energy transfer rate of the energy transfer-based chemosensor on basis of the molecular structure is proposed. The results are intended to give the structure-property relationships for these probes, providing useful knowledge for designing more efficient two-photon fluorescent sensors geared toward biological applications.

Acknowledgments: This work was supported by the National Natural Science Foundation of China (Grant Nos. 11247307, 11304172), Shandong Province Higher Educational Science and Technology Program (Grant No. J12LJ04).

Author Contributions: Jiancai Leng suggested the project; Yujin Zhang performed the theoretical simulations; Jiancai Leng and Yujin Zhang analyzed the data; Yujin Zhang wrote the paper.

Conflicts of Interest: The authors declare no conflict of interest.

References

1. Tsukamoto, K.; Shinohara, Y.; Iwasaki, S.; Maeda, H. A coumarin-based fluorescent probe for Hg^{2+} and Ag^+ with an N'-acetylthioureido group as a fluorescence switch. *Chem. Commun.* **2011**, *47*, 5073–5075. [CrossRef] [PubMed]

2. Ma, Q.J.; Zhang, X.B.; Zhao, X.H.; Jin, Z.; Mao, G.J.; Shen, G.L.; Yu, R.Q. A highly selective fluorescent probe for Hg^{2+} based on a rhodamine-coumarin conjugate. *Anal. Chim. Acta* **2010**, *663*, 85–90. [CrossRef] [PubMed]

3. Xuan, W.; Chen, C.; Cao, Y.; He, W.; Jiang, W.; Liu, K.; Wang, W. Rational design of a ratiometric fluorescent probe with a large emission shift for the facile detection of Hg^{2+}. *Chem. Commun.* **2012**, *48*, 7292–7294. [CrossRef] [PubMed]

4. Sahana, S.; Mishra, G.; Sivakumar, S.; Bharadwaj, P.K. A 2-(2′-hydroxyphenyl)benzothiazole (HTB)—Quinoline conjugate: A highly specific fluorescent probe for Hg^{2+} based on ESIPT and its application in bioimaging. *Dalton Trans.* **2015**, *44*, 20139–20146. [CrossRef] [PubMed]

5. Lin, W.Y.; Cao, X.W.; Ding, Y.D.; Yuan, L.; Long, L.L. A highly selective and sensitive fluorescent probe for Hg^{2+} imaging in live cells based on a rhodamine-thioamide-alkyne scaffold. *Chem. Commun.* **2010**, *46*, 3529–3531. [CrossRef] [PubMed]

6. Karakus, E.; Ucuncu, M.; Emrullahoglu, M. A rhodamine/BODIPY-based fluorescent probe for the differential detection of Hg(II) and Au(III). *Chem. Commun.* **2014**, *50*, 1119–1121. [CrossRef] [PubMed]

7. Kong, F.P.; Ge, L.H.; Pan, X.H.; Xu, K.H.; Liu, X.J.; Tang, B. A highly selective near-infrared fluorescent probe for imaging H_2Se in living cells and in vivo. *Chem. Sci.* **2016**, *7*, 1051–1056. [CrossRef]

8. Xu, K.H.; Qiang, M.M.; Gao, W.; Su, R.X.; Li, N.; Gao, Y.; Xie, Y.X.; Kong, F.P.; Tang, B. A near-infrared reversible fluorescent probe for real-time imaging of redox status changes in vivo. *Chem. Sci.* **2013**, *4*, 1079–1086. [CrossRef]

9. Guo, Z.Q.; Park, S.; Yoon, J.; Shin, I. Recent progress in the development of near-infrared fluorescent probes for bioimaging applications. *Chem. Soc. Rev.* **2014**, *43*, 16–29. [CrossRef] [PubMed]

10. Wang, X.; Sun, J.; Zhang, W.H.; Ma, X.X.; Lv, J.Z.; Tang, B. A near-infrared ratiometric fluorescent probe for rapid and highly sensitive imaging of endogenous hydrogen sulfide in living cells. *Chem. Sci.* **2013**, *4*, 2551–2556. [CrossRef]

11. Anand, T.; Sivaraman, G.; Mahesh, A.; Chellappa, D. Aminoquinoline based highly sensitive fluorescent sensor for lead(II) and aluminum(III) and its application in live cell imaging. *Anal. Chim. Acta* **2015**, *853*, 596–601. [CrossRef] [PubMed]

12. Lee, M.H.; Kim, J.S.; Sessler, J.L. Small molecule-based ratiometric fluorescence probes for cations, anions, and biomolecules. *Chem. Soc. Rev.* **2015**, *44*, 4185–4191. [CrossRef] [PubMed]

13. Yang, W.J.; Zhang, Y.J.; Wang, C.K. Optical properties and responsive mechanism of 4-amino-1,8-naphthalimide-based two-photon fluorescent probe for sensing hydrogen sulfide. *Acta Phys. Chim. Sin.* **2015**, *31*, 2303–2309.

14. Srivastava, P.; Ali, R.; Razi, S.S.; Shahid, M.; Patnaik, S.; Misra, A. A simple blue fluorescent probe to detect Hg^{2+} in semiaqueous environment by intramolecular charge transfer mechanism. *Tetrahedron Lett.* **2013**, *54*, 3688–3693. [CrossRef]

15. Razi, S.S.; Ali, R.; Gupta, R.C.; Dwivedi, S.K.; Sharma, G.; Koch, B.; Misra, A. Phenyl-end-capped-thiophene (P-T type) based ict fluorescent probe (D-π-A) for detection of Hg^{2+} and Cu^{2+} ions: Live cell imaging and logic operation at molecular level. *J. Photochem. Photobiol. A* **2016**, *324*, 106–116. [CrossRef]

16. Xu, Z.C.; Baek, K.H.; Kim, H.N.; Cui, J.N.; Qian, X.H.; Spring, D.R.; Shin, I.; Yoon, J. Zn^{2+}-triggered amide tautomerization produces a highly Zn^{2+}-selective, cell-permeable, and ratiometric fluorescent sensor. *J. Am. Chem. Soc.* **2010**, *132*, 601–610. [CrossRef] [PubMed]

17. Han, Z.X.; Zhang, X.B.; Li, Z.; Gong, Y.J.; Wu, X.Y.; Jin, Z.; He, C.M.; Jian, L.X.; Zhang, J.; Shen, G.L.; et al. Efficient fluorescence resonance energy transfer-based ratiometric fluorescent cellular imaging probe for Zn^{2+} using a rhodamine spirolactam as a trigger. *Anal. Chem.* **2010**, *82*, 3108–3113. [CrossRef] [PubMed]

18. Li, C.Y.; Zhang, X.B.; Qiao, L.; Zhao, Y.; He, C.M.; Huan, S.Y.; Lu, L.M.; Jian, L.X.; Shen, G.L.; Yu, R.Q. Naphthalimide-porphyrin hybrid based ratiometric bioimaging probe for Hg^{2+}: Well-resolved emission spectra and unique specificity. *Anal. Chem.* **2009**, *81*, 9993–10001. [CrossRef] [PubMed]

19. Yuan, L.; Lin, W.Y.; Zheng, K.B.; Zhu, S.S. FRET-based small-molecule fluorescent probes: Rational design and bioimaging applications. *Accounts Chem. Res.* **2013**, *46*, 1462–1473. [CrossRef] [PubMed]

20. Shao, J.Y.; Sun, H.Y.; Guo, H.M.; Ji, S.M.; Zhao, J.Z.; Wu, W.T.; Yuan, X.L.; Zhang, C.L.; James, T.D. A highly selective red-emitting FRET fluorescent molecular probe derived from BODIPY for the detection of cysteine and homocysteine: An experimental and theoretical study. *Chem. Sci.* **2012**, *3*, 1049–1061. [CrossRef]

21. Fan, J.L.; Hu, M.M.; Zhan, P.; Peng, X.J. Energy transfer cassettes based on organic fluorophores: Construction and applications in ratiometric sensing. *Chem. Soc. Rev.* **2013**, *42*, 29–43. [CrossRef] [PubMed]

22. Kumar, M.; Kumar, N.; Bhalla, V.; Singh, H.; Sharma, P.R.; Kaur, T. Naphthalimide appended rhodamine derivative: Through bond energy transfer for sensing of Hg^{2+} ions. *Org. Lett.* **2011**, *13*, 1422–1425. [CrossRef] [PubMed]

23. Liu, Y.L.; Lv, X.; Zhao, Y.; Chen, M.L.; Liu, J.; Wang, P.; Guo, W. A naphthalimide-rhodamine ratiometric fluorescent probe for Hg^{2+} based on fluorescence resonance energy transfer. *Dyes Pigments* **2012**, *92*, 909–915. [CrossRef]

24. Suresh, M.; Mishra, S.; Mishra, S.K.; Suresh, E.; Mandal, A.K.; Shrivastav, A.; Das, A. Resonance energy transfer approach and a new ratiometric probe for Hg^{2+} in aqueous media and living organism. *Org. Lett.* **2009**, *11*, 2740–2743. [CrossRef] [PubMed]

25. Zhang, X.L.; Xiao, Y.; Qian, X.H. A ratiometric fluorescent probe based on fret for imaging Hg^{2+} ions in living cells. *Angew. Chem. Int. Ed.* **2008**, *47*, 8025–8029. [CrossRef] [PubMed]

26. Gong, Y.J.; Zhang, X.B.; Zhang, C.C.; Luo, A.L.; Fu, T.; Tan, W.H.; Shen, G.L.; Yu, R.Q. Through bond energy transfer: A convenient and universal strategy toward efficient ratiometric fluorescent probe for bioimaging applications. *Anal. Chem.* **2012**, *84*, 10777–10784. [CrossRef] [PubMed]

27. Wang, M.; Wen, J.; Qin, Z.H.; Wang, H.M. A new coumarin-rhodamine FRET system as an efficient ratiometric fluorescent probe for Hg^{2+} in aqueous solution and in living cells. *Dyes Pigments* **2015**, *120*, 208–212. [CrossRef]

28. Mao, G.J.; Wei, T.T.; Wang, X.X.; Huan, S.Y.; Lu, D.Q.; Zhang, J.; Zhang, X.B.; Tan, W.H.; Shen, G.L.; Yu, R.Q. High-sensitivity naphthalene-based two-photon fluorescent probe suitable for direct bioimaging of H_2S in living cells. *Anal. Chem.* **2013**, *85*, 7875–7881. [CrossRef] [PubMed]

29. Zhou, L.Y.; Zhang, X.B.; Wang, Q.Q.; Lv, Y.F.; Mao, G.J.; Luo, A.; Wu, Y.X.; Wu, Y.; Zhang, J.; Tan, W.H. Molecular engineering of a TBET-based two-photon fluorescent probe for ratiometric imaging of living cells and tissues. *J. Am. Chem. Soc.* **2014**, *136*, 9838–9841. [CrossRef] [PubMed]

30. Sarkar, A.R.; Kang, D.E.; Kim, H.M.; Cho, B.R. Two-photon fluorescent probes for metal ions in live tissues. *Inorg. Chem.* **2014**, *53*, 1794–1803. [CrossRef] [PubMed]

31. Yu, H.B.; Xiao, Y.; Jin, L.J. A lysosome-targetable and two-photon fluorescent probe for monitoring endogenous and exogenous nitric oxide in living cells. *J. Am. Chem. Soc.* **2012**, *134*, 17486–17489. [CrossRef] [PubMed]

32. Zhang, Y.J.; Zhang, Q.Y.; Ding, H.J.; Song, X.N.; Wang, C.K. Responsive mechanism of 2-(2'-hydroxyphenyl)benzoxazole-based two-photon fluorescent probes for zinc and hydroxide ions. *Chin. Phys. B* **2015**, *24*, 023301. [CrossRef]

33. Jia, H.H.; Zhao, K.; Wu, X.L. Effects of torsional disorder and position isomerism on two-photonabsorption properties of polar chromophore dimers. *Chem. Phys. Lett.* **2014**, *612*, 151–156. [CrossRef]

34. Gaussian09. Available online: http://www.gaussian.com/ (accessed on 22 December 2015).

35. Dalton2013. Available online: http://www.kjemi.uio.no/software/dalton/ (accessed on 10 November 2013).

36. Beljonne, D.; Curutchet, C.; Scholes, G.D.; Silbey, R.J. Beyond Förster resonance energy transfer in biological and nanoscale systems. *J. Phys. Chem. B* **2009**, *113*, 6583–6599. [CrossRef] [PubMed]

37. Xu, Z.; Ren, A.M.; Guo, J.F.; Liu, X.T.; Huang, S.; Feng, J.K. A theoretical investigation of two typical two-photon ph fluorescent probes. *Photochem. Photobiol.* **2013**, *89*, 300–309. [CrossRef] [PubMed]

38. Ji, S.M.; Yang, J.; Yang, Q.; Liu, S.S.; Chen, M.D.; Zhao, J.Z. Tuning the intramolecular charge transfer of alkynylpyrenes: Effect on photophysical properties and its application in design of off-on fluorescent thiol probes. *J. Org. Chem.* **2009**, *74*, 4855–4865. [CrossRef] [PubMed]

sensors

MDPI

Article

Towards the Development of a Low-Cost Device for the Detection of Explosives Vapors by Fluorescence Quenching of Conjugated Polymers in Solid Matrices

Liliana M. Martelo [1,2,*], Tiago F. Pimentel das Neves [3], João Figueiredo [3], Lino Marques [3], Alexander Fedorov [2], Ana Charas [4], Mário N. Berberan-Santos [2] and Hugh D. Burrows [1]

[1] Department of Chemistry, University of Coimbra, 3004-535 Coimbra, Portugal; burrows@ci.uc.pt
[2] Centro de Química-Física Molecular (CQFM) and the Institute of Nanoscience and Nanotechnology (IN), Instituto Superior Técnico, University of Lisbon, 1049-001 Lisbon, Portugal; berberan@tecnico.ulisboa.pt
[3] Institute of Systems and Robotics (ISR), University of Coimbra, 3030-290 Coimbra, Portugal; tiago.pimenteldasneves@epfl.ch (T.F.P.d.N.); jfigueiredo@assystem.com (J.F.); lino@isr.uc.pt (L.M.); aleksander@mail.ist.utl.pt (A.F.)
[4] Instituto de Telecomunicações, Instituto Superior Técnico, Av. Rovisco Pais, 1049-001 Lisbon, Portugal; ana.charas@lx.it.pt
*   Correspondence: liliana.martelo@tecnico.ulisboa.pt; Tel.: +351-218-419-259

Received: 28 September 2017; Accepted: 30 October 2017; Published: 3 November 2017

Abstract: Conjugated polymers (CPs) have proved to be promising chemosensory materials for detecting nitroaromatic explosives vapors, as they quickly convert a chemical interaction into an easily-measured high-sensitivity optical output. The nitroaromatic analytes are strongly electron-deficient, whereas the conjugated polymer sensing materials are electron-rich. As a result, the photoexcitation of the CP is followed by electron transfer to the nitroaromatic analyte, resulting in a quenching of the light-emission from the conjugated polymer. The best CP in our studies was found to be poly[(9,9-dioctylfluorenyl-2,7-diyl)-co-bithiophene] (F8T2). It is photostable, has a good absorption between 400 and 450 nm, and a strong and structured fluorescence around 550 nm. Our studies indicate up to 96% quenching of light-emission, accompanied by a marked decrease in the fluorescence lifetime, upon exposure of the films of F8T2 in ethyl cellulose to nitrobenzene (NB) and 1,3-dinitrobenzene (DNB) vapors at room temperature. The effects of the polymeric matrix, plasticizer, and temperature have been studied, and the morphology of films determined by scanning electron microscopy (SEM) and confocal fluorescence microscopy. We have used ink jet printing to produce sensor films containing both sensor element and a fluorescence reference. In addition, a high dynamic range, intensity-based fluorometer, using a laser diode and a filtered photodiode was developed for use with this system.

Keywords: conjugated polymers; explosives detection; trace analysis; optical sensor; luminescence sensor

1. Introduction

Part of the extensive research in conjugated polymers (CPs) and conjugated polyelectrolytes (CPEs) is motivated by their capacity as sensitive fluorescent materials for chemo- and biosensing. They offer a broad range of possibilities for transforming analyte receptor interactions, as well as nonspecific interactions, into observable (transducible) responses [1,2]. Amplified quenching in fluorescent CP was introduced by Swager and Zhou [3] and opened the way for novel sensory materials using this important class of conjugated polymers. In 1998, Yang and co-workers [4] used a fluorescence quenching transduction mechanism together with the amplifying nature of conjugated polymers to develop a material highly sensitive to 2,4,6-trinitrotoluene (TNT) vapors, the major explosive component of landmines. One peculiarity of nitroaromatics which may be used in detection based on fluorescence techniques is their electron-accepting capability. CPs are promising for redox sensing because they are normally electron donors. This donor behavior is further enhanced in their delocalized pi excited states. This excited state delocalization is crucial because the resulting exciton migration along the polymer chain increases the frequency of interaction with a bound quencher, in this case the nitroaromatic analytes, which contributes to improve detection sensitivity. For these reasons, nitroaromatic analytes can efficiently quench the emission of CP by photoinduced electron transfer process. As a practical result, photoexcitation of the conjugated polymer is followed by electron transfer to the nitrated organic compounds, resulting in a quenching of the CP fluorescence. Fluorescence quenching sensing methods are promising for rapid and sensitive detection of explosives vapors, and possess major advantages, including high sensitivity signal output and operational simplicity [5].

The detection of explosives is a major quest for security in many civilian and military environments, and is usually carried out through the sensing of the vapor emitted by the explosives, or of markers present with them. These sensors must satisfy several criteria, such as sensitivity, reversibility and the capability for real-time signal processing. For nitroaromatic explosives, sensing of a few parts per billion or less of the analyte vapor is mandatory, and should be accompanied with rapid and, ideally, reversible changes in the sensor output. Some assessments of explosives containing soils have been performed, and it has been indicated that the concentration of TNT is around 10–100 ng/kg. The vapor concentration is even lower, around the 100 pg/kg to 100 fg/kg level [6]. For in-field detection of such materials, a portable system would be highly beneficial.

In order to address these issues, we have developed a new conjugated polymer-based optical sensor of trace explosives vapors. For the chemosensory material, we have used hairy-rod polymers [7], an important class of π-conjugated polymers, such as poly(fluorene-2,7-diyl)s (PFs). These have excellent photoluminescence quantum yields, good thermal stability, and good solubility in several solvents [8]. The linear side chain poly[9,9-dioctylfluorene-2,7-diyl] (PFO) and its homologue poly[9,9-dioctylfluorenyl-2,7-diyl)-co-bithiophene]) (F8T2) were used in this study (Figure 1). Detailed spectroscopic and photophysical properties of these polymers have been presented elsewhere [9].

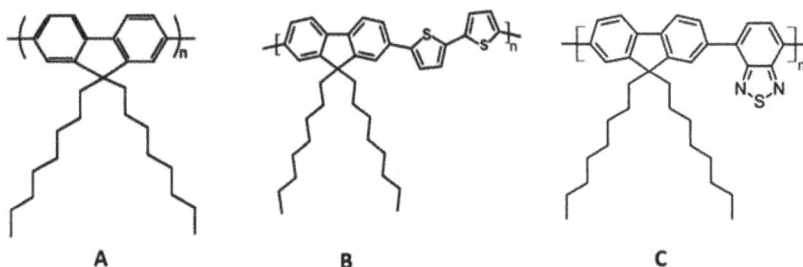

Figure 1. Structures of the CPs used: (**A**) poly[9,9-dioctylfluorene-2,7-diyl] (PFO); (**B**) poly[(9,9-dioctylfluorenyl-2,7-diyl)-co-bithiophene] (F8T2); and (**C**) poly[(9,9-dioctylfluorenyl-2,7-diyl)-alt-co-(1,4-benzo-2,10,3- thiadiazole)] (F8BT).

For many practical applications, it is desirable to incorporate CPs in an appropriate porous inert matrix. In this work, we used ethyl cellulose (EC) to incorporate the CPs. Both CPs exhibit high sensitivity in ethyl cellulose films when exposed to nitrobenzene (NB) and 1,3-dinitrobenzene (DNB) vapors. These are chosen as models or markers of more common nitroaromatic explosives, such as TNT or RDX. EC is the most common insoluble cellulose derivative used and is available in a variety of viscosity grades, according to the molecular weight range of the products. The molecular weight affects the mechanical properties, which have fundamental importance for producing intact films, depending on the application [10]. Plasticizers are generally used to improve the mechanical properties of a polymer matrix. This occurs because the plasticizer can decrease the intramolecular forces between the polymer chains, reducing the glass transition temperature and increasing the permeability of the polymer matrix to gases or other analytes [11]. In this work, we use polyethylene glycol and polypropylene glycol with different molecular weights as plasticizers to improve the mechanical properties and permeability of our polymer matrices.

This contribution is divided in two parts. First, we report absorption, emission spectra, and fluorescence lifetimes of the PFO and F8T2 in ethyl cellulose films, the structural characterization of the thin films, and then discuss the ability of these materials to sense TNT model compounds. In the second part, we study the improved polymer matrices produced by the introduction of plasticizers which increase the sensitivity to TNT-like compounds when compared with the non-plasticized ones. We also develop other methods for CP device preparation in the solid matrix, such as ink jet printing technology: in this case we added an internal reference, a CP whose fluorescence is not quenched by the TNT-like molecules, to provide potential for ratiometric sensing. In this condition, we print different zones with the two CPs, and use as "paper" the non-plasticized ethyl cellulose matrix.

2. Materials and Methods

2.1. Materials

Ethyl cellulose of viscosity grade 100 cP, was acquired from Sigma-Aldrich (St. Louis, MO, USA) and used without any treatment. The conjugated polymers, poly[9,9-dioctylfluorene-2,7-diyl] (PFO, Mw $\geq$ 20,000) poly[9,9-dioctylfluorenyl-2,7-diyl)-co-bithiophene]) (F8T2, Mn > 20,000) and poly(9,9-dioctylfluorene-alt-benzothiadiazole) (F8BT, Mn $\pm$ 17,000-23,000) were from Sigma-Aldrich. Solutions for film preparation were made by dissolving ethyl cellulose and the CPs (200–500 ppm) in toluene (GPS grade, Carlo Erba Reagents) at room temperature. Nitrobenzene (ACS reagent, 99%) and 1,3-dinitrobenzene (99%) were from Sigma-Aldrich.

2.2. Film Preparation and Ethyl Cellulose Plasticization

Films were prepared by solution casting of a mixture of ethyl cellulose and the CPs (200–500 ppm) from toluene at room temperature. To ensure good optical quality of the films, the solvent was evaporated slowly at room temperature (72 h) and the last traces removed in an oven at 60 °C for 10 min. No differences in fluorescence behavior were observed when samples were left at this temperature for longer times. Plasticized films were obtained by the addition of 1–10 wt % of polyethylene glycol (600 and 3400) and polypropylene glycol (average Mw = 1000) to EC, followed by its dissolution in the solvent. The thicknesses of the films were 400–600 µm as measured with a micrometer (Etalon Rolle, Switzerland).

2.3. Ink Jet Printing

A FUJIFILM Dimatix Materials printer DMP-2800 Series was used for printing films. This is suitable for printing the CPs on an appropriate matrix. Toluene solutions of the CPs were used to fill disposable cartridges that have 16 individually-tunable, piezo-actuated nozzles. Cartridges are available for dispensing 10 pL or 1 pL drops. Drops were printed by voltage-driven deformations of a membrane wall of a chamber behind each nozzle. The segments of this action make up a waveform that is optimized for each ink, as well as the intended print job.

2.4. Luminescence Characterization

The UV spectroscopic measurements were performed at room temperature with a Shimadzu UV-3101PC UV-VIS-NIR spectrometer, using cells of 1.0 cm optical path length for solution measurements. The emission and excitation spectra were recorded with a Horiba Jobin-Ivon SPEX Fluorolog 3–22 fluorescence spectrometer. The Fluorolog consists of a modular spectrofluorimeter with double-grating monochromators for excitation (200–950 nm range, optimized in the UV with a blaze angle at 330 nm) and emission (200–950 nm range, optimized in the visible and with a blaze angle at

500 nm). The bandpass for excitation and emission was 5 nm with a wavelength accuracy of ± 0.5 nm. The excitation source consisted of an ozone-free 450 W xenon lamp. The emission detector employed was a Hamamatsu R928 photomultiplier, with a photodiode as the reference detector. The fluorescence quenching of the film was measured in a sealed cuvette containing the nitrobenzene or 1,3-dinitrobenzene vapors at room temperature (293 K).

Time-resolved picosecond fluorescence intensity decays were obtained by the single-photon timing method with laser excitation, with the set-up described elsewhere [12]. Decay data analysis with a sum of exponentials was achieved by means of a Microsoft Excel spreadsheet specially designed for lifetime analysis that considers deconvolution with the instrument response function (IRF) [13].

2.5. Structural Characterization

A confocal laser scanning microscope (Leica TCS-SP5) equipped with a CW Ar ion laser (458, 465, 488, 496, and 514 nm) and a pulsed Ti:sapphire (Spectra-Physics Mai Tai BB, 710–990 nm, 100 fs, 80 MHz) was used to obtain images of the films.

SEM was performed with a Hitachi S2400 microscope and the images were recorded by software Quantax (Bruker; Billerica, MA, USA). Samples were coated with gold and registered at 50× and 500× magnification.

2.6. Sensor Prototype

A portable device (Figure 2) was developed and tested for the study of the fluorescence quenching of the CP films by the nitroaromatic vapors. A Sony SLD3135 laser diode (405 nm; 50 mW) with integrated monitoring photodiode was used as excitation source and a Vishay BPW34 Silicon PIN photodiode covered with a green filter (Edmund Optics #43-934) to avoid excitation light from reaching the measuring photodiode was used for the detection. The general architecture is described elsewhere [14].

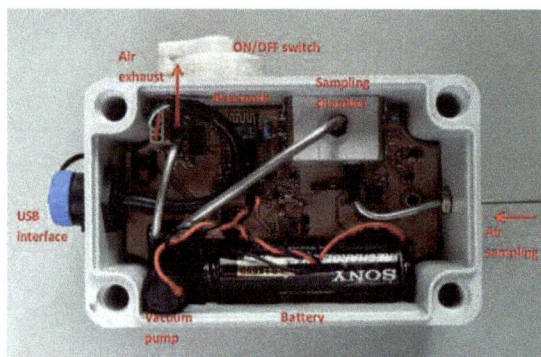

Figure 2. Prototype sensor device.

3. Results and Discussion

The ground-state absorption spectrum of F8T2 was obtained in toluene solution and in ethyl cellulose films (Figure 3). In the case of the thin film a new band is observed at longer wavelengths (492 nm). A number of possible explanations exist for this new band. However, detailed analysis suggests that it is associated with the so-called β phase of the F8T2 [15]. The β phase is a metastable state with part of the CP in a rigid extended structure, and can be formed through evaporation of an appropriate solvent, by treatment of the film or one of the other phases by solvent vapor, or by keeping the polymer in a restricted environment. The β-phase peak was observed originally by Bradley and co-workers [16] in PFO polystyrene films and PFO solutions in poor solvents (such as methylcyclohexane) [17]. We observed this new band at 492 nm in the absorption spectrum of the PFO ethyl cellulose films (Figure 4), even though F8T2 is a less rigid polymer than PFO, strong support has been presented from steady-state and time-resolved fluorescence and fluorescence anisotropy measurements for the formation of the β-phase with this polymer [18].

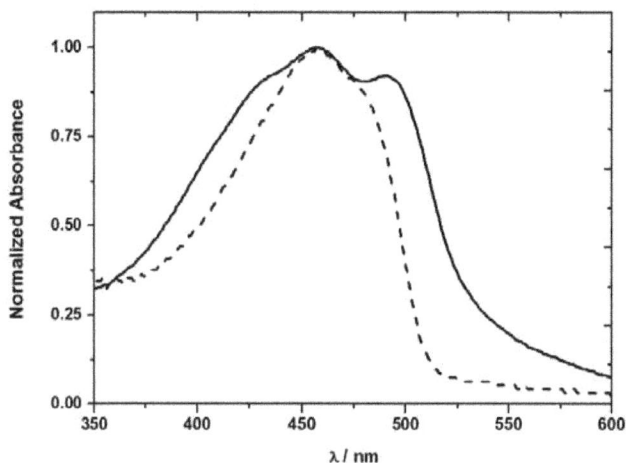

Figure 3. Absorption spectrum of the F8T2 ethyl cellulose film (solid line) and in toluene solution (dashed line).

The presence of NB vapor does not change significantly the absorption spectrum (Figure 4) of the PFT2 or PFO ethyl cellulose films, suggesting the absence of ground-state complexation.

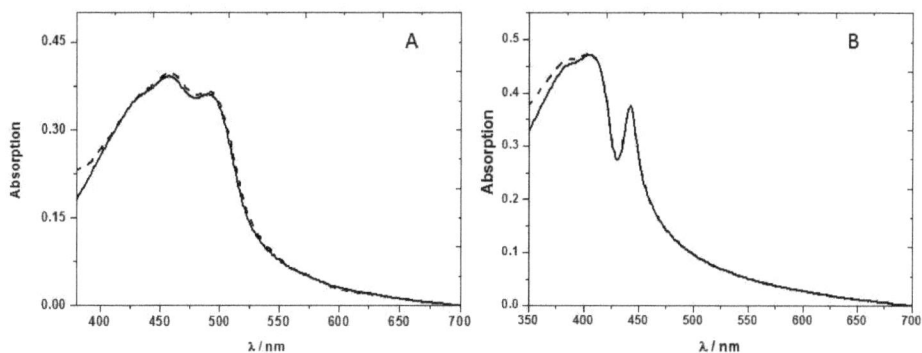

Figure 4. (**A**) Absorption spectra of the F8T2 film in the presence (dashed line) and absence (solid line) of nitrobenzene vapor; (**B**) Absorption spectra of the PFO film in the presence (dashed line) and absence (solid line) of nitrobenzene vapor. Both CPs are incorporated in an ethyl cellulose matrix.

The F8T2 emission spectrum has a maximum at 545 nm (Figure 5). In the presence of NB vapor we observed a decrease of the emission intensity of about 42%. In the case of the PFO, the emission showed a structured fluorescence spectrum between 400 and 600 nm, attributed to at least three vibronic components. We observed only a 34% drop in the fluorescence emission intensity in the presence of NB vapor at 445 nm. As mentioned, the CPs are good electron donors and their fluorescence is quenched by NB through photoinduced electron transfer. The amplifying nature of the exciton delocalization in the conjugated polymers makes them highly sensitive materials to quenching by nitroaromatic vapors.

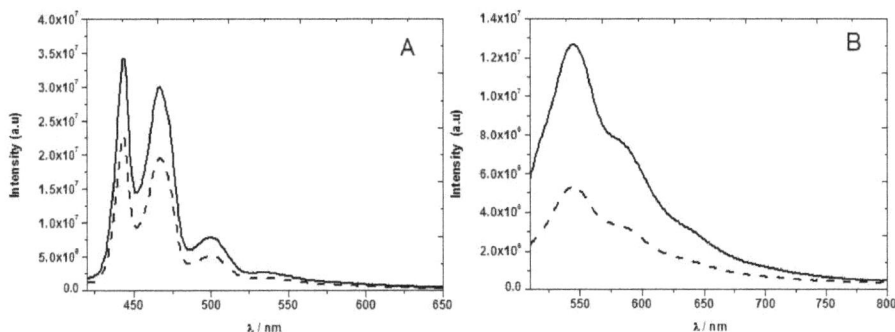

Figure 5. Fluorescence spectra in the absence (solid line) and presence of saturated NB vapors at room temperature (dashed line) of: (**A**) PFO in ethyl cellulose film, λ_{exc} = 410 nm, and (**B**) F8T2 in ethyl cellulose film, λ_{exc} = 490 nm.

Time-resolved fluorescence decays of CPs in thin films of ethyl cellulose recorded at the maximum emission wavelength are well fitted with sums of three exponentials (Table 1). This agrees with previous studies [18] showing triple exponential decays of F8T2 in methylcyclohexane (MCH), with lifetimes of 650 ps, 440 ps, and 20 ps. The longest decay time (650 ps) is assigned to the β-conformation and the intermediate lifetime (440 ps) to the α-conformation [18]. The shortest time (20 ps) may result from solvent/conformational relaxation or intramolecular energy transfer from non-ordered to ordered chain segments. Studies of PFO in toluene solution [19] also show a complex decay, a sum of two or even three exponentials are being required to obtain good fits. A fast component of about 20 ps is found, and is more important (with greater amplitude) at the onset of the emission band. An intermediate component is also observed around 90 ps and a predominant decay time around 360 ps is observed independent of the emission wavelength and attributed to the PFO intrinsic fluorescence lifetime. In thin films, the decay is again described by a of sum of three exponentials, however, at long wavelengths, it is dominated by a long component of 3 ns, attributed to the presence of photooxidized species, such as keto defects and other emissive defects, which are easily populated by efficient energy migration [20].

Table 1. Decay times and amplitudes of the PFO and F8T2 in ethyl cellulose thin film without NB vapors and in the presence of NB vapors.

	NB	τ_1/ns (f_1 *)	τ_2/ns (f_2 *)	τ_3/ns (f_3 *)	$\tau_{average}$/ns	τ_{fluor} change (%) **
F8T2	Without	0.03 (0.10)	0.43 (0.55)	0.62 (0.35)	0.40	65
	With	0.005 (0.07)	0.04 (0.51)	0.29 (0.42)	0.14	
PFO	Without	0.16 (0.26)	0.31 (0.63)	1.70 (0.11)	0.43	72
	With	0.05 (0.37)	0.08 (0.52)	0.55 (0.11)	0.12	

* computed from the individual lifetimes and pre-exponential factors: α_1, α_2 and α_3, $f_1 = \alpha_1\tau_1/(\alpha_1\tau_1 + \alpha_2\tau_2 + \alpha_3\tau_3)$, $f_2 = \alpha_2\tau_2/(\alpha_1\tau_1 + \alpha_2\tau_2 + \alpha_3\tau_3)$ and $f_3 = 1 - (f_1 + f_2)$. ** computed as: $(\frac{\tau_{average} \text{ without NB vapors} - \tau_{average} \text{ with NB vapors}}{\tau_{average} \text{ without NB}}) \times 100$.

The drop in the average lifetimes (τ/τ_0) of F8T2 and PFO, resulting from the presence of NB vapors, is 35% and 28%, respectively. Comparing these values with those measured in steady-state conditions, 42% and 34% for F8T2 and PFO, respectively, it is concluded that the quenching induced by the nitroaromatics is predominantly a dynamic process. This is important as it favors reversibility of the sensing system.

To increase the sensitivity towards detection of nitroaromatic vapors of the P8T2 ethyl cellulose films, we have tested the effect of adding different plasticizers to the ethyl cellulose matrix. Plasticized films show a very high sensitivity towards nitroaromatics when compared with the non-plasticized ones, probably due to higher permeability (Figure 6). For example, using only 1% (w/w) of the plasticizer PEG 3400, we obtained in a short time (three minute) of 2,3-dinotrobenzene, DNB, vapor exposure a fluorescence quenching of 95% (Figure 6B).

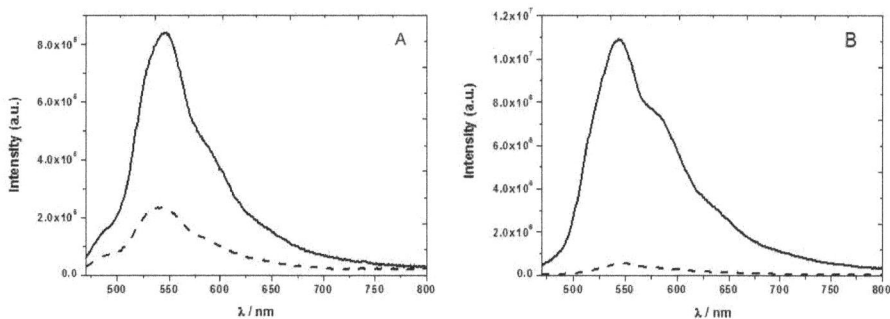

Figure 6. (**A**) Emission spectra of the F8T2 ethyl cellulose film with 1% (w/w) of PEG 3400 in the absence (solid line) and after three minutes of exposure to nitrobenzene (8.74×10^{-6} M) vapor (dashed lines); (**B**) Emission spectra of the F8T2 ethyl cellulose film with 1% (w/w) of PEG 3400 in the absence (solid line) and after three minutes of exposure to 2,3-dinitrobenzene (1.83×10^{-5} M) vapor (dashed lines). Excitation wavelength was 450 nm.

We can see from Figure 6 a stronger fluorescence quenching caused by the DNB vapors. This higher quenching efficiency may arise from the higher electron affinity of DNB, outweighing its lower vapor pressure [21]. In all the studied plasticized films, the sensitivity towards nitroaromatic vapors increases when compared with the neat ethyl cellulose films, as can be seen from the decrease in fluorescence lifetimes in Table 2.

Table 2. Decay times of F8T2 in ethyl cellulose thin films with different plasticizers in the presence and in the absence of nitroaromatic vapors.

	NB			DNB		
	$\tau_{\text{average without}}$ NB vapors/ns	$\tau_{\text{average with}}$ NB vapors/ns	$\tau_{\text{fluor.}}$ Decrease (%) *	$\tau_{\text{average without}}$ NB vapors/ns	$\tau_{\text{average with}}$ DNB vapors	$\tau_{\text{fluor.}}$ Decrease (%) *
Neat Ethyl Cellulose	400	340	15	400	300	25
1% PEG 3400	380	330	13	375	200	47
1% PEG 600	380	300	21	380	180	53

* computed as: ($\frac{\tau_{\text{average without NB or DNB vapors}} - \tau_{\text{average with NB or DNB vapors}}}{\tau_{\text{average without NB or DNB vapors}}}$) $\times$ 100.

By comparing the values in the lifetime attenuation for the plasticized and non-plasticized films (Table 2), we observe that the addition of only 1% by weight of plasticizer to the neat ethyl cellulose matrix increases the sensitivity of F8T2 towards these nitroaromatic vapors by ca. 5% (NB), 29%, and 35% (DNB). The molecular weight of the plasticizer does not appear to have a significant influence on the increase in the sensitivity of the CP, but it is noted that a lower molecular weight one seems to facilitate a slightly stronger quenching with both NB and DNB. This may be a result of better compatibility with the EC matrix, and, hence, production of a more amorphous structure.

The thermal behavior of these plasticizers films was studied. Neat ethyl cellulose has a glass transition temperature (Tg) at 130–133 °C [22]. As expected, the addition of plasticizer decreases the Tg of the samples. For example, using 25% (w/w) PEG 400 the Tg drops to 70 °C [21]. This strong decrease means that there is a good compatibility between the polymer matrix and the plasticizer. In general, plasticizers reduce polymer interchain interactions by distributing themselves homogeneously within the polymer, hence increasing the free volume. However, a reduction in the Tg value down to near room temperature will result in an increase in chain mobility and, consequently, could enhance the crystallization of films by reducing the energy required for this process. This phenomenon would lead to structural changes resulting in loss of transparency. The thermal stability of the films was studied by monitoring the fluorescence intensity with an increase of the temperature. These experiments (Figure 7) were performed in the absence and in the presence of DNB vapors.

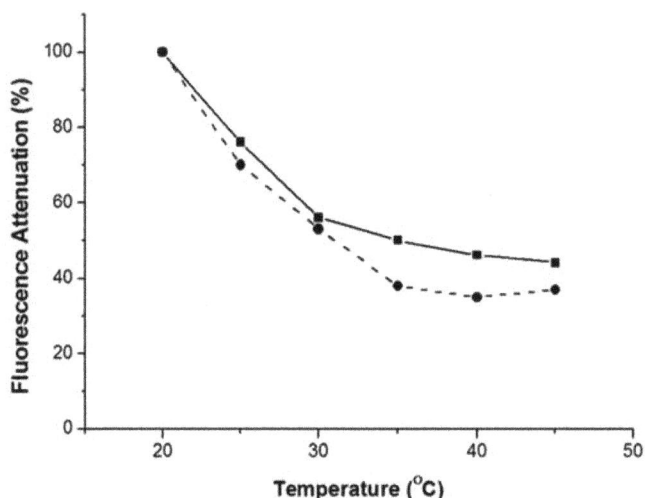

Figure 7. Fluorescence attenuation of the F8T2 ethyl cellulose film with 1% (w/w) of PEG 3400 in the absence (solid line) and presence (dashed line) of DNB vapor with as a function of the temperature. Excitation wavelength was 450 nm.

It can be seen that the increase of DNB vapor pressure with the temperature is not the major factor involved. Instead, the temperature dependence appears to result from the decrease in Tg upon PEG addition.

The morphology of the CP films was studied by confocal fluorescence microscopy. Typical images obtained from this technique are shown in Figure 8, in which the green spots represent the emission of the CPs. These films do not exhibit bulk phase separation at the magnifications studied, but some polymer aggregation can be observed, especially in the ethyl cellulose film containing F8T2.

The surface morphology of the films was studied using scanning electron microscopy (SEM) (Figure 9). The surface of neat ethyl cellulose film (not shown) is rather smooth, compact, and featureless. However, in the case of ethyl cellulose, F8T2 blends, phase separated zones are observed for concentrations above the incorporation capacity of CPEs into ethyl cellulose films, Figure 9A. The addition of a plasticizer to the neat ethyl cellulose can be seen in Figure 9B to introduce some porosity. This film has pores with diameters between 1.5 and 3 μm, randomly distributed. The formation of these pores in the plasticized films may explain, in part, the increased sensitivity to the nitroaromatic vapors observed by fluorescence quenching of the CPs.

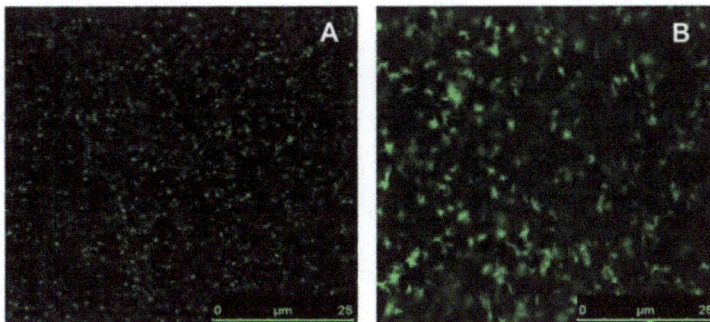

Figure 8. Typical images of (**A**) PFO and (**B**) F8T2 in neat ethyl cellulose film, observed by confocal microscopy. Excitation wavelength was 458 nm and emission wavelength in the range of 510–700 nm. These pictures have 512 × 512 pixels, using a pinhole of 1 AU, zoom 4×, and 400 Hz. The scale bars are 25 μm.

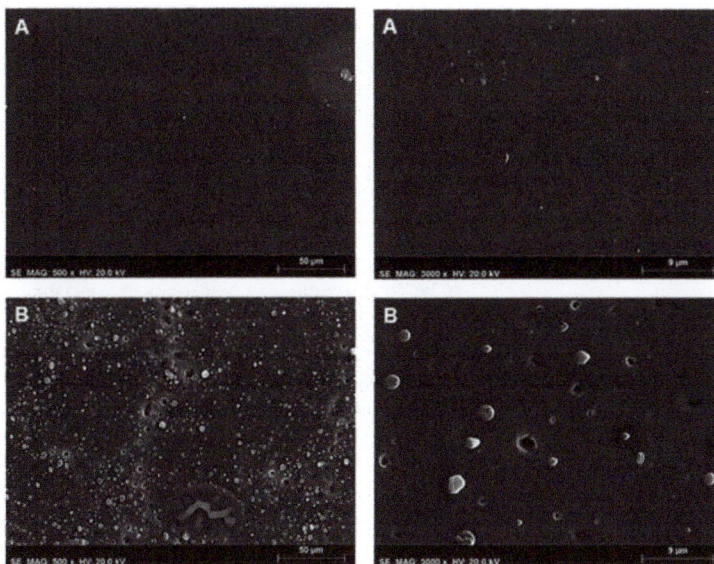

Figure 9. SEM images of ethyl cellulose films containing (**A**) F8T2 and (**B**) F8T2 and 1% PEG 600. The scale bars are 50 μm and 9 μm.

Fluorescence quenching based sensors normally require a reference material, such that the degree of quenching measured by the ratio of signals from sensor and reference materials. We have produced such a ratiometric system using ink-jet microprinting (IJMP) (Figure 10). The IJMP allows direct deposition of minuscule quantities of the CPs onto the ethyl cellulose film substrate (thickness 240 μm). The diameter and uniformity of the microdot can be controlled by modifying substrate surface chemistry and ink preparation [21].

We have used F8T2 as sensor material, and have incorporated another CP, F8BT (Figure 1), which is not readily oxidizable and does not exhibit any fluorescence changes in the presence of DNB and NB vapors, thereby serving as an internal reference. Table 3 shows the analysis of fluorescence decays of the imprinted ethyl cellulose sensor by IJMP. In the presence and absence of concentrated

DNB vapors, we observed that F8BT did not exhibit any significant attenuation of its fluorescence lifetime, in contrast to the quenching observed with F8T2. In this case, we can use this system as a sensor with an internal reference for nitroaromatic vapors.

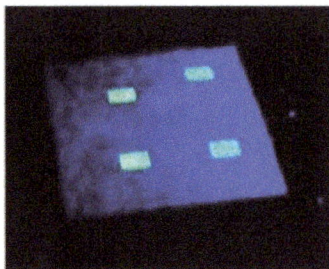

Figure 10. Photograph of the P8T2 IJMP imprinted film on ethylcellulose under UV light.

Table 3. Decay times of the F8T2 and F8BT imprinted by IJMP in ethyl cellulose film in presence or absence of nitroaromatics vapors.

F8T2			F8BT		
$\tau_{average\ without}$ DNB vapors/ns	$\tau_{average\ with}$ DNB vapors/ns	$\emptyset_{fluor.}$ Decrease (%) *	$\tau_{average\ without}$ NB vapors/ns	$\tau_{average\ with}$ DNB vapors	$\tau_{fluor.}$ Decrease (%) *
1. 8	0.86	52	0.79	0.78	1

$$* \text{ computed as: } \left(\frac{\tau_{average\ without\ NB\ vapors} - \tau_{average\ with\ NB\ vapors}}{\tau_{average\ without\ NB\ vapors}} \right) \times 100.$$

If we compare the values in the lifetime attenuation for the F8T2 and F8BT (Table 3), we observe that the F8T2 lifetime decreases by 52%, whereas there are insignificant changes in the F8BT lifetime. The measured sensitivity is equivalent to that obtained in plasticized ethyl cellulose films.

The sensor was tested in a dynamic setup composed of two mass flow controllers (MFCs Dwyer GFC-2102), one controlling the flow of clean air and the other controlling the flow of saturated nitrobenzene vapor, obtained from a bubbler at constant temperature and constant pressure. Both MFCs are controlled from MATLAB through a microcontroller (PIC24FV16KM202) that sets the references to the MFCs, defining the composition of the nitrobenzene-clean air mixture. As expected, the proposed differential approach guaranteed very good sensitivity and fast response from the electronics side while providing high-level input signals. Figure 11 shows the sensor response after a 20-second exposure to minute NB vapors.

For in-field applications, the fast response to the presence of a small concentration of nitroaromatics is a very positive characteristic. Although the recovery time is rather long (several minutes), this can be acceptable for scenarios where the detection of frequent changes in the analyte level is not required.

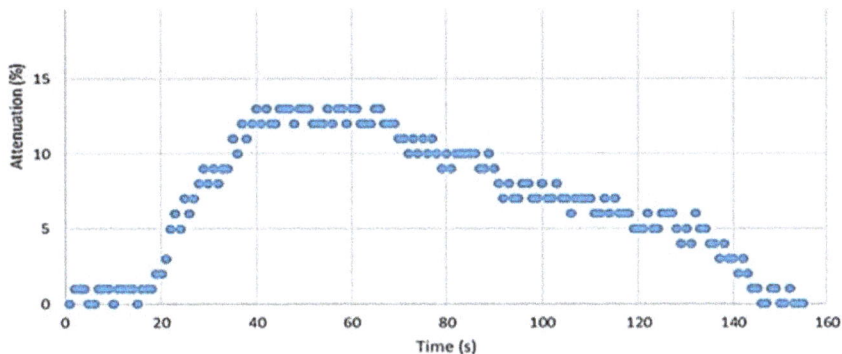

Figure 11. Attenuation of the fluorescence emission of the F8T2 in ethyl cellulose when exposed to NB $(1.7 \times 10^{-5}$ M) vapors for 20 s.

4. Conclusions

We have studied fluorene-based conjugated polymers as fluorescence sensing materials for nitroaromatic vapors with the overall goal of detecting explosives using such polymers. The best conjugated polymer in our studies was found to be poly(9,9-dioctylfluorenyl-2,7-diyl]-co-bithiophene] (F8T2). It is stable, has a good absorption between 400 and 450 nm, a strong and structured fluorescence around 550 nm. A 96% quenching of fluorescence, accompanied by a corresponding decrease in the fluorescence lifetime, is seen on exposure of the plasticizer film of F8T2 ethylcellulose to the model compounds nitrobenzene (NB) and 1,3-dinitrobenzene (DNB) vapors, both from the family of common explosives vapors.

Furthermore, it was demonstrated that ink-jet microprinting can be used as a convenient approach to easily and rapidly fabricate films containing these sensors and inert reference materials, with the same sensitivity of plasticized ethyl cellulose films towards nitroaromatic vapors.

A sensor prototype based on the F8T2 conjugated polymer was developed and tested. The ability of the sensor to detect small quantities of nitrobenzene was confirmed. The sensor prototype showed very fast response (a few seconds) to the presence of small concentrations of the target analyte, but also showed a large recovery time, which limits its potential applications. This slow recovery may result from the designed sampling chamber and also from the time required for desorption of the nitrobenzene molecules from the polymer surface. Both of the above aspects will be addressed in a future designs, optimizing polymer thickness and porosity, and optimizing the shape and arrangement of the sampling chamber.

Acknowledgments: This work was partially carried out in the framework of TIRAMISU project. This project is funded by the European Community's Seventh Framework Program (FP7/2007-2013) under grant 284747. L Martelo was supported by Fundação para a Ciência e Tecnologia (FCT, Portugal) with a postdoctoral fellowship (SFRH/BPD/121728/2016). HDB is grateful for funding from the Coimbra Chemistry Centre (CQC), which is supported by FCT through the programs UID/QUI/UI0313/2013 and COMPETE. AC also thanks FCT for funding through the program UID/EEA/50008/2013. AF also thanks to the FCT for funding through (SFRH/BPD/111301/2015).

Author Contributions: L. Martelo conceived and performed the experiments, under the supervision of H. Burrows and M. N. Berberan-Santos; A. Fedorov performed the time-resolved fluorescence experiments; A. Charas helps in the conception of the ink jet printing films; J. Figueiredo designed and implemented the sensor prototype shown in Figure 2; T. Neves characterized the sensor's response and Lino Marques supervised the works relative to the prototype design, implementation and characterization.

Conflicts of Interest: The authors declare no conflict of interest.

References

1. McQuade, T.D.; Pullen, E.A.; Swager, T.M. Conjugated polymer-based chemical sensors. *Chem. Rev.* **2000**, *100*, 2537–2574. [CrossRef] [PubMed]

2. Gu, C.; Huang, N.; Gao, J.; Xu, Y.; Jiang, D. Controlled synthesis of conjugated microporous polymer films: versatile platforms for highly sensitive and label-free chemo- and biosensing. *Angewandte* **2014**, *53*, 4850–4855. [CrossRef] [PubMed]

3. Thomas, W.S., III; Joly, D.G.; Swager, M.T. Chemical sensors based on amplifying fluorescent conjugated polymers. *Chem. Rev.* **2007**, *107*, 1339–1386. [CrossRef] [PubMed]

4. Gu, C.; Huang, N.; Wu, Y.; Xu, Y.; Jiang, D. Design of highly photofunctional porous polymer films with controlled thickness and prominent microporosity. *Angewandte* **2015**, *54*, 11540–11544. [CrossRef] [PubMed]

5. Yang, J.; Swager, T.M. Fluorescent porous polymer films as TNT chemosensors: Electronic and structural effects. *J. Am. Chem. Soc.* **1998**, *120*, 11864–11873. [CrossRef]

6. Hancock, F.L.; Deans, R.; Moon, J.; Swager, M.T. Amplifying fluorescent polymer detection of bioanalytes. In Proceedings of the Chemical and Biological Early Warning Monitoring for Water, Food, and Ground, Boston, MA, USA, 2001. Available online: https://www.spiedigitallibrary.org/conference-proceedings-of-spie/4575/1/Amplifying-fluorescent-polymer-detection-of-bioanalytes/10.1117/12.456910.short?SSO=1 (accessed on 29 October 2017).

7. Knaapila, M.; Garamus, V.M.; Dias, F.B.; Almásy, L.; Galbrecht, F.; Charas, A.; Morgado, J.; Burrows, H.D.; Monkman, A.P. Influence of solvent quality on the self-organization of archetypical hairy rods−branched and linear side chain polyfluorenes: Rodlike chains versus "Beta-Sheets" in solution. *Macromolecules* **2006**, *39*, 6505–6512. [CrossRef]

8. Wegner, G. Nanocomposites of hairy-rod macromolecules: Concepts, constructs, and materials. *Macromol. Chem. Phys.* **2003**, *204*, 347–357. [CrossRef]

9. Fonseca, S.M.; Pina, A.L.G.; Seixas de Melo, J.; Burrows, H.D.; Chattopadhyay, N.; Alcácer, L.; Charas, A.; Morgado, J.; Monkman, A.P.; Asawapirom, U.; et al. Triplet-state and singlet oxygen formation in fluorene-based alternating copolymers. *J. Phys. Chem. B* **2006**, *110*, 8278–8283. [CrossRef] [PubMed]

10. Andersson, H.; Hjärtstam, J.; Stading, M.; Corswant, V.C.; Larsson, A. Effects of molecular weight on permeability and microstructure of mixed ethyl-hydroxypropyl-cellulose films. *Eur. J. Pharm. Sci.* **2013**, *48*, 240–248. [CrossRef] [PubMed]

11. Martelo, L.; Jiménez, A.; Valente, A.J.M.; Burrows, H.D.; Marques, A.T.; Forster, M.; Scherf, U.; Peltzer, M.; Fonseca, M.S. Incorporation of polyfluorenes into poly(lactic acid) films for sensor and optoelectronics applications. *Polym. Int.* **2012**, *61*, 1023–1030. [CrossRef]

12. Menezes, F.; Fedorov, A.; Baleizão, C.; Valeur, B.; Berberan-Santos, M.N. Methods for the analysis of complex fluorescence decays: Sum of Becquerel functions versus sum of exponentials. *Meth. Appl. Fluoresc.* **2013**, *1*, 015002. [CrossRef]

13. Berberan-Santos, M.N. Unpublished work, 2009.

14. Neves, T.; Marques, L.; Martelo, L.; Burrows, H.D. Conjugated polymer-based explosives sensor: Progresses in the design of a handheld device. *IEEE Sens.* **2014**, *11*, 1415–1418.

15. Knaapila, M.; Dias, F.B.; Garamus, V.M.; Almásy, L.; Torkkeli, M.; Leppänen, K.; Galbrecht, F.; Preis, E.; Burrows, H.D.; Scherf, U.; et al. Influence of side chain length on the self-assembly of hairy-rod poly(9,9-dialkylfluorene)s in the poor solvent methylcyclohexane. *Macromolecules* **2007**, *40*, 9398–9405. [CrossRef]

16. Bradley, D.D.C.; Grell, M.; Long, X.; Mellor, H.; Grice, W.A.; Inbasekaran, M.; EWoo, P.E. Influence of aggregation on the optical properties of a polyfluorene. *Proc. SPIE* **1997**, *3145*, 254–259.

17. Dias, F.B.; Morgado, J.; Maçanita, A.L.; Da Costa, F.P.; Burrows, H.D.; Monkman, A.P. Kinetics and thermodynamics of poly(9,9-dioctylfluorene) β-phase formation in dilute solution. *Macromolecules* **2006**, *39*, 5854–5864. [CrossRef]

18. Rodrigues, R.F.; Charas, A.; Morgado, J.; Maçanita, A. Self-organization and excited-state dynamics of a fluorene−bithiophene copolymer (F8T2) in solution. *Macromolecules* **2010**, *43*, 765–771. [CrossRef]

19. Monkman, A.; Rothe, C.; King, S.F.; Dias, F. Polyfluorene photophysics. In *Polyfluorenes*; Springer: Berlin/Heidelberg, Germany, 2008.

20. Dias, F.B.; Knaapila, M.; Monkman, A.P.; Burrows, H.D. Fast and slow time regimes of fluorescence quenching in conjugated polyfluorene—fluorenone random copolymers: The role of exciton hopping and dexter transfer along the polymer backbone. *Macromolecules* **2006**, *39*, 1598–1606. [CrossRef]

21. Salinas, Y.; Martinéz-Máñez, R.; Marcos, D.M.; Sancenón, F.; Costero, M.A.; Parra, M.; Gil, S. Optical chemosensors and reagents to detect explosives. *Chem. Soc. Rev.* **2012**, *41*, 1261–1296. [CrossRef] [PubMed]

22. Pabst, O.; Perelaer, J.; Beckert, E.; Schubert, U.S.; Eberhardt, R.; Tünnermann, A. All inkjet-printed piezoelectric polymer actuators: Characterization and applications for micropumps in lab-on-a-chip systems. *Org. Electron.* **2013**, *14*, 3423–3429. [CrossRef]

sensors

MDPI

Article

Fluorescent Polymer Incorporating Triazolyl Coumarin Units for Cu^{2+} Detection via Planarization of Ict-Based Fluorophore

Jean Marie Vianney Ngororabanga, Jacolien Du Plessis and Neliswa Mama *

Department of Chemistry, Nelson Mandela Metropolitan University, Port Elizabeth 6031, South Africa;
s212438700@live.nmmu.ac.za (J.M.V.N.); s215011619@live.nmmu.ac.za (J.D.P.)
* Correspondence: neliswa.mama@nmmu.ac.za; Tel.: +27-41-5042-368

Received: 26 June 2017; Accepted: 4 August 2017; Published: 30 August 2017

Abstract: A novel fluorescent polymer with pendant triazolyl coumarin units was synthesized through radical polymerization. The polymer showed reasonable sensitivity and selectivity towards Cu^{2+} in acetonitrile in comparison to other tested metal ions with a significant quenching effect on fluorescence and blue shifting in the range of 20 nm. The blue shift was assigned to the conformation changes of the diethylamino group from the coumarin moiety which led to planarization of the triazolyl coumarin units. The possible binding modes for Cu^{2+} towards the polymer were determined through the comparison of the emission responses of the polymer, starting vinyl monomer and reference compound, and the triazole ring was identified as one of the possible binding sites for Cu^{2+}. The detection limits of the polymer and vinyl monomer towards Cu^{2+} were determined from fluorescence titration experiments and a higher sensitivity (35 times) was observed for the polymer compared with its starting monomer.

Keywords: pendant group; triazolyl coumarin; chemosensor; aggregate; planarization

1. Introduction

The cupric ion (Cu^{2+}) is considered as one of the trace elements in human and other mammal systems due to its essentiality and very limited quantities in the body [1,2]. Being ranked third after Fe^{2+} and Zn^{2+} in the human body as essential transition metal ions, Cu^{2+} concentrations beyond the necessary for biological functions can lead to oxidative stress and cellular toxicity, which are often associated with serious neurodegenerative diseases such as Alzheimer's, Parkinson's, and Wilson's diseases [3–5]. In addition, the accumulation of copper in the environment due to the increased release of this metal from industrial activities presents ecological and human health threats [6,7]. For instance, an over-ingestion of copper can lead to serious health problems such as gastrointestinal disturbances and liver or kidney damage [7]. For these reasons, a limited concentration of 1.3 ppm (~20 µM) was set by the US Environmental Protection Agency (EPA) for copper in drinking water. Due to the widespread use of copper and the toxicity associated with its higher concentration, there is a strong need for reliable, inexpensive and simple methods for detecting and quantifying copper ions in different media for real-time monitoring of the environment, as well as biological and industrial samples. Typical methods that have been developed and employed for the detection of copper ions are mainly based on atomic absorption spectroscopy (AAS) [8], inductively coupled plasma-mass spectroscopy (ICP-MS), inductively coupled plasma-atomic emission spectroscopy (ICP-AES) [9], and electrochemical sensing methods [10,11]. However, these techniques are costly, extremely tedious and destructive. Furthermore, sample preparation requires large amounts and the methods are not suitable for continuous monitoring.

Due to their simplicity and high sensitivity, fluorescent-based methods showed some potential to address some problems faced by these methods, and several fluorescent chemosensors with different fluorophores have been developed. Owing to the excellent properties of the coumarin motif, such as biocompatibility, high fluorescence quantum yield, and relative ease of synthesis, absorption and emission tunability via substituent manipulation [12], the coumarin fluorophore is one of the widely used fluorophores in the synthesis of chemosensors for Cu^{2+}. Nevertheless, most of the currently developed Cu^{2+} chemosensors are of low molecular weight and only a few qualify to be in the macromolecular range.

Owing to low chemical and thermal resistance as well as difficulties in separation and recovery associated with the low-molecular-weight chemosensors, a physical immobilization support is often needed for their application [13]. A physical support does not only improve the mechanical properties, but also minimizes the tendency of the sensing molecules to migrate. To avoid complications associated with synthesizing probes and immobilizing them on a physical support, polymers with host binding sites as part of their backbone or as part of their pendant group were found to be better alternatives. These materials have good thermal and mechanical properties and they also offer an outstanding and permanent immobilization method which allows them to be processed into end-user materials such as coatings and films [14–18]. Furthermore, fluorescent polymer-based chemosensors offer crucial advantages such as higher sensing performance levels (sensitivity and selectivity) compared to their small-molecule counterparts. So far, several polymer-based chemosensors for various metal ions have been developed [19–21], but very few showed selective sensitivity towards Cu^{2+} [22,23].

Herein we describe the chemosensing capability of a novel fluorescent triazolyl coumarin-based polymer (**P1**) towards metal ions. The 1,2,3-triazole ring as a receptor was firstly incorporated in a polymerizable vinyl monomer through Cu(I)-catalyzed azide-alkyne 1,3-dipolar cycloaddition (CuAAC), archetypal click reaction [24,25]. The absorption and emission properties of **P1** were studied in the presence of different metal ions and selective recognition with a fluorescence quenching response was observed in the presence of Cu^{2+}. In addition to this, the presence of Cu^{2+} induced a remarkable blue shift in the emission spectrum of **P1**. Emission studies of the starting vinyl monomer and a reference coumarin-based molecule in the presence of Cu^{2+} were used to investigate the binding mode of **P1** towards Cu^{2+}.

2. Experimental

2.1. Materials

All chemicals and solvents were purchased from Sigma-Aldrich or Merck and were used as received without further purification. The 7-(diethylamino)-3-nitro-2H-chromen-2-one (1), 3-amino-7-(diethylamino)-2H-chromen-2-one (2), 3-azido-7-(diethylamino)-2H-chromen-2-one (3), [26] compound **5** [26] and 7-(diethylamino)-2H-chromen-2-one [27] were prepared according to the literature. A stock solution of the polymer was prepared by dissolving the polymer in 25 mL acetonitrile to afford a solution of 3×10^{-2} g/mL. The solution was further diluted to 7×10^{-5}. Deionized water was used to prepare solutions of metal ions and a concentration of 0.05 mol/L was obtained. All metal ion solutions were prepared using nitrate salts except for Fe^{2+} where sulphate was used. The titration experiments were performed using 3 mL of diluted acetonitrile polymer solution in a 3 mL quartz cuvette. In these experiments, spectroscopic measurements were taken after addition of an aliquot of selected metal ion solution.

2.2. Measurement

^{1}H NMR and ^{13}C NMR spectra were recorded on a Bruker Avance DPX 400 (400 MHz) using TMS as an internal standard. FT-IR spectra were recorded in the range 4000–500 cm^{-1} using Opus software (version 6.5.6) on a Bruker Platinum Tensor 27 ATR-IR spectrophotometer. Size exclusion chromatography experiments were performed in DMAc at 40 °C (flow rate of 0.5 mL/min) using

PMMA as a standard for calibration. The elemental analysis for carbon, hydrogen and nitrogen was performed using a Vario EL (Elementar Analysensystem GmbH) instrument. UV-Vis absorption and emission spectra were recorded at room temperature on a Perkin Elmer Lambda 35 and Perkin Elmer LS 45 respectively.

2.3. Polymer Synthesis

2.3.1. Synthesis of Compound 5

A mixture of 7-azido-4-methyl-2H-chromen-2-one (**4**) (0.1 g, 0.5 mmol), 3-butyn-2-ol (0.06 g, 0.5 mmol), $CuSO_4 \cdot 5H_2O$ (0.02 g, 0.054 mmol), sodium ascorbate (0.03 g, 0.15 mmol) and *N,N,N',N'',N''*-pentamethyldiethylenetriamine (PMDETA) (0.03 g, 0.15 mmol) in tertahydrofuran (THF) (20 mL) was stirred at room temperature for 72 h. THF was removed under reduced pressure and the crude product dissolved in chloroform. The mixture was washed with water and then concentrated to afford a yellow solid product in 80% yield. m.p. 170–175 °C. IR ν_{max} (cm^{-1}): 3152 (C=C-H), 1719.73 (C=O), 1618 (C=C). ^{1}H NMR (CDCl$_3$, 400 MHz): δ = 8.83 (s, 1H), 8–7.98 (m, 2H), 6.47 (s, 1H), 5.46 (d, J = 4.72 Hz, 1H), 4.82 (q, J = 5.92 Hz, 1H), 2.48 (s, 3H), 1.49 (d, J = 6.48 Hz, 1H). ^{13}C NMR (CDCl$_3$, 400 MHz): δ = 159.95, 154.60, 154.13, 153. 24, 139.18, 127.68, 120.39, 119.73, 115.79, 115.03, 107.66, 61.98, 24.05, 18.56. Anal. Calc. for $C_{17}H_{20}N_4O_3$: C: 62.18, H: 6.14, N: 17.06. Found: C: 62.17, H: 6.30, N: 17.59.

2.3.2. Synthesis of Monomer 6

In a two-necked round-bottomed flask equipped with a Dean Stark apparatus, a mixture of 7-(diethylamino)-3-(4-(1-hydroxyethyl)-1H-1,2,3-triazol-1-yl)-2H-chromen-2-one (**5**) (0.51 g, 1.54 mmol) and 20% para-toluenesulfonic acid (0.053 g, 0.31 mmol) in toluene (100 mL) were refluxed at 130 °C for 24 h. The reaction mixture was cooled to room temperature and then neutralized with sodium hydroxide solution (30 mL, 3 M). The organic layer was collected, washed with water (3 × 30 mL) and dried over anhydrous Na$_2$SO$_4$. The solvent was removed under reduced pressure and the crude product purified by column chromatography over silica gel (Hexane: EtOAc, 70:30) to afford a yellow solid in 50% yield. m.p. 149–151 °C. ^{1}H NMR (CDCl$_3$, 400 MHz): δ = 8.33 (s, 1H), 7.33 (d, J = 8.52 Hz, 1H), 6.61 (d, J = 8.48 Hz, 1H), 6.48 (s, 1H), 5.92 (d, J = 17.6 Hz, 1H), 5.32 (d, J = 11 Hz, 1H), 8.48 (s, 1H), 3.38 (q, J = 6.36 Hz, 4H), 1.36 (d, J = 6.44 Hz, 3H), 1.17 (t, J = 6.75 Hz, 6H). ^{13}C NMR (CDCl$_3$, 400 MHz): δ = 156.92, 155.79, 151.56, 146.19, 134.48, 129.98, 125.45, 121, 116.84, 116.39, 110.08, 107.10, 97.03, 45, 12.4. Anal. Calc. for $C_{17}H_{20}N_4O_3$: C: 65.79, H: 5.85, N: 18.05. Found: C: 65.76, H: 6.24, N: 18.35.

2.3.3. Synthesis of **P1**

To a 100 mL Schlenk flask, a mixture of vinyl monomer 6 (0.50 g, 1.61 mmol) and azobisisobutyronitrile (AIBN) (2×10^{-2} mmol) in dimethylformamide (DMF) (4 mL) was degassed using a freeze-thaw method (five cycles) followed by flushing with argon. The mixture was heated for 48 h at 70 °C and then poured to ethanol (50 mL). The precipitated polymer, after dropwise addition of a minimum amount of water, was filtered to afford polymers **P1** in 45% yield.

3. Results and Discussion

3.1. Synthesis and Characterization P1

In order to incorporate a conjugated triazolyl coumarin unit into a polymer chain, a monomer with polymerizable functionality and a triazolyl coumarin unit which may act as both receptor and reporter was prepared. The synthesis was accomplished in five steps from 5-(diethylamino)-2-hydroxybenzaldehyde as shown in Scheme 1. The polymerization of monomer 6 was achieved through controlled radical polymerization, in which a degassed system was required to avoid oxygen interferences with the free radical species [28]. The structures of **P1** were confirmed by ^{1}H NMR, as shown in Figure 1. Notable is the disappearance of the proton signals between 5.2 and

6.1 ppm in the monomer spectra (characteristic for the vinyl functionality), which are compensated by the appearance of the alkyl proton signals from the alkyl polymer backbone. The average molecular weight and polydispersity index of **P1** were determined to be 2.17×10^3 and 1.92, respectively.

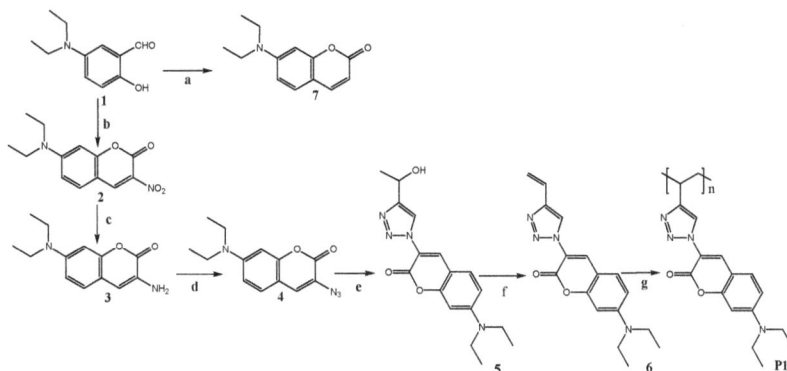

Scheme 1. Synthesis of polymer **P1**; (**a**) Piperidine, ethanol, HCl, AcOH, reflux; (**b**) Ethyl 2-nitroacetate, AcOH, piperidine, Butanol, reflux; (**c**) SnCl$_2$, HCl$_{aq}$, rt; (**d**) NaNO$_2$, KOAc, NaN$_3$, HCl$_{aq}$; (**e**) 3-Butyn-2-ol, CuSO$_4$·5H$_2$O, PMDETA, NaAsc, THF, rt; (**f**) PTSA, Toluene, 110 °C; (**g**) AIBN, DMF.

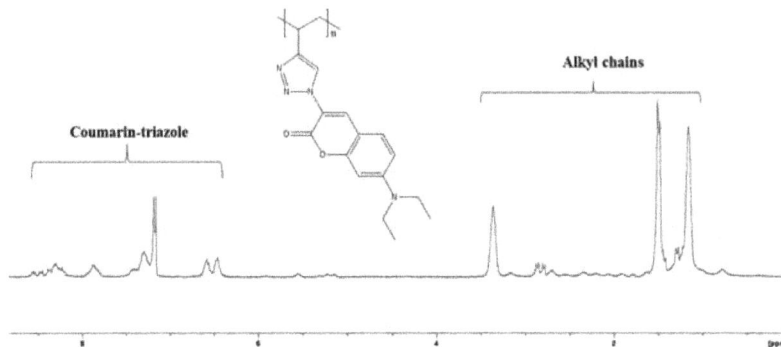

Figure 1. ^{1}H NMR spectrum for **P1** in CDCl$_3$.

P1 was partially soluble in less polar organic solvents and completely soluble in polar organic solvents such as acetonitrile, dimethyl sulfoxide (DMSO) and DMF. In order to avoid the interference of carbonyl and sulfinyl oxygen from DMF and DMSO solvents with the carbonyl group of the coumarin units, all spectroscopic measurements were performed in acetonitrile solvent. Detailed information regarding **P1** is summarized in Table 1 below.

Table 1. Radical polymerization data and photophysical properties of polymer **P1**.

Polymer	M$_w$ *	M$_n$ *	PDI *	λ_{abs} (nm)	λ_{ex} (nm)	λ_{emit} (nm)	Stock Shift (nm)
P1	2.17×10^3	1.19×10^3	1.92	265 and 396	395	484	90

* Determined from SEC experiments (eluent: DMAc, PMMA standard).

3.2. Absorption Spectra Analysis

The chemosensing capability of **P1** towards metal ions was initially investigated by UV-vis spectral analysis. This was achieved at room temperature in the presence of metal ions (monovalent, divalent and trivalent) such as Na^+, Li^+, Ca^{2+}, Ag^+, Al^{3+}, Cr^{3+}, Cu^{2+}, Fe^{2+}, Hg^{2+}, Mn^{2+}, Co^{2+}, Zn^{2+}, Cd^{2+}, Ni^{2+} and Pb^{2+}. As shown in Figure 2, all the metal ions except Cu^{2+} did not show any significant changes in the absorption peaks of **P1** at 265 nm and 396 nm. The addition of a Cu^{2+} aliquot induced an increase in the absorption peak intensity at 265 nm from 0.17 to 0.34 and a red shift to 287 nm. The absorption peak at 396 nm also showed a blue shift to 381 nm.

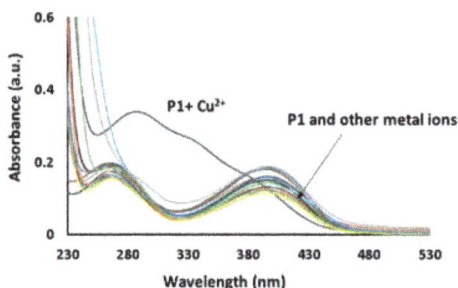

Figure 2. Absorption spectra of **P1** (4×10^{-3} g/L) in the presence of various metal ions (1.5×10^{-4} M aliquots) in acetonitrile.

The titration of Cu^{2+} with **P1** (Figure 3a) showed a blue shift in the absorption peak at 396 nm and a gradual increase in the intensity, accompanied by a red shift in the absorption peak at 265 nm. A clear isosbestic point was observed at 390 nm, indicating a complex formation between **P1** and Cu^{2+}. Using UV-vis titration data, a plot of A/A_o (where A_o and A are absorbance intensities in the absence and in the presence of Cu^{2+}) against the concentration of Cu^{2+} was plotted (Figure 3b). Linearity between A/A_o and $[Cu^{2+}]$ was found in the 130–190 μmol/L range, with a correlation coefficient of $R^2 = 0.9989$. The detection limit of **P1** was calculated according to the literature and was found to be 9×10^{-6} M [29].

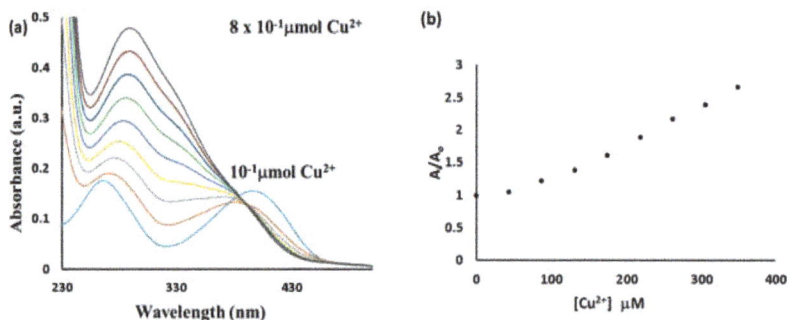

Figure 3. (**a**) Changes in the absorption spectrum of **P1** (4×10^{-3} g/L) in acetonitrile upon the addition of Cu^{2+} aliquots (0.05 M); (**b**) Plots of A/A_o against $[Cu^{2+}]$.

3.3. Emission Spectra Analysis

The selectivity of **P1** towards various metal ions was also investigated using emission spectral analysis under identical conditions as the UV-vis spectral analysis. However, Hg^{2+} and Cu^{2+} were the only metal ions that caused significant changes in the emission spectrum of **P1** when an aliquot of each tested metal ion was added to the acetonitrile solution of **P1** (Figure 4). The addition of Cu^{2+} caused a blue shift in the **P1** emission band from 484 to 469 nm, and a 68% intensity decrease, while the Hg^{2+} addition caused a 32% intensity decrease with no observable spectral shift. The fluorescent quenching in the presence of Hg^{2+} and Cu^{2+} could be respectively attributed to the heavy metal effect [30,31] and the transfer of excitation energy from the fluorophores to the metal *d*-orbital or charge transfer from the fluorophores to the metal ion [32].

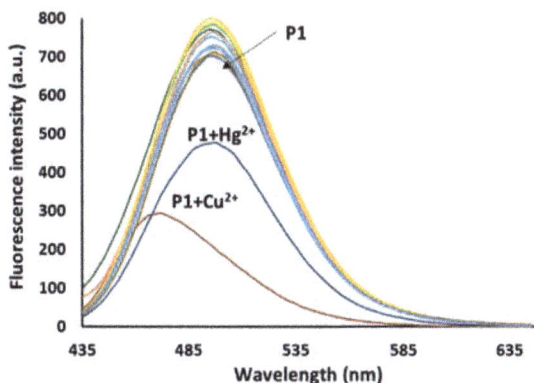

Figure 4. Emission spectra of **P1** (7×10^{-5} g/L) in the presence various metal ions (10 µL of 0.05 M solution) in acetonitrile. Excitation was performed at 395 nm.

The addition of Cu^{2+} aliquots to the **P1** solution from 0.8 to 4 µM led to a gradual decrease in the intensity of the emission peak at 484 nm which was accompanied by a blue shift from 484 to 470 nm (Figure 5a). Surprisingly, additional aliquots beyond 4 µM reversed the fluorescence response with a small increase in the intensity of the shifted emission band. The fluorescence increase was attributed to the aggregation-induced fluorescence mechanism (AIE) [33], from the aggregated triazolyl coumarin units resulting from the addition of Cu^{2+}. This aggregation initiated the AIE process by restricting the intramolecular rotation of the diethylamino group which deactivates the excited states of fluorophores [34–36]. However, the process was not strongly expressed due to the presence of the highly quenching effect from Cu^{2+}. The saturation point was achieved when the concentration of Cu^{2+} exceeded 10 µM. The maximum emission wavelength of the shifted peak at the saturation point was observed at 466 nm, suggesting a total shift of ca. 20 nm. A blue shift in the emission spectrum of **P1** suggests that the presence of Cu^{2+} induces the planarization in the triazolyl coumarin unit. In fact, dyes with electron donor, such as dialkylamino, and electron acceptor groups on the same aromatic ring form planar intermolecular charge transfer (ICT) structures with partial electron transfer upon electronic excitation. In polar environments, the dialkylamino group undergoes twisting which results in perpendicularity between the donor and acceptor orbitals [37]. This allows for the complete transfer of electrons and the mechanism is known as twisted intermolecular charge transfer (TICT) [38].

In **P1**, the addition of Cu^{2+} imposes greater spatial restriction in the triazolyl coumarin unit which leads to the planarization of the donor (diethylamino) and acceptor (carbonyl-triazole) groups in the excited state. This restricted planarization reduces the extent of conjugation which results in blue shifting of the **P1** emission band. Furthermore, it was reported that 7-dialkylaminocoumarins show a drastic reduction in the fluorescence quantum yields and fluorescence lifetimes in highly polar media due to the increased rate of the TICT process [34–36]. Since the addition of Cu^{2+} decreases the rate of TICT through planarization in triazoryl coumarin units, fluorescence quenching due to Cu^{2+} was opposed by the increasing quantum yield of triazoryl coumarin units which prevented the total quenching of the **P1** emission intensity upon the addition of Cu^{2+} aliquots. The detection limit was calculated from the plot of F/F_0 (where F_0 and F are fluorescence intensities in the absence and in the presence of Cu^{2+}) against the concentration of Cu^{2+} (Figure 5b) in the linear range between 0 and 3.2 μM and was found to be 0.75 μM. This low detection limit compared to the detection limit obtained from absorption experiments highlights a higher sensitivity associated with fluorescence-based sensing systems [39,40], and is sufficiently low for the detection of Cu^{2+} at the sub-millimolar level.

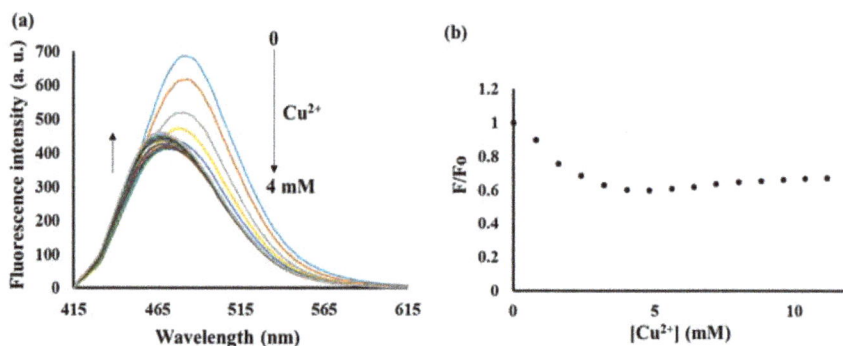

Figure 5. (a) Changes in the emission spectrum of **P1** (7×10^{-5} g/L) in acetonitrile upon the addition of Cu^{2+} aliquots (1.25×10^{-3} M); (b) plots of F/F_0 against $[Cu^{2+}]$. Excitation was performed at 395 nm.

To investigate the effect of other metal ions on the interaction between the **P1** and Cu^{2+}, we conducted competitive studies with other metal ions in the presence of Cu^{2+}. As shown in Figure 6, the fluorescence quenching and blue shift induced by the presence of Cu^{2+} ions in **P1** were not significantly affected by the presence of other metal ions. This indicates that **P1** and Cu^{2+} form a stable complex which cannot be interfered with the presence of other metal ions.

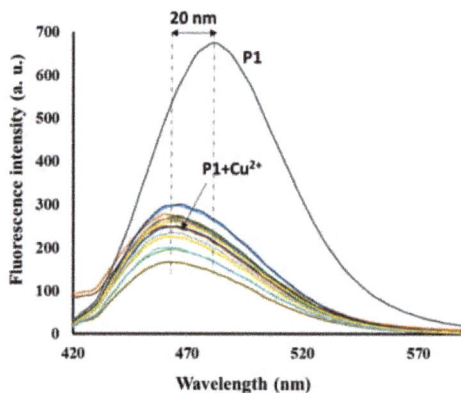

Figure 6. Fluorescence responses of **P1** (7×10^{-5} g/L) in the presence of a mixture of Cu^{2+} (1.5×10^{-4} M) and other metal ions (1.5×10^{-4} M). Excitation was performed at 395 nm.

3.4. Sensing Mechanism of **P1** with Cu^{2+}

To study the binding mode between **P1** and Cu^{2+}, the fluorescence responses of monomer 6 and compound 7 were investigated in the presence of increasing amounts of Cu^{2+} under the same conditions as **P1** spectral investigations. Similar to **P1**, the addition of Cu^{2+} aliquots to the monomer 6 solution induced a gradual decrease in intensity of the emission peak at 489 nm, accompanied by a blue shift from 489 to 458 nm (Figure 7a), while in compound 7 only a gradual decrease in emission intensity was observed (Figure 7b). These observations indicate that the triazole ring interacts with Cu^{2+} which leads to the planarization of the triazolyl coumarin unit in both monomer 6 and **P1**. Since compound 7 has only one possible binding site (the carbonyl group), the fluorescence quenching responses upon the addition of the increasing amount of Cu^{2+} indicate that the carbonyl group also takes part in Cu^{2+} binding. From the fluorescence titration experiments of monomer 6, a detection limit of 26 μM was calculated. This higher detection limit (~35 times higher than the **P1** detection limit) clearly indicates how the collective properties of the fluorophores in the polymer enhance the sensitivity in comparison to their monomer counterparts.

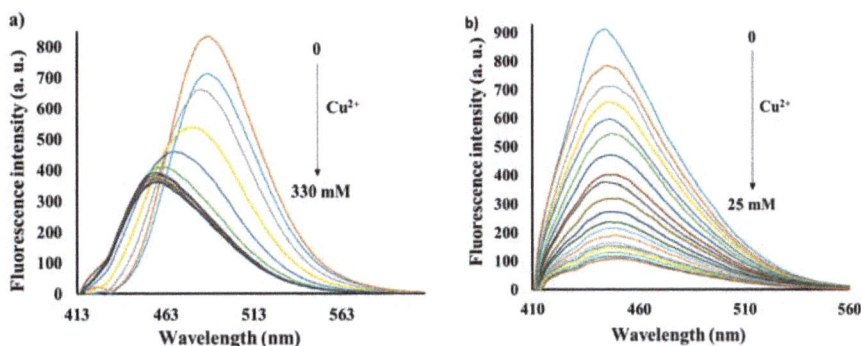

Figure 7. Changes in the emission spectra of (**a**) monomer 6 (5×10^{-12} M) and (**b**) compound 7 (3×10^{-8} M) in acetonitrile upon the addition of Cu^{2+} aliquots (monomer 6: 3×10^{-5} M Cu^{2+}; compound 7: 1.3 μM Cu^{2+}). Excitation was performed at 395 nm.

In order to determine the maximum number of triazole rings which take part in Cu^{2+} binding during complexation, a job plot analysis was carried out on monomer 6 using a continuous variation method [41]. The total concentration of added Cu^{2+} and monomer 6 was kept constant (6.5×10^{-11} M), and the plot of the emission intensity versus the molar fraction of Cu^{2+} at 395 nm is shown in Figure 8. Notably, the minimum emission intensity was achieved when the molar fraction was 0.5, which suggests a 1:1 stoichiometry of Cu^{2+}:monomer 6 complexation. The association constant (K_a) of **P1** with Cu^{2+} was evaluated graphically using the Benesi-Hildebrand equation (absorption method) [42–44]. The calculated value of K_a from the slope and intercept of the line was 2.6×10^3 M.

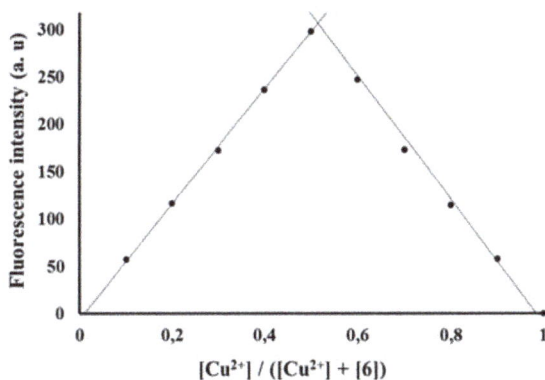

Figure 8. A job plot for monomer 6-Cu^{2+} complexation with a constant total concentration of 6.5×10^{-11} M in acetonitrile. Excitation was performed at 385 nm.

Since the job plot analysis indicated a 1:1 stoichiometry of the Cu^{2+}:monomer 6 complex, the interaction between Cu^{2+} and the triazolyl coumarin, which leads to blue shifting in the emission spectrum of **P1** and monomer 6, can be summarized in Scheme 2. To further analyze the Cu^{2+} interactions in monomer 6 and **P1**, UV-vis spectral analysis of monomer 6 in the presence of increasing amounts of Cu^{2+} was carried out (Figure 9). The addition of Cu^{2+} aliquots to the monomer 6 solution induced a gradual decrease in the intensity of the absorption peak at 412 nm, accompanied by a blue shift to 402 nm and a small absorption enhancement at ~300 nm. The blue shifting of the charge transfer absorption bands at 412 nm as in **P1** confirms the planarization of the triazoryl coumarin units. Furthermore, the comparison of the absorption spectra of **P1** and monomer 6 in the presence of increasing amounts of Cu^{2+} suggests that the planarization of triazoryl coumarin units in **P1** results in aggregation between the triazolyl coumarin units due to the increased π–π stacking. This was supported by a gradual increase in absorbance accompanied by the red shifting in the **P1** absorption band at 265 nm upon the addition of increasing amounts of Cu^{2+}, which is characteristic of J-aggregate behavior [45]. This feature can be used to distinguish Cu^{2+} from other metal ions, as no other tested metal ions induced such a change in the absorption spectrum of **P1**.

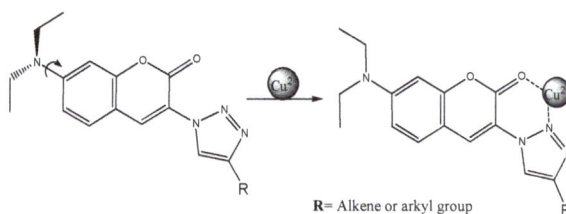

Scheme 2. A schematic representation of the Cu^{2+}-triazolyl coumarin unit interaction.

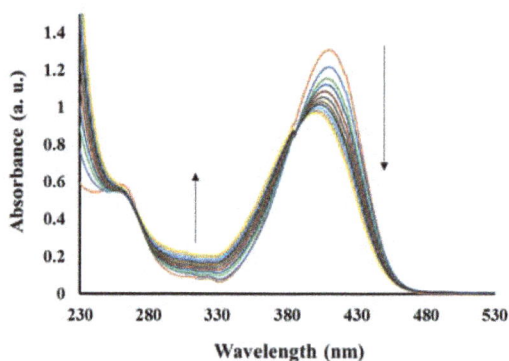

Figure 9. Changes in the absorption spectrum of monomer 6 (2×10^{-5} M) in acetonitrile upon the addition of Cu^{2+} aliquots (0.025 M).

3.5. Reversibility and Response Time

In order to examine the reversibility of **P1** complexation, the addition of EDTA to the acetonitrile solution of **P1** and Cu^{2+} was conducted (Figure 10). After the addition of an excess of EDTA to the solution, the fluorescence at 475 nm increased significantly and was again quenched upon the addition of Cu^{2+}. These findings indicate that **P1** reversibly coordinates to Cu^{2+}. Thus, it is plausible that **P1** may be recovered and reused in the laboratory and industrial settings. Although the fluorescence intensity of **P1** was regenerated upon the addition of EDTA, the blue spectral shift caused by the presence of Cu^{2+} could not be restored. This suggests that the complex that resulted from EDTA and

Cu^{2+} also promotes the aggregation process of the triazolyl coumarin units. The course of the response of **P1** (7×10^{-5} to 1.7×10^{-4} g/L) in acetonitrile was also investigated. The results showed that the recognition interaction takes place in 30 s after the addition of Cu^{2+}. Thus **P1** can be used for real-time monitoring of Cu^{2+} in practical analyses.

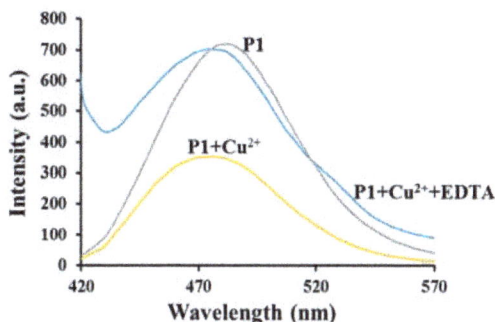

Figure 10. Reversibility of the interaction between **P1** and Cu^{2+} upon EDTA addition.

3.6. Preliminary Analytical Application

The applicability of the developed sensor was investigated in practical sample analysis using tap and lake water samples. The lake water samples were obtained from North End Lake (Eastern Cape, South Africa). In all samples analyzed using AAS, only tap water showed the presence of Cu^{2+} with the concentration of 0.275 ppm. The analytical studies of tap water samples using the developed sensor produced satisfactory results, as shown in Table 2. This indicates that **P1** could be used for the detection of Cu^{2+} even in complex media.

Table 2. Determination of Cu^{2+} (ppm) in tap water.

Sample	1	2	3
Concentration (ppm)	0.255	0.317	0.328

4. Conclusions

A novel polymer **P1** with triazolyl coumarin units as pendant groups was synthesized through multiple step syntheses. **P1** showed reasonable sensitivity and good selectivity towards Cu^{2+} in acetonitrile over a wide range of metal ions with remarkable quenching and blue shifting of the emission band. The blue shift in the emission band of **P1** was assigned to the planarization induced by the presence of Cu^{2+} in the triazolyl coumarin side chains and it was not interfered with by the presence of other metal ions. The investigation of the emission spectra of the starting monomer and reference compound 7 in the presence of Cu^{2+} indicated that the triazole ring and carbonyl functional groups were potential binding sites for Cu^{2+}. From the comparison of the absorption behaviors of monomer 6 and **P1** in the presence of increasing amounts of Cu^{2+}, it was noted that the planarization of triazolyl coumarin units in **P1** results in aggregation between the triazolyl coumarin units, which was supported by a red shift in the absorption spectrum of **P1**. The proposed sensor was also successfully applied in the determination of Cu^{2+} in real water samples. We believe that the design strategy, the remarkable photophysical properties of **P1** and the underlying mechanisms in the detection of Cu^{2+} will be useful in the development of novel polymer-based chemosensors which exploit new emerging signaling mechanisms.

Acknowledgments: For this work we acknowledge the National Research Foundation (NRF) and Nelson Mandela Metropolitan University (NMMU) for funding and facilities to carry out this project.

Author Contributions: Jean Marie Vianney Ngororabanga and Neliswa Mama conceived and designed the experiments; Jean Marie Vianney Ngororabanga and Jacolien Du Plessis performed the experiments; Jean Marie Vianney Ngororabanga and Neliswa Mama analyzed the data; Neliswa Mama contributed reagents/materials/analysis tools; Jean Marie Vianney Ngororabanga wrote the paper.

Conflicts of Interest: The authors declare no conflict of interest.

References

1. Cowan, J.A. *Inorganic Biochemistry: An Introduction*, 2nd ed.; Wiley-VCH: New York, NY, USA, 1997; pp. 133–134.
2. Linder, M.C.; Azam, M.H. Copper biochemistry and molecular biology. *Am. J. Clin. Nutr.* **1996**, *63*, 797S–811S. [PubMed]
3. Multhaup, G.; Schlicksupp, A.; Hesse, L.; Beher, D.; Ruppert, T.; Masters, C.L.; Beyreuther, K. The amyloid precursor protein of Alzheimer's disease in the reduction of copper(II) to copper(I). *Science* **1996**, *271*, 1406–1409. [CrossRef] [PubMed]
4. Deraeve, C.; Boldron, C.; Maraval, A.; Mazarguil, H.; Gornitzka, H.; Vendier, L.; Pitie, M.; Meunier, B. Preparation and study of new poly-8-hydroxyquinoline chelators for an anti-Alzheimer strategy. *Chem. Eur. J.* **2008**, *14*, 682–696. [CrossRef] [PubMed]
5. Bruijn, L.I.; Miller, T.M.; Cleveland, D.W. Unraveling the mechanisms involved in motor neuron degeneration in ALS. *Annu. Rev. Neurosci.* **2004**, *27*, 723–749. [CrossRef] [PubMed]
6. Waldermar, M.; Ryszard, R.; Teresa, U. Effect of excess Cu on the photosynthetic apparatus of runner bean leaves treated at two different growth stages. *Physiol. Plant.* **1994**, *91*, 715–721.
7. Georgopoulos, P.G.; Roy, A.; Yonone-Lioy, M.J.; Opiekun, R.E.; Lioy, P.J. Environmental copper: Its dynamics and human exposure issues. *J. Toxicol. Environ. Health Part B* **2001**, *4*, 341–394. [CrossRef] [PubMed]
8. Gonzales, A.P.S.; Firmino, M.A.; Nomura, C.S.; Rocha, F.R.P.; Oliveira, P.V.; Gaubeur, I. Peat as a natural solid-phase for copper preconcentration and determination in a multicommuted flow system coupled to flame atomic absorption spectrometry. *Anal. Chim. Acta* **2009**, *636*, 198–204. [CrossRef] [PubMed]
9. Liu, Y.; Liang, P.; Guo, L. Nanometer titanium dioxide immobilized on silica gel as sorbent for preconcentration of metal ions prior to their determination by inductively coupled plasma atomic emission spectrometry. *Talanta* **2005**, *68*, 25–30. [CrossRef] [PubMed]
10. Pathirathna, P.; Yang, Y.; Forzley, K.; McElmurry, S.P.; Hashemi, P. Fast-scan deposition-stripping voltammetry at carbon-fiber microelectrodes: Real-time, subsecond, mercury free measurements of copper. *Anal. Chem.* **2012**, *84*, 6298–6302. [CrossRef] [PubMed]
11. Mülazımoğlu, İ.E. Electrochemical determination of copper(II) ions at naringenin-modified glassy carbon electrode: Application in lake water sample. *Desalin. Water Treat.* **2012**, *44*, 161–167. [CrossRef]
12. Wheelock, C.E. The Fluorescence of Some Coumarins1. *Am. Chem. Soc.* **1959**, *81*, 1348–1352. [CrossRef]
13. Yoshimura, I.; Miyahara, Y.; Kasagi, N.; Yamane, H.; Ojida, A.; Hamachi, I. Molecular recognition in a supramolecular hydrogel to afford a semi-wet sensor chip. *J. Am. Chem. Soc.* **2004**, *126*, 12204–12205. [CrossRef] [PubMed]
14. Lv, F.; Feng, X.; Tang, H.; Liu, L.; Yang, Q.; Wang, S. Development of film sensors based on conjugated polymers for copper(II) ion detection. *Adv. Funct. Mater.* **2011**, *21*, 845–850. [CrossRef]
15. García, J.M.; García, F.C.; Serna, F.; de la Peña, J.L. High-performance aromatic polyamides. *Prog. Polym. Sci.* **2010**, *35*, 623–686.
16. Anzenbacher, P.; Liuand, Y.; Kozelkova, M.E. Hydrophilic polymer matrices in optical array sensing. *Curr. Opin. Chem. Biol.* **2010**, *14*, 693–704. [CrossRef] [PubMed]
17. Rotello, V.; Thayumanavan, S. *Molecular Recognition and Polymers: Control of Polymer Structure and Self-Assembly*; Rotello, V., Thayumanavan, S., Eds.; Wiley: Hoboken, NJ, USA, 2008.
18. Gu, C.; Huang, N.; Wu, Y.; Xu, H.; Jiang, D. Design of Highly Photofunctional Porous Polymer Films with Controlled Thickness and Prominent Microporosity. *Angew. Chem. Int. Ed.* **2015**, *54*, 11540–11544. [CrossRef] [PubMed]
19. Cui, W.; Wang, L.; Xiang, G.; Zhou, L.; An, X.; Cao, D. A colorimetric and fluorescence "turn-off" chemosensor for the detection of silver ion based on a conjugated polymer containing 2, 3-di(pyridin-2-yl) quinoxaline. *Sens. Actuators B Chem.* **2015**, *207*, 281–290. [CrossRef]

20. Luo, J.; Jiang, S.; Qin, S.; Wu, H.; Wang, Y.; Jiang, J.; Liu, X. Highly sensitive and selective turn-on fluorescent chemosensor for Hg^{2+} in pure water based on a rhodamine containing water-soluble copolymer. *Sens. Actuators B Chem.* **2011**, *160*, 1191–1197. [CrossRef]

21. Wang, B.; Liu, X.; Hu, Y.; Su, Z. Synthesis and photophysical behavior of a water-soluble coumarin-bearing polymer for proton and Ni^{2+} ion sensing. *Polym. Int.* **2009**, *58*, 703–709. [CrossRef]

22. Guo, Z.Q.; Zhu, W.H.; Tian, H. Hydrophilic copolymer bearing dicyanomethylene-4 H-pyran moiety as fluorescent film sensor for Cu^{2+} and pyrophosphate anion. *Macromolecules* **2010**, *43*, 739–744. [CrossRef]

23. Dong, Y.; Koken, B.; Ma, X.; Wang, L.; Cheng, Y.; Zhu, C. Polymer-based fluorescent sensor incorporating 2, 2'-bipyridyl and benzo[2,1,3]thiadiazole moieties for Cu^{2+} detection. *Inorg. Chem. Commun.* **2011**, *14*, 1719–1722. [CrossRef]

24. Rostovtsev, V.V.; Green, L.G.; Fokin, V.V.; Sharpless, K.B. A stepwise huisgen cycloaddition process: Copper(I)-catalyzed regioselective "ligation" of azides and terminal alkynes. *Angew. Chem. Int. Ed.* **2002**, *41*, 2596–2599. [CrossRef]

25. Meldal, M.; Tornøe, C.W. Cu-catalyzed azide-alkyne cycloaddition. *Chem. Rev.* **2008**, *108*, 2952–3015. [CrossRef] [PubMed]

26. Sivakumar, K.; Xie, F.; Cash, B.M.; Long, S.; Barnhill, H.N.; Wang, Q. A fluorogenic 1, 3-dipolar cycloaddition reaction of 3-azidocoumarins and acetylenes. *Org. Lett.* **2004**, *6*, 4603–4606. [CrossRef] [PubMed]

27. Wu, J.; Liu, W.; Zhuang, X.; Wang, F.; Wang, P.; Tao, S.; Zhang, X.; Wu, S.; Lee, S. Fluorescence turn on of coumarin derivatives by metal cations: A new signaling mechanism based on C=N isomerization. *Org. Lett.* **2007**, *9*, 33–36. [CrossRef] [PubMed]

28. Hasirci, V.; Yilgor, P.; Endogan, T.; Eke, G.; Hasirci, N. Polymer fundamentals: Polymer synthesis. *Compr. Biomater.* **2011**, *1*, 349–371.

29. Goswami, S.; Das, A.K.; Maity, S. 'PET' vs. 'push–pull' induced ICT: A remarkable coumarinyl-appended pyrimidine based naked eye colorimetric and fluorimetric sensor for the detection of Hg^{2+} ions in aqueous media with test trip. *Dalton Trans.* **2013**, *42*, 16259–16263. [CrossRef] [PubMed]

30. McClure, D.S. Spin-orbit interaction in aromatic molecules. *J. Chem. Phys.* **1952**, *20*, 682–686. [CrossRef]

31. Masuhara, H.; Shioyama, H.; Saito, T.; Hamada, K.; Yasoshima, S.; Mataga, N. Fluorescence quenching mechanism of aromatic hydrocarbons by closed-shell heavy metal ions in aqueous and organic solutions. *J. Phys. Chem.* **1984**, *88*, 5868–5873. [CrossRef]

32. Wang, R.; Wan, Q.; Feng, F.; Bai, Y. A novel coumarin-based fluorescence chemosensor for Fe^{3+}. *Chem. Res. Chin. Univ.* **2014**, *30*, 560–565. [CrossRef]

33. Hong, Y.; Lam, J.W.; Tang, B.Z. Aggregation-induced emission: Phenomenon, mechanism and applications. *Chem. Commun.* **2009**, *29*, 4332–4353. [CrossRef] [PubMed]

34. Satpati, A.K.; Kumbhakar, M.; Nath, S.; Pal, H. Photophysical Properties of Coumarin-7 Dye: Role of twisted intramolecular charge transfer state in high polarity protic solvents. *Photochem. Photobiol.* **2009**, *85*, 119–129. [CrossRef] [PubMed]

35. Nag, A.; Bhattacharyya, K. Role of twisted intramolecular charge transfer in the fluorescence sensitivity of biological probes: Diethylaminocoumarin laser dyes. *Chem. Phys. Lett.* **1990**, *169*, 12–16. [CrossRef]

36. Ramakrishna, G.; Ghosh, H.N. Efficient electron injection from twisted intramolecular charge transfer (TICT) state of 7-diethyl amino coumarin 3-carboxylic acid (D-1421) dye to TiO2 nanoparticle. *J. Phys. Chem. A* **2002**, *106*, 2545–2553. [CrossRef]

37. Sasaki, S.; Drummen, G.P.; Konishi, G.I. Recent advances in twisted intramolecular charge transfer (TICT) fluorescence and related phenomena in materials chemistry. *J. Mater. Chem. C* **2016**, *4*, 2731–2743. [CrossRef]

38. Nad, S.; Kumbhakar, M.; Pal, H. Photophysical properties of coumarin-152 and coumarin-481 dyes: Unusual behavior in nonpolar and in higher polarity solvents. *J. Phys. Chem. A* **2003**, *107*, 4808–4816. [CrossRef]

39. Kikuchi, K.P.K. Design, synthesis and biological application of chemical probes for bio-imaging. *Chem. Soc. Rev.* **2010**, *39*, 2048–2053. [CrossRef] [PubMed]

40. Carter, K.P.; Young, A.M.; Palmer, A.E. Fluorescent sensors for measuring metal ions in living systems. *Chem. Rev.* **2014**, *114*, 4564–4601. [CrossRef] [PubMed]

41. Renny, J.S.; Tomasevich, L.L.; Tallmadge, E.H.; Collum, D.B. Method of continuous variations: Applications of job plots to the study of molecular associations in organometallic chemistry. *Angew. Chem. Int. Ed.* **2013**, *52*, 11998–12013. [CrossRef] [PubMed]

42. Benesi, H.A.; Hildebrand, J.H. A spectrophotometric investigation of the interaction of iodine with aromatic hydrocarbons. *J. Am. Chem. Soc.* **1949**, *71*, 2703–2707. [CrossRef]
43. Barra, M.; Bohne, C.; Scaiano, J.C. Effect of cyclodextrin complexation on the photochemistry of xanthone. Absolute measurement of the kinetics for triplet-state exit. *J. Am. Chem. Soc.* **1990**, *112*, 8075–8579. [CrossRef]
44. Wang, L.; Ye, D.; Cao, D. A novel coumarin Schiff-base as a Ni(II) ion colorimetric sensor. *Spectrochim. Acta Part A* **2012**, *90*, 40–44. [CrossRef] [PubMed]
45. Würthner, F.; Kaiser, T.E.; Saha-Möller, C.R. J-Aggregates: From Serendipitous Discovery to Supramolecular Engineering of Functional Dye Materials. *Angew. Chem. Int. Ed.* **2011**, *50*, 3376–3410. [CrossRef] [PubMed]

MDPI

Article

Carbon Nanotubes as Fluorescent Labels for Surface Plasmon Resonance-Assisted Fluoroimmunoassay

Hiroki Ashiba [1],*, Yoko Iizumi [2], Toshiya Okazaki [2], Xiaomin Wang [3],† and Makoto Fujimaki [1]

[1]   Electronics and Photonics Research Institute, National Institute of Advanced Industrial Science and Technology (AIST), Tsukuba, Ibaraki 305-8565, Japan; m-fujimaki@aist.go.jp

[2]   CNT-Application Research Center, National Institute of Advanced Industrial Science and Technology (AIST), Tsukuba, Ibaraki 305-8565, Japan; iizumi-yoko@aist.go.jp (Y.I.); toshi.okazaki@aist.go.jp (T.O.)

[3]   Nanoelectronics Research Institute, National Institute of Advanced Industrial Science and Technology (AIST), Tsukuba, Ibaraki 305-8565, Japan; wang_x105@optoquest.co.jp

*   Correspondence: h.ashiba@aist.go.jp; Tel.: +81-29-851-4739

†   Present address: Research and Development Division, Optoquest Co., Ltd., Ageo, Saitama 362-0021, Japan.

Received: 15 September 2017; Accepted: 3 November 2017; Published: 7 November 2017

Abstract: The photoluminescence properties of carbon nanotubes (CNTs), including the large Stokes shift and the absence of fluorescent photobleaching, can be used as a fluorescent label in biological measurements. In this study, the performance of CNTs as a fluorescent label for surface plasmon resonance (SPR)-assisted fluoroimmunoassay is evaluated. The fluorescence of (8, 3) CNTs with an excitation wavelength of 670 nm and an emission wavelength of 970 nm is observed using a sensor chip equipped with a prism-integrated microfluidic channel to excite the SPR. The minimum detectable concentration of a CNT dispersed in water using a visible camera is 0.25 μg/mL, which is equivalent to 2×10^{10} tubes/mL. The target analyte detection using the CNT fluorescent labels is theoretically investigated by evaluating the detectable number of CNTs in a detection volume. Assuming detection of virus particles which are bound with 100 CNT labels, the minimum number of detectable virus particles is calculated to be 900. The result indicates that CNTs are effective fluorescent labels for SPR-assisted fluoroimmunoassay.

Keywords: biosensor; carbon nanotube; fluorescent probe; surface plasmon resonance; virus detection

---

1. Introduction

Carbon nanotubes (CNTs) have attracted considerable attention from researchers because of their superior electrical, mechanical, and optical properties [1,2]. In recent years, the photoluminescence of CNTs in the near-infrared region (1000–1400 nm) [3,4] has been utilized to develop fluorescent labels for biological experiments [5,6]. These CNT labels exhibit a large Stokes shift between the excitation and emission wavelengths, thus, the noises arising from the stray excitation light and autofluorescence of sensor chip substrates can be effectively removed using optical filters, and a high signal-to-noise ratio can be achieved. In addition, CNTs exhibit no fluorescent photobleaching that allows stable fluorescent measurements. However, a low quantum yield of CNTs, at most several per cent [3,7–9], limits the performance of fluorescent labels.

Surface plasmon resonance-assisted fluoroimmunoassay (SPRF), or surface plasmon-enhanced fluorescence spectroscopy (SPFS), is a sensitive biosensing technique that uses a fluorescent label and surface plasmon resonance (SPR) [10–13]. In SPRF, the luminescence of fluorescent labels is enhanced by two effects: (i) enhancement of the electric field intensity of the excitation light and (ii) enhancement of the quantum yield of fluorescent labels. The quantum yield enhancement has a greater effect when the original quantum yield of a fluorescent label is lower [14]. In addition, for conventional SPR sensors [15,16], carbon nanomaterials including CNTs and graphene has been used to enhance the resonance shift signal [17,18]; the combination of CNT and SPR is considered to be promising. Therefore, the use of CNT fluorescent labels with SPRF is effective in improving the low quantum yield.

The authors reported a norovirus detection system based on SPRF using a CdSe quantum dot fluorescent label, which exhibits a large Stokes shift similar to CNTs, and demonstrated sensitive detection of norovirus virus-like particles (VLPs) [19]. Although CdSe quantum dots possess good optical properties as a fluorescent label, the disposal of quantum dots is problematic due to the presence of toxic heavy metals. This problem, in some cases, would be critical for practical use. In contrast, CNTs can be incinerated, and therefore, are more suitable for practical use. CNT fluorescent labels are thus worth investigating with SPRF.

In this study, the performance of CNTs as a fluorescent label for target analyte detection using SPRF was evaluated. A V-trench biosensor, which was developed as a miniature and simple SPRF apparatus [19–21], was used for the evaluation. The sensor is equipped with a V-shaped microfluidic channel, which functions as a prism to excite the SPR, and it can perform sensitive biosensing based on SPRF. SPR-induced fluorescence emission on the V-trench was confirmed for fluorescent molecules and quantum dot fluorescent labels in previous studies [19–21]. Herein, the evaluation of a CNT fluorescent label using a V-trench biosensor designed for CNTs is presented.

2. Materials and Methods

2.1. CNT Dispersion Preparation

An aqueous dispersion of CNTs was prepared using single-strand deoxyribonucleic acid (DNA; Sonicated Salmon Sperm DNA, Agilent Technologies, Santa Clara, CA, USA). A 10 mg/L aqueous solution of the DNA was boiled for 5 min and cooled with ice. CNTs (CoMoCAT SG65, SouthWest NanoTechnologies, Canton, MA, USA) with a total mass of 20 mg were pre-dispersed in 10 mL of ethanol by bath sonication for 2 min. The CNTs were washed by repetitive vacuum filtering and sonic dispersion with 10 mL of ethanol 3 times and 10 mL of boiled water 16 times. The washed CNTs with a mass of 1 mg and 0.4 mL of the DNA solution were mixed with 10 mL of 50 mM phosphate buffer (pH 8.0). The mixture was sonicated using a VCX 500 instrument (Sonics & Materials, Newtown, CT, USA) for 10 min and centrifuged at 171000G for 60 min. In the present experiment, the supernatant containing monodisperse CNTs was used.

In this study, we focused on (8, 3) CNTs with excitation and emission wavelengths of 670 nm and 970 nm, respectively. (8, 3) CNTs were chosen for the following reasons: (i) the emission at a wavelength of 970 nm is detectable using a visible charge-coupled device (CCD) camera with a relative response of several percent; (ii) the excitation at a wavelength of 670 nm is suitable for SPR excitation by gold, which is stable and is the most commonly used material for SPR. A photoluminescence map of the dispersion of CNTs measured using a spectrometer (Fluorolog-3-2-iHR320, Horiba Jobin Yvon, Kyoto, Japan) is shown in Figure 1. From the integrated photoluminescence intensity, the content ratio of the (8, 3) CNTs in the dispersion was estimated to be approximately 15%. The length of the CNTs contained in the dispersion was measured using an atomic force microscope (AFM). The AFM image is shown in Figure S1 in the Supplementary Material. The typical length of the CNT was 1 μm. The monodispersity of the CNTs was also evaluated using the AFM image, and over 90% of CNTs was monodisperse in the prepared dispersion.

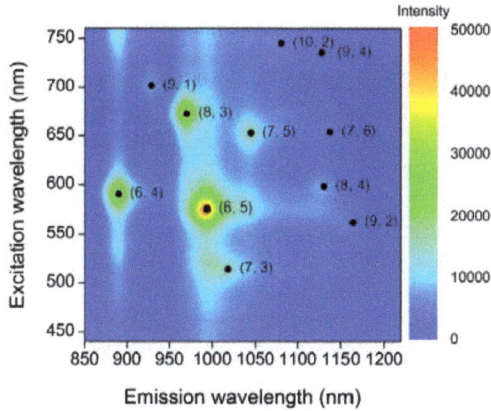

Figure 1. Photoluminescence map of the CNT dispersion used here.

2.2. Experimental Apparatus

A schematic diagram of the V-trench biosensor is shown in Figure 2. The optical system consists of a light source, sensor chip, and camera that are aligned in a straight line. The sensor chip is equipped with a microfluidic channel having a V-shaped cross section, which functions as a prism to excite the SPR. The vertex angle of the V-shaped trench determines the incident angle of the excitation light, and is designed using electric field simulation to maximize the electric field intensity at the surface (Section 2.3). The analyte is captured at the surface of the V-shaped trench and labeled with a fluorescent dye. The excitation light enters the sensor chip from the backside and illuminates the V-trench to excite the SPR, and the fluorescence excited by the SPR is detected using the camera.

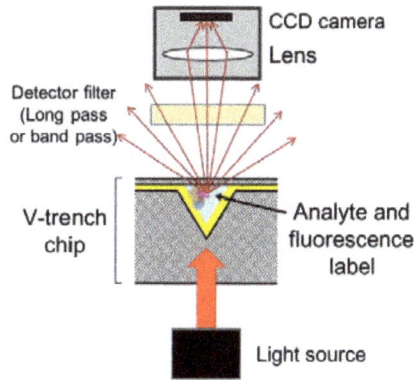

Figure 2. Schematic diagram of the V-trench biosensor.

For the measurement of fluorescence from (8, 3) CNTs, a fluorescence imaging instrument (Light-Capture II, Atto, Tokyo, Japan) equipped with a laser diode with a wavelength of 670 nm (Shibasaki, Chichibu, Japan) is used. The excitation light, which is 6 mm in diameter, is p- or s-polarized to perform the fluorescence measurement with or without SPR excitation, respectively. The emitted light from the CNTs excited by the SPR passes through a long pass filter with a cut-off wavelength of 900 nm (FEL0900, Thorlabs, Newton, NJ, USA) and is detected using a cooled visible CCD camera (pixel size: 8.4 μm (H) × 9.8 μm (V)).

2.3. Sensor Chip

The V-trench sensor chip was designed using electric field simulation based on the transfer matrix method [22]. The vertex angle of the V-trenches and the thickness of a gold film as an SPR excitation layer were determined to maximize the electric field intensity against the excitation light with a wavelength of 670 nm. A multi-layer model was used to represent the V-trench sensor chip, as shown in Figure 3a, which includes, from bottom to top, a polystyrene substrate, an adhesive layer of chromium with a thickness of 0.6 nm, a gold layer, a protein layer with a thickness of 20 nm, and water. The protein layer represents the surface modifications to the sensor chip, including immobilized antibodies and self-assembled monolayers for protein immobilization. The complex refractive indices at a wavelength of 670 nm were 1.5848 for polystyrene [23], $3.7176 + i4.3666$ for chromium [24], $0.16316 + i3.4625$ for gold [25], 1.45 for protein, and 1.3302 for water [26]. An electric field enhancement factor ($|E/E_0|^2$) map against the p-polarized excitation light with a wavelength of 670 nm is shown in Figure 3b, where the values of $|E/E_0|^2$ are calculated at the boundary of the protein and water layers. The maximum value of $|E/E_0|^2$ is 23.8, which is obtained at a vertex angle of 45.6° and a gold thickness of 49 nm. Thus, the sensor chip having the V-shaped trench with a vertex angle of 45.6° was prepared.

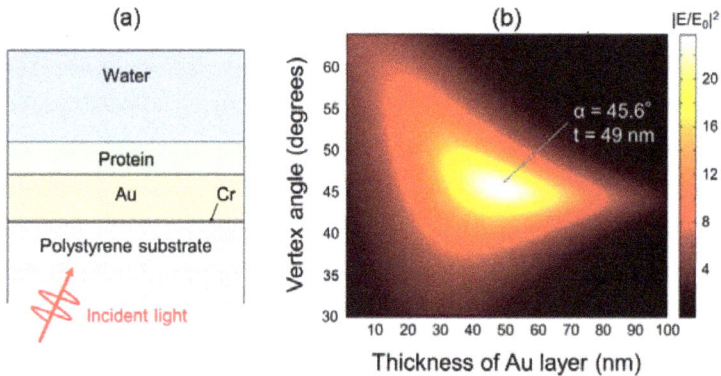

Figure 3. (**a**) Schematic of the cross section of the V-trench sensor chip surface used in the electric field simulation of a multi-layer model based on the transfer matrix method; (**b**) Calculated electric field enhancement factor ($|E/E_0|^2$) against the thickness of the gold layer (t) and the vertex angle of the V-trench (α). The excitation wavelength is 670 nm, and $|E/E_0|^2$ at the boundary between the protein and water layers is shown. The arrow indicates the point of maximum $|E/E_0|^2$.

The substrate of the V-trench sensor chip was fabricated using injection molding (Cluster Technology, Higashi-Osaka, Japan) of polystyrene (CR-3500, DIC, Tokyo, Japan). The width of the opening and bottom and the length of the V-trench were 0.3, 0.02, and 10 mm, respectively. The chromium and the gold layers were formed on the substrate using a vacuum deposition system (Biemtron, Shirosato, Japan). The setting thicknesses of the deposited chromium and gold films were 0.6 nm and 126 nm, respectively. Under this condition, the thickness of the gold film in the direction perpendicular to the surface of the 45.6° V-trenches becomes 49 nm.

2.4. Fluorescence Measurement of CNTs

The CNT dispersion was directly applied to the V-trenches without any surface modifications. The intensities of fluorescence of the CNT dispersion samples at various concentrations were measured.

Before applying the samples, the V-trench sensor chips were cleaned using a plasma cleaner (PDC-32G, Harrick Plasma, Ithaca, NY, USA) at an RF power of 18 W for 30 s. To acquire "Blank" fluorescent images, the sensor chip filled with 15 µL of Milli-Q water (Merck, Darmstadt, Germany) was placed in the V-trench biosensor system, the excitation light was irradiated, and the images were acquired with an exposure time of 1 min. Then, the Milli-Q water was removed from the V-trench by blowing nitrogen, and 15 μL of the CNT dispersions samples was applied. The concentration of the original CNT dispersion was 25 μg/mL, and the CNT dispersion with various concentrations ranging from 0.063 to 25 μg/mL were prepared by diluting the original dispersion with Milli-Q water. Under the excitation light irradiation, fluorescent images of the samples were acquired with an exposure time of 1 min. The luminescence intensity was obtained by integrating the brightness of each pixel in a defined area (80 × 20 pixels; approximately 7 × 2 mm^2 on the actual dimension) of the fluorescent images. The power density of the excitation light illuminated on the sensor chip was 11 mW/cm^2. It is known that CNTs release heat by the photothermal effect [27,28]. The effect of heat to the sensor chip and fluorescent signal was not observed for the power density used here (see Figure S2 in the Supporting Material).

3. Results

The luminescent intensities of the CNT dispersion sample with a concentration of 25 μg/mL measured using the V-trench biosensor under the p- and s-polarized excitation light are shown in Figure 4. The bars in the figure indicate the average luminescent intensities obtained from three fluorescent images, and the error bars indicate the standard errors. The luminescent intensity measured under the p-polarized light irradiation, where SPR was excited, was 4.4-folds greater than that measured under the s-polarized light. This result indicates that the SPR was excited on the V-trench and that the fluorescence of CNTs was excited by the enhanced electric field of SPR.

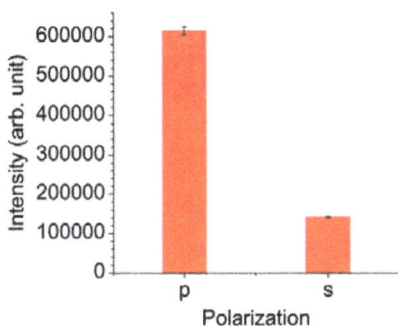

Figure 4. Luminescent intensities of the CNT dispersion sample with a concentration of 25 μg/mL measured using a V-trench biosensor. The excitation light was p- or s-polarized. The error bars indicate the standard errors.

The luminescent intensities of the CNT dispersion samples with various concentrations measured under the p-polarized excitation light are shown in Figure 5. The symbols in the figure indicate the average luminescent intensities obtained from three fluorescent images, and the error bars indicate the standard errors. The fluorescence intensities were positively correlated with the concentration of the CNTs. By comparing the luminescent intensities of "Blank" and "CNT," the minimum detectable concentration of the CNTs in this experiment was 0.25 μg/mL. The intensity is linear to the concentration at the range of approximately over 1 μg/mL, whereas the gradient of the intensity decreases below 1 μg/mL. This would be due to the attachment of CNTs to the sensing surface. Assuming that a density of CNTs at the sensing surface increases due to the attachment, it would result in increased luminescent intensity and the decreased gradient when the concentration is low. When the concentration is high, the attachment to the surface would be saturated and the luminescent intensities become linear.

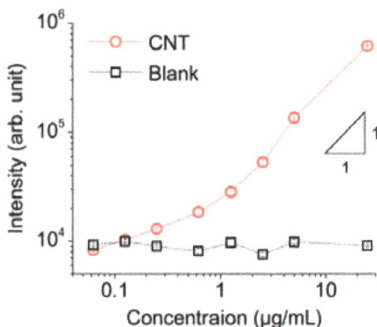

Figure 5. Luminescent intensities of the CNT dispersion samples with various concentrations measured using a V-trench biosensor. The excitation light was p-polarized. "CNT" and "Blank" were measured by applying the CNT dispersion samples and Milli-Q water into the V-trench, respectively. The error bars indicate the standard errors.

4. Discussion

For evaluating the potential of CNT fluorescent label, the minimum detectable number of CNTs is calculated herein, and the minimum detectable number of target analyte, including protein molecules and virus particles, is discussed accordingly. In the fluorescent measurement described above, CNTs were considered to be uniformly distributed in the dispersion. Then, the number of CNTs contained in a detection volume, N_{CNT}, is calculated as:

$$N_{CNT} = D_{CNT} A H \tag{1}$$

where D_{CNT} is the number of CNTs in a unit volume, A is the measurement area, and H is the height of a detection volume. In general, H is related with the height of the illumination by excitation light, since the dimension of a fluorescent label is smaller than the height of illumination. However, in this study, the CNT length is considered to be larger than the height of illumination by SPR. Specifically, the CNT length was typically 1 μm as mentioned in Section 2.1, and from the electric field simulation described in Section 2.3, the decay length of the electric field of SPR was derived to be 176 nm for the developed system. In this case, if the height of illumination is employed as H, the number of CNTs is underestimated. Therefore, the CNT length (1 μm) is herein employed as H. Next, D_{CNT} is calculated from the concentration of CNT dispersion as follows. For example, when the concentration of dispersion is 0.25 μg/mL, the concentration of luminous (8, 3) CNTs is 3.8×10^{-2} μg/mL, since the content ratio of (8, 3) CNTs was approximately 15%, as mentioned in Section 2.1. The structural model of (8, 3) CNTs yielded the number of carbon atoms in its unit length as 93 atoms/nm. Using the number of carbon atoms and the CNT length (1 μm), a concentration of 3.8×10^{-2} μg/mL yields $D_{CNT} = 2.0 \times 10^{10}$ tubes/mL. A is calculated to be 4.3 mm^2 for the developed V-trench biosensor using the excitation light with a diameter of 6 mm and an opening width, a bottom width, and a vertex angle of the V-trench of 0.3 mm, 0.02 mm, and 45.6°, respectively. When 0.25 μg/mL is employed as the minimum detectable concentration, using Equation (1) with $D_{CNT} = 2.0 \times 10^{10}$ tubes/mL, $A = 4.3$ mm^2, and $H = 1$ μm, the minimum detectable number of CNTs is calculated to be 9×10^4 tubes.

As mentioned in Section 3, the luminescent intensities shown in Figure 5 reveal non-linearity at the low concentration region, which would be due to the attachment of CNTs to the sensing surface. This means that, if the attachment to the sensing surface is avoided, the minimum detectable concentration would be larger than 0.25 μg/mL. Considering the gradient of the intensities over 1 μg/mL, the minimum detectable concentration without CNT attachment is considered to be greater up to twice. Thus, the minimum detectable number of CNTs is considered to be in between 9×10^4 and 2×10^5 tubes.

When considering the detection of target analyte using a CNT fluorescent label, since the CNT length is larger than the height of illumination and the dimension of target analyte, the detectable number of target analyte, N_T, is simply calculated as $N_T = N_{CNT}/B$, where B is the number of CNT labels bound to one target analyte. The value of B varies depending on various factors such as the structure of target analyte, binding effectivity of a CNT label, etc. Figure 6 shows N_T against B for $N_{CNT} = 9 \times 10^4$ and 2×10^5 tubes. Typical ranges of B for proteins and virus particles are also indicated. For example, when the CNT label is used for the detection of virus with $B = 100$, the minimum detectable number of virus particles is 900–2000. The calculation indicates that CNTs possess the potential as a fluorescent label for sensitive biosensing using SPRF sensors.

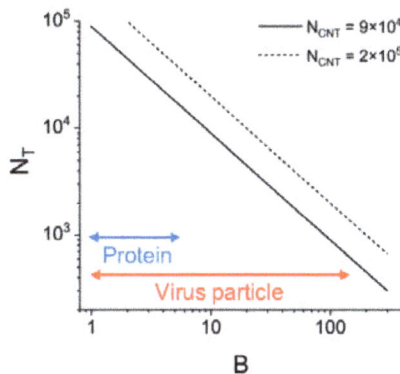

Figure 6. Detectable number of target analyte (N_T) against the number of CNT labels bound to one target analyte (B) for the detectable number of CNTs (N_{CNT}) of 9×10^4 and 2×10^5 tubes. Blue and red arrows indicates typical ranges of B for proteins and virus particles.

The effect of quenching in the fluorescence measurements should also be discussed. Since the surface of the V-trench is bare gold, the photoluminescence of fluorescent labels near the surface is likely quenched. Firstly, it has been reported that CNTs coated with organic molecules are less affected by quenching [29]. The CNTs used in this study were coated with DNA molecules and thus considered tolerant against quenching. Secondly, the minimum detectable concentration of CNTs was measured by including the quenching effect. That is to say, the minimum detectable number of target analyte evaluated above also includes the quenching effect. Since the CNTs are long, as mentioned above, even if a part of the CNT is inside the quenching distance, the other part of the CNT would likely be outside. Other fluorescent labels such as fluorescent molecules and quantum dots are small (<20 nm), and if these fluorescent labels come close to the surface, their luminescence intensities are significantly affected by quenching. The tolerability against quenching is an advantage of the CNT compared with other fluorescent labels.

In the fluorescence measurement, the CNT dispersions were directly applied onto the gold layer of the V-trenches, whereas, as shown in Section 2.3, the structure of the V-trench sensor chip was optimized assuming that a protein layer was formed on the gold layer to capture the target. $|E/E_0|^2$ of the V-trenches used for the CNT dispersion measurement is considered to be lower than the optimal condition; the simulation for the V-trenches without the protein layer yields $|E/E_0|^2 = 12.6$, which is half the optimal value. Therefore, in practical target detection using the V-trenches with a protein layer, the minimum detectable number of CNTs can be better than evaluated above. On the other hand, under practical conditions, various factors would depress the performance of target detection by lowering the value of B and generating noise. These factors includes labeling efficiency of the CNT to the antibody, binding efficiency of the antibody to the target analyte, nonspecific binding of the CNT-labeled antibody, and content ratio of luminous CNTs. In particular, labeling with antibodies is a major challenge for CNTs, because of poor reactivity. Although several techniques have been reported for CNT labeling [5,30,31], the efficiency, stability, and controllability of the labeling were insufficient. The binding of the unlabeled antibodies to the target analyte compete with that of the CNT-labeled antibodies, resulting in depressed sensitivity of target detection. Further investigation of the labeling techniques would be beneficial for CNT fluorescent labels. Regarding the content ratio of the CNTs, the CNT dispersion used in this study contained 15% of (8, 3) CNTs that emitted the observed fluorescence. It is desirable that only (8, 3) CNTs are contained in the fluorescent labels. In recent years, effective methods of selective growth and purification of CNTs have been reported [32,33]. These methods are useful to prepare efficient CNT fluorescent labels. In addition, the performance of CNT fluorescent labels can be further improved by modifying the CNT itself. For example, the luminescent intensity of CNTs is greatly enhanced by introducing defects using oxidization or covalent functionalization [34–36]. These techniques improve the potential of CNT labels even further.

5. Conclusions

The performance of (8, 3) CNTs as a fluorescent label for SPRF was evaluated. The fluorescent measurement of CNTs was performed using a V-trench biosensor, and its capability for the influenza virus detection was estimated. A sensor chip optimized for (8, 3) CNTs was designed to enhance fluorescence with an excitation wavelength of 670 nm. The minimum detectable concentration of the CNTs was 0.25 µg/mL. The minimum detectable number of CNT labels was evaluated to be 9×10^4–2×10^5 tubes. The detectable number of target analyte was evaluated as the function of the minimum detectable number of CNTs and the number of CNT labels bound to one target analyte. Once the challenges such as labeling efficiency are overcome, CNTs are considered as a promising candidate for use as fluorescent labels for SPRF.

Supplementary Materials: The following are available online at http://www.mdpi.com/1424-8220/17/11/2569/s1, Figure S1: Atomic force microscope image of the CNT dispersion. Figure S2: Luminescent intensities of the CNT dispersion sample measured with various powers of the excitation light.

Author Contributions: H.A., T.O. and M.F. conceived and designed the experiments; H.A. and Y.I. performed the experiments; H.A. analyzed the data; Y.I. contributed materials; X.W. performed the electric field simulation; H.A. wrote the paper.

Conflicts of Interest: The authors declare no conflict of interest.

References

1. Dresselhaus, M.S.; Dresselhaus, G.; Avouris, P. (Eds.) *Carbon Nanotubes: Synthesis, Structure, Properties, and Applications*; Springer: Berlin, Germany, 2001; ISBN 978-3-540-41086-7.
2. Jorio, A.; Dresselhaus, G.; Dresselhaus, M.S. (Eds.) *Carbon Nanotubes: Advanced Topics in the Synthesis, Structure, Properties and Applications*; Springer: Berlin, Germany, 2008; ISBN 978-3-540-72864-1.
3. O'Connel, M.J.; Bachilo, S.M.; Huffman, C.B.; Moore, V.C.; Strano, M.S.; Haroz, E.H.; Rialon, K.L.; Boul, P.J.; Noon, W.H.; Kittrell, C.; et al. Band Gap Fluorescence from Individual Single-Walled Carbon Nanotubes. *Science* **2002**, *297*, 593–596. [CrossRef] [PubMed]
4. Bachilo, S.M.; Strano, M.S.; Kittrell, C.; Hauge, R.H.; Smalley, R.E.; Weisman, R.B. Structure-Assigned Optical Spectra of Single-Walled Carbon Nanotubes. *Science* **2002**, *298*, 2361–2366. [CrossRef] [PubMed]
5. Liu, Z.; Tabakman, S.; Welsher, K.; Dai, H. Carbon nanotubes in biology and medicine: In vitro and in vivo detection, imaging and drug delivery. *Nano Res.* **2009**, *2*, 85–120. [CrossRef] [PubMed]
6. Hong, G.; Antaris, A.L.; Dai, H. Near-infrared fluorophores for biomedical imaging. *Nat. Biomed. Eng.* **2017**, *1*. [CrossRef]
7. Lefebvre, J.; Austing, D.G.; Bond, J.; Finnie, P. Photoluminescence imaging of suspended single-walled carbon nanotubes. *Nano Lett.* **2006**, *6*, 1603–1608. [CrossRef] [PubMed]
8. Crochet, J.; Clemens, M.; Hertel, T. Quantum Yield Heterogeneities of Aqueous Single-Wall Carbon Nanotube Suspensions. *J. Am. Chem. Soc.* **2007**, *129*, 8058–8059. [CrossRef] [PubMed]

9. Hertel, T.; Himmelein, S.; Ackermann, T.; Stich, D.; Crochet, J. Diffusion Limited Photoluminescence Quantum Yields in 1-D Semiconductors: Single-Wall Carbon Nanotubes. *ACS Nano* **2010**, *4*, 7161–7168. [CrossRef] [PubMed]

10. Attridge, J.W.; Daniels, P.B.; Deacon, J.K.; Robinson, G.A.; Davidson, G.P. Sensitivity enhancement of optical immunosensors by the use of a surface plasmon resonance fluoroimmunoassay. *Biosens. Bioelectron.* **1991**, *6*, 201–214. [CrossRef]

11. Liebermann, T.; Knoll, W. Surface-plasmon field-enhanced fluorescence spectroscopy. *Colloids Surf. A* **2000**, *171*, 115–130. [CrossRef]

12. Roy, S.; Kim, J.-H.; Kellis, J.T.; Poulose, A.J.; Robertson, C.R.; Gast, A.P. Surface Plasmon Resonance/Surface Plasmon Enhanced Fluorescence: An Optical Technique for the Detection of Multicomponent Macromolecular Adsorption at the Solid/Liquid Interface. *Langmuir* **2002**, *18*, 6319–6323. [CrossRef]

13. Toma, K.; Vala, M.; Adam, P.; Homola, J.; Knoll, W.; Dostálek, J. Compact surface plasmon-enhanced fluorescence biochip. *Opt. Express* **2013**, *21*, 10121–10132. [CrossRef] [PubMed]

14. Kajikawa, K.; Okamoto, T.; Takahara, J.; Okamoto, K. *Active Plasmonics*; Chap. 5; Corona Publishing: Tokyo, Japan, 2013; ISBN 978-4-339-00836-4. (In Japanese)

15. Nylander, C.; Liedberg, B.; Lind, T. Gas detection by means of surface plasmon resonance. *Sens. Actuators* **1982**, *3*, 79–88. [CrossRef]

16. Homola, J.; Yee, S.S.; Gauglitz, G. Surface plasmon resonance sensors: Review. *Sens. Actuators B* **1999**, *54*, 3–15. [CrossRef]

17. Zeng, S.; Baillargeat, D.; Ho, H.-P.; Yong, K.-T. Nanomaterials enhanced surface plasmon resonance for biological and chemical sensing applications. *Chem. Soc. Rev.* **2014**, *43*, 3426–3452. [CrossRef] [PubMed]

18. Zeng, S.; Sreekanth, K.V.; Shang, J.; Yu, T.; Chen, C.-K.; Yin, F.; Baillargeat, D.; Coquet, P.; Ho, H.-P.; Kabashin, A.V.; et al. Graphene-Gold Metasurface Architectures for Ultrasensitive Plasmonic Biosensing. *Adv. Mater.* **2015**, *27*, 6163–6169. [CrossRef] [PubMed]

19. Ashiba, H.; Sugiyama, Y.; Wang, X.; Shirato, H.; Higo-Moriguchi, K.; Taniguchi, K.; Ohki, Y.; Fujimaki, M. Detection of norovirus virus-like particles using a surface plasmon resonance-assisted fluoroimmunosensor optimized for quantum dot fluorescent labels. *Biosens. Bioelectron.* **2017**, *93*, 260–266. [CrossRef] [PubMed]

20. Nomura, K.; Gopinath, S.C.B.; Lakshmipriya, T.; Fukuda, N.; Wang, X.; Fujimaki, M. An angular fluidic channel for prism-free surface-plasmon-assisted fluorescence capturing. *Nat. Commun.* **2013**, *4*, 2855. [CrossRef] [PubMed]

21. Ashiba, H.; Fujimaki, M.; Wang, X.; Awazu, K.; Tamura, T.; Shimizu, Y. Sensor chip design for increasing surface-plasmon-assisted fluorescence enhancement of the V-trench biosensor. *Jpn. J. Appl. Phys.* **2016**, *55*, 067001. [CrossRef]

22. Born, M.; Wolf, E. *Principles of Optics: Electromagnetic Theory of Propagation, Interference and Diffraction of Light*, 6th ed.; Pergamon: Oxford, UK, 1980; ISBN 978-0-08-026482-0.

23. Refractive Index Database. Available online: http://refractiveindex.info (accessed on 17 April 2017).

24. Lynch, D.W.; Hunter, W.R. An Introduction to the Data for Several Metals. In *Handbook of Optical Constants of Solids II*; Parik, E.D., Ed.; Academic Press: San Diego, CA, USA, 1998; pp. 374–385, ISBN 0-12-544422-2.

25. Lynch, D.W.; Hunter, W.R. Comments on the Optical Constants of Metals and an Introduction to the Data for Several Metals. In *Handbook of Optical Constants of Solids*; Parik, E.D., Ed.; Academic Press: San Diego, CA, USA, 1998; pp. 286–295, ISBN 0-12-544420-6.

26. Querry, M.R.; Wieliczka, D.M.; Segelstein, D.J. Water (H_2O). In *Handbook of Optical Constants of Solids II*; Parik, E.D., Ed.; Academic Press: San Diego, CA, USA, 1998; pp. 1059–1077, ISBN 0-12-544422-2.

27. Kam, N.W.S.; O'Connell, M.; Wisdom, J.A.; Dai, H. Carbon nanotubes as multifunctional biological transporters and near-infrared agents for selective cancer cell destruction. *Proc. Natl. Acad. Sci. USA* **2005**, *102*, 11600–11605. [CrossRef] [PubMed]

28. Murakami, T.; Nakatsuji, H.; Inada, M.; Matoba, Y.; Umeyama, T.; Tsujimoto, M.; Isoda, S.; Hashida, M.; Imahori, H. Photodynamic and Photothermal Effects of Semiconducting and Metallic-Enriched Single-Walled Carbon Nanotubes. *J. Am. Chem. Soc.* **2012**, *134*, 17862–17865. [CrossRef] [PubMed]

29. Hong, G.; Tabakman, S.M.; Welsher, K.; Wang, H.; Wang, X.; Dai, H. Metal-enhanced fluorescence of carbon nanotubes. *J. Am. Chem. Soc.* **2010**, *132*, 15920–15923. [CrossRef] [PubMed]

30. Liu, Z.; Tabakman, S.M.; Chen, Z.; Dai, H. Preparation of carbon nanotube bioconjugates for biomedical applications. *Nat. Protoc.* **2009**, *4*, 1372–1382. [CrossRef] [PubMed]

31. Iizumi, Y.; Okazaki, T.; Ikehara, Y.; Ogura, M.; Fukata, S.; Yudasaka, M. Immunoassay with Single-Walled Carbon Nanotubes as Near-Infrared Fluorescent Labels. *ACS Appl. Mater. Interfaces* **2013**, *5*, 7665–7670. [CrossRef] [PubMed]
32. Yang, F.; Wang, X.; Zhang, D.; Yang, J.; Luo, D.; Xu, Z.; Wei, J.; Wang, J.-Q.; Xu, Z.; Peng, F.; et al. Chirality-specific growth of single-walled carbon nanotubes on solid alloy catalysts. *Nature* **2014**, *510*, 522–524. [CrossRef] [PubMed]
33. Yomogida, Y.; Tanaka, T.; Zhang, M.; Yudasaka, M.; Wei, X.; Kataura, H. Industrial-scale separation of high-purity single-chirality single-wall carbon nanotubes for biological imaging. *Nat. Commun.* **2016**, *7*, 12056. [CrossRef] [PubMed]
34. Ghosh, S.; Bachilo, S.M.; Simonette, R.A.; Beckingham, K.M.; Weisman, R.B. Oxygen Doping Modifies Near-Infrared Band Gaps in Fluorescent Single-Walled Carbon Nanotubes. *Science* **2010**, *330*, 1656–1659. [CrossRef] [PubMed]
35. Miyauchi, Y.; Iwamura, M.; Mouri, S.; Kawazoe, T.; Ohtsu, M.; Matsuda, K. Brightening of excitons in carbon nanotubes on dimensionality modification. *Nat. Photonics* **2013**, *7*, 715–719. [CrossRef]
36. Piao, Y.; Meany, B.; Powell, L.R.; Valley, N.; Kwon, H.; Schatz, G.C.; Wang, Y. Brightening of carbon nanotube photoluminescence through the incorporation of sp^3 defects. *Nat. Chem.* **2013**, *5*, 840–845. [CrossRef] [PubMed]

sensors

MDPI

Article

Design and Fabrication of a Ratiometric Planar Optode for Simultaneous Imaging of pH and Oxygen

Zike Jiang, Xinsheng Yu * and Yingyan Hao

Key Lab of Submarine Geosciences and Prospecting Techniques, Ministry of Education,
College of Marine Geosciences, Ocean University of China, Qingdao 266100, China;
jiangzike2011@126.com (Z.J.); 15634219925@163.com (Y.H.)
* Correspondence: xsyu@ouc.edu.cn; Tel.: +86-532-667-82913

Academic Editor: Sheshanath Bhosale
Received: 9 April 2017; Accepted: 27 May 2017; Published: 7 June 2017

Abstract: This paper presents a simple, high resolution imaging approach utilizing ratiometric planar optode for simultaneous measurement of dissolved oxygen (DO) and pH. The planar optode comprises a plastic optical film coated with oxygen indicator Platinum(II) octaethylporphyrin (PtOEP) and reference quantum dots (QDs) embedded in polystyrene (PS), pH indicator 5-Hexadecanoylamino-fluorescein (5-Fluorescein) embedded in Hydromed D4 matrix. The indicator and reference dyes are excited by utilizing an LED (Light Emitting Diode) source with a central wavelength of 405 nm, the emission respectively matches the different channels (red, green, and blue) of a 3CCD camera after eliminating the excitation source by utilizing the color filter. The result shows that there is low cross-sensitivity between the two analytes dissolved oxygen and pH, and it shows good performance in the dynamic response ranges of 0–12 mg/L and a dynamic range of pH 6−8. The optode has been tested with regard to the response times, accuracy, photostability and stability. The applied experiment for detecting pH/Oxygen of sea-water under the influence of the rain drops is demonstrated. It is shown that the planar optode measuring system provides a simple method with low cross-talk for pH/Oxygen imaging in aqueous applications.

Keywords: pH; dissolved oxygen; planar optode; ratiometric; sea-water; rain drops

---

1. Introduction

For decades, microelectrodes have been widely used in underwater observation, enabling the further understanding of chemical gradients of seawater [1]. However, they can only provide point measurements with high spatial resolution. Recently, a better technique based on the use of an optical oxygen indicator for spatial mapping (called planar optode) has been introduced for real time applications [2,3]. The planar optode based on luminescence of specific indicators allowed multiple measurement of analytes in a variety of formats, and imaging analysis of larger areas.

So far, the majority of planar optode studies have been focused on pH and oxygen dynamics in rhizospheres [4–6]; measurements in sea-ice for resolving physical and biologically induced oxygen dynamics [7]; marine sediments including the investigation of microbial mats and biofilms [8,9]; effects of bioturbating/irrigating fauna [10–12]; and studies of pH two-dimensional in benthic substrates [13,14]. There is a broad interest in applying a multi-analyte planar optode to environmental sciences, biological, marine sediments, and medical science benthic communities [15–18], but the requirements of relatively complicated multi-analyte planar optode, and measuring systems, have limited the number of users.

The first planar optode systems introduced for aquatic applications were based on pure intensity measurements [19]. Pure intensity measurements for a planar optode suffer from several disadvantages and limitations such as the fluorescence intensity being sensitive to variations in excitation light, background interference, and inhomogeneous indicator [19–21]. Holst et al. [22] introduced the superior luminescence life-time-based method to overcome the main disadvantage of the pure intensity measurements. While the lifetime method was superior and overcame the main limitations of the intensity-based systems, it required a relatively complex trigger control circuit and high precision industrial camera [23]. One alternative to the two systems is the ratiometric approach, it can eliminate the deficiencies caused by varied excitation light, background interference, and inhomogeneous indicator, while not requiring complex available hardware [24–26].

The previous studies of the ratiometric sensors are mostly based on the DSLR (Digital Single Lens Reflex) or CCD (Charge Coupled Device) cameras with single CCD. The majority of digital cameras make use of the bayer color filter, which allows them to separate the incident light into the three primary colors: red, green, and blue. However, it detects only one-third of the color information for each pixel. The other two-thirds must be interpolated with a demosaicing algorithm to 'fill in the gaps', resulting in a much lower effective resolution [27]. In this scheme, it could cause two-thirds of light intensity loss, and affect the sensitivity of the planar optode. In this paper we present an imaging observation system based on 3CCD camera. The imaging system of 3CCD camera uses three separate charge-coupled devices (CCDs), each one taking a separate measurement of the primary colors, red, green, or blue light. The incident light coming into the lens is split by a trichroic prism assembly, which directs the appropriate wavelength ranges of light to their respective CCDs. Compared to cameras with single CCD, 3CCD cameras generally provide superior image quality through enhanced resolution and lower noise. By taking separate readings of red, green, and blue values for each pixel, 3CCD cameras achieve much better precision than single-CCD cameras, and ensure the spectral intensity.

In order to simultaneously realize the imaging of pH and oxygen chemical parameters, multiple indicators can be combined within a single planar optode [28,29]. It is very difficult to separate the emission of the dyes, and this would cause cross-sensitivity between the red, green and blue channel of a 3CCD camera. Moßhammer et al. introduced a new optical dual-analyte sensor for imaging. It is a simple way to use a 2CCD camera with near-infrared (NIR) sensitive camera chip to expand the potential of ratiometric readout [30].

The 3CCD camera can be used to read out one reference dye and two indicator dyes. However, the principle is limited by the spectral overlap of the traditional dyes possessing broad emission spectrum. This calls for simpler, lower overlap, good antijamming capability and user-friendly dyes and measuring setups. Owing to the characteristics of insensitivity, broad spectrum excitation, tunable emission wavelength, narrow half peak width, good light stability and high quantum yield [31], quantum dots (QDs) is an excellent reference. Wang et al. introduced a sensing film doped with modified water-soluble CdTe/CdS quantum dots and PtF20-TPP imbed in sol-gel with two layers respectively on slides. The thickness of the sensing film is about 70 μm, and it achieved colorimetric oxygen determination with precise, distinct, and tunable color [32]. In order to improve homogeneity, mechanical strength and applicability of the sol-gel in extreme condition, we synthesized the oleic acid modified QDs which was lipophilic and could be imbed in polystyrene (PS) layer without leakage in aquatic conditions.

In this study, we synthesized the oleic acid modified QDs [33,34] with narrow spectral emission. It was lipophilic, high-efficiency and pH/oxygen insensitive. By doping the QDs as the reference dye, a new ratiometric pH/Oxygen planar optode combination with a 3CCD camera was designed for simultaneous imaging of pH and oxygen. The planar optode used PtOEP as the oxygen indicator with red emission, 5-Fluorescein as the pH indicator with green emission, and QDs as the reference dye with blue emission. These three dyes were respectively embedded in polystyrene (PS) and Hydromed D4 matrix. The performance of the planar optode was validated for detecting pH/ Oxygen of sea-water under the influence of the rain drops [35,36].

2. Experimental

2.1. Materials Preparation

The oxygen indicator dye PtOEP (CAS number : 31248-39-2), pH indicator dye 5-Fluorescein (CAS number: 73024-80-3), chloroform, ethanol, octadecene (ODE), TOP (trioctylphosphine) and oleic acid (OA) were all analytical grade, they were purchased from Aladdin Chemical Company (Shanghai, China); The matrix PS, D4 were purchased from J&K Chemical Company (Shanghai, China).

2.2. Synthesis of Lipophilic QDs

The oleic acid modified QDs was synthesized according to the literature [37]. Briefly, The 70 mL clear octadecene (ODE) solution was prepared by mixing t 5.54 g OA, 0.51 g CdO and was heated at 180 °C under N_2 atmosphere. We could obtain different sizes of QDs by controlling the temperature respectively, and herein the mixture was heated to 275 °C. Fully stirred TOP-Se solution containing 1.3 mmol Se powder, 0.5 g trioctylphosphine and 10 mL ODE, was quickly injected into the three-neck round-bottom flask. Finally, the resulting stable QDs solution was dispersed in chloroform, and stored at a temperature of 8 °C.

2.3. pH/Oxygen Planar Optode Fabrication

For the oxygen layer, the solution was fabricated by mixing 200 mg of polymer PS, 1 mg of QDs, 1 mg of PtOEP dye in 0.4 mL chloroform. The sensor film was fabricated by utilizing a knife coating which was a fast and simple method [3]. The solution was coated onto a 125 μm thick polyester support foil to form the a ~5 μm thick oxygen sensitive layer after solvent evaporation. The solution containing 200 mg of polymer D4 (ethanol: water, 9:1 w/w), 1 mg of 5-Fluorescein dye in chloroform was knife coated on top of the oxygen layer, obtaining ~15 μm thick pH sensitive layer. For the optical isolation layer, the solution containing 1.00% w/w of carbon black in 10% w/w of Hydromed D4 (ethanol: water, 9:1 w/w) was coated onto the pH layer after solvent evaporation (Figure 1).

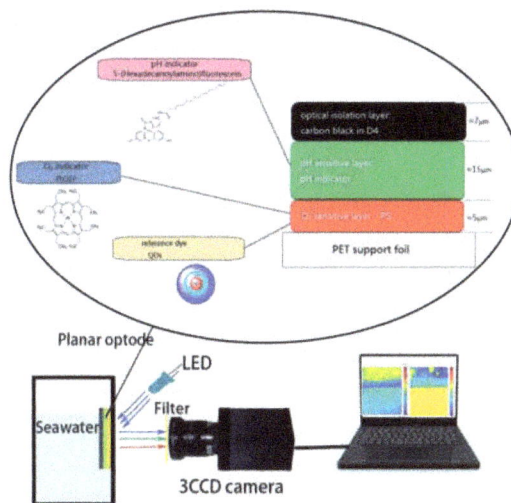

Figure 1. Schematic drawing of the new ratiometric multiple pH/Oxygen planar optode (the oval part) and measuring system with JAI AT-200GE 3CCD camera. The oval part indicates the drawing of the pH/Oxygen sensor composed of a three layer system containing the depicted indictor and reference dyes. Filter: 455 nm longpass. LED: high power LED (peak wavelength 405 nm) as a light source.

The ratiometric measuring scheme was sensitive to wavelength-dependent scattering and reflection from any background behind the transparent sensor. However, a thin translucent D4 layer imbedded with carbon powder solved the problem. Thus, the pH/Oxygen sensitive film was coated with D4 matrix doped with carbon powder. The dry isolation layer was ~7 μm thick and semitransparent with a light transmission of ~20%. This ensured that any structures behind the sensor were still visible during oxygen measurements, without affecting the ratiometric approach.

2.4. Instrument and Method

The pH/Oxygen measuring system is illustrated in Figure 1. For this present work, the luminescence excitation was provided by 6 high power 405 nm LED (λ-peak = 405 nm, Φ = 400 mW at I_F = 1100 mA) from Tianyao companies, and the LED was powered by a DH171 5A-3 DC stabilized power supply (Dahua Company, Beijing, China). The relative luminescence intensity was measured using the Ocean Optics spectrometer USB2000+ and JAI AT-200GE 3CCD camera (Daheng Image Company, Beijing, China). Emissions filters which were fabricated in front of the 3CCD camera were 440 nm long pass filters (Nantong Optical glass Company, Beijing, China).

The JAI AT-200GE camera provided 2-megapixel resolution (1628 × 1236) based on three 1/1.8-inch CCDs, 1624 (h) × 1236 (v) effective pixels (4.40 μm square) for each CCD. With 3-CCD technology, a specific R, G, and B value was captured for each pixel. The camera has a GigE Vision interface and its output can be either 24-bit or 32-bit RGB in TIFF format. The 32-bit RGB TIFF images were automatically saved onto the computer (exposure time of 15535 μs) and used for further calculations. Processing of the recorded images was performed with the free software ImageJ (http://rsbweb.nih.gov/ij/) and analyzed in MATLAB software.

Bayer mosaic color cameras used a pattern of color filters and an interpolation process to estimate the approximate RGB value of a given pixel. With 3-CCD technology, a specific R, G, and B value was captured for each pixel (Figure 2). This inherently produced higher color precision in the 3-CCD output. Furthermore, the spectral curves resulting from the hard dichroic prism coatings were much steeper than the curves from the soft polymer dyes used in Bayer filters. This enabled the 3-CCD cameras to produce exceptionally accurate color data without the uncertainty that comes from the overlap regions. In addition to reducing color precision, the overlap in the color filter response also causes part of each pixel's well capacity to fill with photons resulting from the crosstalk, thus decreasing the available well capacity. Precision response from the dichroic coatings enables each channel to efficiently use the full well capacity of the pixel, allowing the maximum possible dynamic range.

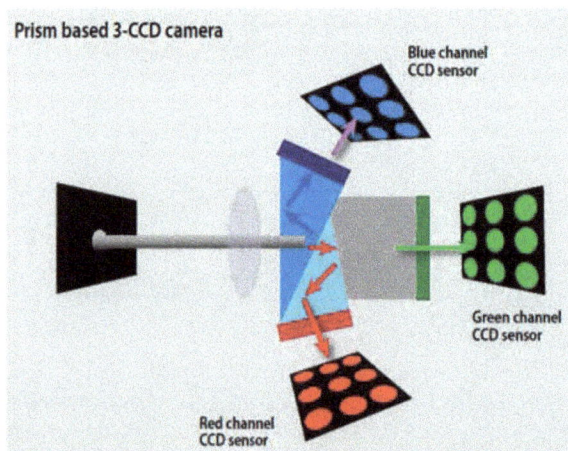

Figure 2. Schematic drawing of prism based JAI AT-200GE 3CCD camera.

The oxygen concentrations in 30‰ salinity seawater was controlled by mixing oxygen and nitrogen, and this process was controlled by gas flow-meters. The pH calibration solution was confected by the 30‰ artificial seawater containing 10 mM TRIS-HCL buffer. The pH of the seawater was adjusted by adding either 1 M HCl or 1 M NaOH to the glass tank. The dissolved oxygen concentration was measured by O_2 microelectrode (Unisense O_2 Microsensor, Aarhus, Denmark), and the temperature-compensated pH microelectrode (YSI pH 100A, Shanghai, China) was used to continuously monitor the pH level of the seawater.

2.5. Simulated Rainfall Experiments

For sea-water measurements, the pH/Oxygen sensitive film was transferred to a custom-made quartz aquarium (H × L × W: 12 × 12 × 12 cm) that was subsequently filled with natural seawater collected in the Shilaoren Bay, Qingdao, China. The quartz aquarium was allowed to settle for 24 h before the presented images were recorded. During this period, the setup was kept at a constant temperature of 17 °C and exposed to a 12/12 h light/dark cycle. We simulated rainfall at the rate of about 0.5 inch/h for 10 min by experimental rainfall setup in a laboratory. A device with an array of micro holes of constant sizes (3 mm in diameter) was used to produce rain of constant size drops. The rainwater (the oxygen concentrations of 8.31 mg/L, pH value 7.2) was collected and stored in the tank with an air pump, and the air pump is a device for pushing air during this period.

3. Results and Discussion

3.1. Optical Properties of Multiple pH/Oxygen Planar Optode

Figure 3 showed the luminescence spectra of three dyes materials (PtOEP, 5-Fluorescein, and QDs) at room-temperature. When excited with a 405 nm LED, the planar optode exhibited intense luminescence emissions at 650 nm, 520 nm, and 460 nm, respectively. The emissions of QDs served as the reference insensitive to pH and oxygen, and it also acted as an internal donor molecule for the indicator. Since the emissions of the QDs overlapped with the absorption of 5-Fluorescein, the brightness of the two indicators, especially the 5-Fluorescein was enhanced by making use of QDs. The emissions of indicator PtOEP and 5-Fluorescein are respectively sensitive to oxygen and pH, thus the pH and oxygen can be measured. In addition, by the ratio process, the deficiencies of the intensity was eliminated by the reference QDs.

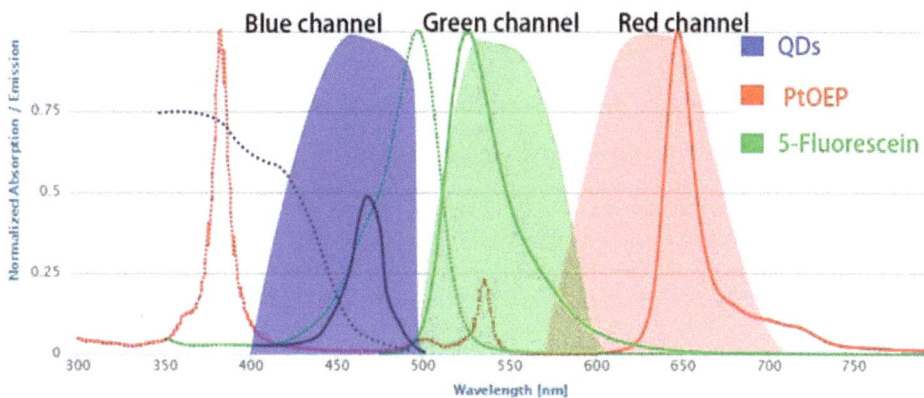

Figure 3. Spectra of the dyes (absorption in dashed and emission in solid lines), and the spectral range of three different channels (red, green, and blue) in JAI AT-200GE 3CCD camera.

The emission intensity of the three dye materials (PtOEP, 5-Fluorescein, and QDs) respectively was recorded by the corresponding three spectrum channels (red, green, and blue) of the 3CCD camera. The pixel intensity of the red, green channel was dominated by the luminescence from the PtOEP and the 5-Fluorescein, respectively. The blue channel represented the luminescence of QDs and blue LED excited light, after eliminating the excitation source by utilizing the longpass filter (>455 nm). There was a negligible spectral overlap between the three dye materials.

3.2. Calibration and Accuracy Evaluation of the pH/Oxygen Planar Optode

Figure 4 showed the fit equations and calibration plots for pH and oxygen in artificial sea water at a concentration of 30‰ (salinity). For the oxygen calibration, the relationship between intensity and oxygen concentration was reflected by the modified Stern–Volmer equation [38]:

$$\frac{I}{I_0} = \left(A + \frac{1-A}{1 + K_{sv}[O_2]} \right) \tag{1}$$

where, I was the intensity ratios of the red and blue channels (R/B) of the planar optode at different oxygen concentration, I_0 expressed the corresponding values at oxygen free condition. Ksv and A respectively were the constant of calibration curve and the quenching ratio of indicator. In an ideal quencher system, there is a linear relationship between I_0/I and oxygen concentration, so a linear Stern-Volmer equation can be applied. However, a matrix effected on the quenching properties of luminescent dyes in solid solutions strongly depends on the polymer used. It is necessary to take into account the influence of polymer which was attributed to a distribution of quenching rate constants.

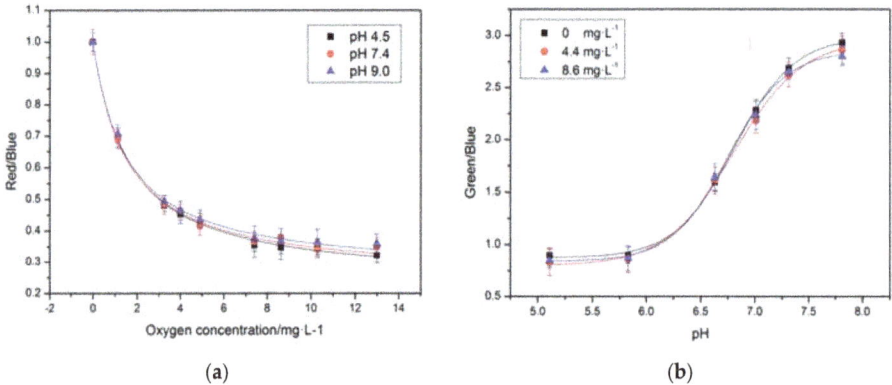

Figure 4. The calibration curves for (**a**) oxygen and pH (**b**) in seawater. (17.0 ± 0.2 °C).

For the pH calibration, the relationship between intensity and pH values was fitted by the four-parametric Boltzmann (sigmoidal) equation [39]:

$$R = b + \frac{a - b}{1 + e^{\frac{pH - pka}{dx}}} \tag{2}$$

where, R was the intensity ratios of the green and blue channels (G/B) of the planar optode at different pH values, a, b and dx respectively express the empirical parameters and the width of the curve, pKa was the coefficient of center of the calibration curve.

In order to evaluate potential cross-talk for oxygen and pH, the experiments were conducted at varying condition (Figure 5). Figure 4a presented the calibration curves for the oxygen sensitivity in the dual pH/Oxygen planar optode system, and these were conducted for three different pH values: one acidic (pH 4.5), alkalescence (pH 7.4) and one alkaline (pH 9.0).

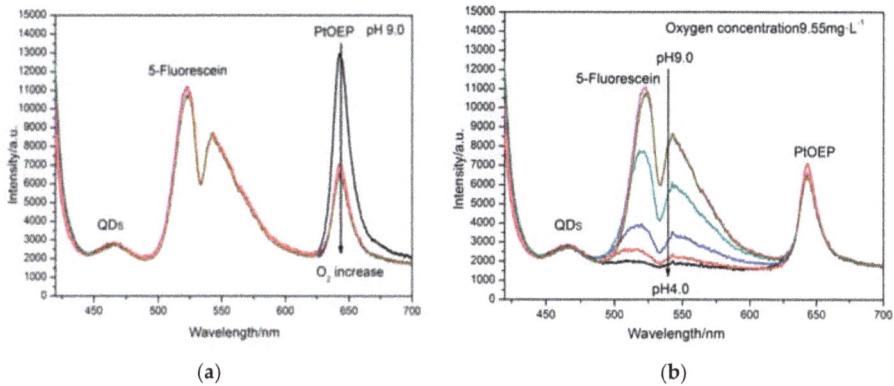

(a)   (b)

Figure 5. The emission spectra of ratiometric pH/oxygen planar optode:(a) Emission spectra in different oxygen concentration at pH 9.0, the different colors of the lines represented the different oxygen concentration, and the arrow indicates the increasing of oxygen concentration; (b) Emission spectra at pH values between 4 and 9 under the oxygen concentration of about 9.55 mg·L^{-1}, the different colors of the lines represented the different pH value, and the arrow indicated the decrease of pH.

The optode's response for the oxygen showed nonlinear calibration curves with maximum sensitivity at low oxygen concentrations. The ratio decreases more than 80% when the oxygen concentration increases from 0 to 8 mg/L. For this oxygen concentration scope, the pH/Oxygen planar optode retained a highly effective large reduction. This was mainly because of the simultaneous presence of static and dynamic quenching [40] and the inequality of the microenvironment of the immobilized dye PtOEP placed within the polymer matrix PS. In the oxygen concentration of 0–6 mg/L, the calibration curves exhibited no cross interferences caused by changes in the pH. However, a slight cross-sensitivity toward oxygen was observed in the oxygen concentration of 8–12 mg/L. This can be explained by the energy transfer to the PtOEP from the luminescence of 5-Fluorescein, the interference increased with decreased luminescence of PtOEP under hyperoxia condition. The pH response of the dual pH/Oxygen planar optode followed a sigmoidal pattern with an apparent pKa of around 6.95. As the Figure 4b showed, a minor cross-sensitivity toward oxygen was observed at pH values around the pKa. In the pH scope between 6.3 and 7.3, the calibration curves exhibited no cross interferences caused by changes in the oxygen.

Tables 1 and 2 illustrated the accuracy evaluation between the pH/Oxygen planar optode and pH/oxygen microelectrode. It was noted that the pH/Oxygen planar optode possessed superior behavior in oxygen depleted condition for the concentration range of 0–6 mg/L than the supersaturated condition. For the pH values, the optode with an apparent pKa of around 7.15 showed superior behavior at pH values between 6 and 8.

Table 1. Comparison of the pH/Oxygen planar optode and electrodes for oxygen measurements. AE: absolute error; RE: relative error; SD: standard deviation.

Oxygen Electrodes		pH/Oxygen Planar Optode		
Measured	Calculated	AE	RE	SD
0.01	0.0096	0.0004	4.00%	0.0038
4	3.884	0.116	2.90%	0.1244
7	7.432	0.432	6.17%	0.3102
9.34	9.974	0.634	6.79%	0.7352
10.59	9.812	0.778	7.35%	0.6321

Table 2. Comparison of the pH/Oxygen planar optode and electrodes for pH measurements. AE: absolute error; RE: relative error; SD: standard deviation.

pH Electrodes		pH/Oxygen Planar Optode		
Measured	Calculated	AE	RE	SD
4.23	3.92	0.31	7.32%	0.0462
4.89	4.63	0.26	5.32%	0.1487
5.56	5.80	0.24	3.72%	0.1292
6.45	6.27	0.18	2.79%	0.3043
7.75	7.45	0.3	3.87%	0.2945

3.3. Long Term Stability and Photostability

The stability and photostability was a key analytical figure of the pH/Oxygen planar optode [26,41]. The photostability was tested by placing the ratiometric pH/Oxygen planar optode into seawater with continuous irradiation by using a USHIO EKE, 150W halogen lamp (Weiguang Company, Shenzhen, China) equipped with an optical filter at room temperature for around 100 min. The LI-250A photometer (ECOTEK Company ,Beijing, China)was used to measure the output of halogen lamp, and the irradiances was 253 μmol photons$\cdot$s$^{-1}\cdot$m^{-2}. The relative luminescence intensities of the reference QDs, 5-Fluorescein, and PtOEP decreased 9.82%, 4.701%, 6.45%, respectively (Figure 6a). The ration of the Green channel/Blue channel changed 4.67%, and ratios of Red channel/Blue channel changed 5.25% (Figure 6c).

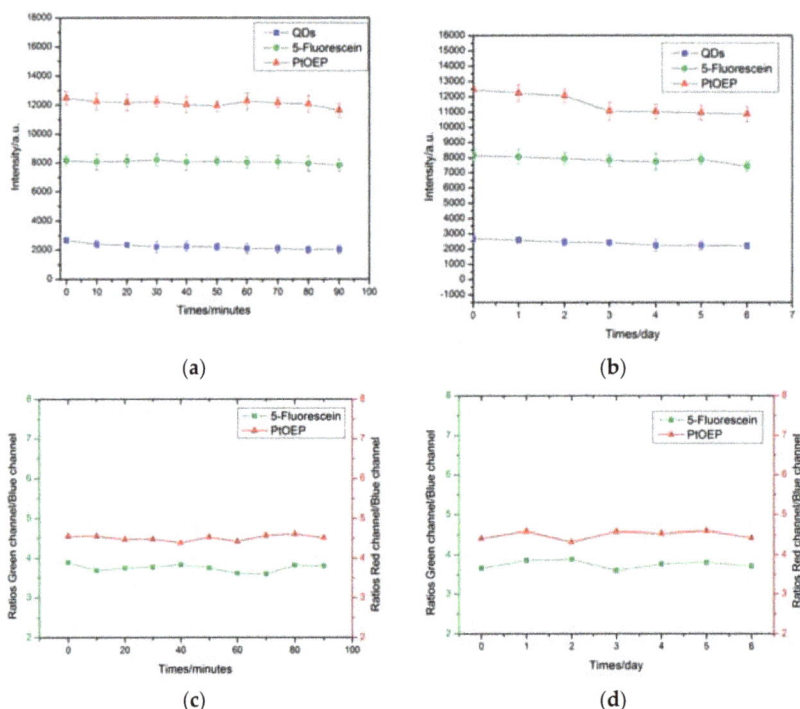

(a)

(b)

(c)

(d)

Figure 6. Photostability and long term stability of the pH/Oxygen planar optode. (**a,b**) The relative luminescence intensities of the QDs, 5-Fluorescein, and PtOEP (Platinum(II) octaethylporphyrin). (**c,d**) The ration of the pH/Oxygen planar optode: ratios of Green channel/Blue channel, and ratios of Red channel/Blue channel.

For the stability test, the experiment was conducted to evaluate the pH/Oxygen planar optode by putting the planar optode in artificial sea-water (salinity 30‰) at 25 °C for a period of 1 week. Figure 6b showed that there was no obvious leakage for the period time, the relative luminescence intensities of the reference QDs, 5-Fluorescein, and PtOEP decreased 12.15%, 9.01%, 13.55%, respectively (Figure 6b). The ration of the Green channel/Blue channel changed 7.43%, and ratios of Red channel/Blue channel changed 8.38% (Figure 6d). The photostability of the pH/Oxygen planar optode was stable. For the stability and photostability, the ratio is less unaffected and more superior than the intensity. In the process of continuous long-term use, in order to assure the accuracy of the measurement result, the pH/Oxygen planar optode needed regular calibration.

In order to evaluate the possibility of migration of the lipophilic fluorescein into the optical isolation layer, and into the seawater, experiments were conducted. We analyzed the fluorescence of seawater used in the quartz aquarium during the stability experiment. The negligible fluorescence was detected when the seawater was excited by a 405 nm LED, so the effect of migration was negligible.

3.4. Response Time of the pH/Oxygen Planar Optode

In real-time applications measurements, response time (time to reach 90% of the full signal) was a critical performance factor. The response times of the pH/Oxygen planar optode (from 9.55 mg/L to 0 mg/L) are around ~35 s, ~2 s (from 0 mg/L to 9.55 mg/L), and ~16 s for pH changes (from 5 to 8). The signal changes of the pH/Oxygen planar optode were fully reversible.

4. Application of the Planar Optode for Imaging oxygen and pH Simultaneously within Seawater

In order to test the applicability of the pH/Oxygen planar optode on the sample, the seawater was collected (in Shilaoren Bay, Qingdao, China) and analyzed. By applying the pH/oxygen planar optode, the results in Figure 7 showed the measurement of the two-dimensional pH/oxygen distribution of sea-water under the influence of the rain drops. The scale of pH and oxygen concentration was respectively expressed with color bars.

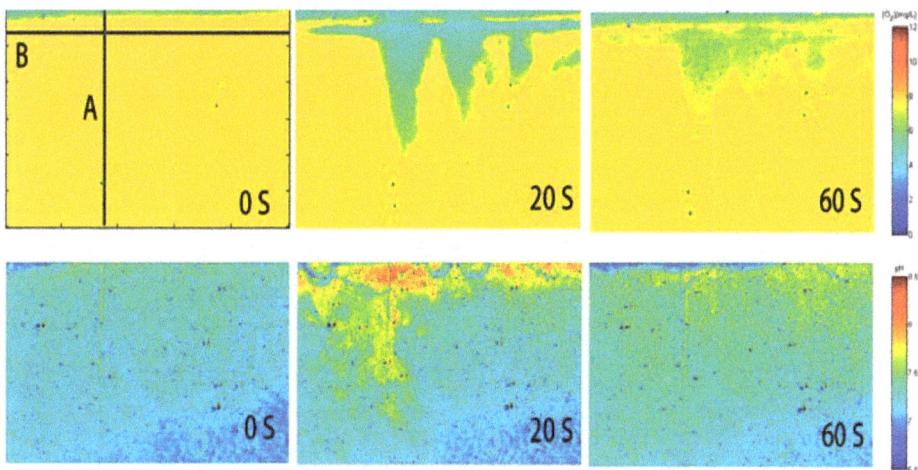

Figure 7. Time series recording of the pH/oxygen distribution. The Line A represented extracted vertical profiles of the raindrops landing area, the Line B indicated the position of air-water interface.

To analyze the pH and oxygen dynamics of the sea-water under the influence of the rain drops in more detail, the vertical and horizontal profiles (corresponds to the marked line A, B in Figure 8) of the pH/Oxygen were extracted from the two-dimensional pH/oxygen figure. The results showed that the rainfall could cause significant changes of dissolved oxygen and pH value of the water surface in a vertical and horizontal direction. On the surface of the water to a vertical depth of 23 mm, the change of dissolved oxygen was the most obvious: the content of oxygen increased 2.3 mg/L within 40 s after rainfall, the pH value decreased to 7.2. To a vertical depth of 12 mm, the diffusion of dissolved oxygen and pH value was slow. It is proved that the change of dissolved oxygen content and pH value in the process of regulating the surface water was affected by the rain drops in the surface water in the region with a small variation of wind speed, temperature or pressure. This study provided a new technical method for understanding the influence of raindrops on the dissolved oxygen concentration and pH of the surface water in low wind impact areas or static water areas.

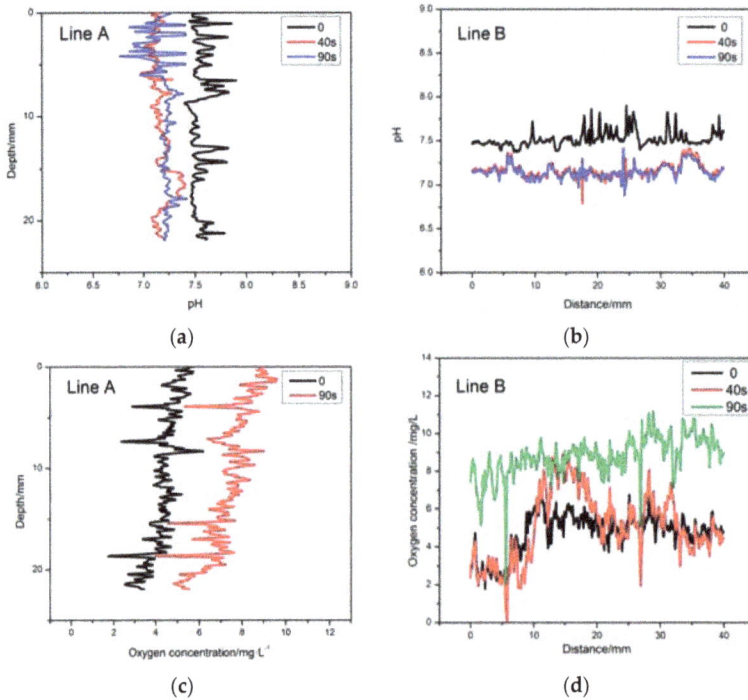

Figure 8. Extraction of time-resolved and depth-resolved dynamics from time series recording pH/oxygen distribution images in Figure 7. (**a,c**) The vertical and horizontal pH/oxygen profiles of Line A in Figure 7; (**b,d**) The horizontal pH/oxygen profiles of Line B in Figure 7.

5. Conclusions

In this work, a novel dual ratiometric pH/Oxygen planar optode combination with a 3CCD camera for simultaneous imaging of pH and oxygen was developed and applied on sea-water under the influence of the rain drops. The planar optode used PtOEP as the oxygen sensitive dye with red emission, 5-Fluorescein as the pH sensitive dye with green emission, and QDs as the pH/Oxygen insensitive reference dye with blue emission. These three dyes were respectively embedded in polystyrene (PS) and Hydromed D4 matrix. The reference QDs and indicator dye was chosen due to its superior optical property, stability, and high lipophilicity. The results indicated that the pH/Oxygen planar optode can be an effective simple, high resolution approach for simultaneous measurement of dissolved oxygen (DO) and pH. This study provides a new technical method for the pH/Oxygen planar optode. It could simultaneously visualize the dynamic changes in pH and oxygen, and have a good nolinear response within dissolved oxygen concentration between 0 and up to 12 mg/L and a pH range from 7 to 9. These specific properties made the pH/Oxygen planar optode suitable for applications in the environmental monitoring field.

Acknowledgments: This work is financially supported by National Natural Science Foundation of China (Grants No. 41176078, No. 41276089), and the National Basic Research Program of China (973 program, Grant No. 2013CB429704).

Author Contributions: Zike Jiang and Xinsheng Yu conceived and designed the experiments; Zike Jiang and Yingyan Hao performed the experiments; Zike Jiang analyzed the data; Xinsheng Yu contributed reagents/materials/analysis tools; Zike Jiang wrote the paper.

Conflicts of Interest: The authors declare no conflict of interest.

Abbreviations

The following abbreviations are used in this manuscript:

DO	dissolved oxygen
PtOEP	Platinum(II) octaethylporphyrin
PS	polystyrene
5-Fluorescein	5-Hexadecanoylamino-fluorescein
D4	Hydromed D4
TOP	trioctylphosphine
OA	oleic acid
DC	direct current
ODE	octadecene
QDs	Quantumdots

Appendix A. Matlab Script to Generate Colormap Images of Spatial Oxygen and pH

```
Im=imread('E:13.tif');
Imcut=imcrop(Im,[650,650,499,499]);
Imd=double(Imcut);

Imdr=Imd(:,:,1);
Imdg=Imd(:,:,2);
Imdb=Imd(:,:,3);
 h=ones(3,3);
 h(1,1) = 0;
    h(1,3) = 0;
    h(3,1) = 0;
 h(1,3) =0;
Imdr = medfilt2(Imdr);
 Imdg = medfilt2(Imdg);
 Imdb = medfilt2(Imdb);
Imd2=zeros(m5,n5);
for i=1:m5
       for j=1:n5
           temp1=Imdr(i,j);
           temp2=Imdb(i,j);
           aver1=mean(temp1(:));
           aver2=mean(temp2(:));
           Imd2(i,j)=aver1/aver2;~end
end
Idmd2=Imd2./I0;
for i=1:m5
   for j=1:n5
       if Imd2(i,j)>1
           Imd2(i,j)=1;
       end
   end
end
do2=((1-A)./(Idmd2-A)-1)./K;
for i=1:m5
```

179

```
  for j=1:n5
      if do2(i,j)>14
          do2(i,j)=14;
      end
      if do2(i,j)<=0
          do2(i,j)=0;
      end
  end
end
imagesc(do2)
h=colorbar;
set(get(h,'Title'),'string','[O_2](mg/L)');
axis off

clear; clc;
Im=imread('E:12.tif');
Imcut=imcrop(Im,[590,420,500,500]);
Imd=double(Imcut);
m5=500
n5=500
Imdr=Imd(:,:,1);
Imdg=Imd(:,:,2);
Imdb=Imd(:,:,3);

 h=ones(3,3);
 h(1,1) = 0;
   h(1,3) = 0;
  h(3,1) = 0;
 h(1,3) =0;
Imdr = medfilt2(Imdr);
Imdg = medfilt2(Imdg);
Imdb = medfilt2(Imdb);

Imd2=zeros(m5,n5);
for i=1:m5
      for j=1:n5
          temp1=Imdg(i,j);
          temp2=Imdb(i,j);
          aver1=mean(temp1(:));
          aver2=mean(temp2(:));
          Imd2(i,j)=temp1./temp2;
      end
end
for i=1:m5
   for j=1:n5
      if Imd2(i,j)<0
          Imd2(i,j)=0;
      end
   end
end
pH=7.13719+0.2310*log(-1.47284./(Imd2-2.67781)-1);
pH=real(pH)
imagesc(pH)
h=colorbar;
set(get(h,'Title'),'string','pH');
axis off
```

References

1. Glud, R.N.; Stahl, H.; Berg, P.; Wenzhofer, F.; Oguri, K.; Kitazato, H. In situ microscale variation in distribution and consumption of O_2: A case study from a deep ocean margin sediment (Sagami Bay, Japan). *Limnol. Oceanogr.* **2009**, *54*, 723–734. [CrossRef]
2. Glud, R.N.; Tengberg, A.; Kühl, M.; Hall, P.O.J.; Klimant, I. An in situ instrument for planar O_2 optode measurements at benthic interfaces. *Limnol. Oceanogr.* **2001**, *46*, 2073–2080. [CrossRef]

3. Larsen, M.; Borisov, S.M.; Grunwald, B.; Klimant, I.; Glud, R.N. A simple and inexpensive high resolution color ratiometric planar optode imaging approach: Application to oxygen and pH sensing. *Limnol. Oceanogr. Methods* **2011**, *9*, 348–360. [CrossRef]

4. Elgetti Brodersen, K.; Koren, K.; Lichtenberg, M.; Kühl, M. Nanoparticle-based measurements of pH and O_2 dynamics in the rhizosphere of *Zostera marina* L.: Effects of temperature elevation and light-dark transitions. *Plant Cell Environ.* **2016**, *39*, 1619–1630. [CrossRef] [PubMed]

5. Koren, K.; Brodersen, K.E.; Jakobsen, S.L.; Kühl, M. Optical sensor nanoparticles in artificial sediments–a new tool to visualize O_2 dynamics around the rhizome and roots of seagrasses. *Environ. Sci. Technol.* **2015**, *49*, 2286–2292. [CrossRef] [PubMed]

6. Ingemann Jensen, S.; Kühl, M.; Glud, R.N.; Jørgensen, L.B.; Priemé, A. Oxic microzones and radial oxygen loss from roots of *Zostera marina*. *Mar. Ecol.-Prog. Ser. Online* **2005**, *293*, 49–58. [CrossRef]

7. Rysgaard, S.; Glud, R.N.; Sejr, M.K.; Blicher, M.E.; Stahl, H.J. Denitrification activity and oxygen dynamics in Arctic sea ice. *Polar Biol.* **2008**, *31*, 527–537. [CrossRef]

8. Glud, R.N.; Santegoeds, C.M.; De Beer, D.; Kohls, O.; Ramsing, N.B. Oxygen dynamics at the base of a biofilm studied with planar optodes. *Aquat. Microb. Ecol.* **1998**, *14*, 223–233. [CrossRef]

9. Glud, R.N.; Kühl, M.; Kohls, O.; Ramsing, N.B. Heterogeneity of oxygen production and consumption in a photosynthetic microbial mat as studied by planar optodes. *J. Phycol.* **1999**, *35*, 270–279. [CrossRef]

10. Behrens, J.W.; Stahl, H.J.; Steffensen, J.F.; Glud, R.N. Oxygen dynamics around buried lesser sandeels Ammodytes tobianus (Linnaeus 1785): mode of ventilation and oxygen requirements. *J. Exp. Biol.* **2007**, *210*, 1006–1014. [CrossRef] [PubMed]

11. Volkenborn, N.; Polerecky, L.; Wethey, D.S.; Woodin, S.A. Oscillatory porewater bioadvection in marine sediments induced by hydraulic activities of Arenicola marina. *Limnol. Oceanogr.* **2010**, *55*, 1231–1247. [CrossRef]

12. Pischedda, L.; Poggiale, J.; Cuny, P.; Gilbert, F. Imaging oxygen distribution in marine sediments. The importance of bioturbation and sediment heterogeneity. *Acta Biotheor.* **2008**, *56*, 123–135. [CrossRef] [PubMed]

13. Hulth, S.; Aller, R.C.; Engström, P.; Selander, E. A pH plate fluorosensor (optode) for early diagenetic studies of marine sediments. *Limnol. Oceanogr.* **2002**, *47*, 212–220. [CrossRef]

14. Zhu, Q.; Aller, R.C.; Fan, Y. High-performance planar pH fluorosensor for two-dimensional pH measurements in marine sediment and water. *Environ. Sci. Technol.* **2005**, *39*, 8906–8911. [CrossRef] [PubMed]

15. Byrne, R.H.; Breland, J.A. High precision multiwavelength pH determinations in seawater using cresol red. *Deep Sea Res. Part A Oceanogr. Res. Pap.* **1989**, *36*, 803–810. [CrossRef]

16. Diaz, R.J.; Rosenberg, R. Spreading dead zones and consequences for marine ecosystems. *Science* **2008**, *321*, 926–929. [CrossRef] [PubMed]

17. Jovanovic, Z.; Pedersen, M.Ø.; Larsen, M. Rhizosphere O_2 dynamics in young *Zostera marina* and Ruppia maritima. *Mar. Ecol.* **2015**, *518*, 95–105. [CrossRef]

18. Wang, X.D.; Wolfbeis, O.S. Optical methods for sensing and imaging oxygen: materials, spectroscopies and applications. *Chem. Soc. Rev.* **2014**, *43*, 3666–3761. [CrossRef] [PubMed]

19. Glud, R.N.; Ramsing, N.B.; Gundersen, J.K.; Klimant, I. Planar optrodes: A new tool for fine scale measurements of two-dimensional O_2 distribution in benthic communities. *Mar. Ecol. Prog. Ser.* **1996**, *140*, 217–226. [CrossRef]

20. Rudolph-Mohr, N.; Vontobel, P.; Oswald, S.E. A multi-imaging approach to study the root-soil interface. *Ann. Bot.* **2014**, *114*, 1779–1787. [CrossRef] [PubMed]

21. Zhu, Q.; Aller, R.C. Planar fluorescence sensors for two-dimensional measurements of H_2S distributions and dynamics in sedimentary deposits. *Mar. Chem.* **2013**, *157*, 49–58. [CrossRef]

22. Holst, G.; Kohls, O.; Klimant, I.; Konig, B.; Kuhl, M.; Richter, T. A modular luminescence lifetime imaging system for mapping oxygen distribution in biological samples. *Sens. Actuators B Chem.* **1998**, *51*, 163–170. [CrossRef]

23. Holst, G.; Grunwald, B. Luminescence lifetime imaging with transparent oxygen optodes. *Sens. Actuators B Chem.* **2001**, *74*, 78–90. [CrossRef]

24. Lu, H.; Jin, Y.; Tian, Y.; Zhang, W.; Holl, M.R.; Meldrum, D.R. New ratiometric optical oxygen and pH dual sensors with three emission colors for measuring photosynthetic activity in Cyanobacteria. *J. Mater. Chem.* **2011**, *2011*, 19293–19301. [CrossRef] [PubMed]

25. Staal, M.; Prest, E.I.; Vrouwenvelder, J.S.; Rickelt, L.F.; Kuhl, M. A simple optode based method for imaging O_2 distribution and dynamics in tap water biofilms. *Water Res.* **2011**, *45*, 5027–5037. [CrossRef] [PubMed]

26. Koren, K.; Kühl, M. A simple laminated paper-based sensor for temperature sensing and imaging. *Sens. Actuators B* **2015**, *210*, 124–128. [CrossRef]

27. Wootton, C. *A Practical Guide to Video and Audio Compression: From Sprockets and Rasters to Macroblocks*; Taylor & Francis: Crowborough, UK, 2005.

28. Wang, X.D.; Stolwijk, J.A.; Lang, T.; Sperber, M.; Meier, R.J.; Wegener, J.; Wolfbeis, O.S. Ultra-small, highly stable, and sensitive dual nanosensors for imaging intracellular oxygen and pH in cytosol. *J. Am. Chem. Soc.* **2012**, *134*, 17011–17014. [CrossRef] [PubMed]

29. Borisov, S.M.; Vasylevska, A.S.; Krause, C.; Wolfbeis, O.S. Composite luminescent material for dual sensing of oxygen and temperature. *Adv. Funct. Mater.* **2006**, *16*, 1536–1542. [CrossRef]

30. Moßhammer, M.; Strobl, M.; Kühl, M.; Klimant, I.; Borisov, S.M.; Koren, K. Design and Application of an Optical Sensor for Simultaneous Imaging of pH and Dissolved O_2 with Low Cross-Talk. *ACS Sens.* **2016**, *1*, 681–687. [CrossRef]

31. Danek, M.; Jensen, K.F.; Murray, C.B.; Bawendi, M.G. Synthesis of luminescent thin-film CdSe/ZnSe quantum dot composites using CdSe quantum dots passivated with an overlayer of ZnSe. *Chem. Mater.* **1996**, *8*, 173–180. [CrossRef]

32. Wang, X.D.; Chen, X.; Xie, Z.X.; Wang, X.R. Reversible optical sensor strip for oxygen. *Angew. Chem.* **2008**, *120*, 7560–7563. [CrossRef]

33. Han, M.; Gao, X.; Su, J.Z.; Nie, S. Quantum-dot-tagged microbeads for multiplexed optical coding of biomolecules. *Nat. Biotechnol.* **2001**, *19*, 631–635. [CrossRef] [PubMed]

34. Uyeda, H.T.; Medintz, I.L.; Jaiswal, J.K.; Simon, S.M.; Mattoussi, H. Synthesis of compact multidentate ligands to prepare stable hydrophilic quantum dot fluorophores. *J. Am. Chem. Soc.* **2005**, *127*, 3870–3878. [CrossRef] [PubMed]

35. Turk, D.; Zappa, C.J.; Meinen, C.S.; Christian, J.R.; Ho, D.T.; Dickson, A.G.; McGillis, W.R. Rain impacts on CO_2 exchange in the western equatorial Pacific Ocean. *Geophys. Res. Lett.* **2010**, *37*, L23610. [CrossRef]

36. Cunliffe, M.; Engel, A.; Frka, S.; Gasparovic, B.; Guitart, C.; Murrell, J.C.; Salter, M.; Stolle, C.; Upstill-Goddard, R.; Wurl, O. Sea surface microlayers: A unified physicochemical and biological perspective of the air-ocean interface. *Prog. Oceanogr.* **2013**, *109*, 104–116. [CrossRef]

37. Bullen, C.R.; Mulvaney, P. Nucleation and growth kinetics of CdSe nanocrystals in octadecene. *Nano Lett.* **2004**, *4*, 2303–2307. [CrossRef]

38. Stern, O.; Volmer, M. Über die abklingzeit der fluoreszenz. *Phys. Z.* **1919**, *20*, 183–188.

39. Schroder, C.R.; Polerecky, L.; Klimant, I. Time-resolved pH/pO_2 mapping with luminescent hybrid sensors. *Anal. Chem.* **2007**, *79*, 60–70. [CrossRef] [PubMed]

40. Lee, S.; Okura, I. Photoluminescent determination of oxygen using metalloporphyrin-polymer sensing systems. *Spectrochim. Acta Part A Mol. Biomol. Spectrosc.* **1998**, *54*, 91–100. [CrossRef]

41. Tang, Y.; Tehan, E.C.; Tao, Z.Y.; Bright, F.V. Sol-gel-derived sensor materials that yield linear calibration plots, high sensitivity, and long-term stability. *Anal. Chem.* **2003**, *75*, 2407–2413. [CrossRef] [PubMed]

Article

Low Cost Lab on Chip for the Colorimetric Detection of Nitrate in Mineral Water Products

Mohammad F. Khanfar [1], Wisam Al-Faqheri [2] and Ala'aldeen Al-Halhouli [2,*]

[1] Department of Pharmaceutical and Chemical Engineering, School of Applied Medical Sciences, German Jordanian University, P.O. Box 35247, Amman 11180, Jordan; Mohammad.Khanfar@gju.edu.jo

[2] NanoLab, School of Applied Technical Sciences, German Jordanian University, P.O. Box 35247, Amman 11180, Jordan; Wisam.AlFakhri@gju.edu.jo

* Correspondence: alaaldeen.alhalhouli@gju.edu.jo; Tel.: +962-6-429-4500

Received: 7 August 2017; Accepted: 27 September 2017; Published: 14 October 2017

Abstract: The diagnostics of health status and the quality of drinking water are among the most important United Nations sustainable development goals. However, in certain areas, wars and instability have left millions of people setting in refugee camps and dangerous regions where infrastructures are lacking and rapid diagnostics of water quality and medical status are critical. In this work, microfluidic testing chips and photometric setups are developed in cheap and portable way to detect nitrate concentrations in water. The performed test is designed to work according to the Griess procedure. Moreover, to make it simple and usable in areas of low resource settings, commercially available Arduino mega and liquid crystal display (LCD) shield are utilized to process and display results, respectively. For evaluation purposes, different local products of tap water, bottled drinking water, and home-filter treated water samples were tested using the developed setup. A calibration curve with coefficient of determination (R^2) of 0.98 was obtained when absorbance of the prepared standard solutions was measured as a function of the concentrations. In conclusion, this is the first step towards a compact, portable, and reliable system for nitrate detection in water for point-of-care applications.

Keywords: colorimetric; microfluidic; sensor; LOC

1. Introduction

The photometric determination of chemical species is one of the key techniques in chemical analysis. If the target species does not absorb radiations in the ultraviolet-visible region, it could be introduced as a limiting reactant in a reaction to produce a colored product. The concentration of the target analyte could be deduced from its absorbance, which follows the Beer-Lambert law over a wide, useful range of concentrations [1–3].

Nitrate is an anion of significant interest, since it could be hazardous to public health if accumulated in human body in concentrations higher than 500 ppm. Through a series of chemical reactions in the body, nitrate is reduced to nitrous acid, which oxidizes the ferrous ion of the hemoglobin to the 3+ oxidation state, converting the hemoglobin to methemoglobin (brown) which does not transport oxygen as efficiently. This causes a disease known as methemoglobinemia or "blue baby syndrome" that is accompanied with a fast heart rate and shortness of breath, and could result in death [4–6].

The photometric detection of nitrate is based on the Griess test. In brief, nitrate is reduced to nitrite by means of cadmium in an acidic solution. Sulfanilamide is added to the nitrite to produce a cation known as diazonium salt. The last step in the test procedure is coupling the salt with N-alpha-naphthyl-ethylenediamine to yield the azo dye, which has a pink color. The intensity of the pink color is correlated to the original nitrate concentration. Photometric (or colorimetric) measurements are based on the Beer-Lambert law; in brief, the absorbing ingredient, also known as chromophore, absorbs portion of the radiant energy emitted by a light source. As a consequence, the transmitted light is attenuated to an extent that is directly proportional to the amount of the absorbing species. The key parameters, the absorbance (A) and the concentration (c), are connected via the relationship: $A = \varepsilon \cdot b \cdot c$, where ε is the absorptivity coefficient and b is the distance the light passes across the analyte solution [7–10].

Colorimetric determination of chemicals could be more versatile if conducted on a smaller scale, rather than an industrial scale. For field measurements, it is practical to use portable measuring devices that could provide the operator with a rapid quantitative determination of the target compounds. Miniaturization of the lab-scale equipment and measuring devices could be achieved through the utilization of lab on a chip technology, where mixing processes take place in grooves and/or channels patterned on paper or plastic templates and the desirable quantitative analysis of the target species is performed with the assistance of compact electronic circuits that function in a manner similar to that of the lab-scale measuring instruments [11–13]. In addition, miniaturizing detection systems provides the advantages of fast analysis, parallelization, low cost, portability, minute reagent consumption, the possibility of running the system by non-trained public workers, and the option of use in areas of low resource settings [14].

The detection of nitrate concentration at the micro molar level has been investigated by different research groups. Optimization of the detection conditions include adjusting the pH of the analyte, its flow rate, the utilized substrates, and the performance of the light-emitting diode and the photodiode array systems used for the detection purposes. The fabricated setups have been mainly used for the determination of nitrate and nitrite concentrations in aqueous systems such as sea water and wastewater for environmental monitoring purposes [15–20].

In this work, a home-made photometric miniaturized detection system was fabricated and its performance toward the detection of nitrate in mineral water products supplied locally was examined and utilized for the colorimetric detection of nitrate in local drinking waters. The fabricated system is simple in design and operation, and its reported results are reproducible and precise.

2. Microfluidic Platform Design and Fabrication

For this project, a microfluidic chip with a straight-forward design was designed and fabricated. As can be seen in Figure 1a, the microfluidic chip consists of a curved channel with a 1-mm width and 2-mm depth. The spiral channel has a single inlet hole for sample load and pumping. The channel is connected to a 10-mm diameter detection chamber where the final chemical reaction and detection will be performed. The detection chamber has two venting holes to allow air movement (in and out) during fluid flow process.

The final microfluidic chip design was fabricated on polymethyl methacrylate (PMMA) plastic (Moden Glas, Bangkok, Thailand). The fabricated chip has three PMMA layers: two 1-mm layers on the top and bottom of the chip, as well as a 2-mm layer in the middle (see Figure 1b). The main microfluidic features (the channel and the detection chamber) were cut into the middle 2-mm PMMA layers. The inlet hole and two venting holes were cut into the top layer, while the bottom PMMA layer has no features (implemented as a cover only). All the microfluidic features were introduced into the PMMA layers using Bodor CO_2 laser cutter (Bodor, Shandong, China). Afterwards, the machined PMMA layers pass through different steps of cleaning, washing, and drying before the bonding process starts. The cleaned PMMA layers were aligned and bonded together using two pressure-sensitive adhesive (PSA) layers (FLEXcon, Spencer, MA, USA). A 100-μm, texture- and color-free PSA is implemented in this work to avoid any signal and/or absorbance interruption.

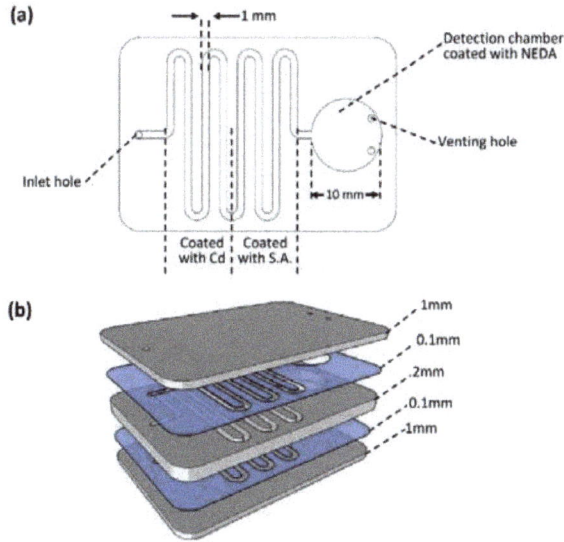

Figure 1. Microfluidic chip for nitrate detection. (**a**) Chip design where a 1 mm × 2 mm spiral channel with one inlet is connected to a final detection chamber; (**b**) microfluidic chip layers.

2.1. Detection Setup

In this work, a simple and cost-effective colorimetric detection setup that can read the final reaction results directly from the detection chamber has been designed and developed. The developed detection setup can facilitate the main goal of the development of a simple, portable, on-site, and cost-effective method for the detection of chemical compounds. Figure 2 shows a 3D view of the proposed detection setup. The main setup is located inside a black semi-cubic box that is 80 mm, 84 mm, and 90 mm in dimension. This black cover has a slot in the front wall for the microfluidic chip to slide in for the final result-reading process. The box is made of 3-mm PMMA sheets and painted black. The inside view shows that the used detection setup consists of three layers: a top layer with a green LED (light-emitting diode) (Farnell, Aschheim, Germany) emits light in the 520–530 nm wavelength range through the detection chamber. The bottom layer is made with a photodiode (Farnell, Aschheim, Germany) which is aligned directly under the LED for light observation and analysis. Finally, the middle layer is made of a specific design to hold and align the detection chamber exactly between the LED and photodiode. For signal read-out and display, Arduino mega, along with LCD shield were utilized. The Arduino was also used as a power source for the LED and the photodiode. For each test, the utilized Arduino takes the voltage signal of the photodiode and displays it on the LCD screen. It was also programmed to display the average of 100 readings each 5 s. Elevation of the nitrate concentration decreases the amount of light transmitted to the photodiode and, as a consequence, the corresponding electrical signal-voltage in this case—decreases. With that, the absorbance (or the transmittance) is correlated to the measured nitrate concentration.

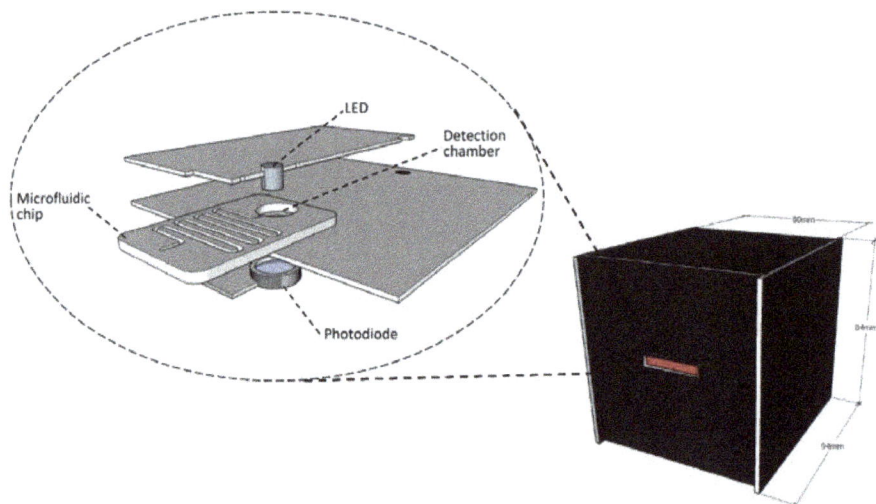

Figure 2. Colorimetric detection setup (on the right) external view of the detection setup (on the left) internal view, top layer with LED, bottom layer with photodiode, and middle layer is the holder for microfluidic chip

2.2. Chemicals Preparation

Cadmium powder, potassium nitrate, and dihydrogen sodium phosphate were purchased from Sigma Aldrich, St. Louis, MO, USA. Phosphoric acid was provided by Riedel de haen (now Honeywell-Riedel de Haen, Morristown, NJ, USA). Benzensulfanylamide (S.A.) was purchased from Applichem GmbH, Darmstadt, Germany, and N-1-naphthylethylenediamin dihydrochloride (NEDA) from Carlo Erba reagents, Peypin, France. Potassium hydroxide was ordered from S.D. Fine Chem Limited, Mumbai, India.

All of the tested solutions were prepared using HPLC grade water obtained from UltraMax 372 Yonglin Water Purification System, Anyang, Korea. Phosphate pH 2.0 buffer solution was employed as the working solution; the solution was prepared by mixing appropriate amounts of dihydrogen sodium phosphate and phosphoric acid. The pH of the prepared solution was adjusted using 0.1 M NaOH aqueous solution.

A 0.5 mM solution of potassium nitrate was prepared and passed through a titration pipette with the cadmium powder located at the bottom of the pipette atop a cotton bed. The reduced nitrate was then mixed with 1.0 mM buffer solutions of S.A. and NEDA to produce the desired standard pink solution. The prepared solution was diluted serially to make standard solutions of lower concentrations. Absorbance of the prepared solutions was measured using the fabricated detection system.

For the detection of the nitrate content in water samples, suspensions of the three ingredients were immobilized on the regions shown in Figure 1 using a micropipette. Masses of the ingredients were selected so that they did exist in excess amounts when compared to the target analyte; the nitrate and the three ingredients, Cd, S.A., and NEDA, were immobilized on the walls of the microfluidic channel as aqueous solutions, then dried at 45 °C for 1 h. The chips were then sealed and became ready for the desired detection procedure.

2.3. Operational Concept

The novel microfluidic design which was discussed earlier eliminates the need for any valving or gating elements to perform the chemical reaction. Instead, the design simply requires a pumping method (using syringe pump in our case) to drive the sample through the microfluidic channel while accurately controlling the flow rate. Figure 3 shows how the experimental process works step-by-step. The upper part of the graph presents the syringe pump status over time as it changes from OFF (no liquid flow) to ON (pump the liquid to the next part of the chip). Under each pump status, the process status and liquid position is presented. For the first time cycle (0 min to 4 min), the syringe pump is OFF and the sample is located within the first part of the chip, which is coated with Cd. Afterward, syringe pump is activated for 1 min to drive the sample from Cd-coated part to S.A.-coated part. It can be noted that we implemented a low flow rate to allow for accurate liquid flow control without the need for the integration of valving mechanism. At 5 min, the pump is deactivated and liquid is allowed to react with the S.A. coating for 4 min. At 9 min, the pump is activated again to drive the liquid to its final destination (the detection chamber). In the detection chamber, the liquid reacts with the NEDA coating for 4 min, where liquid color turns pink. At the end of the reaction time, the pink color intensity presents the concentration of nitrate compounds in the processed sample. The microfluidic chip is then slid into the detection setup for the final result reading. The results will be displayed and saved on a PC dedicated for this process.

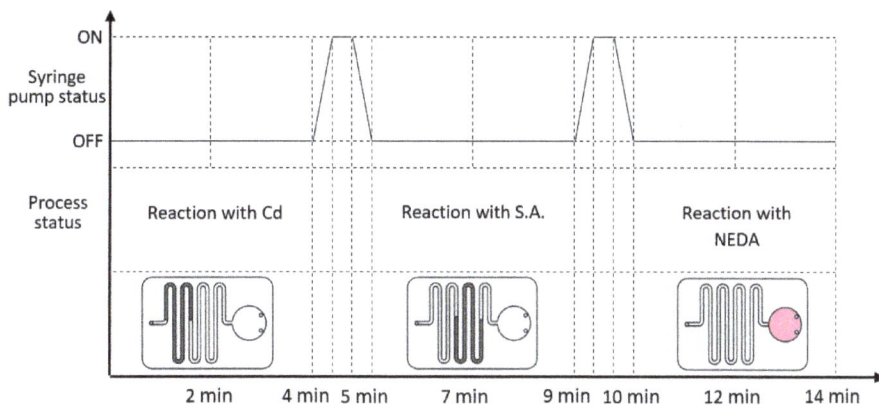

Figure 3. Experimental steps (top part) syringe pump status over time (bottom part) sample position inside the microfluidic chip over time.

3. Results and Discussion

3.1. The Calibration Curve

The main objective of this part is to establish a calibration curve that correlates the produced dye absorbance to the nitrate concentration in parts per million (ppm). As shown in Table 1, the extent of light absorption increases as the nitrate concentration (standard solutions) increases.

Table 1. Dependence of absorbance of the produced dye on the concentration of the standard nitrate solutions.

Standard Solution	Concentration in ppm	Voltage in mV (N = 6)	Absorbance
Blank	0	2845.386667	0.0000
Solution 1	0.033271	2823.436667	0.003363293
Solution 2	0.066541	2803.05	0.006510498
Solution 3	0.166353	2772.118333	0.011329578
Solution 4	0.415882	2689.8	0.024421354
Solution 5	1.039704	2569.236667	0.044337232
Solution 6	2.079408	2416.596667	0.070937171
Solution 7	5.198519	2060.125	0.140247771

Absorbance of the standard solutions was measured in six repeats using the phosphate buffer solution as the blank and the fabricated system as mentioned earlier. The reported results are shown in Figure 4.

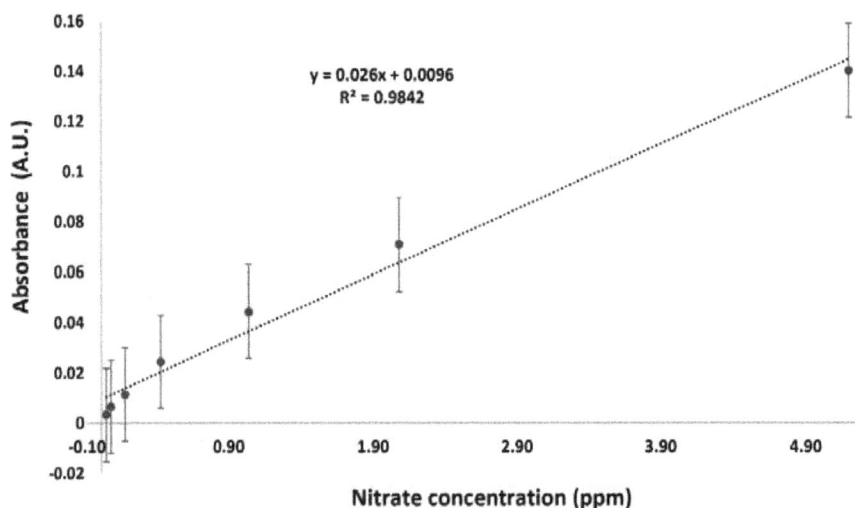

Figure 4. Calibration curve for the prepared standard solutions.

The results were obtained based on the 535 nm absorbance maxima of the azo dye. As shown in Figure 4, the reported R^2 value is equal to 0.9842. The limit of detection (LOD) and limit of quantitation (LOQ) were found to be 0.0782 and 0.237 ppm, respectively. The LOD is estimated to be 3 σ/m, while the LOQ is calculated as $10\sigma/m$ where σ is the standard deviation of the lowest measured concentration (measured six times) and m is the slope of the calibration curve. These values indicate that the performed work needs further improvement when compared to the previously reported value of 0.0016 mg/L [21]. Issues related to quality of the materials used for the chip fabrication, intensity of the employed incident light, and sensitivity of the used detector must be considered in order to optimize the experimental conditions.

3.2. The Water Samples

In this work, the nitrate content in eight water samples of different origins was measured using the fabricated system. The samples were tap water, bottled drinking water, and home-filter treated water. With the utilization of the fabricated system, the total nitrate content (nitrate and nitrite) was quantified and the corresponding results are listed in Figure 5. In the tested samples, the nitrate content was less than the maximum concentration limit (MCL) of 10 ppm (or 1.176×10^{-4} M).

The nitrate content in each sample was measured six times, and the listed results present the average nitrate content in the water samples. As shown in Figure 5, the nitrate content is below the MCL value. Unexpectedly, the maximum reported nitrate content is that of a bottled water sample (Bottled 2), which is even higher than that of the tap water. That reported, and unexpected, value could be attributed to missing key step during course of the purification, which is the anions removal process, allowing nitrate along with other anions to exist as species dissolved in the aqueous matrix with relatively high concentrations. The home-filter treated waters demonstrated moderate nitrate concentrations. The obtained results are of acceptable credibility, since the lowest detected nitrate concentration (Bottled 3) is almost twice as high as the limit of quantification.

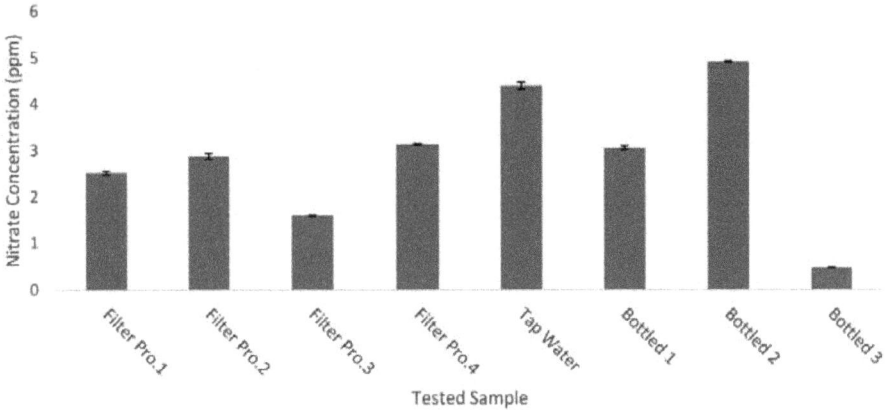

Figure 5. Total nitrate concentration in selected water samples.

The reported results point to the total nitrate concentrations, including nitrate (NO^{3-}) and nitrite (NO^{2-}). The nitrite content could be estimated if the experiments were repeated without the utilization of cadmium, which is responsible for the nitrate to nitrite reduction. The detection conditions could be improved with the utilization of a glass light transparent cover that transmits the incident radiations more efficiently than the PMMA used in this work; with that, adherence to Beer's law could be expected in a manner more linear than that shown in Figure 4 (i.e., at $R^2 = 0.9842$).

Compared to the previously reported works, our developed setup is simple in design and operation, and its reported results are reproducible and precise. In term of fabrication, the developed microfluidic chip can be fabricated very easily with a basic milling machine or laser cutter with a doable dimension of microfluidic channel (one that is not very small and does not require highly accurate machining). The developed chip is cheap and disposable, without any integrated electronic components such as LED or fiber optics [15,17]. Finally, the developed setup does not need bulky laboratory equipment and/or components such as a controlling PC, data processors, a bulky AC or DC power source (a battery is enough), or a micro-valve for flow control. Even sample pumping can be perfumed manually with a normal pipette or syringe. This makes our setup applicable in low-resource and extreme point-of-care areas.

4. Conclusions

In this work, microfluidic testing chips and a colorimetric setup were developed in a cheap and portable way to detect nitrate concentrations in water. The performed test was developed to work according to the Griess procedure. The microfluidic chip was designed to have a long-coated channel fabricated using PMMA layers of different thicknesses. On the other hand, the colorimetric setup mainly consisted of an LED light source, photodiode, Arduino mega, and LCD shield. The developed setup was evaluated using different water samples including bottled water products. The concept of colorimetric detection at the micro level and its viability have been proven, and a useful system for the detection of nitrate in local mineral water products has been fabricated. This chip is easy to fabricate and use, disposable, cheap, and can be operated by non-trained personnel. However, much still needs to be done to improve the detection conditions. In the near future, an effort will be paid to enhance the sensitivity of the spectroscopic detector, thus improving the detection performance of the fabricated setup.

Acknowledgments: The authors would like to thank the Abdul Hameed Shoman Foundation for funding the project entitled "Utilization of Lab on a Chip Technology for the Colorimetric Determination of Chemical

Compounds". The authors would like to show their high appreciation for Engineers Obada Idhoon, Mohsen Diraneyya, and Ahmed Al-Baghdadi for their efforts and help in this work.

Author Contributions: Mohammad F. Khanfar and Ala'aldeen Al-Halhouli contributed to the design of basic idea, setup design, project funding, experiments, supervision, and paper writing. Wisam Al-Faqheri contributed to the design of microfluidic chips, experimental testing, and paper writing

Conflicts of Interest: The authors declare no conflicts of interest.

References

1. Skoog, D.A.; Holler, F.J.; Crouch, S.R. *Principles of Instrumental Analysis*, 6th ed.; Thomson Brooks/Cole: Belmont, CA, USA, 2007; Chapter 13.
2. Jeffery, C.H.; Bassett, J.; Mendham, J.; Denny, R.C. *Vogel's Textbook of Quantitative Chemical Analysis*, 5th ed.; Longman Scientific & Technical: Harlow, Essex, UK; John Wiely and Sons: Hoboken, NJ, USA, 1989; Chapter 14.
3. Jiménez-Colmenero, F.; Solana, J.B. Additives: Preservatives. In *Handbook of Processed Meats and Poultry Analysis*; Taylor and Francis Group: Milton Park, Didcot, UK, 2009; Chapter 5; pp. 91–108.
4. United States Environmental Protection Agency. Inert Ingredient Tolerance Reassessment. Available online: https://www.epa.gov/sites/production/files/2015-04/documents/nitrate_0.png (accessed on 5 February 2017).
5. Randy Merino, L. Nitrate in Foodstuffs: Analytical Standardization and Monitoring and Control in Leafy Vegetables. Licentiate Thesis, Swedish University of Agriculture Sciences, Uppsala, Sweden, 2009; pp. 14–15.
6. Yilong, Z.; Dean, Z.; Daoliang, L. Electrochemical and other methods for detection and determination of dissolved nitrite: A review. *Int. J. Electrochem. Sci.* **2015**, *10*, 1144–1168.
7. Bianchil, E.; Bruschi, R.; Draisci, R.; Lucentini, L. Comparison between ion chromatography and a spectrophotometric method for determination of nitrates in meat products. In *Zeitschrift für Lebensmittel-Untersuchung und Forschung*; Springer: Berlin, Germany, 1995; Volume 200, pp. 256–260.
8. Xiong, Y.; Wang, C.; Tao, T.; Duan, M.; Fang, S.; Zheng, M. A miniaturized fiber-optic colorimetric sensor for nitrite determination by coupling with a microfluidic capillary waveguide. *Anal. Bioanal. Chem.* **2016**, *408*, 3413–3423. [CrossRef] [PubMed]
9. Cardoso, T.M.G.; Garciaa, A.P.T.; Coltro, W.K.T. Colorimetric determination of nitrite in clinical, food and environmental samples using microfluidic devices stamped in paper platforms. *Anal. Methods* **2015**, *7*, 7311–7317. [CrossRef]
10. Ma, J.; Yuan, D.; Lin, K.; Feng, S.; Zhou, T.; Li, Q. Applications of flow techniques in seawater analysis: A review. *Trends Environ. Anal. Chem.* **2016**, *10*, 1–10. [CrossRef]
11. Martinez, A.W.; Phillips, S.T.; Carrilho, E.; Thomas, S.W., III; Sindi, H.; Whitesides, G.M. Simple telemedicine for developing regions: Camera phones and paper-based microfluidic devices for real-time, off-site diagnosis. *Anal. Chem.* **2008**, *80*, 3699–3707. [CrossRef] [PubMed]
12. Yetisen, A.K.; Akram, M.S.; Lowe, C.R. Paper-based microfluidic point of-care diagnostic devices. *Lab. Chip.* **2013**, *13*, 2210–2251. [CrossRef] [PubMed]
13. Peterat, G.; Schmolke, H.; Lorenz, T.; Llobera, A.; Rasch, D.; Al-Halhouli, A.T.; Dietzel, A.; Büttgenbach, S.; Klages, C.P.; Krull, R. Characterization of oxygen transfer in vertical micro bubble columns for biotechnological process intensification. *Biotechnol. Bioeng.* **2014**, *111*, 1809–1819. [CrossRef] [PubMed]
14. Al-Halhouli, A.T.; Demming, S.; Alahmad, L.; LIobera, A.; Büttgenbach, S. In-line photonic biosensor for monitoring of glucose concentrations. *Sensors* **2014**, *14*, 15749–15759. [CrossRef] [PubMed]
15. Petsul, P.H.; Greenway, G.M.; Haswell, S.J. The development of an on-chip micro-flow injection analysis of nitrate with a cadmium reductor. *Anal. Chim. Acta* **2001**, *428*, 155–161. [CrossRef]
16. Baeza, M.; Del, M.; Ibanez-Garcia, N.; Baucells, J.; Bartrolí, J.; Alonso, J. Microflow injection system based on a multicommutation technique for nitrite determination in wastewaters. *Analyst* **2006**, *131*, 1109–1115. [CrossRef] [PubMed]
17. Sieben, V.J.; Floquet, C.F.A.; Ogilvie, I.R.G.; Mowlem, M.C.; Morgan, H. Microfluidic colourimetric chemical analysis system: Application to nitrite detection. *Anal. Methods* **2010**, *2*, 484–491. [CrossRef]
18. Liu, B.; Su, H.; Wang, S.; Zhang, Z.; Liang, Y.; Yuan, D.; Ma, J. Automated determination of nitrite in aqueous samples with an improved integrated flow loop analyzer. *Sens. Actuators B Chem.* **2016**, *237*, 710–714. [CrossRef]

19. Zhang, M.; Yuan, D.; Huang, Y.; Chen, G.; Zhang, Z. Sequential injection spectrophotometric determination of nanomolar nitrite in seawater by on-line preconcentration with HLB cartridge. *Acta Oceanol. Sin.* **2010**, *29*, 100–107. [CrossRef]

20. Bui, D.A.; Hauser, P.C. Analytical devices based on light-emitting diodes—A review of the state-of-the-art. *Anal. Chim. Acta* **2015**, *853*, 46–58. [CrossRef] [PubMed]

21. Beaton, A.D.; Cardwell, C.L.; Thomas, R.S.; Sieben, V.J.; Legiret, F.E.; Waugh, E.M.; Statham, P.J.; Mowlem, M.C.; Morgan, H. Lab-on-chip measurement of nitrate and nitrite for in situ analysis of natural waters. *Environ. Sci. Technol.* **2012**, *46*, 9548–9556. [CrossRef] [PubMed]

sensors

MDPI

Article

Portable Multispectral Colorimeter for Metallic Ion Detection and Classification

Mauro S. Braga [1,2], Ruth F. V. V. Jaimes [3], Walter Borysow [2], Osmar F. Gomes [4] and Walter J. Salcedo [1,*]

1   Laboratório de Microeletrônica, Escola Politécnica da Universidade de São Paulo,
    São Paulo 05508-010, Brazil; msbraga@lme.usp.br
2   Instituto Federal de Educação, Ciência e Tecnologia de São Paulo, Cubatão 11533-160, Brazil;
    wborysow@ifsp.edu.br
3   Centro de Ciências Naturais e Humanas, Universidade Federal do ABC, Santo Andre 09210-580, Brazil;
    rfvillam@iq.usp.br
4   Centro de Capacitação e Pesquisa em Meio Ambiente (Cepema-USP), Cubatão 11540-990, Brazil;
    ofgomes@usp.br
*   Correspondence: wsalcedo@lme.usp.br; Tel.: +55-113091-0720

Received: 12 June 2017; Accepted: 25 July 2017; Published: 28 July 2017

Abstract: This work deals with a portable device system applied to detect and classify different metallic ions as proposed and developed, aiming its application for hydrological monitoring systems such as rivers, lakes and groundwater. Considering the system features, a portable colorimetric system was developed by using a multispectral optoelectronic sensor. All the technology of quantification and classification of metallic ions using optoelectronic multispectral sensors was fully integrated in the embedded hardware FPGA (Field Programmable Gate Array) technology and software based on virtual instrumentation (NI LabView®). The system draws on an indicative colorimeter by using the chromogen reagent of 1-(2-pyridylazo)-2-naphthol (PAN). The results obtained with the signal processing and pattern analysis using the method of the linear discriminant analysis, allows excellent results during detection and classification of Pb(II), Cd(II), Zn(II), Cu(II), Fe(III) and Ni(II) ions, with almost the same level of performance as for those obtained from the Ultravioled and visible (UV-VIS) spectrophotometers of high spectral resolution.

Keywords: portable environmental monitoring systems; metallic ions detection; colorimetric system

1. Introduction

Heavy metal ions have presented strong threats to human health as they have a lot of toxic bio-cumulative properties in the natural environment. Once these ions are thrown into rivers and lakes near cities, they can affect the vegetables and animals, unbalancing the whole food chain [1]. The main health problem caused by heavy metal ions and the threshold level in drinking water according to the World Health Organization (WHO) are summarized in Table 1. To this end, great effort has been made by the scientific community in order to develop devices and systems for metal ions detection. Electrochemical devices were initially offered, which have achieved high accuracy and, in some special cases, also high selectivity by using nanomaterials such as active electrodes [2–4]. However, these types of devices normally suffer interference from electromagnetic noise sources. In order to avoid this, many dye molecules or also bio-indicator molecules have been successfully used for signal detection of the absorption (colorimetric) and fluorescent emission spectra. For example, Anabas testudines were applied as a bio-indicator for Hg and Pb metal ions detection using an ion exchange chromatography spectrometer. The experimental setup of this assay needs complex procedures for the separation and purification of samples [5]. Aminopyridine shift base molecules were also used for Ni(II), Zn(II), Fe(III) and UO_2(II) ions detection by colorimetric and flourogenic methods as a conventional spectrometer.

All the previous procedures have showed that the colorimetric method allows selectivity for Ni(II) and Zn(II) ions, the selectivity was only for Zn(II) when applying the fluorescence technique [6]. A sucessful review paper showed that the colorimetric technique is a good suitable procedure for metal ions detection, especially when functionalized gold is used in a nanoparticle absorption spectra shift for ions detection [7]. The dyad biodipy–rhodamine molecule was utilized for three-valent ions (Al(III), Cr(III), Fe(III)) detection by monitoring changes in fluorescence emission due to energy transfer from biodipy to rhodamine moiety, this dyad molecule did not show any selectivity between these ions, as was reported. The dyad could also be manipulated as an imaging indicator in the biological cell culture [8]. A carbon dot ending with carboxylate groups was applied as a chemo sensor for the detection of many metallic ions, by using the photoluminescence quenching of these dots. However, only selectivity to Fe(III), Pb(II) and Hg(II) was achieved after buffer solution switching for each kind of ion [9]. A single pyridine-linked anthracene-based molecule was taken for the detection of various metallic ions; the change of the photoluminescence emission and its dye also presented selectivity to Pd(II) ion when an Sodium Dodecyl Sulfate (SDS) surfactant was additionally used in the sample solution [10]. A portable microfluidic system for a microbial biosensor was reported to detect Pb(II) and Cd(II) ions, for this purpose the authors used an inverted fluorescence microscopy spectrometer [11]. The Plasmon resonance fiber-optic-based sensor was chosen for metal ions detection, using the peak resonance shift [12]. A review paper reported many different carbon nanoparticle structures to detect Hg(II), Cu(II) and Fe(III) ions by fluorescent off, fluorescence on and ratiometric detection mechanisms. Even though these structures were shown to be a good potential material for metallic ion detection, they could not be precise in the selection and detection of all the mechanisms described here, as they were very sensible to buffer usage in the metallic ions solution [13]. The benzothiazolium-derived molecules were proposed as a colorimetric and fluorescent chemosensor to detect Hg(II) ions; these molecules showed high selectivity for Hg(II) ions and the colorimetric and fluorescent calibration curves were achieved by monitoring the peak position changes (i.e., a specific spectral point) of the absorption spectra and fluorescent spectra respectively [14]. The main challenge in metallic ion detection is to develop a recoverable system, it was reported for a photonic colorimetric device based in the Bragg diffraction process. In this work, the sensor was doped with hydroxyquilonine molecules and the sensor had a selective response for Pb(II) and Cu(II) ions with good reversibility [15]. The selectivity of optical sensors for metallic ion detection is still an issue to solve. Some authors reported the matrix array indicators to overcome this problem, the array of 12 different thiophene-based compounds were used to detect and classify various metallic ions, the authors suggest that 100% classification was possible when they used the fluorescence signal from the phiophene-based molecules and these signals were processed by linear discriminant analyses [16].

As we described above, all the systems that used colorimetric or fluorescence techniques used the conventional test bench spectrophotometer and the selectivity of these systems was specific for some type of metallic ions. In these contexts, this article presents the development of a portable colorimetric and fluorescent chemical detection system, for the detection of metallic ions in liquid media. The system is based on the optoelectronic multispectral sensor as the detector and the white light emitting diode has been used as an excitation source. All components of the system such as excitation, detection and test calibration curves, have been controlled by a real-time embedded national board acquisition system programed with LabVIEW software from the National Instrument Company. The system was tested using the 1-(2-pyridylazo)-2-naphthol (PAN) molecules as the colorimetric indicator and the achieved results showed that this system could detect and classify many metallic ions at the same time (Pb(II), Cd(II), Zn(II), Cu(II), Fe(III) and Ni(II)). The portable system proposed, together with signal processing technique, could apply to metallic ion detection in situ environments such as rivers and lakes.

Table 1. Limit of various heavy metal ions in drinking water according to World Health Organiztion (WHO).

Metal Ions	WHO Limit mg/L (ppm) [17]	Effects [1]
Cu(II)	2	Alliergies, anaemia, kidney disorder
Zn(II)	3	Respiratore disorder, neuronal disorder, prostate cancer
Ni(II)	0.07	At hig level may be toxic, even carcinogenic
Cd(II)	0.003	Renal toxicity, hypertension, lymphocytosis, pulmonary fibrosis, lung cancer, osteoporosis, hyperuricemia
Pb(II)	0.05	Penetrates through protective blood brain barrier, Alzheimer's disease and senile dementia, neuro degenerative diseases, kigney damage
Fe(III)	3	At high level may be originated hemochromatosis, damage cell in the hear liver
As(III)	0.05	Causes effect on central nervous system, cardio vascular and pulmonary diseases, anorexia, gastrointestinal disease, hyper pigmentation, skin cancer
Ag(I)	0.1	Argyria, gastroenteritis, neuronal disorder, mental fatigue, rheumatism
Cr(VI)	0.05	Reproductuve toxicity, embryotoxicity, mutagenicity, carcinogenicity, lung cancer, dermatitis, skin ulcers
Hg(II)	0.001	Impared neurologic development, effects on digestive system, immune system, hypertension

2. Experimental Procedures

The portable embedded system for the detection of different metallic ions by the colorimetric method used a photodetector optoelectronic chip, composed of 18 sets of photodiodes (3×6) encapsulated in a same enclosure, MMCS6CS type, manufactured by the MAZeT company (Jena, Germany). In this device, there were three groups of six photodiodes symmetrically distributed in a circular structure of 2 mm diameter. Each group of photodiodes had a spectral dielectric filter that selects the specific wavelength band so that the complete array of photodiodes covers the spectral region from 380 nm to 780 nm, where each group with specific filters is sensible to the band centered at 425, 475, 525, 625, 575 and 675 nm respectively. Additionally, there was one group of six photodiodes that did not have any filter, i.e., unfiltered array (PW). The photodiodes were connected directly to two integrated transimpedance amplifiers of MTI04CS type, which have four channels with programmable gains. The amplifiers chips were manufactured by MAZeT company (Jena, Germany). The gain selection was achieved by combining the three-bit binary entrance of the MTI04CS integrated circuit, allowing up to eight different stages of amplification levels. After the amplification step, the signals from the photodiodes are multiplexed and directed to a processing and signal acquisition module in order to get the electrical signal (V_{DC}) that corresponds to a light intensity that arrived at each groups of photodiode array of the multispectral sensor. As light source, a white light-emitting diode (LED) (P_{MAX} = 120 mW, IF = 30 mA), manufactured by the company Laser Roithner Technik (B3B-440-JB) was used. This source was set up at the front side of the quartz cuvette that contains the sample solution. The LED was fed with constant current source. The acquisition, control and processing of the signals was performed based on Field Programmable Gate Array (FPGA) technology, which was developed based on virtual instrumentation software (NI LabView®), manufactured by National Instruments, NI model myRIO-1900 (Austin, TX, USA). Figure 1 shows a schematic diagram of the portable embedded system for the detection of the heavy metal ions (Cu(II), Zn(II), Ni(II), Cd(II), Pb(II), and Fe(III)) by the colorimetric method using a chromogen reagent and multispectral optical sensor.

The system manufactured in this way is a portable system that can easily be plugged and played to a computer. The physical picture of the system is depicted in Figure A7 of the Appendix A.

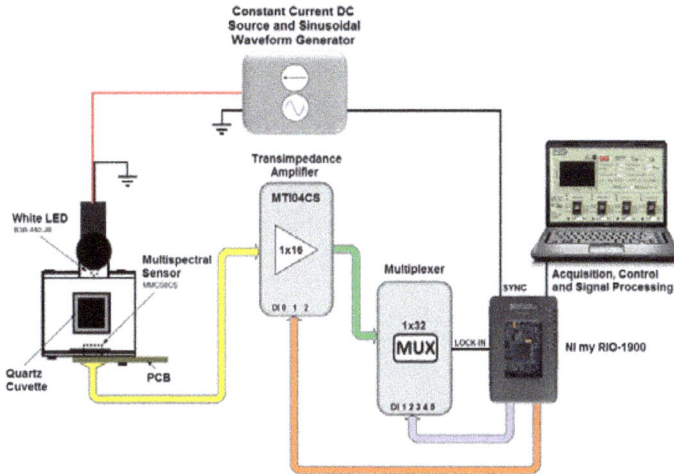

Figure 1. Schematic diagram of the portable colorimetric system built with a multispectral optoelectronic sensor.

The solution with different metal ions was prepared with reagents of 99.9% purity. All reagents were acquired from the Sigma-Aldrich Chemistry (São Paulo, Brazil) and deionized (DI) water was purified with a Milli-Q system Gradient. Standard solutions for the different metals were prepared in water DI, with a suitable dilution of 250 ppm of salts of copper sulphate ($CuSO_4$), zinc sulfate ($ZnSO_4$), nickel chloride ($NiCl_2$), cadmium chloride ($CdCl_2$), lead nitrate ($Pb(NO_3)_2$), iron(III) nitrate (FeN_3O_9). The pH values of the ionic solutions were read by a pH meter, LUCA-210 model, manufacturer Lucadema and unmodified according to those values obtained after the process of the dilution of salts in the water, as is shown in Table 2.

Table 2. The pH values of prepared ionic solutions for colorimetric assays.

Ionic Solution of 250 ppm	pH
Cu(II)	4.0
Zn(II)	4.5
Ni(II)	4.5
Cd(II)	4.5
Pb(II)	4.5
Fe(III)	3.0

The chromogen reagent of 1-(2-pyridylazo)-2-naphthol (PAN) was diluted with methanol in order to get a concentration of 100 μM. Before each data acquisition, a volume of 2.5 mL of the prepared PAN solution was added into a quartz cuvette and then small additions of appropriate volumes of the metal ions Cu(II), Zn(II), Ni(II), Cd(II), Pb(II), Fe(III) were performed in order to get concentrations of 1 to 10 ppm, respectively. It is important to point out that all ion concentrations were authenticated by the EPA SW-846 Test Method 6010D: Inductively Coupled Plasma-Optical Emission Spectrometry using a standard of Pb-CGPB1-1 (1000 μg/mL) in 0.5% HNO_3 (v/v) Inorganic Ventures—CAS No.: 7439-92-1, Cd-CGCD1-1 (1000 μg/mL) in 2.0% HNO_3 (v/v) Inorganic Ventures—CAS No.: 7440-43-9, Zn-CGZN1-1 (1000 μg/mL) in 2.0%

HNO$_3$ (v/v) Inorganic Ventures—CAS No.: 7440-66-6, Cu-CGCU1-1 (1000 µg/mL) in 2.0% HNO$_3$ (v/v) Inorganic Ventures—CAS No.: 7440-50-8, Ni-CGNI1-1 (1000 µg/mL) in 2.0% HNO$_3$ (v/v) Inorganic Ventures—CAS No.: 7440-02-0, Fe-CGFE1-1 (1000 µg/mL) in 2.0% HNO$_3$ (v/v) Inorganic Ventures—CAS No.: 7439-89-6.

The response of the multispectral sensor MMCS6CS in the presence of metal ions of Pb(II), Cd(II), Zn(II), Cu(II), Fe(III) and Ni(II) was based on spectral change measurement of the optical transmittance spectra of a PAN solution, due to the action of different ions. In this case, the light intensity transmitted and received by the array of photodiodes was converted by transimpedance amplifiers (MTI04CS) into V_{DC} voltage values and stored by the acquisition, control and processing module (myRIO-1900). The transmittance was determined relative to the reference signal which corresponded to the response of each photodiode in different arrays to the transmitted white light through the solvent used in the preparation of the solution samples.

Before each signal reading, a volume of 2.5 mL of prepared PAN solution (100 µM) was added into a cuvette of quartz which has square shape of 10 mm each side. After this, a calibrated pipette, model P100 (20–100 µL), Gilson Pipetman, was used to add a small volume of metal ions diluted in water in order to get a concentration of ions in the range of 1 to 10 ppm. During the experiment, the ambient temperature was kept at 26 °C.

3. Results and Discussion

In order to compare the performance of our proposed portable colorimeter first, the transmittance spectra (T) was obtained of PAN (sensitive molecule) and of the solutions of this molecule in environments containing metal ions of Pb(II), Cd(II), Zn(II), Cu(II), Fe(III) and Ni(II), respectively. The spectra were obtained by a UV-VIS spectrometer Cary 50 model, Varian and are presented in Figures A1a, A2a, A3a, A4a, A5a and A6a, which can be seen in the supplementary information (Appendix A).

Then, the portable colorimeter which was built with the multispectral sensor MMCS6CS was used to obtain the transmittance spectra of the PAN solution containing the different metallic ions so that the solutions had the same condition as the ones used with the UV-VIS spectrometer. The transmitted light signals were detected with the six photodetector output terminals (MAZeT) that correspond to the responses of the array of photodiodes with band pass optical filters centered at 425, 475, 525, 575, 625 and 675 nm, respectively. These signals were conditioned using the digital LOCK-IN amplification process. The transmittance spectra were determined comparing the signal from the PAN solution relative to the signal corresponding to the solvent (methanol) used for the PAN solution preparation. Equation (1) gives the transmittance relation that was obtained using the detected signals on the photodiode array.

$$T = \frac{I_{sample}(\lambda)}{I_{solvent}(\lambda)} \tag{1}$$

where $I_{sample}(\lambda)$ is the current generated by the photodiode array with an optical filter centered at the wavelength λ when the samples were the PAN solution without or with metal ions, respectively. $I_{solvent}$ is the current generated by the photodiode array with an optical filter centered at the wavelength λ when the sample only corresponds to a solvent (methanol).

The transmittance spectra for different metal ions and at different concentrations obtained this way are depicted in the Figures A1b, A2b, A3b, A4b, A5b and A6b in the supplementary information (Appendix A).

The spectra results with the UV-VIS spectrometer and portable system clearly show that the presence of metal ions in the solutions of the PAN molecules changes the profile of the transmittance bands and these changes are related to the change in color of the original solution (PAN solution free of ions). The color change mechanism could be explained as follows: the PAN molecule is composed of two aromatic groups, the pyridyl group and naphthol group, joined by azonitrogen. The aromatic groups act as an optical antenna in the UV-VIS region. The interaction of the PAN molecule and the metallic ion in a solution promoted a reaction such that the PAN acts as a tridentate ligand complexing with metal ions through the ortho-hydroxyl group of naphthol rings and the azonitrogen approach hetrocyclic nitrogen atom. This reaction promotes changes in the electronic orbital of pyridyl and naphthol groups which are responsible for the absorption spectrum of the PAN molecule in the UV-VIS region. Thus, the PAN molecule chelation with metal ions changes its spectral band absorption shape and these band changes are used as indicators to identify different types of metallic ions [18,19].

The spectra, obtained with the multispectral sensor MMCS6CS, certainly have lower quality than the spectra obtained with the conventional UV-VIS spectrometer, since the multispectral sensor system has a discrete number of spectral points (six filtered sensors). However, it can be observed in the Figures A1b, A2b, A3b, A4b, A5b and A6b that the profiles of the discrete spectra follow the same trend in the change of spectra that were obtained with the UV-VIS spectrophotometer. It is important to point out that the spectral range of the set of six filtered sensors was limited to a range between 380 and 780 nm. In this sense, in order to obtain a more accurate comparison, the region of the wavelength bands in the ultraviolet region (275–375 nm), seen in the spectra with the UV-VIS spectrometer, were suppressed for the quantitative analyses.

Before the colorimetric analyses, the sensitivity response of our proposed system was compared with that obtained with conventional spectrometers. For this proposal, the transmittance coefficient was analyzed at 525 nm, which is a sensible spectral point that changes significantly with metal ion concentrations. Thus, we define a response function at this point to both the spectrometer and the multispectral MMCS6CS system, using the following Equation (2).

$$\text{Response} = \frac{T_0 - T}{T_0} \qquad (2)$$

where T_0 and T are the transmittance coefficients of the PAN solution without and with metallic ions, respectively.

Figures 2 and 3 depict the calibration curves of the responses obtained by UV-VIS spectrometer and multispectral sensor MMCS6CS, in different concentrations of metals ions, at a wavelength of 525 nm, respectively. It is observed that, for both systems, the response curves for this spectral point (525 nm) present the same profile, showing the compatibility of the sensitive results of our proposed system with the results obtained by a conventional UV-VIS spectrometer.

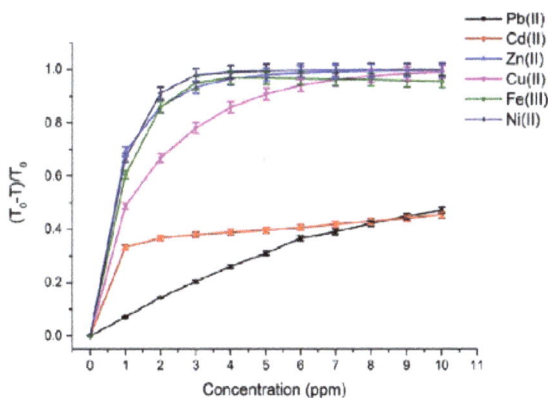

Figure 2. The response curves obtained from the transmittance spectra at 525 nm which were achieved with the Ultraviolet-visible (UV-VIS) stectrophotometer for different metallic ions at different concentrations. The measurements were repeated ten times and the fluctuations of each experimental point were about 0.01%. The error bars were calculated considering the transmittance error of the spectrometer 2.5%.

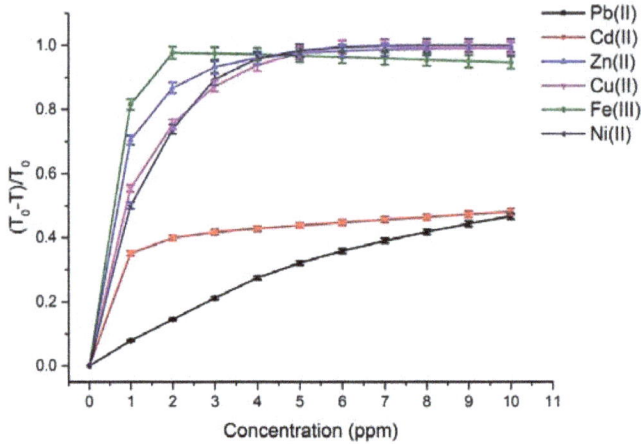

Figure 3. The response curves obtained from the transmittance spectra at 525 nm which were achieved with a portable colorimetric system based in a multispectral sensor for different metallic ions at different concentrations. The measurements were repeated ten times and the fluctuations of each experimental point were about 0.1%. The error bars were calculated considering the error in the photocurrent measurement of the multispectral sensor 1.7% (in the worst case).

On the other hand, Figures 2 and 3 show that the response curves saturate early, showing the high sensitivity of PAN molecules to detecting the metal ions studied in this work, except for Pb and Cd ions. Considering that the response error in our proposed system (MMCS6CS) was about 1.7% (in the worst case), the limit of detection of our system was estimated by using the slope of the linear part of the response curves (Figure 3) in Equation (3) [20]. These limits of detection for all the ions studied in this work are showed in the Table 3.

$$DL = \frac{3.3\sigma}{S},\tag{3}$$

where: S is the slope of response curves (linear region) and σ is the imprecision of the detection system (error).

Table 3. Detection limit (DL) of the MMCS6CS system.

Metal Ions	DL (ppm)
Fe(III)	0.0068
Zn(II)	0.0079
Cu(II)	0.010
Ni(II)	0.011
Cd(II)	0.016
Pb(II)	0.077

The limit of detection for the Pb(II) ion is really close to the limit level for drinking water (Table 1). However, the limit of detection of the Cd(II) ion is greater in one order of magnitude than to the limit level for drinking water. The limit of detection for the Fe(III), Z(II), Cu(II) and Ni(II) ions are much smaller than the limit levels of these ions in drinking water (Table 1). These results showed that the multispectral portable system proposed in this work could be used successfully to control the water quality.

The most relevant results reported in this work are related to the classification power of different metallic ions achieved with the proposed portable colorimeter system. The classification procedure was achieved by using the Fisher linear discriminant analysis. For this procedure we used a set of

20 data for each type of ion, of which ten data were used for the training process and the other ten data were used for the testing process. In the case of the spectra from the UV-VIS spectrometer, first the transmittance curves were fitted with seven harmonic functions (Equation (4)).

$$T(\lambda) = A_0 + \sum_{j=1}^{7}\left[A_j sin(jK\lambda) + B_j cos(jK\lambda)\right], \tag{4}$$

where λ is the wavelength of excited light; and K is the fundamental frequency of the harmonic series.

The sixteen parameters K, A_j and B_j ($j = 1, 2, \ldots, 7$) were used as the input data for the linear discriminant analyses (training and testing process). It is important to point out that different authors proposed a classification process using the colorimetric technique by using the spectral point where the significant variation of the transmittance (or absorbance) coefficient happened [9]. This strategy certainly loses the profile change of all the transmittance bands. In this regard, the fitting process proposed in this work preserved the intensity and shape variation of the spectral bands on the classification process.

In the case of the multispectral MMCS6CS portable colorimeter, we have the six spectral points for the transmittance spectra, so these six transmittance coefficients were directly used as the input data for the linear discriminant analyses.

The canonical score plots for the training and testing process are depicted in the Figures 4 and 5, respectively.

Figure 4 shows that the training process achieved an excellent classification for the spectra data obtained with the UV-VIS spectrometer, since the different classes were clearly separated between them by hyperplanes. This figure also shows that the testing results and the error rates for all metallic ion recognition were 0%, as is shown in Table 4.

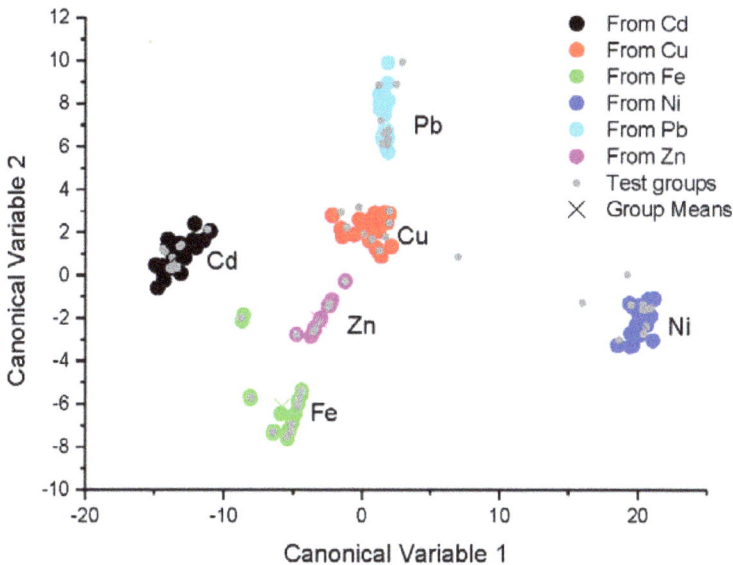

Figure 4. The canonical score plots for the training and testing processes built from the set of spectra data which were obtained with the Ultravioled-visible (UV-VIS) bench spectrometer. The classification procedure was obtained by using the linear discriminant analyses.

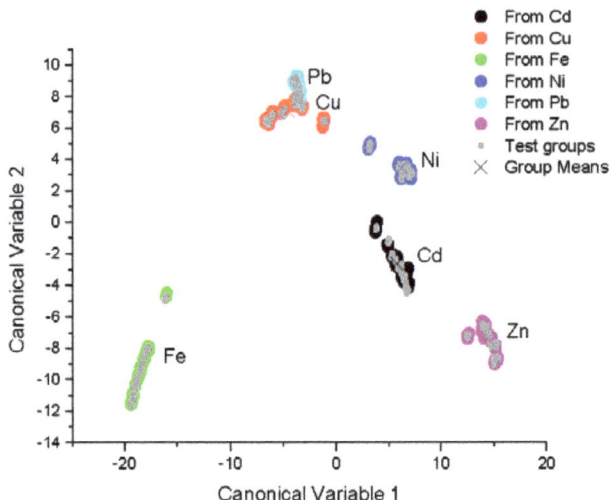

Figure 5. The canonical score plots for the training and testing processes built from the set of spectra data which were obtained with a portable colorimetric system based in multispectral sensors. The classification procedure was obtained by using linear discriminant analyses.

Table 4. Classification counts and error rates of different metallic ions obtained after processing and analyzing the fitted parameter of transmittance spectra (obtained with UV-VIS spectrometer) by the linear discriminant analysis method.

	Predicted Group						
	Cd(II)	**Cu(II)**	**Fe(III)**	**Ni(II)**	**Pb(II)**	**Zn(II)**	**Total**
Cd	20	0	0	0	0	0	20
	100.00%	0.00%	0.00%	0.00%	0.00%	0.00%	100.00%
Cu	0	20	0	0	0	0	20
	0.00%	100.00%	0.00%	0.00%	0.00%	0.00%	100.00%
Fe	0	0	20	0	0	0	20
	0.00%	0.00%	100.00%	0.00%	0.00%	0.00%	100.00%
Ni	0	0	0	20	0	0	20
	0.00%	0.00%	0.00%	100.00%	0.00%	0.00%	100.00%
Pb	0	0	0	0	20	0	20
	0.00%	0.00%	0.00%	0.00%	100.00%	0.00%	100.00%
Zn	0	0	0	0	0	20	20
	0.00%	0.00%	0.00%	0.00%	0.00%	100.00%	100.00%
Total	20	20	20	20	20	20	120

	Error Rate						
	Cd(II)	**Cu(II)**	**Fe(III)**	**Ni(II)**	**Pb(II)**	**Zn(II)**	**Total**
Prior	0.16667	0.16667	0.16667	0.16667	0.16667	0.16667	
Rate	0.00%	0.00%	0.00%	0.00%	0.00%	0.00%	0.00%

The score plot of the training and testing process, which were obtained from the proposed portable colorimeter, is depicted in Figure 5. The clusters of different classes were almost totally separated by hyperplanes, except for the clusters corresponding to Cu(II) and Pb(II) ions, where it was not possible to draw a hyperplane which could separate these clusters. The testing process also shows an error rate for Cu(II) ion recognition of 10%, as can be seen in Table 3—i.e., 10% of Cu(II) ion samples were misunderstood as Pb(II) ions. Even though it was not possible to understand the samples for Cu(II) and Pb(II) ions, all the other metallic ions used in this work were successfully classified with an error rate of 0% (Table 5). The classification results obtained with the portable multispectral colorimetric system almost showed equivalent performance with those obtained with the conventional UV-VIS spectrometer, the 10% of misunderstood Cu(II) and Pb(II) ions must be due to discrete spectral points of the multispectral detector of our system, which loses fine details of band shape changes.

Table 5. Classification counts and error rates of different metallic ions obtained after processing and analyzing the output signal from the multispectral sensor by the linear discriminant analysis method.

	Predicted Group						
	Cd(II)	Cu(II)	Fe(III)	Ni(II)	Pb(II)	Zn(II)	Total
Cd	50	0	0	0	0	0	50
	100.00%	0.00%	0.00%	0.00%	0.00%	0.00%	100.00%
Cu	0	45	0	0	5	0	50
	0.00%	90.00%	0.00%	0.00%	10.00%	0.00%	100.00%
Fe	0	0	50	0	0	0	50
	0.00%	0.00%	100.00%	0.00%	0.00%	0.00%	100.00%
Ni	0	0	0	50	0	0	50
	0.00%	0.00%	0.00%	100.00%	0.00%	0.00%	100.00%
Pb	0	0	0	0	50	0	50
	0.00%	0.00%	0.00%	0.00%	100.00%	0.00%	100.00%
Zn	0	0	0	0	0	50	50
	0.00%	0.00%	0.00%	0.00%	0.00%	100.00%	100.00%
Total	50	45	50	50	55	50	300
	Error Rate						
	Cd(II)	Cu(II)	Fe(III)	Ni(II)	Pb(II)	Zn(II)	Total
Prior	0.16667	0.16667	0.16667	0.16667	0.16667	0.16667	
Rate	0.00%	10.00%	0.00%	0.00%	0.00%	0.00%	1.67%

4. Conclusions

In the present work, a portable device system applied in the detection of different metallic ions was proposed and developed, aiming at its application in the monitoring of hydrological systems like rivers, lakes and groundwater. A portable colorimetric system was designed and developed, embedded in the board acquisition of National Instruments. The system functioned as a colorimeter by using the chromogen reagent of 1-(2-pyridylazo)-2-naphthol (PAN) as an indicator, along with signal processing and pattern analysis using the linear discriminant analysis method, allowing us to obtain excellent results in the detection and classification of Pb(II), Cd(II), Zn(II), Cu(II), Fe(III) and Ni(II) ions, with almost the same level of performance as those obtained from UV-VIS spectrometers with high spectral resolution. All the technology for the quantification and classification of metallic ions using optoelectronic multispectral sensors was fully integrated into the embedded hardware FPGA technology and software based on virtual instrumentation (NI LabView®).

The portable system developed in this work suggests its application for environmental control in situ and in real time, in such a way that it can be integrated into a network of sensors that can provide data continuously and receive commands to control environmental monitoring centers. In addition, the proposed system can be applied for the detection of various types of gases simultaneously, since the different dye molecules sensitive to different types of gas and with different spectral responses could be integrated into the active area of multispectral sensors. In this case it will be used for the absorption or photoluminescence spectra of dye molecules since our portable system provided an easy process for switching the source of a white-light-emitting diode (used for absorption spectra obtention) by an emitting laser diode at a specific wavelength, which can be used as a source to excite the dye molecules for photoluminescence emission.

Acknowledgments: The authors thank CNPq, CAPES, FAPESP, INCT, CEPEMA-USP and IFSP campus Cubatão for resources and technical support offered. In addition, the German company MAZeT for the free supply of photodetectors and transimpedance amplifiers to carry out the project.

Author Contributions: M.S.B. and W.J.S. conceived and designed the experiment; M.S.B. performed the experiments; R.F.V.V.J. gave important suggestions for experimental process and helped in the results discussion; M.S.B. and W.B. analyzed the data; O.F.G. contributed reagents/materials/analysis tools; W.J.S. and M.S.B. wrote the paper.

Conflicts of Interest: The authors declare no conflict of interest.

Appendix A

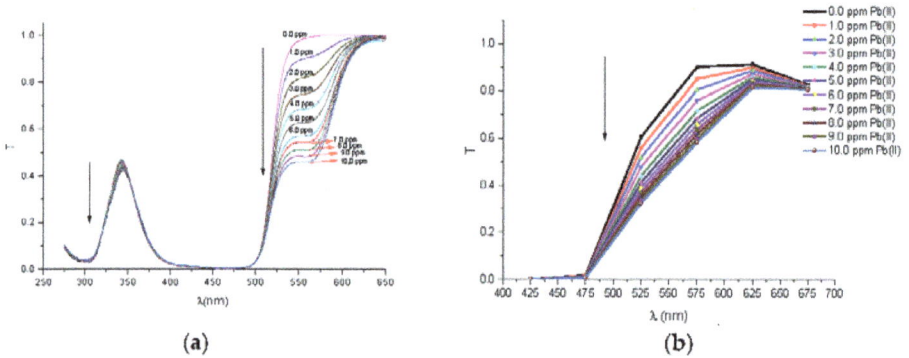

Figure A1. The transmittance spectra of the chromogen reagent of 1-(2-pyridylazo)-2-naphthol (PAN) solution containing Pb(II) ions at different concentrations corresponding to: (**a**) The UV-VIS Cary® 50-Varian spectrometer (**b**) the multiespectral MMCS6CS sensor.

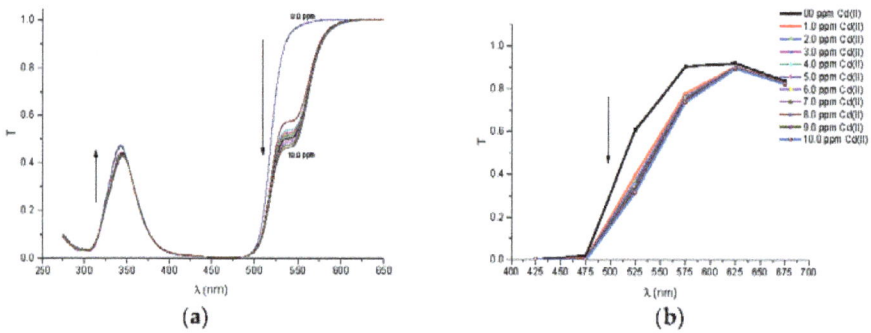

Figure A2. The transmittance spectra of the chromogen reagent of 1-(2-pyridylazo)-2-naphthol (PAN) solution containing Cd(II) ions at different concentrations corresponding to: (**a**) The UV-VIS Cary® 50-Varian spectrometer (**b**) the multiespectral MMCS6CS sensor.

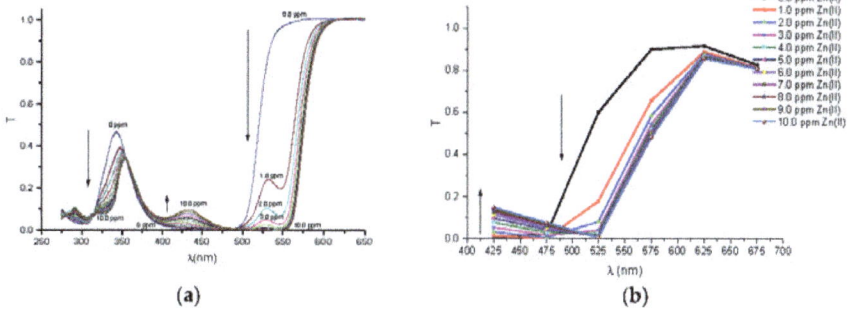

Figure A3. The transmittance spectra of the chromogen reagent of 1-(2-pyridylazo)-2-naphthol (PAN) solution containing Zn(II) ions at different concentrations corresponding to: (**a**) The UV-VIS Cary® 50-Varian spectrometer (**b**) the multiespectral MMCS6CS sensor.

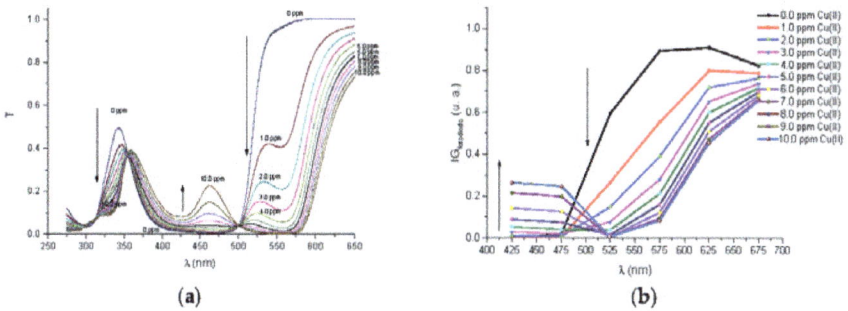

Figure A4. The transmittance spectra of the chromogen reagent of 1-(2-pyridylazo)-2-naphthol (PAN) solution containing Cu(II) ions at different concentrations corresponding to: (**a**) The UV-VIS Cary® 50-Varian spectrometer (**b**) the multiespectral MMCS6CS sensor.

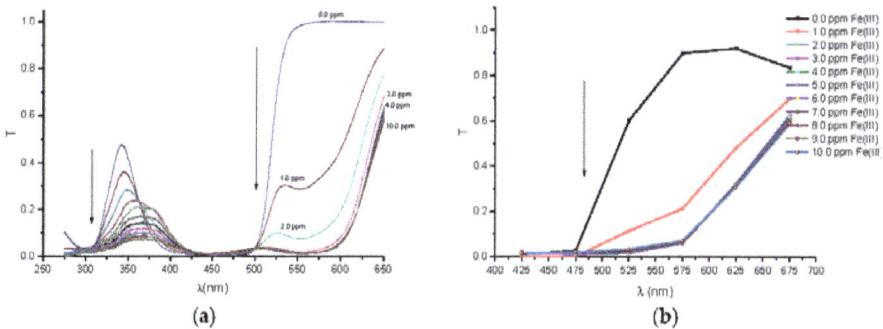

Figure A5. The transmittance spectra of the chromogen reagent of 1-(2-pyridylazo)-2-naphthol (PAN) solution containing Fe(III) ions at different concentrations corresponding to: (**a**) The UV-VIS Cary® 50-Varian spectrometer (**b**) the multiespectral MMCS6CS sensor.

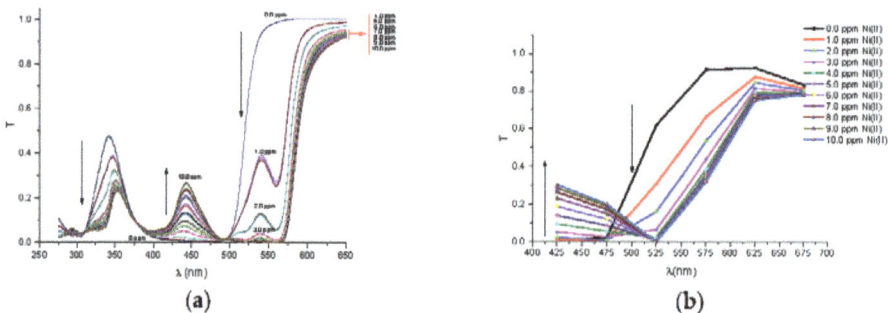

Figure A6. The transmittance spectra of the chromogen reagent of 1-(2-pyridylazo)-2-naphthol (PAN) solution containing Ni(II) ions at different concentrations corresponding to: (**a**) The UV-VIS Cary® 50-Varian spectrometer (**b**) the multiespectral MMCS6CS sensor.

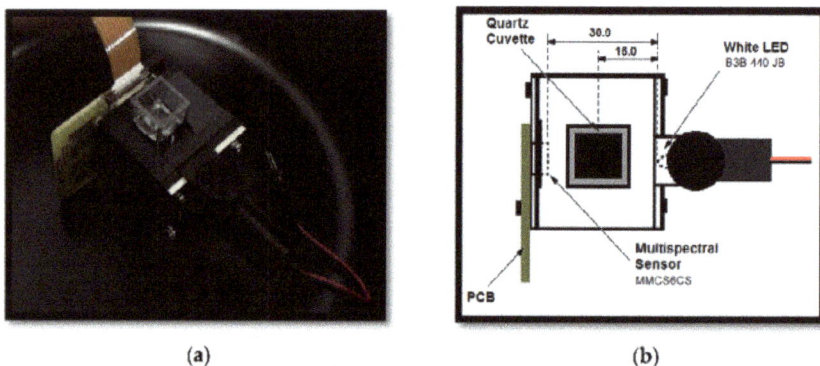

Figure A7. (**a**) Picture of the portable system with the detection sensor MMCS6CS set at the front of the source of a white light-emitting diode (LED). (**b**) Schematic representation of the portable system.

References

1. Järup, L. Hazards of heavy metal contamination. *Br. Med. Bull.* **2003**, *68*, 167–182. [CrossRef] [PubMed]
2. Gumpua, M.B.; Sethuramanb, S.; Krishnanb, U.M.; Rayappan, J.B.B. A review on detection of heavy metal ions in water—An electrochemical approach. *Sens. Actuators B Chem.* **2015**, *213*, 515–533. [CrossRef]
3. Mayne, L.J.; Christie, S.D.R.; Platt, M. A tunable nanopore sensor for the detection of metal ions using translocation velocity and biphasic pulses. *Nanoscale* **2016**, *8*, 19139–19147. [CrossRef] [PubMed]
4. Ting, S.L.; Ee, S.J.; Ananthanarayanan, A.; Leong, K.C.; Chen, P. Graphene quantum dots functionalized gold nanoparticles for sensitive electrochemical detection of heavy metal ions. *Electrochim. Acta* **2015**, *172*, 7–11. [CrossRef]
5. Ahmad, S.A.; Wong, Y.F.; Shukor, M.Y.; Sabullah, M.K.; Yasid, N.A.; Hayat, N.M.; Shamaan, N.A.; Khalid, A.; Syed, M.A. An alternative bioassay using Anabas testudineus (Climbing perch) colinesterase for metal ions detection. *Int. Food Res. J.* **2016**, *23*, 1446–1452.
6. Guptaa, V.K.; Singha, A.K.; Kumawata, L.K.; Mergu, N. An easily accessible switch-on optical chemosensor for the detection of noxious metal ions Ni(II), Zn(II), Fe(III) and UO_2(II). *Sens. Actuators B Chem.* **2016**, *222*, 468–482. [CrossRef]
7. Priyadarshinia, E.; Pradhan, N. Gold nanoparticles as efficient sensors in colorimetric detection of toxic metal ions: A review. *Sens. Actuators B Chem.* **2017**, *238*, 888–902. [CrossRef]

8. Chereddya, N.R.; Rajua, M.V. N.; Reddya, B.M.; Krishnaswamyb, V.R.; Korrapatib, P.S.; Reddyc, B.J.M.; Rao, V.J. A TBET based BODIPY-rhodamine dyad for the ratiometric detection of trivalent metal ions and its application in live cell imaging. *Sens. Actuators B Chem.* **2016**, *237*, 605–612. [CrossRef]

9. Li, C.; Liu, W.; Ren, Y.; Sun, X.; Pan, W.; Wang, J. The selectivity of the carboxylate groups terminated carbon dots switched by buffer solutions for the detection of multi-metal ions. *Sens. Actuators B Chem.* **2017**, *240*, 941–948. [CrossRef]

10. Baiga, M.Z.K.; Pawara, S.; Tulichalab, R.N.P.; Naga, A.; Chakravarty, M. A single fluorescent probe as systematic sensor for multiple metal ions: Focus on detection and bio-imaging of Pd^{2+}. *Sens. Actuators B Chem.* **2017**, *243*, 226–233. [CrossRef]

11. Kim, M.; Lim, J.W.; Kim, H.J.; Lee, S.K.; Lee, S.J.; Kim, T. Chemostat-like microfluidic platform for highly sensitive detection of heavy metal ions using microbial biosensors. *Biosens. Bioelectron.* **2015**, *65*, 257–264. [CrossRef] [PubMed]

12. Verma, R.; Gupta, B.D. Detection of heavy metal ions in contaminated water by surface plasmon resonance based optical fibre sensor using conducting polymer and chitosan. *Food Chem.* **2015**, *166*, 568–575. [CrossRef] [PubMed]

13. Guo, Y.; Zhang, L.; Zhang, S.; Yang, Y.; Chen, X.; Zhang, M. Fluorescent carbon nanoparticles for the fluorescent detection of metalions. *Biosens..Bioelectron.* **2015**, *63*, 61–71. [CrossRef] [PubMed]

14. Nhan, D.T.; Nhung, N.T.A.; Vien, V.; Trung, N.T.; Cuong, N.D.; Bao, N.C.; Huong, D.Q.; Hien, N.K.; Quang, D.T. A benzothiazolium-derived colorimetric and fluorescent chemosensor for detection of Hg^{2+} ions. *Chem. Lett.* **2016**, *46*, 135–138. [CrossRef]

15. Yetisen, A.K.; Montelongo, Y.; Qasim, M.M.; Butt, H.; Wilkinson, T.D.; Monteiro, M.J.; Yun, S.H. Photonic Nanosensor for Colorimetric Detection of Metal Ions. *Anal. Chem.* **2015**, *87*, 5101–5108. [CrossRef] [PubMed]

16. Smith, D.G.; Sajid, N.; Rehn, S.; Chandramohan, R.; Carney, I.J.; Khan, M.A.; New, E.J. A library-screening approach for developing a fluorescence sensing array for the detection of metal ions. *Analyst* **2016**, *141*, 4608–4613. [CrossRef] [PubMed]

17. World Health Organization. Guidelines for Drinking-Water Quality. Available online: http://www.who.int/water_sanitation_health/dwq/gdwq0506.png (accessed on 27 July 2017).

18. Cheng, K.L.; Bray, R.H. 1-(2-Pyridylazo)-2-Naphthol as Possible Analytical Reagent. *Anal. Chem.* **1955**, *27*, 782–785. [CrossRef]

19. Malik, A.K.; Sharma, V.; Sharma, V.K.; Rao, A.L.J. Column Preconcentration and Spectrophotometric Determination of Ziram and Zineb in Commercial Samples and Foodstuffs Using 1-(2-Pyridylazo)-2-naphthol (PAN)-Naphthalene as Adsorbate. *J. Agric. Food Chem.* **2004**, *52*, 7763–7767. [CrossRef] [PubMed]

20. Logaranjan, K.; Devasena, T.; Pandian, K. Quantitative Detection of Aloin and Related Compounds Present in Herbal Products and Aloe veraPlant Extract Using HPLC Method. *Am. J. Anal. Chem.* **2013**, *4*, 600–605. [CrossRef]

sensors

MDPI

Article

Multichannel Discriminative Detection of Explosive Vapors with an Array of Nanofibrous Membranes Loaded with Quantum Dots

Zhaofeng Wu [1,2], Haiming Duan [1], Zhijun Li [1,*], Jixi Guo [2], Furu Zhong [2], Yali Cao [2,*] and Dianzeng Jia [2,*]

1   School of Physics Science and Technology, Xinjiang University, Urumqi 830046, China;
    wzf911@mail.ustc.edu.cn (Z.W.); dhm@xju.edu.cn (H.D.)
2   Key Laboratory of Energy Materials Chemistry, Ministry of Education, Key Laboratory of Advanced
    Functional Materials, Xinjiang University, Urumqi 830046, China; jxguo1012@163.com (J.G.);
    zhfuru@shzu.edu.cn (F.Z.)
*   Correspondence: lizhjun@xju.edu.cn (Z.L.); caoyali@xju.edu.cn (Y.C.); jdz@xju.edu.cn (D.J.);
    Tel.: +86-991-858-2401 (Z.L.)

Received: 12 October 2017; Accepted: 17 November 2017; Published: 20 November 2017

Abstract: The multichannel fluorescent sensor array based on nanofibrous membranes loaded with ZnS quantum dots (QDs) was created and demonstrated for the discriminative detection of explosives. The synergistic effect of the high surface-to-volume ratio of QDs, the good permeability of nanofibrous membranes and the differential response introduced by surface ligands was played by constructing the sensing array using nanofibrous membranes loaded with ZnS QDs featuring several surface ligands. Interestingly, although the fluorescence quenching of the nanofibrous membranes is not linearly related to the exposure time, the fingerprint of each explosive at different times is very similar in shape, and the fingerprints of the three explosives show different shapes. Three saturated vapors of nitroaromatic explosives could be reliably detected and discriminated by the array at room temperature. This work is the first step toward devising a monitoring system for explosives in the field of public security and defense. It could, for example, be coupled with the technology of image recognition and large data analysis for a rapid diagnostic test of explosives. This work further highlights the power of differential, multichannel arrays for the rapid and discriminative detection of a wide range of chemicals.

Keywords: discriminative detection; explosive vapors; sensor array; nanofibrous membranes; quantum dots

1. Introduction

One pressing concern in antiterrorism and homeland security is explosive detection [1–3]. Among the current explosive detection methods, fluorescent sensing represents one of the most promising approaches for trace explosives detection due to possible short response time, excellent sensitivity, simplicity and low cost [1,4,5]. Great efforts regarding fluorescent materials have been made in order to conveniently, quickly and effectively detect explosives. Conjugated polymers [1,6,7], organic dyes such as porphyrinoid and dendrimer [3,8], and microporous metal-organic frameworks [9–11] are proven to be high-performance fluorescent sensing materials, but their application is always limited by costly and cumbersome syntheses [4,12]. In comparison with organic dyes such as rhodamine, the fluorescent quantum dots (QDs) are 20 times as bright and 100 times as stable against photobleaching, showing better potential applications in various fields [13]. Recently, considerable progress has been made in the field of explosive detection based on the fluorescent sensors of QDs [14–16] with a high surface area-to-volume ratio. For example, Itamar Willner reported on the use of chemically modified CdSe/ZnS QDs as fluorescent probes for the detection of trinitrotoluene (TNT) or trinitrotriazine (RDX). The sensitivities of the QDs sensors are controlled by the electron donating properties of the capping layer that modifies the particles, thus allowing the quantitative analysis of the explosive substrates. Bingxin Liu [17] constructed the dual-luminescence-emission probe of CdTe/ZnS QDs and a novel ligand containing 8-hydroxyquinoline for selective sensing of the picric acid (PA) in aqueous solution. The study provides a new insight into highly selective fluorescent sensing of the nitroaromatic explosive PA with a detection limit of 9 nm. Leyu Wang [18] prepared ZnS:Mn^{2+}@allyl mercaptan nanocomposites through novel light-induced in situ polymerization to detect sensitively and selectively nitroaromatic explosives. The fluorescent probe can linearly detect TNT and PA in the range of 0.01–0.5 µg/mL and 0.05–8.0 µg/mL, respectively, barely interfered with by other nitroaromatics such as 2,4-dinitrotoluene (DNT) and nitrobenzene (NB). Neelotpal Sen Sarma [19] synthesized Poly(vinyl alcohol)-grafted polyaniline (PPA) and its nanocomposites with 2-mercaptosuccinic acid (MSA)-capped CdTe QDs and with MSA-capped CdTe/ZnS QDs via a single step free radical polymerization reaction. The detection limits of PPA, MSA-capped CdTe, and MSA-capped CdTe/ZnS QDs for PA in aqueous solution are found to be 23, 1.6, and 0.65 nm, respectively, which are remarkably low. William J. Peveler [7] reported a multichannel array based on multicolored, fluorescent CdTe/ZnS QDs with surface functionalities for the detection of explosives in a rapid single fluorometric test. Pattern analysis of the fluorescence quenching data allows for explosive detection and identification, and five explosives, DNT, TNT, tetryl, RDX and pentaerythritol tetranitrate (PETN), are detected and differentiated in a multichannel fluorescent platform. Zhongping Zhang [20] embedded the red-emitting CdTe QDs in silica nanoparticles and covalently linked the green-emitting CdTe QDs to the silica surface, respectively, to form a dual-emissive fluorescent hybrid nanoparticle. The fluorescence of red QDs in the silica nanoparticles stays constant, whereas the green QDs functionalized with polyamine can selectively bind TNT, leading to the green fluorescence quenching due to resonance energy transfer. The variations of the two fluorescence intensity ratios display continuous color changes from yellow-green to red upon exposure to different amounts of TNT.

As can be seen from the above representative works, a series of fluorescent probes containing QDs have also been applied for the detection of explosives, but mainly limited to explosive solutions and/or particulates (through direct contact) [2,3,21]. Moreover, the vast majority of studies have not tried to discriminate between multiple types of explosive. Compared to the detection in solution and solid phases, the detection of explosives in vapor phase is more challenging and desired since most of them have substantially low volatility [5,22]. Furthermore, QDs in solutions are easy to agglomerate and precipitate, and the long-term stability is poor, which is not conducive to the effective detection of explosives. Electrospinning has become a simple, cost-effective and versatile technique for the preparation of nanomaterial films with high porosity and flexibility, which has great potential for enhanced explosive detection [23–25]. The combination of QDs and electrospun fibers may provide a novel possibility for the low-cost, sensitive, discriminative detection of explosives in vapor phase. Lastly, the preparation of some QDs, such as CdTe and InP, usually needs oil bath with high reaction temperature and they are easily oxidized in air, while ZnS QDs can be prepared in aqueous solution at room temperature and have good oxidation resistance, which can be easily combined with electrospun polymer fibers for the detection of gaseous explosives.

Herein we report a fluorescent sensor array based on nanofibrous membranes loaded with ZnS QDs followed by the modification of several surface ligands for the detection of explosive vapors. The synergistic effect of the high surface-to-volume ratio of QDs, the good permeability of nanofibrous membranes and the differential response owing to the surface ligands was played in the sensing system. The sensing system is designed to respond to a range of explosives through supramolecular interactions, such as host-guest binding and electrostatics, causing fluorescence quenching of the QDs, to create an analytical fingerprint for the sensitive, quick, discriminative detection of explosive vapors at room temperature.

2. Materials and Methods

2.1. Materials

Analytically pure sodium sulfide, zinc acetate, manganese acetate, lysine, L-cysteine, trifluoroacetyl lysine, L-cysteine hydrochloride were purchased from Aladdin Reagent Co., Ltd., (Los Angeles, CA, USA). Nitrobenzene (NB) was purchased from Sinopharm Chemical Reagent Co., Ltd., (Shanghai China). Picric acid and 2, 4-dinitrotoluene were used as received from national standard substance Center. Dimethylformamide (DMF) and methanol solution of 2, 4, 6-Trinitrotoluene (1000 μg/mL, TNT) was purchased from Sinopharm Chemical Reagent Co., Ltd., (Shanghai China), and TNT was recrystallized to produce saturated vapor at room temperature. Polyurethane (PU, Mw = 200,000) was supplied by Anhui Amway synthetic leather Co., Ltd., (Hefei China).

Caution: The highly explosive TNT and PA should be used with extreme caution and handled only in small quantities.

2.2. Synthesis of Mn^{2+}-Doped ZnS QDs

Sodium sulfide was used at the mole amount equal to that of zinc acetate. Typically, 50 mmol of zinc acetate was dissolved in mixed solution of ethanol and deionized water (volume ratio = 1:1). An amount of 2.5 mmol of manganese acetate was added into the above solution, and the mixture was ultrasonicated for 20 min at room temperature. After the mixture solution was refluxed in a flask under nitrogen, 10 mL of aqueous solution containing 50 mmol of sodium sulfide was added dropwise into the reaction system, and the mixture was vigorously stirred for 5 h. The resultant Mn^{2+} doped ZnS QDs were centrifuged and washed with deionized water several times.

2.3. Preparation of Electrospun Nanofibrous Membrane

The preparation of PU electrospun nanofibrous membrane was performed as follows: an electrospun solution was initially prepared by dissolving 1.2 g of PU and 0.18 g of zinc acetate in 10 mL of DMF. The prepared solution was transferred into a 10 mL syringe mounted on the electrospinning apparatus (SS-2535H). A high-voltage power supply generated direct-current voltage up to 18 kV. The electrospun solution was fed at a constant rate of 0.8 mL/h by a syringe pump. Nanofibers were collected on aluminum foil with a collection time of 120 min, and were then dried in an oven at 50 °C for 12 h to remove the residual organic solvent. The electrospun membranes were recorded as PU-0.

2.4. Preparation of Fluorescent Nanofibrous Membranes

The preparation of membranes loaded with ZnS QDs was prepared as follows.

First, 50 mmol of zinc acetate was dissolved in 50 mL mixed solution of ethanol and deionized water (volume ratio = 1:1). Second, 2.5 mmol of manganese acetate was added into the above mixed solution, followed by a ultrasonic dispersion process for 20 min and a reflux for 1 h under nitrogen protection. Third, the PU-0 membranes were immersed in the mixed solution with slow stirring and the reaction device was kept in an ice water mixture for 5 h. Fourth, 10 mL of aqueous solution containing 50 mmol of sodium sulfide was added dropwise into the reaction system, and stirred for 4, 8, and 12 h, respectively. Last, the PU-0 membranes loaded with ZnS QDs were washed with deionized water several times and the resultant electrospun membranes corresponding to 4, 8, and 12 h were recorded as PU-1, PU-2 and PU-3, respectively.

The preparation of membranes including ZnS QDs was performed as follows.

First, an electrospun solution was initially prepared by dissolving 0.25 g of ZnS QDs as fluorescence probes and 1.2 g of PU as the supporting polymer in 10 mL of DMF. Second, the prepared solution was transferred into a 10 mL syringe mounted on the electrospinning apparatus (SS-2535H). Third, the electrospun solution was fed at a constant rate of 0.8 mL/h by a syringe pump with a 20 kV direct-current voltage. At last, nanofibers were collected on aluminum foil with a collection time of

120 min, and then dried in an oven at 50 °C for 12 h to remove the residual organic solvent. The sample was recorded as PU-4.

2.5. Surface Modification of Fluorescent Nanofibrous Membranes

For the further surface modification, an amount of 0.25 mmol of lysine, cysteine, trifluoroacetyl lysine, cysteine hydrochloride (Structural formula shown in Figure S1) was dissolved in 50 mL of mixed solution of ethanol and deionized water (volume ratio = 1:1), respectively. Then the fluorescent membranes (PU-1) were immersed in the above mixture solutions with slow stirring for 24 h, respectively. Mercapto groups of cysteine and cysteine hydrochloride tightly attached onto the surface of the QDs due to the excess of metal ions with respect to sulfide ions at the surface of the QDs [26]. As shown in Figure S2, the infrared bands located at about 3500 cm^{-1} are due to the –OH stretching on the surfaces of ZnS QDs. The strong bands located at about 3500 cm^{-1} indicates the presence of a large number of hydroxyl on the surface of ZnS QDs. Similarly, lysine and trifluoroacetyl lysine also tightly attached onto the surface of the QDs due to the interaction between carboxyl groups of lysine and trifluoroacetyl lysine and hydroxyl groups at the surface of the QDs [27]. The modified fluorescent membranes were washed with ethanol several times to remove the residue and were dried at 50 °C for use. PU-1 membranes treated by lysine, cysteine, trifluoroacetyl lysine and cysteine hydrochloride were recorded as PU-1L, PU-1C, PU-1TL and PU-1CH, respectively.

2.6. Quenching Tests of Fluorescent Nanofibrous Membranes towards Nitroaromatic Explosives

Before quenching tests, the fluorescent nanofibrous membranes were placed under an ultraviolet (UV) lamp for 12 h to make the fluorescence intensity stable. Small granules of nitro analytes were placed on the bottom of a sealed testing box with four quartz windows and a slot. Meanwhile, a small amount of cotton gauze was tucked into the testing box to help maintain a constant vapor pressure of analyte. After the slot was sealed, the vapor of nitro compound was up to saturation in the testing box for 48 h at room temperature (25 °C). The nanofibrous membranes were fixed onto the sample shelf tightly matched with the slot of the testing box and promptly inserted into the testing box filled with saturated vapor through the slot. The evolution of fluorescence spectra was recorded for specific time intervals after exposing the films to the vapor of analytes.

2.7. Characterization

Steady-state luminescence spectra were acquired under excitation at 300 nm on a Hitachi F4600 luminescence spectrometer (Hitachi, Tokyo Japan). The UV-Vis absorbance spectra were recorded with a UV-3900H spectrometer (Hitachi, Japan). Thermal gravimetric analysis (TGA) was carried out using a 200F3 thermal gravimetric analyzer (Netzsch, Gebrüder Germany) at a heating rate of 10 °C/min under air condition. Transmission electron microscope (JEM-2100F) and field emission scanning electron microscopy (FE-SEM, S-4800, Hitachi, Japan) was used to characterize morphology of samples. Contact angles of the films with deionized water drop were measured with a contact angle goniometer (G-1, Erma) at room temperature. Infrared spectra of the ZnS QDs was recorded on a VERTEX 70 Fourier transform infrared spectrometer (Bruker, Karlsruhe Germany).

3. Results and Discussion

3.1. Preparation and Surface Modification of Fluorescent Nanofibrous Membranes

As shown in Figure 1a,b, the PU-0 nanofibrous membrane is consisted of fibers between 150 and 200 nm in diameter, and their surfaces are basically smooth. Similarly, the PU-4 nanofibrous membrane is consisted of smooth fibers between 50 and 250 nm in diameter, and there are no ZnS QDs on the fiber surfaces (Figure 1c,d), indicating that the ZnS QDs are mainly distributed in the inner part of the nanofibers. To further analyze the surface properties of two kinds of fiber membranes, the static contact angle test was performed. Their static contact angles of PU-0 and PU-4 are 96 ± 2.2° and

$123 \pm 2.4°$, respectively (inserts in Figure 1). While the static contact angle of pure PU nanofibrous membrane is about 125° (Figure S3), which is very close to that of PU-4. The result indicates that the incorporation of ZnS QDs has not changed significantly the surface properties of PU-4 compared to the pure PU nanofibrous membrane. The surface properties of PU-0 are changed significantly because of the incorporation of hydrophilic zinc acetate, which is conducive to the growth or load of the hydrophilic QDs on the PU-0 membrane.

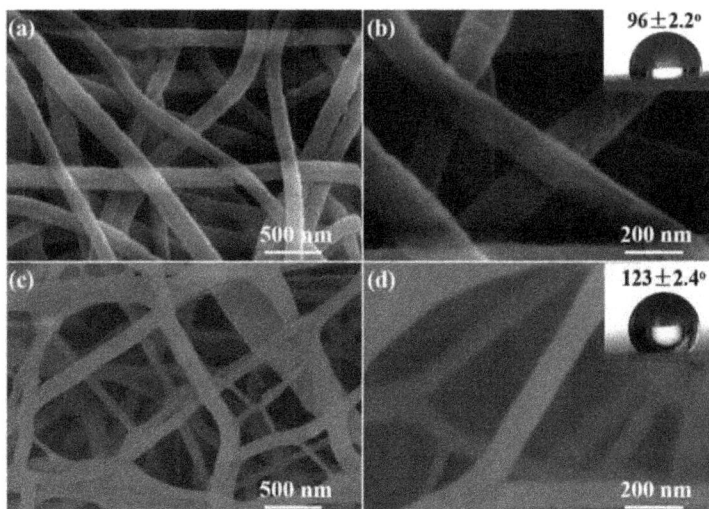

Figure 1. Scanning electron microscopy (SEM) images with different magnification of (**a,b**) PU-0, (**c,d**) polyurethane (PU)-4 nanofibrous membrane, the insets in (**b,d**) show images of water drops and the corresponding contact angles.

As shown in Figure 2a–f, compared with the PU-0, the fiber diameter of PU-1, PU-2 and PU-3 membranes loaded with ZnS QDs did not change significantly because of the tiny particle size, but the surfaces of fibers become rough. Further observations show that the surfaces of the fibers are loaded with a layer of ZnS QDs and more and more ZnS QDs are loaded on the fiber surface from PU-1 to PU-3. The static contact angles of PU-1, PU-2 and PU-3 also have obviously changed, reaching 93, 82 and 68°, respectively. The significant change of surface properties of nanofibers should be attributed to the hydrophilicity of the hydroxyl groups of ZnS surfaces (Figure S2). These results indicate that ZnS QDs are effectively loaded on the fiber surfaces, which could be illustrated by a schematic diagram. As a result, the rough topography of fibers loaded with ZnS QDs shown in Figure 2 is formed. As shown in Figure S4, when PU-0 membranes are immersed in the mixture of manganese and zinc ions with slow stirring, the zinc ions in the solution are adsorbed on the surface of the nanofibers owing to the high surface-to-volume ratio of nanofibers, forming a large number of reactive sites. With the increase of soaking time, the zinc ions in the nanofiber inner migrate to the surface of the fiber and also form the reactive sites. At the same time, the ice bath is used to control the reaction temperature, thus controlling the reaction rate. When the aqueous solution of sodium sulfide is slowly added, the sulfide ion combines with the reactive sites of the fiber surface to form ZnS QDs and ZnS QDs continue to grow or agglomerate together with the increase of reaction time, forming larger QDs agglomerates on the fiber surfaces.

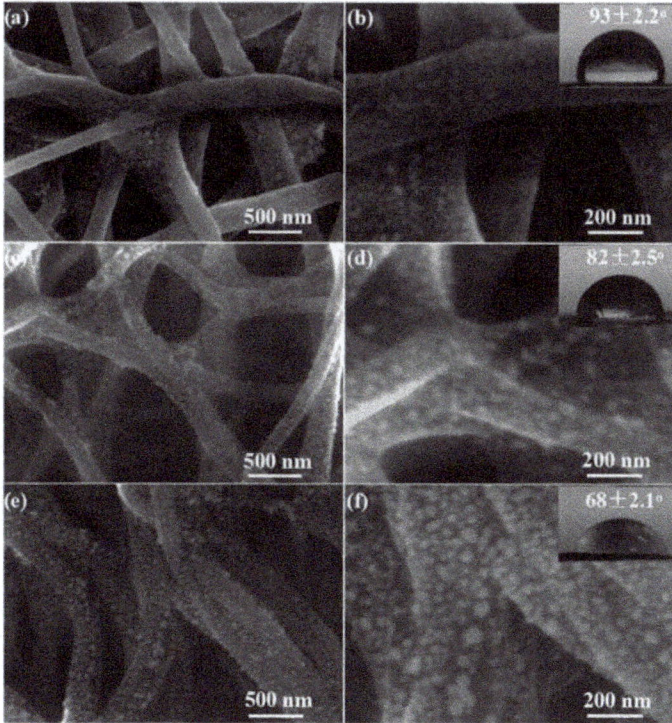

Figure 2. SEM images with different magnification of (**a,b**) PU-1, (**c,d**) PU-2, and (**e,f**) PU-3 nanofibrous membrane, the insets in (**b,d,f**) show images of water drops and the corresponding contact angles.

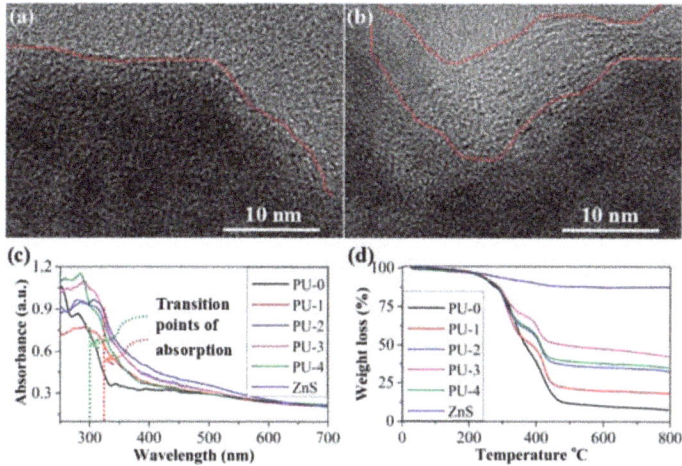

Figure 3. Transmission electron microscope (TEM) images of ZnS quantum dots (QDs) (**a**) on the fiber surfaces of PU-2, (**b**) in the fiber inner of PU-4 (the red dotted curves in Figure 4a,b show the boundary between ZnS and the polymer layer), (**c**) ultraviolet (UV)-vis absorption spectrum (the green and red dotted lines show the transition points of UV-Vis absorption of the PU-0 and ZnS), (**d**) thermal gravimetric analysis (TGA) curves of samples.

In order to qualitatively and quantitatively determine the amount of ZnS QDs on the nanofiber surfaces, TEM, UV-Vis absorption and TGA were performed. As shown in Figure 3a, ZnS QDs are in a state of agglomeration and there is no polymer layer on the surface of ZnS QDs for PU-2. For PU-4, although ZnS QDs are also in a state of agglomeration because of the large amount of ZnS addition, ZnS QDs are coated with a polymer layer of about several nanometers in thickness (as shown by the red dotted curve in Figure 3b). The results also show that ZnS QDs are distributed in the inner part of the PU-4 nanofibers, which is consistent with the observation of SEM in Figure 1. As shown by the red and green dotted lines in Figure 3c, the transition points of UV-Vis absorption of PU-0 and ZnS QDs are about 300 and 325 nm, respectively. Also, the absorption intensity of ZnS QDs is obviously higher than that of the PU-0 in the range of 240-500 nm. It is worth pointing out that the transition points of UV-Vis absorption of PU-1, PU-2, PU-3 and PU-4 are very close to that of ZnS, showing the similar absorption characteristics to ZnS. The increase of absorption intensity in the ultraviolet region from PU-1 to PU-3 indicate that the amount of ZnS loaded on the nanofiber surface also increase with the increase of reaction time. This result is further demonstrated by TGA tests. TGA was performed in a nitrogen atmosphere to prevent the oxidation of ZnS at elevated temperatures. As shown in Figure 3d, the residue of all the samples is basically stable after 500 °C, and the residue weight of PU-0, PU-1, PU-2, PU-3, PU-4 and ZnS is 7.8, 18.5, 33.3, 42.5, 35.2 and 87.6% at 800 °C, respectively. Compared with PU-0, the residue weight of PU-1, PU-2, PU-3 and PU-4 is increased by about 10.7, 25.5, 34.7 and 27.4%, respectively, due to the introduction of ZnS QDs. The TGA results also illustrate the effective load of ZnS on the nanofiber surfaces, facilitating the preparation of fluorescent nanofibrous membrane array for explosive sensing.

3.2. Fluorescent Nanofibrous Membrane Array for Explosive Sensing

In order to evaluate the quenching effect of fluorescent nanofibrous membranes with different loading amounts of ZnS towards nitroaromatic explosive vapors, time-dependent fluorescence emission spectra were performed. As shown in Figure 4, the fluorescence intensity of PU-1, PU-2 and PU-3 increase significantly with the increase of loaded ZnS before exposure to TNT vapor, reaching about 1180, 1900 and 3000. Because the loading amounts of ZnS in PU-4 is close to that of PU-2, the fluorescence intensity of PU-4 is close to that of PU-2, reaching about 1700 due to the coating of polymer layers. And one can see that from Figure 4 and Figure S5, the fluorescence quenching rate gradually decreases from PU-1 to PU-3 due to the increase of loaded ZnS. It is worth pointing out that although PU-4 does not have the most loads of ZnS, the fluorescence quenching rate of PU-4 is the slowest, which should be attributed to the coating of ZnS by the polymer layer. The coating of polymer layer effectively prevents the contact of ZnS and TNT vapor, causing the slow quenching rate. Based on the quenching performances of the above fluorescent membranes, PU-1 is selected as an ideal fluorescent sensing membrane for the detection of nitroaromatic explosive vapors. Furthermore, the reproducibility of fluorescent nanofibrous membranes were evaluated. The fluorescence quenching efficiencies (presented as $(1-I/I_o)$, where I_o is the initial fluorescence intensity in the absence of analyte, I is the fluorescence intensity in the presence of analytes) of the above four kinds of membranes towards TNT vapor as a function of time among three batches were comparable (Figure S5). The results indicated that the fluorescent nanofibrous membranes show satisfactory reproducibility for application to the detection of explosives.

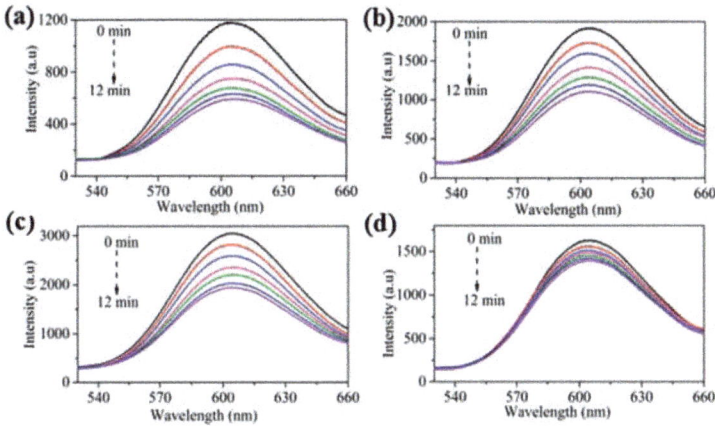

Figure 4. Time-dependent fluorescence emission spectra of the nanofibrous membranes (**a**) PU-1, (**b**) PU-2, (**c**) PU-3, and (**d**) PU-4 upon exposure to 4 ppb of trinitrotoluene (TNT) vapor.

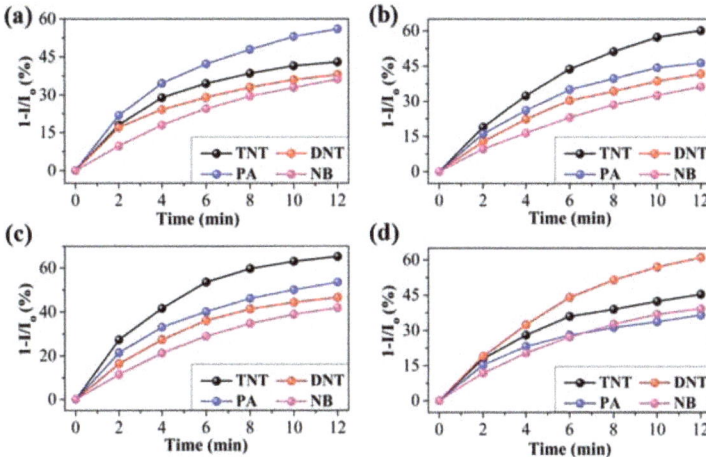

Figure 5. Variation of the quenching percentage for (**a**) PU-1C, (**b**) PU-1CH, (**c**) PU-1L and (**d**) PU-1TL as a function of the exposure time to the saturated air of TNT, 2,4-dinitrotoluene (DNT), picric acid (PA) and nitrobenzene (NB) at room temperature.

In order to realize the differential quenching of fluorescent membranes towards explosives, PU-1 membranes are modified by lysine, cysteine, trifluoroacetyl lysine and cysteine hydrochloride, recorded as PU-1L, PU-1C, PU-1TL and PU-1CH, respectively. After modification, mercapto groups of cysteine and cysteine hydrochloride tightly attached onto the surface of the QDs due to the excess of metal ions with respect to sulfide ions at the surface of the QDs [26]. Similarly, lysine and trifluoroacetyl lysine also tightly attached onto the surface of the QDs due to the interaction between carboxyl groups of lysine and trifluoroacetyl lysine and hydroxyl groups at the surface of the QDs [27]. When fluorescent membranes are exposed to the explosive vapors, the ligand should bind the explosive, and an electron-transfer mechanism between the QD and the electron-deficient explosive causes QD fluorescence quenching. Figure 5 shows the variation of the quenching percentage as a function of the exposure time to TNT, DNT, PA and NB. The response curves of fluorescent membranes towards different analytes are displayed in Figure S6 and S7. As for PU-1C, after about 12 min, the quenching percentage is ~43, 38, 56 and 36% (Figure 5) for saturated TNT (4 ppb) [26], DNT (180 ppb [28] or 200 ppb [26]), PA (0.0077 ppb) [26,28] and NB (3 × 10^5 ppb [28] or 4 × 10^5 ppb [26]) vapors at 25 °C, respectively. Because the vapor pressures of TNT and DNT are about 520 and 2.3~2.6 × 10^4-fold that of PA, respectively, the quenching percentage for PA is thus surprisingly larger than that expected from the relative vapor pressure of these analytes. In terms of molecular structure (Figure S8), PA is a stronger acid than TNT, and a stronger acid-base pairing interaction thus occurs between PA and amino ligands, resulting in the formation of PA anions at the surface of amine-capped ZnS QDs [26]. However, it is well known that the DNT molecules with two nitro groups are much weaker Lewis acids and electron acceptors than TNT molecules. This suggests that it is less likely to form a Mesienheimer complex with the amine for DNT by the relatively weak basic amine groups. Therefore, the enhanced sensitivity toward PA vapor originates from the extremely strong adsorption of PA species at the amino of QDs and the larger quenching efficiency due to the high electron-accepting ability. Moreover, the high surface-to-volume ratio of QDs [29,30] and the good permeability of nanofiber membranes [5,31,32] are further advantageous to the enhancement of the interaction between nitroaromatic explosive vapors and the amino ligands, helping maximize the quenching efficiency. The fluorescent membranes of PU-1CH show the differential responses towards explosives owing to the different molecular structures of ligands, after 12 min the quenching percentage was ~60, 42, 47 and 36% for saturated TNT, DNT, PA and NB vapors at 25 °C, respectively. Lysine has two amino groups, but the response of PU-1L is not as we expected it to be. The quenching percentage towards TNT, DNT and PA vapors does not increase simultaneously, only reaching 36 and 45% for DNT and PA, respectively. Trifluoroacetyl lysine has an amino group, an amino group and three substituted fluorine atoms and the quenching percentage of PU-1TL towards saturated TNT, DNT, PA and BN vapors at 25 °C is ~45, 61, 37 and 39%, respectively. It is possible that the high quenching percentage towards DNT should be attributed to the polarity interaction of DNT and trifluoroacetyl lysine brought by the fluorine atoms. It is worth noting that although the vapor pressure of NB is much higher than TNT and PA, the fluorescence quenching efficiency is still lower than TNT and PA. This should be attributed to the weaker electron-withdrawing ability of BN with only one electron-withdrawing nitro group, compared with TNT and PA with three electron-withdrawing nitro groups. As a result, the differential quenching of fluorescent membranes towards nitroaromatic explosives is basically achieved by the surface modification of QDs, which lays a good foundation for the recognizable detection of nitroaromatic explosives.

3.3. Discriminative Detection Based on Fluorescent Nanofibrous Membrane Array

In order to evaluate the recognizable detection of the fluorescent sensor array towards explosives, the quenching responses at different times of the sensor array including four fluorescent sensors were used to obtain the fingerprints of four nitroaromatic explosives. Figure 6 shows the transformation of fingerprints of four nitroaromatic explosives along with time. Interestingly, although the quenching of the fluorescent membranes is not linearly related to the exposure time (Figure 6), the fingerprint of each explosive at different times is very similar in shape and the fingerprints of the four explosives show different shapes. For example, the fingerprints of TNT, DNT, PA and NB corresponding to 2 min are not significantly different from those corresponding to 6 or 8 min, so it is possible to discriminate the explosives based on the fingerprints at 2 min or even shorter time. Thus, the fingerprints of the four explosives are clearly different because of the differential quenching of fluorescent membranes towards nitroaromatic explosives, which means that the recognizable detection of explosives by the fluorescent sensor array has been realized. More importantly, because of the stable relationship between fingerprints and time, the detection time of explosives can be effectively shortened, so as to achieve the recognizable detection of explosives in a shorter period of time. Or it can be used to determine unidentified explosives by monitoring the changes in the fingerprints along with time. It could, for example, be coupled with the technology of image recognition and large data analysis for a rapid diagnostic test of explosives, which is very important for the rapid and recognizable detection of explosives.

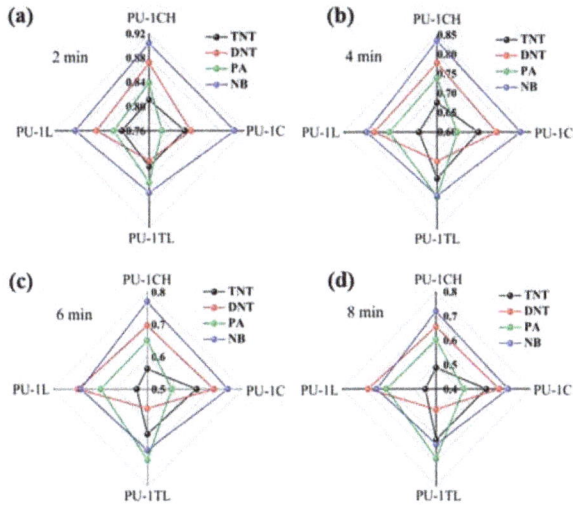

Figure 6. Fingerprints of three nitroaromatic explosives according to the quenching percentage of the sensing array based on fluorescent membranes as a function of time (**a**) 2 min, (**b**) 4 min, (**c**) 6 min, (**d**) 8 min.

3.4. Homogeneity, Stability and Recoverability of Fluorescent Nanofibrous Membranes

Homogeneity, stability and recoverability are known to be important for fluorescence sensors, therefore, these properties of fluorescent nanofibrous membranes are also evaluated. As shown in Figure 7a, the fluorescent homogeneity of 10 test points of fluorescent nanofibrous membranes is also investigated and both the modified nanofibrous membranes (PU-1C, PU-1CH, PU-1L and PU-1TL) and the pre-modified nanofibrous membranes (PU-1, PU-2, PU-3) show good fluorescent uniformity. The time stability of PU-1 and PU-2 membranes was tested at room temperature with the 40% relative humidity and the peak values at about 603 nm were collected every minute for 1 h (Figure 7b). Figure 7b shows that the fluorescent intensity of PU-1 and PU-2 membranes decreased by 4.3% and 3.4% after 1 h, respectively, showing the good time stability. The good time stability should be attributed to the UV irradiation treatment of the fluorescent nanofibrous membranes for 12 h to make the fluorescence intensity stable before quenching tests. Figure 7c illustrates the normalized fluorescence reversible investigation of PU-1C membrane with five cycles of PA detection between the saturated PA vapor at room temperature and the air at 50 °C. The PU-1L membrane shows a reversible fluorescent performance after five cycles of being immersed in the saturated PA vapor at room temperature for 10 min, heated in an oven at 50 °C for 20 min to detach the PA molecules. Evidently, this good recovery cycles demonstrate that the fluorescent nanofibrous membranes are reusable after a simple desorption process by heating.

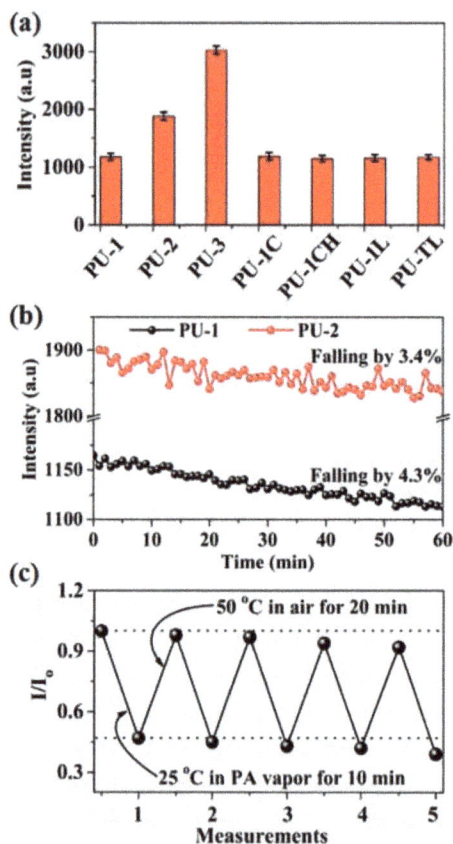

Figure 7. (**a**) Fluorescent homogeneity of 10 test points of fluorescent nanofibrous membranes (error bars represent the standard deviation of 10 test points of fluorescent membranes), (**b**) time stability of PU-1 and PU-2 membranes (relative humidity is 40%), (**c**) the normalized fluorescence recovery cycles of PU-1C membrane.

4. Conclusions

We have created and demonstrated the multichannel fluorescent sensor array based on nanofibrous membranes loaded with ZnS QDs for the discriminative detection of explosives. The array was constructed from nanofibrous membranes loaded with ZnS QDs featuring several surface ligands, playing a synergistic effect of the high surface-to-volume ratio of QDs, the good permeability of nanofiber membranes and the differential quenching introduced by surface ligands. The almost invariant transformation of fingerprints of four nitroaromatic explosives along with time was discovered. Four saturated vapors of nitroaromatic explosives could be reliably detected and discriminated by the array at room temperature. This work is the first step toward devising a monitoring system for explosives in the field of public security and defense. It could, for example, be coupled with the technology of image recognition and large data analysis for a rapid diagnostic test of explosives. This work further highlights the power of differential, multichannel arrays for the rapid and recognizable detection of a wide range of chemicals.

Supplementary Materials: The following are available online at http://www.mdpi.com/1424-8220/17/11/2676/s1, Figure S1:Structural formula of lysine, cysteine, trifluoroacetyl lysine and cysteine hydrochloride, Figure S2: The infrared spectrum of ZnS QD, Figure S3: SEM images with different magnification of (a, b) pure PU nanofibrous membrane, the inset in (b) shows image of water drops and the corresponding contact angles, Figure S4: Schematic diagram of ZnS QDs loading on nanofiber surfaces, Figure S5: Reproducibility of the quenching sensitivity of the nanofibrous membranes, Figure S6: Time-dependent fluorescence curves of the sensing array based on fluorescent membranes towards saturated TNT, DNT and PA vapors at room temperature, Figure S7: Time-dependent fluorescence curves towards the saturated NB vapor at room temperature (a) PU-1C, (b) PU-1CH, (c) PU-1L and (d) PU-1TL, Figure S8: Structural formula of TNT, PA, DNT and NB.

Acknowledgments: The authors thank the financial support from the Research Program of Natural Science Foundation of Xinjiang Uygur Autonomous Region (2015211B033).

Author Contributions: Z.W., Z.L., Y.C., D.J. and H.D. conceived and designed the experiments; Z.W. and F.Z. performed the experiments; Z.W. and Z.L. and D.J. analyzed the data; J.G. and F.Z. contributed reagents/materials/analysis tools; Z.W. and Z.L. wrote the paper.

Conflicts of Interest: The authors declare no conflict of interest.

References

1. Rose, A.; Zhu, Z.; Madigan, C.F.; Swager, T.M.; Bulovi, V. Sensitivity gains in chemosensing by lasing action in organic polymers. *Nature* **2005**, *434*, 876–879. [CrossRef] [PubMed]
2. Lichtenstein, A.; Havivi, E.; Shacham, R.; Hahamy, E.; Leibovich, R.; Pevzner, A.; Krivitsky, V.; Davivi, G.; Presman, I.; Elnathan, R. Supersensitive fingerprinting of explosives by chemically modified nanosensors arrays. *Nat. Commun.* **2014**, *5*, 4195. [CrossRef] [PubMed]
3. Geng, Y.; Ali, M.A.; Clulow, A.J.; Fan, S.; Burn, P.L.; Gentle, I.R.; Meredith, P.; Shaw, P.E. Unambiguous detection of nitrated explosive vapours by fluorescence quenching of dendrimer films. *Nat. Commun.* **2015**, *6*, 8240. [CrossRef] [PubMed]
4. Sun, X.; Brückner, C.; Nieh, M.P.; Lei, Y. A fluorescent polymer film with self-assembled three-dimensionally ordered nanopores: preparation, characterization and its application for explosives detection. *J. Mater. Chem. A* **2014**, *2*, 14613–14621. [CrossRef]
5. Wang, Y.; La, A.; Ding, Y.; Liu, Y.X.; Lei, Y. Novel Signal-Amplifying Fluorescent Nanofibers for Naked-Eye-Based Ultrasensitive Detection of Buried Explosives and Explosive Vapors. *Adv. Funct. Mater.* **2012**, *22*, 3547–3555. [CrossRef]
6. Zahn, S.; Swager, T.M. Three-dimensional electronic delocalization in chiral conjugated polymers. *Angew. Chem. Int. Edit.* **2002**, *41*, 4225–4230. [CrossRef]
7. Peveler, W.J.; Roldan, A.; Hollingsworth, N.; Porter, M.J.; Parkin, I.P. Multichannel Detection and Differentiation of Explosives with a Quantum Dot Array. *ACS Nano.* **2015**, *10*, 1139–1146. [CrossRef] [PubMed]
8. Paolesse, R.; Nardis, S.; Monti, D.; Stefanelli, M.; Natale, C.D. Porphyrinoids for Chemical Sensor Applications. *Chem. Rev.* **2016**, *10*, 1139–1146. [CrossRef] [PubMed]
9. Sarkar, S.; Dutta, S.; Chakrabarti, S.; Bairi, P.; Pal, T. Redox-switchable copper(I) metallogel: a metal-organic material for selective and naked-eye sensing of picric acid. *Acs Appl. Mater. Interfaces* **2014**, *6*, 6308–6316. [CrossRef] [PubMed]
10. Ye, J.; Zhao, L.; Bogale, R.F.; Gao, Y.; Wang, X.; Qian, X.; Guo, S.; Zhao, J.; Ning, G. Highly Selective Detection of 2,4,6-Trinitrophenol and Cu^{2+} Ion Based on a Fluorescent Cadmium-Pamoate Metal-Organic Framework. *Chem. Eur. J.* **2014**, *21*, 2029–2037. [CrossRef] [PubMed]
11. Hu, Z.; Deibert, B.J.; Li, J. Luminescent metal-organic frameworks for chemical sensing and explosive detection. *Chem. Soc. Rev.* **2014**, *43*, 5815–5840. [CrossRef] [PubMed]
12. Yang, Y.; Wang, H.; Su, K.; Long, Y.; Peng, Z.; Li, N.; Liu, F. A facile and sensitive fluorescent sensor using electrospun nanofibrous film for nitroaromatic explosive detection. *J. Mater. Chem.*, **2011**, *21*, 11895–11900. [CrossRef]
13. Chan, W.C.; Nie, S. Quantum dot bioconjugates for ultrasensitive nonisotopic detection. *Science* **1998**, *281*, 2016–2018. [CrossRef] [PubMed]
14. Enkin, N.; Sharon, E.; Golub, E.; Willner, I. Ag Nanocluster/DNA Hybrids: Functional Modules for the Detection of Nitroaromatic and RDX Explosives. *Nano Lett.* **2014**, *14*, 4918–4922. [CrossRef] [PubMed]
15. Freeman, R.; Willner, I. NAD^+/NADH-Sensitive Quantum Dots: Applications To Probe NAD^+-Dependent Enzymes and To Sense the RDX Explosive. *Nano Lett.* **2009**, *9*, 322–326. [CrossRef] [PubMed]

16. Freeman, R.; Finder, T.; Bahshi, L.; Gill, R.; Willner, I. Functionalized CdSe/ZnS QDs for the Detection of Nitroaromatic or RDX Explosives. *Adv. Mater.* **2012**, *24*, 6416–6421. [CrossRef] [PubMed]

17. Liu, B.; Tong, C.; Feng, L.; Wang, C.; He, Y.; Lü, C. Water-soluble polymer functionalized CdTe/ZnS quantum dots: a facile ratiometric fluorescent probe for sensitive and selective detection of nitroaromatic explosives. *Chem. Eur. J.* **2014**, *20*, 2132–2137. [CrossRef] [PubMed]

18. Bai, M.; Huang, S.; Xu, S.; Hu, G.; Wang, L. Fluorescent nanosensors via photoinduced polymerization of hydrophobic inorganic quantum dots for the sensitive and selective detection of nitroaromatics. *Anal. Chem.* **2015**, *87*, 2383–2388. [CrossRef] [PubMed]

19. Dutta, P.; Saikia, D.; Adhikary, N.C.; Sarma, N.S. Macromolecular Systems with MSA-Capped CdTe and CdTe/ZnS Core/Shell Quantum Dots as Superselective and Ultrasensitive Optical Sensors for Picric Acid Explosive. *ACS Appl. Mater. Interfaces* **2015**, *7*, 24778. [CrossRef] [PubMed]

20. Zhang, K.; Zhou, H.; Mei, Q.; Wang, S.; Guan, G.; Liu, R.; Zhang, J.; Zhang, Z. Instant visual detection of trinitrotoluene particulates on various surfaces by ratiometric fluorescence of dual-emission quantum dots hybrid. *J. Am. Chem. Soc.* **2011**, *133*, 8424–8427. [CrossRef] [PubMed]

21. Wu, Z.; Zhou, C.; Zu, B.; Li, Y.; Dou, X. Contactless and Rapid Discrimination of Improvised Explosives Realized by Mn^{2+} Doping Tailored ZnS Nanocrystals. *Adv. Funct. Mater.* **2016**, *26*, 4578–4586. [CrossRef]

22. Ostmark, H.; Wallin, S.; Ang, H.G. Vapor Pressure of Explosives: A Critical Review. *Propell. Explos. Pyrot.* **2012**, *37*, 12–23. [CrossRef]

23. Zhang, C.L.; Yu, S.H. Nanoparticles meet electrospinning: recent advances and future prospects. *Chem. Soc. Rev.* **2014**, *43*, 4423–4448. [CrossRef] [PubMed]

24. Sun, B.; Long, Y.Z.; Zhang, H.D.; Li, M.M.; Duvail, J.L.; Jiang, X.Y.; Yin, H.L. Advances in three-dimensional nanofibrous macrostructures via electrospinning. *Prog. Polym. Sci.* **2014**, *39*, 862–890. [CrossRef]

25. Ramakrishna, S.; Fujihara, K.; Teo, W.E.; Yong, T.; Ma, Z.; Ramaseshan, R. Electrospun nanofibers: Solving global issues. *Mater. Today* **2006**, *9*, 40–50. [CrossRef]

26. Tu, R.Y.; Liu, B.H.; Wang, Z.Y.; Gao, D.M.; Wang, F.; Fang, Q.L.; Zhang, Z.P. Amine-capped ZnS-Mn^{2+} nanocrystals for fluorescence detection of trace TNT explosive. *Anal. Chem.* **2008**, *80*, 3458–3465. [CrossRef] [PubMed]

27. Borges, J.; Mano, J.F. Molecular Interactions Driving the Layer-by-Layer Assembly of Multilayers. *Chem. Rev.* **2014**, *114*, 8883–8942. [CrossRef] [PubMed]

28. Lv, Y.Y.; Xu, W.; Lin, F.W.; Wu, J.; Xu, Z.K. Electrospun nanofibers of porphyrinated polyimide for the ultra-sensitive detection of trace TNT. *Sens. Actuators B Chem.* **2013**, *184*, 205–211. [CrossRef]

29. Liu, H.; Li, M.; Voznyy, O.; Hu, L.; Fu, Q.Y.; Zhou, D.X.; Xia, Z.; Sargent, E.H.; Tang, J. Physically Flexible, Rapid-Response Gas Sensor Based on Colloidal Quantum Dot Solids. *Adv. Mater.* **2014**, *26*, 2718–2724. [CrossRef] [PubMed]

30. Sanderson, K. Quantum dots go large. *Nature* **2009**, *459*, 760–761. [CrossRef] [PubMed]

31. Long, Y.Y.; Chen, H.B.; Wang, H.M.; Peng, Z.; Yang, Y.F.; Zhang, G.Q.; Li, N.; Liu, F.; Pei, J. Highly sensitive detection of nitroaromatic explosives using an electrospun nanofibrous sensor based on a novel fluorescent conjugated polymer. *Anal. Chim. Acta* **2012**, *744*, 82–91. [CrossRef] [PubMed]

32. Sun, X.C.; Liu, Y.X.; Shaw, G.; Carrier, A.; Dey, S.; Zhao, J.; Lei, Y. Fundamental Study of Electrospun Pyrene-Polyethersulfone Nanofibers Using Mixed Solvents for Sensitive and Selective Explosives Detection in Aqueous Solution. *ACS Appl. Mater. Interfaces* **2015**, *7*, 13189–13197. [CrossRef] [PubMed]

sensors

MDPI

Article

Graphene-Supported Spinel CuFe$_2$O$_4$ Composites: Novel Adsorbents for Arsenic Removal in Aqueous Media

Duong Duc La [1], Tuan Anh Nguyen [2], Lathe A. Jones [1,3] and Sheshanath V. Bhosale [1,*]

[1] School of Science, RMIT University, GPO Box 2476, Melbourne, VIC 3001, Australia;
    duc.duong.la@gmail.com (D.D.L.); lathe.jones@rmit.edu.au (L.A.J.)
[2] Applied Nanomaterial Laboratory, ANTECH, Hanoi 100000, Vietnam; Tuananhnguyendhb@gmail.com
[3] Centre for Advanced Materials and Industrial Chemistry (CAMIC), School of Science, RMIT University,
    GPO Box 2476, Melbourne, VIC 3001, Australia
*   Correspondence: sheshanath.bhosale@rmit.edu.au; Tel.: +61-3-9925-2680; Fax: +61-3-9925-3747

Academic Editor: W. Rudolf Seitz
Received: 8 May 2017; Accepted: 2 June 2017; Published: 5 June 2017

Abstract: A graphene nanoplate-supported spinel CuFe$_2$O$_4$ composite (GNPs/CuFe$_2$O$_4$) was successfully synthesized by using a facile thermal decomposition route. Scanning electron microscopy (SEM), high resolution transmission electron microscopy (HRTEM), Electron Dispersive Spectroscopy (EDS), X-ray diffraction (XRD) and X-ray Photoelectron Spectroscopy (XPS) were employed to characterize the prepared composite. The arsenic adsorption behavior of the GNPs/CuFe$_2$O$_4$ composite was investigated by carrying out batch experiments. Both the Langmuir and Freundlich models were employed to describe the adsorption isotherm, where the sorption kinetics of arsenic adsorption by the composite were found to be pseudo-second order. The selectivity of the adsorbent toward arsenic over common metal ions in water was also demonstrated. Furthermore, the reusability and regeneration of the adsorbent were investigated by an assembled column filter test. The GNPs/CuFe$_2$O$_4$ composite exhibited significant, fast adsorption of arsenic over a wide range of solution pHs with exceptional durability, selectivity, and recyclability, which could make this composite a very promising candidate for effective removal of arsenic from aqueous solution. The highly sensitive adsorption of the material toward arsenic could be potentially employed for arsenic sensing.

Keywords: graphene-supported CuFe$_2$O$_4$ composite; graphene nanoplates; spinel CuFe$_2$O$_4$; arsenic removal; graphene-oxide hybrid material

1. Introduction

Arsenic is highly toxic in the +3 and +5 oxidation state, and is widely present in the environment through leaching from soils, mining activities, fertilizers, industrial wastes, biological activity, and naturally occurring As containing minerals [1,2]. Long-term ingestion and drinking of arsenic contaminated food or water are linked to kidney, skin and lung cancers [3–6]. Therefore, it is of continued importance to remove arsenic from contaminated water, and to provide safe drinking water below the maximum concentration recommended by WHO (<10 ppb). Many approaches have been used for arsenic removal from contaminated water, including adsorption, ion exchange, chemical treatment, reverse osmosis, electrochemical treatment, membrane filtration, and co-precipitation [7–10]. However, due to its simplicity, low cost and high efficiency, adsorption is widely employed and studied as a promising technology for effectively removal of arsenic from contaminated water. The simplicity of these materials is especially important when it is recognized that As contamination is common in the developing world, where treatment processes must be convenient and affordable.

Many adsorbents based on agriculture and industrial waste, surfactants, carbon-base materials, polymers and metal oxides have been employed for arsenic adsorption [11,12]. Among these, metal and metal oxides such as TiO_2 [13–15], nano zero-valent iron [16,17], Fe_2O_3 [3,18,19], Fe_3O_4 [20], CeO_2 [21], CuO [22,23], CaO [24] and ZrO_2 [25,26] have been extensively studied for arsenic treatment in aqueous solution because of their high affinity to arsenic species, low cost, and the tunability of adsorption capacity [12,27]. Recently, considerable attention has been focused on the development of adsorbent composites containing two or more metals as metal oxides, to maximize arsenic adsorption. For instance, Zhang and co-workers synthesized a nanostructured Fe-Cu binary oxide with high adsorption capacity for arsenic [28].

Fe-Mn binary oxides were also successfully fabricated by Shan et al. with a high adsorption capacity toward arsenic [29]. In another report, Yu et al. presented Fe–Ti binary oxide magnetic nanoparticles which combined the photocatalytic oxidation property of TiO_2 with the high adsorption capacity and magnetic properties of γ-Fe_2O_3, for arsenic treatment [30]. Basu et al. found that Fe(III)-Al(III) mixed oxides and Fe(III)-Ce(IV) oxides have a high adsorption capacity toward arsenic [31,32].

Graphene, a two-dimensional (2D) material, has been attracting significant interest in the past decade, due to its exceptional chemical and physical properties which can be applied to many different areas including, but not limited to, electronic devices, energy storage and conversion, sensors, adsorption, and composites [33–38]. Most recently, graphene has gained tremendous interest as a supporting material for enhancement of adsorption properties of adsorbents, due to its large surface area, high conductivity, ionic mobility, and superior mechanical flexibility. For example, Ganesh et al. reported a smart magnetic graphene that removed heavy metals from drinking water [39]. A hybrid of monolithic Fe_2O_3/graphene was also fabricated, and showed favorable properties for arsenic removal [40]. Reduced graphene oxide-supported mesoporous Fe_2O_3/TiO_2 nanoparticles synthesized by a sol-gel route showed high adsorption towards arsenic [41]. Kumar et al. synthesized single-layer graphene oxide with manganese ferrite magnetic nanoparticles for efficient removal of arsenic from contaminated water [42].

In our previous work, we successfully fabricated a graphene nanoplates (GNPs) -supported Fe-Mg binary oxide composite by a simple hydrothermal method. This adsorbent showed a very high adsorption capacity toward arsenic [43]. In continuation of our efforts to this end, herein we report a simple one-pot hydrothermal method to prepare a graphene nanoplates-supported spinel $CuFe_2O_4$ (GNPs/$CuFe_2O_4$) composite. The optimized Cu:Fe molar ratio to fabricate the spinel $CuFe_2O_4$ for arsenic adsorption was adopted from Zhang's work, which is 1:2 [28]. TEM, SEM, EDS, TGA, XPS and XRD were used to characterize the prepared composite. The arsenic adsorption capacity of the material was carefully studied. The effects of parameters including graphene loading, initial arsenic concentration, adsorption time and solution pH on arsenic adsorption, selectivity, and recyclability were investigated through batch experiments and a column test.

2. Materials and Methods

2.1. Materials

Graphene nanoplates (GNPs) were obtained from VNgraphene. Dry acetone, ethanol, sodium hydroxide (NaOH), posstasium hydroxide (KOH), sodium persulfate ($Na_2S_2O_8$), As_2O_5, anhydrous $CuCl_2$ and $FeCl_3$ were purchased from Ajax Finechem. All chemicals were used as received.

2.2. Synthesis of GNPs/CuFe₂O₄ Composite

GNPs/$CuFe_2O_4$ composites were fabricated by a simple one-pot hydrothermal strategy. Firstly, $CuCl_2$ and $FeCl_3$ with various Cu:Fe molar ratios of 1:2 were dissolved in 50 mL of ethanol. Then graphene nanoplates with different loadings were dispersed in the mixture solution by sonication for 10 min, and then stirred for 1 h. Subsequently, a 2 M NaOH solution was added dropwise to the solution under vigorous stirring until a pH of ~8–9 was reached. After 1 h of further stirring, the reaction solution was transferred and sealed in a Teflon-lined autoclave, and placed in an oven pre-heated to 150 °C, for 2 h. Then the solution was cooled to room temperature, and the precipitate was filtered and washed three times each with ethanol and distilled water. The sample was dried overnight at a temperature of 60 °C in air to obtain the GNPs/CuFe$_2$O$_4$ composites.

2.3. Characterization

The morphology and mapping elemental composition of samples were studied by an EDS-equipped (Oxford Instruments plc, Abingdon, Oxfordshire, UK) scanning electron microscope using an FEI Nova NanoSEM (Hillsboro, AL, USA) operating under high vacuum with an accelerating voltage of 30 keV and an Everhart Thornley Detector (ETD). HRTEM images were obtained on a JEOL 2010 TEM instrument operated at an accelerating voltage of 100 kV. A BrukerAXS D8 Discover instrument with a general area detector diffraction system (GADDS) using a Cu Kα source was utilized to obtain XRD patterns. X-ray photoelectron spectra (XPS) were obtained on a K-Alpha XPS instrument using monochromated aluminum as the X-ray source. The C 1s, Fe 2p, Cu 2p, As 3d and O 1s core level spectra were recorded with an overall resolution of 0.1 eV. The core level spectra were background corrected using the Shirley algorithm, and chemically distinct species were resolved using a nonlinear least square fitting procedure.

2.4. Adsorption Studies

A stock solution of 1000 ppm As(V) was prepared by dissolving As$_2$O$_5$ in water. Arsenic concentrations were determined using an Agilent 4200 microwave plasma-atomic emission spectrometer (MP-AES). All samples were analyzed within 24 h of filtration.

2.4.1. Effect of GNPs Loading on Arsenic Sorption

Adsorption experiments were carried out in closed glass vessels. Typically, 24 mg of each adsorbent prepared from different GNP loadings were added into glass vessels containing 50 mL of 10 mg/L arsenic solution at a pH of 4. The solution was kept shaking at 200 rpm at room temperature for 24 h. Then, all samples were filtered by vacuum filtration to remove the adsorbent, and the concentration of arsenic in the residual solutions was analyzed.

2.4.2. Adsorption Isotherm

A total of 10 mg of optimally fabricated adsorbent was added to 50 mL of As(V) solution with initial concentrations ranging from 5 to 90 mg/L in glass vessels. The adsorption was carried out at a solution pH of 4 at room temperature, shaking at a speed of 200 rpm for 24 h. The mixtures were then filtered by vacuum filter and analyzed for residual arsenic by microwave plasma—atomic emission spectrometry (MP-AES).

2.4.3. Adsorption Kinetics

In a typical experiment, 40 mg of GNPs/CuFe$_2$O$_4$ composite was mixed with 200 mL of 40 mg/L arsenic in a glass vessel. The mixed solution was shaken on an orbital shaker at a speed of 200 rpm at room temperature and solution pH of 4. At certain time intervals, 10 mL of the mixture was taken, filtered by vacuum filter and analyzed for arsenic.

2.4.4. Effect of Solution pH

10 mg of GNPs/CuFe$_2$O$_4$ composite was added to 50 mL of 10 mg/L arsenic at various solution pH values ranging from 4 to 11 (the pH values were adjusted by dilute HCl and NaOH solutions). The suspensions were shaken at a speed of 200 rpm at room temperature for 24 h. Then, all samples were filtered by vacuum filter and residual arsenic concentrations were determined.

2.4.5. Selectivity Test

10 mg of GNPs/ CuFe$_2$O$_4$ adsorbent was added to 50 mL of a solution containing 3 mg/L of each ion such as As(V), Na$^+$, K$^+$, Ca^{2+} and Mg^{2+}. The mixed solution was shaken on an orbital shaker with a speed of 200 rpm at room temperature and solution pH of 7 for 12 h. The mixtures were then filtered by vacuum filter and analyzed for residual ions by MP-AES.

2.4.6. Recyclability Test

The reusability of the prepared adsorbent was studied by a column test. The GNPs/CuFe$_2$O$_4$ oxide composite was assembled as a part of a filter column. Other parts included a glass tube with both ends wrapped with a few layers of filter papers and cotton. The filter column was regenerated by washing several times with 2 M NaOH after each arsenic adsorption cycle before implementing the next experiment.

3. Results and Discussion

The morphology of the obtained graphene nanoplates (GNPs) was studied by SEM (Figure 1). It can be clearly seen in Figure 1a,b that the GNPs had a crumpled, wrinkled morphology with a diameter of tens of microns and a thickness of <20 nm [44].

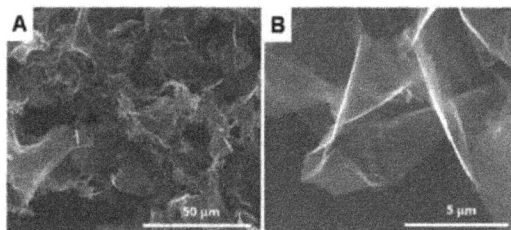

Figure 1. Low (**A**) and high (**B**) scanning electron microscopy (SEM) images of graphene nanoplates.

It is believed that the Fe^{3+} and Cu^{2+} ions firstly physically adsorbed on the GNPs, and then these ions reacted under the hydrothermal conditions to form Fe-Cu binary oxides on the GNPs. The morphology of the as-prepared GNPs/Fe-Cu binary oxides material was investigated by SEM and HRTEM studies. Figure 2a,b and Figure S1 show low and high resolution of SEM images of the composites on a silicon wafer. The SEM images confirmed that the Fe-Cu binary oxides were uniformly dispersed on the surface of the GNPs. The low resolution HRTEM image shown in Figure 2C and Figure S2, also confirmed a good distribution of oxides on the GNPs. When the composite was viewed at the high resolution of HRTEM (Figure 2D), it can be clearly seen that Fe-Cu binary oxides were well-separated, with particle sizes of approximately 5 nm in diameter. The uniform distribution of Fe-Cu binary oxides on the GNPs was further confirmed by EDS mapping (Figure 3). The distribution of Cu and Fe elements on the surface of graphene (elemental C) was uniform and homogeneous.

The EDS study also confirmed that the atomic ratio of Cu:Fe:O is approximately 1:2:4, which was consistent with the theoretical formula of the spinel CuFe$_2$O$_4$, or a mixture of CuO and Fe$_2$O$_3$ oxides formula of obtained binary oxides. In order to further confirm the formation of metal oxides, XRD diffraction patterns were obtained. Figure 4A shows the XRD pattern of the pure CuFe$_2$O$_4$ and GNPs/Fe-Cu binary oxides composite. In the XRD spectrum of the pure CuFe$_2$O$_4$, the diffraction peaks at 2θ = 30.56, 36, 43.7, 51.8, 56.6, and 63.1° could be indexed to the (220), (311), (400), (422), (511), and (440) planes of cubic spinel CuFe$_2$O$_4$ (PDF 06-0545) [45]. When incorporated onto the GNPs surface, the metal oxides mainly formed the cubic spinel structure of CuFe$_2$O$_4$, as the main diffraction peaks matched with the pure CuFe$_2$O$_4$. The peaks with asterisks were attributed to the crystallites of the supporting graphene nanoplates [44,46].

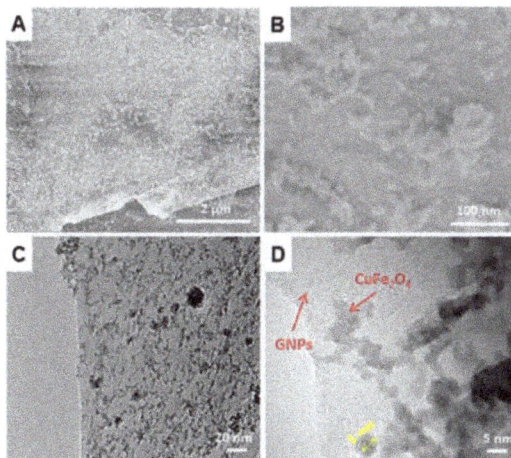

Figure 2. (**A**) and (**B**) SEM images and (**C**) and (**D**) transmission electron microscopy (TEM) images of the GNPs/CuFe$_2$O$_4$ composite.

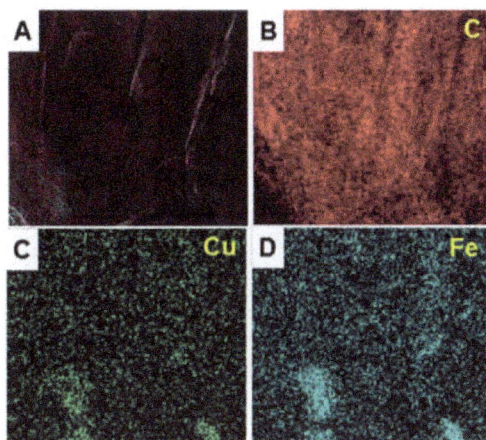

Figure 3. Electron dispersive spectroscopy mappping of the GNPs/CuFe$_2$O$_4$ composite.

The core level XPS spectra of C 1s, Fe 2p and Cu 2p were obtained to probe the chemical environment and oxidation states of C, Fe and Cu in the GNPs/CuFe$_2$O$_4$ composite, as exhibited in Figure 4B–D. The deconvoluted core level of C 1s (Figure 4B) revealed two major peaks at 284.1 and 284.8 eV corresponding to graphenic carbon with C=C (sp2) and C-C (sp3) bonds [47], respectively. In the Figure 4C, the Fe core level XPS spectrum had two dominant peaks at 711.18 and 724.28 eV with small satellite, which was consistent with the Fe 2p3/2 and Fe 2p1/2 of the Fe^{3+} state in the spin-orbit of CuFe$_2$O$_4$, respectively [48]. Figure 4D showed the binding energy of core level Cu 2p. The fitting revealed peaks at around 933.78 and 953.78 eV with a broad satellite at around 942 eV, which corresponded to the Cu2 2p$_{3/2}$ and 2p$_{1/2}$, respectively, of Cu^{2+} in the spinel CuFe$_2$O$_4$ [48]. All of these results further confirmed the formation of CuFe$_2$O$_4$ on the GNPs.

It is of note that the As(V) sorption by the Fe-Cu binary oxide reaches a maximum when the molar ratio of Cu:Fe is 1:2 [28]. Hence, in this study, we have chosen this molar ratio as an optimal condition when preparing the spinel CuFe$_2$O$_4$. We then investigated the effect of graphene loading on the arsenic sorption capacity by the GNPs/CuFe$_2$O$_4$ composite with an initial As(V) concentration of 24 mg/L, adsorbent dose = 200 mg/L, pH = 4, at room temperature (Figure 5A). It was seen from Figure 5 that the As(V) adsorption was enhanced along with an increase in the GNPs loading, and reached a maximal sorption capacity of about 58 mg/g at the GNPs:CuFe$_2$O$_4$ weight ratio of 1:1 (6:6 in the figure). However, the sorption capacity dramatically dropped as GNPs loading increased above 1:1, and without CuFe$_2$O$_4$, the As(V) sorption capacity by pure GNPs was only 2.38 mg/g. These results demonstrated a significant improvement in As(V) adsorption with the incorporation of GNPs with CuFe$_2$O$_4$.

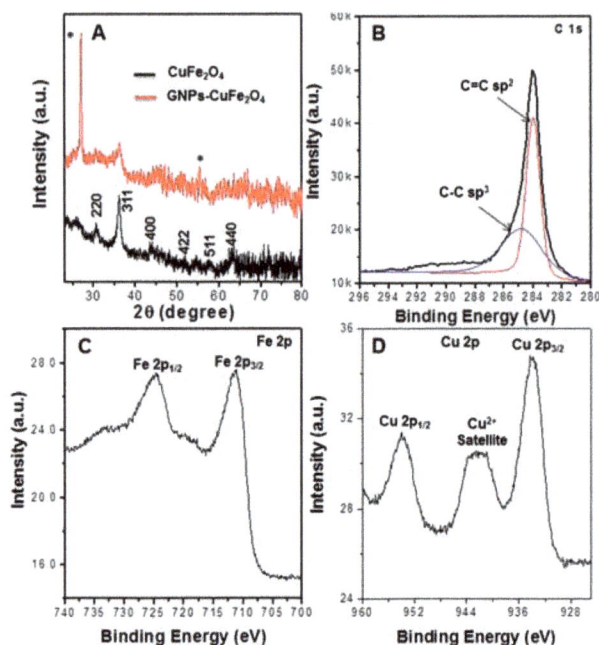

Figure 4. (**A**) X-ray diffraction (XRD) patterns of pure spinel CuFe$_2$O$_4$ and GNPs/CuFe$_2$O$_4$ composites; (**B**–**D**) core level X-ray Photoelectron Spectroscopy pectra of C 1s, Fe 2p and Cu 2p, respectively, obtained from the GNPs/CuFe$_2$O$_4$ composite.

Figure 5B shows the effect of pH on As(V) adsorption by the GNPs/CuFe$_2$O$_4$ composite at an initial As(V) concentration of 10 mg/L, and adsorbent dose = 200 mg/L at room temperature. It is obvious that the sorption capacity of As(V) strongly depended on the solution pH. Arsenic adsorption occurred strongly in acidic conditions with a maximum capacity of 39 mg/g at pH = 4. When the solution pH increased, the adsorption significantly decreased. The change in As(V) adsorption capacity was negligible in the solution pH range of 5 to 9 before declining greatly when the solution pH further increased. This phenomenon may be ascribed to the dependence of adsorption of strong acid anions by metal oxides and hydroxides oxide in solution pH [28]. The pK_a1, pK_a2, pK_a3 of As(V) are 2.1, 6.7, 11.2, which is present in a negative ionic form under most pH conditions. Since the electrostatic attraction is the main force, which is responsible for the adsorption of As(V) on graphene-metal oxide composites [49,50], the change of electrostatic force between As(V) and the GNPs/CuFe$_2$O$_4$ composite may explain the effect of pH on As(V) adsorption. At a low pH, the GNPs/CuFe$_2$O$_4$ adsorbent has a net positive charge due to protonation of –OH groups in the spinel CuFe$_2$O$_4$. As a result, they attract the negatively charged As(V) ions (AsO$_4^{3-}$), which leads to the greater adsorption. When the pH increases, the positive charge decreases, resulting in a decrease of As(V) adsorption.

The leaching of Fe and Cu in the GNPs/CuFe$_2$O$_4$ composite at different pH values was also recorded, as shown in Figure 5B. The release of Fe and Cu was low compared to As(V) adsorption, which indicated that the GNPs/CuFe$_2$O$_4$ composite was a stable and effective adsorbent for arsenic.

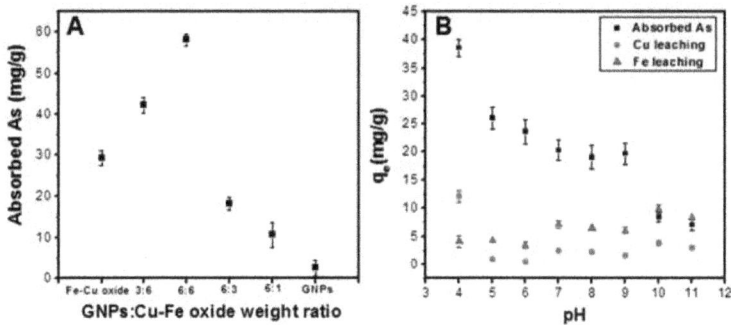

Figure 5. Effect of graphene nanoplates (GNPs) loading (**A**) and solution pH (**B**) on As (V) adsorption by GNPs/CuFe$_2$O$_4$ composites.

The adsorption isotherm was obtained in order to assess the arsenic adsorption and determine the maximum As(V) adsorption capacity by the GNPs/CuFe$_2$O$_4$ composite. The amount of arsenic adsorbed on the composite at equilibrium (q_e) was calculated from different concentrations of arsenic with the following equation:

$$q_e = \frac{(C_0 - C_e) \times V}{m} \tag{1}$$

where C_0 (mg/L) is the initial concentration, C_e (mg/L) is the equilibrium concentration, V (L) is the solution volume, and m (g) is the mass of the GNPs/CuFe$_2$O$_4$ adsorbent.

Figure 6 shows the arsenic adsorption capacity by the composite at equilibrium with various As(V) concentrations in the range of 5–90 mg/L, at an adsorbent dose of 200 mg/L, pH 4 under room temperature. Both adsorption isotherms for the Langmuir and Freundlich models were used to fit the data as expressed in Equations (2) and (3), respectively:

$$q_e = \frac{q_{max} K_L C_e}{1 + K_L C_e} \tag{2}$$

$$q_e = K_F C_e^n \tag{3}$$

where q_e is the amount of arsenic adsorbed on the solid phase at equilibrium (mg/g), q_{max} (mg/g) is the maximum arsenic adsorption capacity per unit weight of adsorbent, C_e is the equilibrium arsenic concentration (mg/L), K_L is the equilibrium adsorption constant represented by the affinity of binding sites (L/mg), K_F is the Freundlich constant, and n is the heterogeneity factor.

The obtained As(V) adsorption constants are presented in Table 1. The higher correlation coefficient (0.966) values of As(V) from the fitted Freundlich plots compared to that of Langmuir plots (0.95) suggested that the Freundlich model was more suitable for representing the adsorption behavior of As(V) by the GNPs/CuFe$_2$O$_4$ composite. The low calculated heterogeneity factor (n = 0.56 for As(V)) also suggests that the Freundlich was the more favorable model. These results indicate that As(V) was heterogeneously adsorbed on the composite surface, suggesting the simultaneous existence of graphene and iron-copper binary oxides in the solid phase. The maximum As(V) adsorption capacity by the GNPs/CuFe$_2$O$_4$ determined from the Langmuir model was 172.27 mg/g, which was a very effective adsorbent for the removal of arsenic. The maximum As(III) adsorption capacity by the GNPs/CuFe$_2$O$_4$ was also determined from the Langmuir model as 236.29 mg/g (Figure S3 and Table S1).

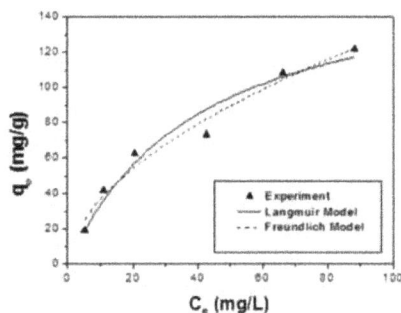

Figure 6. Adsorption isotherm for As(V) by the GNPs/CuFe$_2$O$_4$ composite.

Table 1. Langmuir and Freundlich isotherm parameters for As(V) adsorption on the GNPs/CuFe$_2$O$_4$ composite.

	Langmuir Model			Freundlich Model		
	Q_m (mg/g)	K_L (L/mg)	R^2	K_F	n	R^2
As(V)	172.27	0.02	0.95	10.31	0.56	0.966

Table 2 compares the As(V) adsorption capacity between the GNPs/CuFe$_2$O$_4$ composite with other adsorbents from the literature. It can be seen from the table that the sorption capacity of the GNPs/CuFe$_2$O$_4$ composite was superior to most of the other adsorbents, which could make the GNPs/CuFe$_2$O$_4$ composite a practical adsorbent for arsenic removal.

Table 2. Comparison of maximal arsenic adsorption capacity by various adsorbents.

Absorbates	pH	q_{max} (mg/g)	References
$Mg_{0.27}Fe_{2.5}O_4$	7	83.2	[51]
Fe_3O_4-GO (MGO)	6.5	59.6	[49]
$FeMnO_x$/RGO	7	22.22	[52]
CeO_2-grahene composite	4	1.019	[53]
$GO-ZrO(OH)_2$	5–11	84.89	[50]
nZVI/graphene	7	29	[54]
Magnetic graphene	4	3.26	[39]
Fe_3O_4/graphene/LDH	6	73.1	[39]
Magnetic-GO	4	38	[55]
Magnetic-rGO	4	12	[55]
$MnFe_2O_4$	3	94	[56]
$CoFe_2O_4$	3	74	[56]
$CuFe_2O_4$ binary oxide	7	82.7	[28]
GNPs/Fe-Mg Oxide	7	103.9	[43]
GNPs/CuFe$_2$O$_4$	4	172.7	This work

In order to further understand the adsorption behavior of As(V) on the GNPs/CuFe$_2$O$_4$ surface, the adsorption kinetics of As(V) adsorption were obtained with an initial As(V) concentration of 40 mg/L (adsorbent dose of 200 mg/L, pH 4 and at room temperature), and sorption capacities were determined at different time intervals (Figure 7). The adsorption quickly reached equilibrium within 2 h. The pseudo-second-order model was applied to describe the kinetic data as expressed in Equation (4):

$$q_t = \frac{K q_e^2 t}{1 + K q_e t} \tag{4}$$

where q_t (mg/g) is the amount of arsenic adsorbed on the solid phase at time t (hr), q_e (mg/g) is the amount of arsenic adsorbed on the solid phase at equilibrium, and K is the adsorption rate constant (g mg.h). According to the adsorption kinetic values listed in Table 3, the experimental data was well-fitted, with a correlation coefficient of 0.916. This result implies that the adsorption process may have occurred through chemical adsorption and/or electrostatic attraction, accompanied by electron exchange between the composite and arsenic [57]. The adsorption capacity at equilibrium of the composite with an initial As(V) concentration of 40 mg/L calculated from the pseudo-second-order model was 84.46 mg/g.

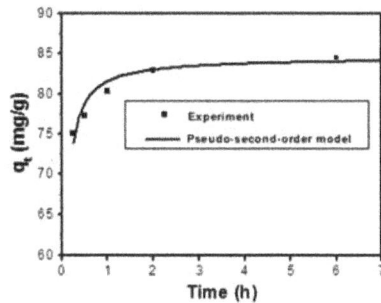

Figure 7. Adsorption kinetics of As(V) on the GNPs/CuFe$_2$O$_4$ composite.

Table 3. Adsorption kinetics parameters for As(V) adsorption on GNPs/CuFe$_2$O$_4$ composite.

Pseudo-Second-Order Model		
q_e (mg/g)	K (h^{-1})	R^2
84.46	0.331	0.916

The XPS As 3d core level spectrum was recorded to verify the presence and chemical state of arsenic on the surface of adsorbent (Figure 8). The appearance of As 3d XPS peak confirmed the presence of arsenic on the surface of the GNPs/CuFe$_2$O$_4$ composite. The core level XPS revealed one dominant peak at 45.5 eV, which was consistent with the binding energy of As(V) [28,58]. As a result, it was obvious that there was no change in oxidation state of As(V) during the sorption process. This further confirmed that arsenic was adsorbed onto the GNPs/CuFe$_2$O$_4$ surface by chemical adsorption and/or an electrostatic attraction mechanism.

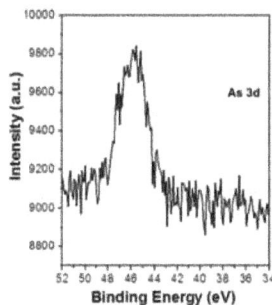

Figure 8. Core level XPS spectra of As 3d obtained from the GNPs/CuFe$_2$O$_4$ composite after adsorption.

Graphene with a large surface area, high conductivity, ionic mobility and superior mechanical flexibility, can be an excellent supporting material for the enhancement of adsorption properties of adsorbents. In this case, $GNPs/CuFe_2O_4$ adsorbent showed enhanced adsorption capacity in comparison with free standing $CuFe_2O_4$ ($q_{max} = 82.7$ mg/g). Based on well-documented understanding and from the discussion above, we proposed a possible adsorption of arsenic by $GNPs/CuFe_2O_4$ (Figure 9). When adding adsorbent into arsenic-containing solution, arsenic is adsorbed on the $CuFe_2O_4$ surface by chemical adsorption and/or electrostatic attraction. The presence of graphene increases the surface area of adsorbent, and as a consequence increases the absorption sites for arsenic.

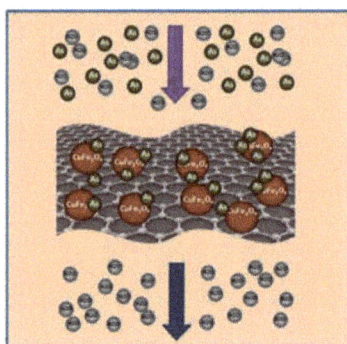

Figure 9. Possible adsorption mechanism of $GNPs/Cu_2FeO_4$ toward arsenic.

Figure 10A shows the selectivity of $GNPs/CuFe_2O_4$ adsorbent towards arsenic in the presence of common positive ions in drinking water such as Na^+, K^+, Ca^{2+} and Mg^{2+}, with an initial concentration of As(V) and other ions at 3 mg/L. It can be seen that while more than 98% of arsenic was adsorbed, there was an insignificant amount of Na^+, K^+ and Ca^{2+} ions adsorbed on the $GNPs/CuFe_2O_4$ composite, indicating that this adsorbent can be effectively and selectively used for arsenic removal. The adsorption capacity of $GNPs/CuFe_2O_4$ composite toward arsenic was also higher than other heavy metals such as lead ions (Figure S4).

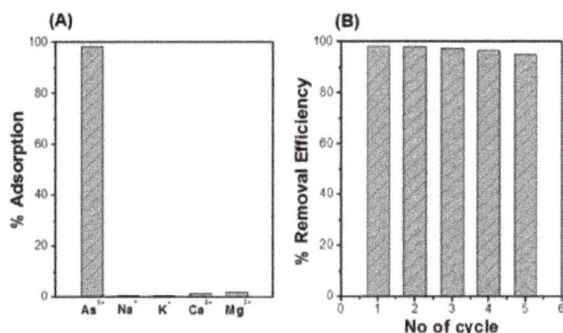

Figure 10. Selectivity (**A**) and recyclability (**B**) of the $GNPs/CuFe_2O_4$ composite for As(V) removal in a column test.

To evaluate recyclability, a filter column with a diameter of 2 cm and a height of 10 cm was assembled, as shown in Figure S5. The mass of the $GNPs/CuFe_2O_4$ composite used was 200 mg. The adsorption process was carried out with 10 mL of flow solution (pH 7) of 3 mg/L As(V). After adsorption, the filter column was washed with 20 mL of 2 M NaOH solution to regenerate the adsorbent before the next test cycle. The adsorption-regeneration process was repeated for five cycles. The result in Figure 10B shows an insignificant decrease of removal efficiency (less than 4%) after 5 cycles, suggesting that the GNPs/CuFe$_2$O$_4$ composite has high durability for arsenic removal.

4. Conclusions

A graphene-supported spinel CuFe$_2$O$_4$ composite was conveniently synthesized by co-precipitating graphene nanoplates with iron and copper ions in ethanol solution. The CuFe$_2$O$_4$ was crystallized and well-dispersed on the graphene surface. The prepared GNPs/CuFe$_2$O composite showed fast, high adsorption capacity toward As(V), with a maximum adsorption capacity of 172.27 mg/g at pH 4, which is superior to the majority of reported adsorbents. This adsorbent showed excellent selectivity toward arsenic ions over common metal ions such as Na$^+$, K$^+$, Ca^{2+} and Mg^{2+}. The arsenic adsorption by the GNPs/CuFe$_2$O composite was very effective over a wide range of solution pHs. Moreover, the absorbent could be readily regenerated and recycled for arsenic removal. With these excellent results, it could be concluded that the GNPs/CuFe$_2$O composite could be considered a promising candidate for practical arsenic removal from aqueous solution. Furthermore, the GNPs/CuFe$_2$O composite can be potentially used as a sensor probe of arsenic, based on its high sensitive adsorption toward arsenic.

Supplementary Materials: The following are available online at http://www.mdpi.com/1424-8220/17/6/1292/s1, Figure S1: SEM images of the GNPs/CuFe$_2$O$_4$ composite, Figure S2: TEM images of the GNPs/CuFe$_2$O$_4$ composite, Figure S3: Adsorption isotherm for As(III) by GNPs/CuFe$_2$O$_4$ composite, Table S1: Langmuir and Freundlich isotherm parameters for As(III) adsorption on GNPs/CuFe$_2$O$_4$ composite, Figure S4: Adsorption of 10 mg GNPs@Fe$_2$CuO$_4$ composite toward 3 mg/L of As^{5+} and Pb^{2+} for 2 hours, Figure S5: Filter column with a diameter of 2 cm and a height of 10 cm for recyclability.

Acknowledgments: D.D.L. thanks RMIT University and Program 165 for financial support. S.V.B. acknowledges the Australian Research Council under a Future Fellowship Scheme (FT110100152). The authors acknowledge the facilities, and the scientific and technical assistance, of the RMIT Microscopy & Microanalysis Facility (RMMF).

Author Contributions: D.D.L. performed all the experiment and T.A.N. help for XRD and XPS analysis, L.A.J. monitors the analysis, S.V.B. supervises the research project. All co-authors contributed for the manuscript preparation.

Conflicts of Interest: The authors declare no conflict of interest.

References

1. Kitchin, K.T.; Conolly, R. Arsenic-Induced Carcinogenesis-Oxidative Stress as a Possible Mode of Action and Future Research Needs for More Biologically Based Risk Assessment. *Chem. Res. Toxicol.* **2009**, *23*, 327–335. [CrossRef] [PubMed]

2. Erdoğan, H.; Yalçınkaya, Ö.; Türker, A.R. Determination of inorganic arsenic species by hydride generation atomic absorption spectrometry in water samples after preconcentration/separation on nano ZrO$_2$/B$_2$O$_3$ by solid phase extraction. *Desalination* **2011**, *280*, 391–396. [CrossRef]

3. Tuzen, M.; Çıtak, D.; Mendil, D.; Soylak, M. Arsenic speciation in natural water samples by coprecipitation-hydride generation atomic absorption spectrometry combination. *Talanta* **2009**, *78*, 52–56. [CrossRef] [PubMed]

4. Babazadeh, M.; Hosseinzadeh Khanmiri, R.; Abolhasani, J.; Ghorbani-Kalhor, E.; Hassanpour, A. Synthesis and Application of a Novel Functionalized Magnetic Metal-Organic Framework Sorbent for Determination of Heavy Metal Ions in Fish Samples. *Bull. Chem. Soc. Jpn.* **2015**, *88*, 871–879. [CrossRef]

5. Ding, S.-Y.; Dong, M.; Wang, Y.-W.; Chen, Y.-T.; Wang, H.-Z.; Su, C.-Y.; Wang, W. Thioether-based fluorescent covalent organic framework for selective detection and facile removal of mercury(II). *J. Am. Chem. Soc.* **2016**, *138*, 3031–3037. [CrossRef] [PubMed]

6. Akamatsu, M.; Komatsu, H.; Matsuda, A.; Mori, T.; Nakanishi, W.; Sakai, H.; Hill, J.P.; Ariga, K. Visual Detection of Cesium Ions in Domestic Water Supply or Seawater using a Nano-optode. *Bull. Chem. Soc. Jpn.* **2017**, *90*, 678–683. [CrossRef]

7. Bissen, M.; Frimmel, F.H. Arsenic—A review. Part I: Occurrence, toxicity, speciation, mobility. *Acta Hydroch. Hydrob.* **2003**, *31*, 9–18. [CrossRef]

8. Mohan, D.; Pittman, C.U. Arsenic removal from water/wastewater using adsorbents—A critical review. *J. Hazard. Mater.* **2007**, *142*, 1–53. [CrossRef] [PubMed]

9. Jadhav, S.V.; Bringas, E.; Yadav, G.D.; Rathod, V.K.; Ortiz, I.; Marathe, K.V. Arsenic and fluoride contaminated groundwaters: A review of current technologies for contaminants removal. *J. Environ. Manag.* **2015**, *162*, 306–325. [CrossRef] [PubMed]

10. Singh, R.; Singh, S.; Parihar, P.; Singh, V.P.; Prasad, S.M. Arsenic contamination, consequences and remediation techniques: A review. *Ecotoxicol. Environ. Saf.* **2015**, *112*, 247–270. [CrossRef] [PubMed]

11. Kurniawan, T.A.; Sillanpää, M.E.; Sillanpää, M. Nanoadsorbents for remediation of aquatic environment: Local and practical solutions for global water pollution problems. *Crit. Rev. Environ. Sci. Technol.* **2012**, *42*, 1233–1295. [CrossRef]

12. Ray, P.Z.; Shipley, H.J. Inorganic nano-adsorbents for the removal of heavy metals and arsenic: A review. *RSC Adv.* **2015**, *5*, 29885–29907. [CrossRef]

13. Jézéquel, H.; Chu, K.H. Enhanced adsorption of arsenate on titanium dioxide using Ca and Mg ions. *Environ. Chem. Lett.* **2005**, *3*, 132–135. [CrossRef]

14. Deedar, N.; Aslam, I. Evaluation of the adsorption potential of titanium dioxide nanoparticles for arsenic removal. *J. Environ. Sci.* **2009**, *21*, 402–408.

15. Xu, Z.; Li, Q.; Gao, S.; Shang, J.K. As(III) removal by hydrous titanium dioxide prepared from one-step hydrolysis of aqueous $TiCl_4$ solution. *Water Res.* **2010**, *44*, 5713–5721. [CrossRef] [PubMed]

16. Bhowmick, S.; Chakraborty, S.; Mondal, P.; Van Renterghem, W.; Van den Berghe, S.; Roman-Ross, G.; Chatterjee, D.; Iglesias, M. Montmorillonite-supported nanoscale zero-valent iron for removal of arsenic from aqueous solution: Kinetics and mechanism. *Chem. Eng. J.* **2014**, *243*, 14–23. [CrossRef]

17. Dong, H.; Guan, X.; Lo, I.M. Fate of As(V)-treated nano zero-valent iron: Determination of arsenic desorption potential under varying environmental conditions by phosphate extraction. *Water Res.* **2012**, *46*, 4071–4080. [CrossRef] [PubMed]

18. Tang, W.; Li, Q.; Gao, S.; Shang, J.K. Arsenic(III, V) removal from aqueous solution by ultrafine α-Fe_2O_3 nanoparticles synthesized from solvent thermal method. *J. Hazard. Mater.* **2011**, *192*, 131–138. [CrossRef] [PubMed]

19. Tang, W.; Li, Q.; Li, C.; Gao, S.; Shang, J.K. Ultrafine α-Fe2O3 nanoparticles grown in confinement of in situ self-formed "cage" and their superior adsorption performance on arsenic(III). *J. Nanopart. Res.* **2011**, *13*, 2641–2651. [CrossRef]

20. Akin, I.; Arslan, G.; Tor, A.; Ersoz, M.; Cengeloglu, Y. Arsenic (V) removal from underground water by magnetic nanoparticles synthesized from waste red mud. *J. Hazard. Mater.* **2012**, *235*, 62–68. [CrossRef] [PubMed]

21. Feng, Q.; Zhang, Z.; Ma, Y.; He, X.; Zhao, Y.; Chai, Z. Adsorption and desorption characteristics of arsenic onto ceria nanoparticles. *Nanoscale Res. Lett.* **2012**, *7*, 1–8. [CrossRef] [PubMed]

22. Reddy, K.; McDonald, K.; King, H. A novel arsenic removal process for water using cupric oxide nanoparticles. *J. Colloid Interface Sci.* **2013**, *397*, 96–102. [CrossRef] [PubMed]

23. Goswami, A.; Raul, P.; Purkait, M. Arsenic adsorption using copper (II) oxide nanoparticles. *Chem. Eng. Res. Des.* **2012**, *90*, 1387–1396. [CrossRef]

24. Olyaie, E.; Banejad, H.; Afkhami, A.; Rahmani, A.; Khodaveisi, J. Development of a cost-effective technique to remove the arsenic contamination from aqueous solutions by calcium peroxide nanoparticles. *Sep. Purf. Technol.* **2012**, *95*, 10–15. [CrossRef]

25. Cui, H.; Su, Y.; Li, Q.; Gao, S.; Shang, J.K. Exceptional arsenic (III, V) removal performance of highly porous, nanostructured ZrO_2 spheres for fixed bed reactors and the full-scale system modeling. *Water Res.* **2013**, *47*, 6258–6268. [CrossRef] [PubMed]

26. Cui, H.; Li, Q.; Gao, S.; Shang, J.K. Strong adsorption of arsenic species by amorphous zirconium oxide nanoparticles. *J. Ind. Eng. Chem.* **2012**, *18*, 1418–1427. [CrossRef]

27. Habuda-Stanić, M.; Nujić, M. Arsenic removal by nanoparticles: A review. *Environ. Sci. Pollut. Res.* **2015**, *22*, 8094–8123. [CrossRef] [PubMed]

28. Zhang, G.; Ren, Z.; Zhang, X.; Chen, J. Nanostructured iron (III)-copper (II) binary oxide: A novel adsorbent for enhanced arsenic removal from aqueous solutions. *Water Res.* **2013**, *47*, 4022–4031. [CrossRef] [PubMed]

29. Shan, C.; Tong, M. Efficient removal of trace arsenite through oxidation and adsorption by magnetic nanoparticles modified with Fe-Mn binary oxide. *Water Res.* **2013**, *47*, 3411–3421. [CrossRef] [PubMed]

30. Yu, L.; Peng, X.; Ni, F.; Li, J.; Wang, D.; Luan, Z. Arsenite removal from aqueous solutions by γ-Fe$_2$O$_3$-TiO$_2$ magnetic nanoparticles through simultaneous photocatalytic oxidation and adsorption. *J. Hazard. Mater.* **2013**, *246*, 10–17. [CrossRef] [PubMed]

31. Basu, T.; Ghosh, U.C. Arsenic(III) removal performances in the absence/presence of groundwater occurring ions of agglomerated Fe(III)-Al(III) mixed oxide nanoparticles. *J. Ind. Eng. Chem.* **2011**, *17*, 834–844. [CrossRef]

32. Basu, T.; Ghosh, U.C. Nano-structured iron(III)-cerium(IV) mixed oxide: Synthesis, characterization and arsenic sorption kinetics in the presence of co-existing ions aiming to apply for high arsenic groundwater treatment. *Appl. Surf. Sci.* **2013**, *283*, 471–481. [CrossRef]

33. Novoselov, K.S.; Geim, A.K.; Morozov, S.; Jiang, D.; Zhang, Y.; Dubonos, S.A.; Grigorieva, I.; Firsov, A. Electric field effect in atomically thin carbon films. *Science* **2004**, *306*, 666–669. [CrossRef] [PubMed]

34. Bunch, J.S.; Van Der Zande, A.M.; Verbridge, S.S.; Frank, I.W.; Tanenbaum, D.M.; Parpia, J.M.; Craighead, H.G.; McEuen, P.L. Electromechanical resonators from graphene sheets. *Science* **2007**, *315*, 490–493. [CrossRef] [PubMed]

35. Katsnelson, M.I. Graphene: Carbon in two dimensions. *Mater. Today* **2007**, *10*, 20–27. [CrossRef]

36. Kopelevich, Y.; Esquinazi, P. Graphene physics in graphite. *Adv. Mater.* **2007**, *19*, 4559–4563. [CrossRef]

37. Morozov, S.; Novoselov, K.; Katsnelson, M.; Schedin, F.; Elias, D.; Jaszczak, J.; Geim, A. Giant intrinsic carrier mobilities in graphene and its bilayer. *Phys. Rev. Lett.* **2008**, *100*, 016602. [CrossRef] [PubMed]

38. Becerril, H.A.; Mao, J.; Liu, Z.; Stoltenberg, R.M.; Bao, Z.; Chen, Y. Evaluation of solution-processed reduced graphene oxide films as transparent conductors. *ACS Nano* **2008**, *2*, 463–470. [CrossRef] [PubMed]

39. Gollavelli, G.; Chang, C.-C.; Ling, Y.-C. Facile synthesis of smart magnetic graphene for safe drinking water: heavy metal removal and disinfection control. *ACS Sustain. Chem. Eng.* **2013**, *1*, 462–472. [CrossRef]

40. Ye, J.-H.; Liu, J.; Wang, Z.; Bai, Y.; Zhang, W.; He, W. A new Fe^{3+} fluorescent chemosensor based on aggregation-induced emission. *Tetrahedron Lett.* **2014**, *55*, 3688–3692. [CrossRef]

41. Babu, C.M.; Vinodh, R.; Sundaravel, B.; Abidov, A.; Peng, M.M.; Cha, W.S.; Jang, H.-T. Characterization of reduced graphene oxide supported mesoporous Fe$_2$O$_3$/TiO$_2$ nanoparticles and adsorption of As(III) and As(V) from potable water. *J. Taiwan Inst. Chem. Eng.* **2016**, *62*, 199–208. [CrossRef]

42. Kumar, S.; Nair, R.R.; Pillai, P.B.; Gupta, S.N.; Iyengar, M.; Sood, A. Graphene oxide-MnFe$_2$O$_4$ magnetic nanohybrids for efficient removal of lead and arsenic from water. *ACS Appl. Mater. Interfaces* **2014**, *6*, 17426–17436. [CrossRef] [PubMed]

43. La, D.D.; Patwari, J.M.; Jones, L.A.; Antolasic, F.; Bhosale, S.V. Fabrication of a GNP/Fe-Mg Binary Oxide Composite for Effective Removal of Arsenic from Aqueous Solution. *ACS Omega* **2017**, *2*, 218–226. [CrossRef]

44. La, M.; Duc, D.; Bhargava, S.; Bhosale, S.V. Improved and A Simple Approach For Mass Production of Graphene Nanoplatelets Material. *ChemistrySelect* **2016**, *1*, 949–952. [CrossRef]

45. Zhao, Y.; He, G.; Dai, W.; Chen, H. High catalytic activity in the phenol hydroxylation of magnetically separable CuFe$_2$O$_4$-reduced graphene oxide. *Ind. Eng. Chem. Res.* **2014**, *53*, 12566–12574. [CrossRef]

46. Zhu, J.; Sadu, R.; Wei, S.; Chen, D.H.; Haldolaarachchige, N.; Luo, Z.; Gomes, J.; Young, D.P.; Guo, Z. Magnetic graphene nanoplatelet composites toward arsenic removal. *ECS J. Solid State Sci. Technol.* **2012**, *1*, M1–M5. [CrossRef]

47. Mateo, D.; Esteve-Adell, I.; Albero, J.; Royo, J.F.S.; Primo, A.; Garcia, H. 111 oriented gold nanoplatelets on multilayer graphene as visible light photocatalyst for overall water splitting. *Nat. Commun.* **2016**, *7*, 1–8. [CrossRef] [PubMed]

48. Nedkov, I.; Vandenberghe, R.; Marinova, T.; Thailhades, P.; Merodiiska, T.; Avramova, I. Magnetic structure and collective Jahn-Teller distortions in nanostructured particles of CuFe$_2$O$_4$. *Appl. Surf. Sci.* **2006**, *253*, 2589–2596. [CrossRef]

49. Sheng, G.; Li, Y.; Yang, X.; Ren, X.; Yang, S.; Hu, J.; Wang, X. Efficient removal of arsenate by versatile magnetic graphene oxide composites. *RSC Adv.* **2012**, *2*, 12400–12407. [CrossRef]

50. Luo, X.; Wang, C.; Wang, L.; Deng, F.; Luo, S.; Tu, X.; Au, C. Nanocomposites of graphene oxide-hydrated zirconium oxide for simultaneous removal of As(III) and As(V) from water. *Chem. Eng. J.* **2013**, *220*, 98–106. [CrossRef]

Sensors **2017**, *17*, 1292

51. Tang, W.; Su, Y.; Li, Q.; Gao, S.; Shang, J.K. Superparamagnetic magnesium ferrite nanoadsorbent for effective arsenic (III, V) removal and easy magnetic separation. *Water Res.* **2013**, *47*, 3624–3634. [CrossRef] [PubMed]
52. Zhu, J.; Lou, Z.; Liu, Y.; Fu, R.; Baig, S.A.; Xu, X. Adsorption behavior and removal mechanism of arsenic on graphene modified by iron–manganese binary oxide (FeMnO x/RGO) from aqueous solutions. *RSC Adv.* **2015**, *5*, 67951–67961. [CrossRef]
53. Yu, L.; Ma, Y.; Ong, C.N.; Xie, J.; Liu, Y. Rapid adsorption removal of arsenate by hydrous cerium oxide–graphene composite. *RSC Adv.* **2015**, *5*, 64983–64990. [CrossRef]
54. Wang, C.; Luo, H.; Zhang, Z.; Wu, Y.; Zhang, J.; Chen, S. Removal of As(III) and As(V) from aqueous solutions using nanoscale zero valent iron-reduced graphite oxide modified composites. *J. Hazard. Mater.* **2014**, *268*, 124–131. [CrossRef] [PubMed]
55. Yoon, Y.; Park, W.K.; Hwang, T.-M.; Yoon, D.H.; Yang, W.S.; Kang, J.-W. Comparative evaluation of magnetite–graphene oxide and magnetite-reduced graphene oxide composite for As(III) and As(V) removal. *J. Hazard. Mater.* **2016**, *304*, 196–204. [CrossRef] [PubMed]
56. Zhang, S.; Niu, H.; Cai, Y.; Zhao, X.; Shi, Y. Arsenite and arsenate adsorption on coprecipitated bimetal oxide magnetic nanomaterials: MnFe$_2$O$_4$ and CoFe$_2$O$_4$. *Chem. Eng. J.* **2010**, *158*, 599–607. [CrossRef]
57. Azizian, S. Kinetic models of sorption: A theoretical analysis. *J. Colloid Interface Sci.* **2004**, *276*, 47–52. [CrossRef] [PubMed]
58. Ouvrard, S.; De Donato, P.; Simonnot, M.; Begin, S.; Ghanbaja, J.; Alnot, M.; Duval, Y.; Lhote, F.; Barres, O.; Sardin, M. Natural manganese oxide: Combined analytical approach for solid characterization and arsenic retention. *Geochim. Cosmochim. Acta* **2005**, *69*, 2715–2724. [CrossRef]

![sensors logo] *sensors*

MDPI

Article

A Simple Assay for Ultrasensitive Colorimetric Detection of Ag⁺ at Picomolar Levels Using Platinum Nanoparticles

Yi-Wei Wang [1], Meili Wang [2], Lixing Wang [1], Hui Xu [1], Shurong Tang [3,*], Huang-Hao Yang [2], Lan Zhang [2,*] and Hongbo Song [1,*]

[1] Key Laboratory of Predictive Microbiology and Chemical Residual Analysis, College of Food Science, Fujian Agriculture and Forestry University, Fuzhou 350002, China; wangyw@fafu.edu.cn (Y.-W.W.); fj150823@163.com (L.W.); xhuifst@163.com (H.X.)

[2] The Key Lab of Analysis and Detection Technology for Food Safety of the MOE, State Key Laboratory of Photocatalysis on Energy and Environment, College of Chemistry, Fuzhou University, Fuzhou 350108, China; wang641132431@163.com (M.W.); hhyang@fzu.edu.cn (H.-H.Y.)

[3] Department of Pharmaceutical Analysis, Faculty of Pharmacy, Fujian Medical University, Fuzhou 350108, China

* Correspondence: srtang@fjmu.edu.cn (S.T.); zlan@fzu.edu.cn (L.Z.); sghgbode@163.com (H.S.); Tel.: +86-591-2286-2738 (S.T.); +86-591-2286-6135 (L.Z.); +86-591-8378-9348 (H.S.)

Received: 17 September 2017; Accepted: 31 October 2017; Published: 2 November 2017

Abstract: In this work, uniformly-dispersed platinum nanoparticles (PtNPs) were synthesized by a simple chemical reduction method, in which citric acid and sodium borohydride acted as a stabilizer and reducer, respectively. An ultrasensitive colorimetric sensor for the facile and rapid detection of Ag⁺ ions was constructed based on the peroxidase mimetic activities of the obtained PtNPs, which can catalyze the oxidation of 3,3′,5,5′-tetramethylbenzidine (TMB) by H_2O_2 to produce colored products. The introduced Ag⁺ would be reduced to Ag⁰ by the capped citric acid, and the deposition of Ag⁰ on the PtNPs surface, can effectively inhibit the peroxidase-mimetic activity of PtNPs. Through measuring the maximum absorption signal of oxidized TMB at 652 nm, ultra-low detection limits (7.8 pM) of Ag⁺ can be reached. In addition to such high sensitivity, the colorimetric assay also displays excellent selectivity for other ions of interest and shows great potential for the detection of Ag⁺ in real water samples.

Keywords: platinum nanoparticles; peroxidase-mimic activity; colorimetric sensor; silver ions detection

---

1. Introduction

Peroxidase is a hemin-containing oxidase that can catalyze the chemical reactions in a variety of biological processes by binding electrons to specific substrates. Since the peroxidase is capable of catalyzing the formation of colored products in very low concentrations, it has become the most frequently used enzyme in enzyme-linked immunosorbent assay and widely used in the detection of various substances through the combination with other enzymes to form multi-enzyme systems [1]. However, the inherent defects of natural enzymes, such as limited source, low stability, complex purification processes, and expensive purification costs, restricted their production and application. Therefore, great efforts have been made to synthesize mimetic enzymes. In virtue of chemical reactions that happen mainly on the surface of nanozymes, different surface modification methods are studied to improve catalytic activity, substrate specificity, and stability [2]. Since the first discovery of Fe_3O_4 nanoparticles [3], many inorganic nanomaterials with enzyme-mimic activities are explored and widely used in biomedical and environmental monitoring, such as glutathione-capped palladium or platinum nanoparticles [4,5], AuPt nanoparticles [6], gold nanoparticles@carbon shells [7], cobalt oxyhydroxide nanoflakes [8], g-C_3N_4/Pt nanoparticles [9], and MoS_2 nanosheets [10].

Silver ions (Ag^+), as one of the heavy metal ions, is highly toxic to bacteria, viruses, algae, and fungi. Due to the unique antibacterial properties, Ag^+ has been widely used in cosmetics, building materials and medical products [11,12]. The excessive uptake of Ag^+ may lead to many serious diseases, including cytotoxicity, organ failure, and mitochondrial dysfunction [13]. Due to the hazardous effects of Ag^+, the maximum allowable level of Ag^+ in drinking water is limited by the U.S. Environmental Protection Agency (EPA) to about 900 nM [14]. The U.S. EPA reported that the concentration of Ag^+ higher than 1.6 nM is toxic to fish and micro-organisms [15]. Hence, it has become increasingly important to develop a simple method for the sensitive detection of Ag^+ in the environment and biological samples.

Over the past decades, many analytical methods have been developed to detect Ag^+ with high sensitivity and selectivity, involving inductively-coupled plasma mass spectrometry (ICP-MS) [16], atomic absorption spectroscopy (AAS) [17], and atomic emission spectrometry (AMS) [18]. The requirements of large instruments, highly-trained operators, and lengthy sample preparation procedures in these methods, impede their capacity to routine and in situ detection. In contrast, chemical sensors provide an excellent platform to make up for the deficiency [19]. A novel silver-specific RNA-cleaving DNAzyme has been selected in vitro for sensitive fluorescence detection of Ag^+ [20]. Colorimetric sensors offer great potential for simple, rapid, low-cost, non-destructive, on-site, and real-time tracking of various analytes, with the advantages of being easy to miniaturize, visual detection results, and lacking the need of expensive equipment, complex pretreatment processes, and toxic fluorescence probes, etc. The variety of enzyme-mimic nanomaterial-based colorimetric sensors have been developed for the detection of heavy metal ions, such as Hg^{2+} [21], Cu^{2+} [22], Ag^+ [23], and Fe^{2+} [24]. However, the detection limits of these sensors are restricted only to micromolar (μM) or nanomolar (nM) levels.

In this paper, a simple chemical reduction method was performed to generate uniform-sized PtNPs using citrate as the capping molecule. An ultrasensitive and selective colorimetric sensor for the rapid detection of Ag^+ was developed with a detection limit down to the picomolar (pM) level based on the peroxidase-mimetic activity of PtNPs. The oxidation of TMB catalyzed by PtNPs could be inhibited by the reduced Ag^0. As a result, the quantitative detection of Ag^+ would be obtained by recording the UV absorption of oxidized TMB. To the best of our knowledge, the proposed sensor showed the highest sensitivity for Ag^+ detection compared to recently-reported colorimetric sensors. The practical application of the colorimetric sensor for the detection of Ag^+ in real water samples was also investigated and satisfactory results were obtained.

2. Materials and Methods

2.1. Materials and Instruments

3,3′,5,5′-tetramethylbenzidine dihydrochloride (TMB·2HCl) was purchased from Beyotime Biotechnology Co., Ltd. (Shanghai, China). All the other chemicals, such as chloroplatinic acid (H_2PtCl_6), $NaBH_4$, citric acid, $AgNO_3$, and ethylene diamine tetraacetic acid (EDTA) etc., were purchased from Sinopharm Chemical Reagent Co., Ltd. (Beijing, China). The reagents were of analytical grade and used as received without further purification. The solutions were prepared using ultrapure water purified by Milli-Q biocel from Millipore China Ltd. (Shanghai, China).

The UV–VIS absorption spectra and kinetic studies were performed on a UV-2450 UV-VIS spectrometer (Shimadzu, Tokyo, Japan). Terephthalic acid (TA) assay was carried out by an F-4600 fluorescence spectrofluorometer (Hitachi, Tokyo, Japan). Transmission electron microscopy (TEM) images were characterized by a high-resolution transmission electron microscopy (HRTEM) on a Philips Tecnai G2 F20 microscope (Philips, Amsterdam, The Netherlands) with an accelerating voltage of 200 kV. Before measurement, samples were prepared by dropping the PtNPs suspension on the surface of carbon-coated copper grid and drying it in air. X-ray photoelectron spectra (XPS) characterization was measured by the ESCALAB 250Xi X-ray photoelectron spectroscopy (Thermo Fisher Scientific, Waltham, MA, USA) using monochromatic Al Ka radiation (hv = 1486.6 eV). X-ray diffraction (XRD) characterization was performed by a Rigaku X-ray diffractometer (D/Max-3C, Tokyo, Japan).

2.2. Synthesis of Citric Acid-Modified PtNPs

Typically, 1 mL of chloroplatinic acid (16 mM), 1 mL of sodium citrate (40 mM) and 38 mL deionized water were added into a 50 mL beaker and stirring for 30 min at room temperature. After that, 200 μL of NaBH$_4$ (50 mM) was introduced to the mixture drop by drop. The solution changed from colorless to brownish-yellow during the reaction process. Finally, citric acid-modified PtNPs were obtained after continuous stirring at room temperature for 1 h.

2.3. Colorimetric Detection of Ag$^+$

Briefly, 8 μL of PtNPs (1.25 mg/L), 40 μL different concentrations of Ag$^+$, and an appropriate amount of deionized water were added into a 0.6 mL centrifuge tube. After reacting for 2 min, 200 μL of TMB (1.6 mM) and 100 μL of H$_2$O$_2$ (2 M) were added into the solution to initiate the chromogenic reaction. The total reaction volume was 400 μL and the reaction was continued for a further 10 min. Finally, the absorption spectra of the resulting solutions were recorded in the range from 500–800 nm, the highest absorption at 652 nm was used as detection signal. The specific experiment was performed as above, except that other ions were used instead of Ag$^+$.

3. Results and Discussion

3.1. Sensing Principle of the Ag$^+$ Colorimetric Sensor

The schematic diagram of the Ag$^+$ colorimetric sensor was shown in Scheme 1. By the use of citric acid as a stabilizer, chloroplatinic acid can be reduced by NaBH$_4$ to generate stable and uniform PtNPs. The obtained PtNPs exhibit excellent peroxidase-mimic activity, which can catalyze the oxidization of TMB by H$_2$O$_2$ to produce a colored product. After addition of Ag$^+$, the introduced Ag$^+$ would be reduced by the capped citrate and deposited on the surface of PtNPs, which led to significant inhibition of the peroxidase-like activity of PtNPs. The specific Ag-Pt interaction provides the excellent selectivity toward Ag$^+$ over other ions. Through measuring the maximum absorption of the oxidized TMB at 652 nm, an ultrasensitive, facile, and rapid colorimetric Ag$^+$ sensor was established.

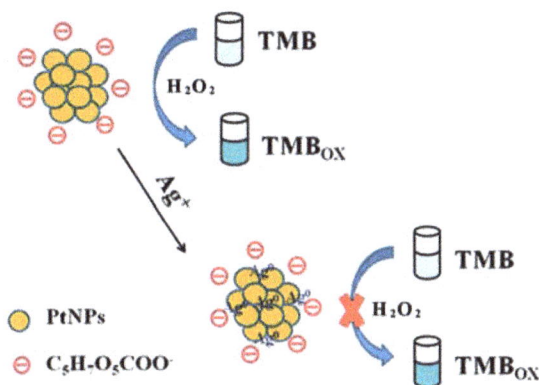

Scheme 1. Schematic of the Ag$^+$ colorimetric sensor based on citrate-modified PtNPs.

3.2. Characterization of the Formed PtNPs

Transmission electron microscopy was used to investigate the morphological characteristics of the synthesized citric acid-modified PtNPs. As shown in Figure 1A, uniformly-dispersed PtNPs with narrow size distribution (diameter~2.5 nm) are observed. From the HRTEM image (inset Figure 1A), obvious lattice fringes of PtNPs can be noticed, which proves that the synthesized PtNPs have good crystal form. The measured lattice spacing is 0.223 nm, corresponding to the (1 1 1) facet of the Pt crystal [25]. The XRD patterns of the PtNPs are shown in Figure 1B. The diffraction peaks at angles of 39.8°, 46.3°, and 67.7° can be assigned to the (1 1 1), (2 0 0), and (2 2 0) facets of the face-centered cubic structures of platinum crystals ((JCPDS No. 4-802)) [26].

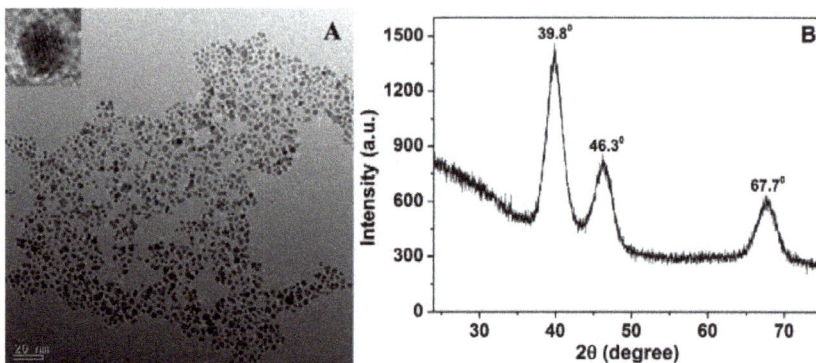

Figure 1. (**A**) TEM image of citric acid-modified PtNPs (Inset: HRTEM image of citric acid-modified PtNPs); and (**B**) XRD patterns of the PtNPs.

XPS spectra were further performed to characterize the citric acid-modified PtNPs. Figure 2A shows the whole XPS spectrum of citrate-capped PtNPs. It can be seen that the elements of C, O, Na and Pt existed, indicating that the citric acid has been successfully modified on the surface of PtNPs. The binding energy of Pt 4f was shown in Figure 2B, the Pt $4f_{7/2}$ peak can be divided into two peaks with binding energy of 71.44 eV and 72.16 eV, corresponding to Pt^0 and Pt^{4+}, respectively. The Pt $4f_{5/2}$ peak also can be divided into two peaks at the binding energies of 74.88 eV and 75.97 eV, corresponding to Pt^0 and Pt^{4+}, respectively [27]. The ratio of Pt^0 (59.7%) and Pt^{4+} (40.3%) on the PtNPs surface is determined as 1.48.

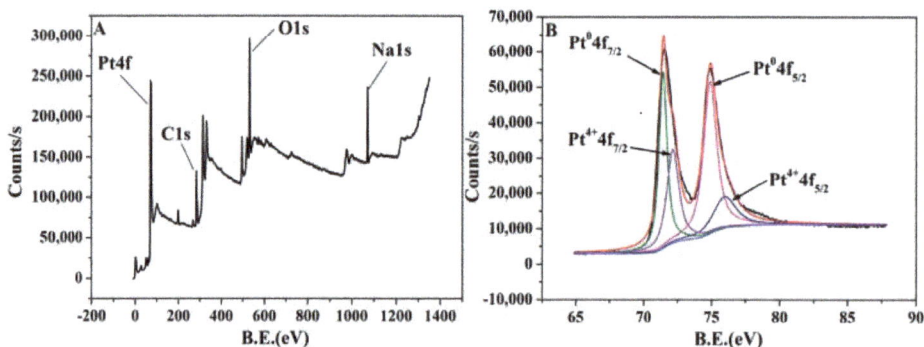

Figure 2. (**A**) The whole XPS spectrum of citrate-modified PtNPs; and (**B**) XPS spectrum showing the binding energy of Pt 4f.

3.3. Catalytic Activity of PtNPs for TMB Oxidation

A series of control experiments were conducted to investigate the catalytic ability of citrate-modified PtNPs for the oxidation of TMB. As shown in Figure 3, the absorption signal was very small in the presence of TMB and H_2O_2 (curve a) or TMB and PtNPs (curve b) only, and the color of the solution is almost colorless (inset a, b). On the contrary, the absorption signal was remarkably increased when PtNPs, TMB, and H_2O_2 were coexistent in the solution (curve c), the color of the solution became deep blue (inset c). These experimental results showed that citrate-modified PtNPs have good peroxidase-mimetic properties to catalyze the oxidation of TMB by H_2O_2 effectively. Terephthalic acid (TA) was used to evaluate the effects of PtNPs on ·OH signal intensity, in which the added TA will react with ·OH to form a highly-fluorescent product, 2-hydroxyterephthalic acid (TAOH) [28]. As shown in Figure 4, a gradual decrease of the fluorescence intensity was observed while increasing the concentration of PtNPs, suggesting that the PtNPs reduced the ·OH radical signal, which is similar to the behavior of reported Co_3O_4 NPs [29], C-Dots [30] and MnO_2 NPs [31]. In addition, experiments showed that the addition of high concentration of PtNPs into the mixture of TA and H_2O_2 resulted in a lot of bubbles (data not shown), indicating the PtNPs accelerated the decomposition of H_2O_2 to produce oxygen. These results and TA assays prove that PtNPs behave analogously to enzymes [32].

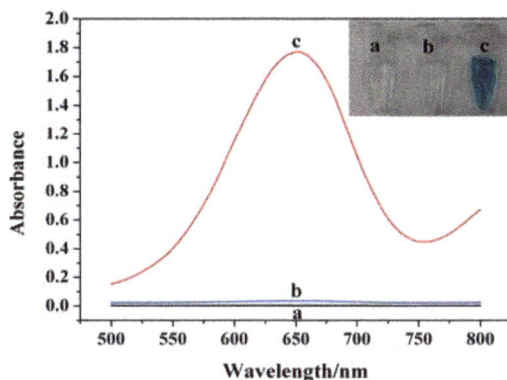

Figure 3. UV-VIS absorption spectrum of (a) TMB + H_2O_2; (b) TMB + PtNPs and (c) TMB + H_2O_2 + PtNPs (inset: the corresponding photographs).

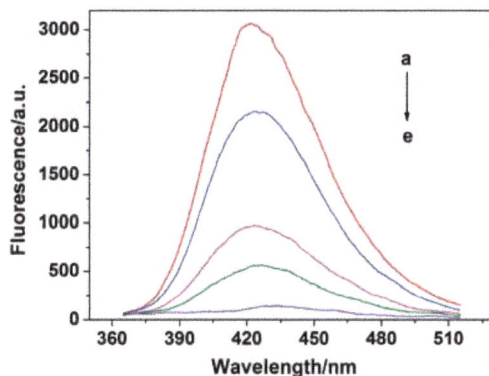

Figure 4. The effect of PtNPs on the formation of hydroxyl radicals in the H_2O_2/TA system. Samples were a mixture of 0.25 mM TA, 10 mM H_2O_2, and various concentrations of PtNPs (a) 0, (b) 5, (c) 12.5, (d) 25 and (e) 125 µg/L.

3.4. Inhibitory Effect of Ag+ on Catalytic Activity

In order to examine the feasibility of the designed colorimetric sensor for the detection of Ag^+, the absorption signals before and after addition of Ag^+ were investigated. From Figure 5, it can be observed that a dark blue color solution was produced (inset a) with a strong absorption signal in the absence of Ag^+ (curve a). After addition of 1.5 nM Ag^+, the absorption signal was significantly decreased (curve b). At the same time, the color of the solution became lighter (inset b). When the Ag^+ concentration was increased to 3.0 nM, the absorption signal was further inhibited (curve c), along with the color of the solution becoming shallower (inset c). These results indicated that the catalytic activities of PtNPs can be effectively inhibited by trace amounts of Ag^+. Thus, a simple colorimetric sensor can be established for Ag^+ detection with high sensitivity. The peroxidase-like activity of the PtNPs in the absence and presence of Ag^+ was further investigated using steady-state kinetics (Figure 6). The apparent kinetic parameters were calculated based on the Michaelis-Menten equation: $v = V_{max} \times [S]/(K_m + [S])$, where v is the initial velocity, V_{max} is the maximal reaction velocity, [S] is the concentration of the substrate, and K_m is the Michaelis constant. K_m is an important parameter to evaluate the enzyme affinity to substrate. As shown in Table 1, the K_m value of the PtNPs increased, while the V_{max} value decreased after interaction with Ag^+. These results indicated that Ag^+-treated PtNPs have lower affinity to the substrates and weaker catalytic activity.

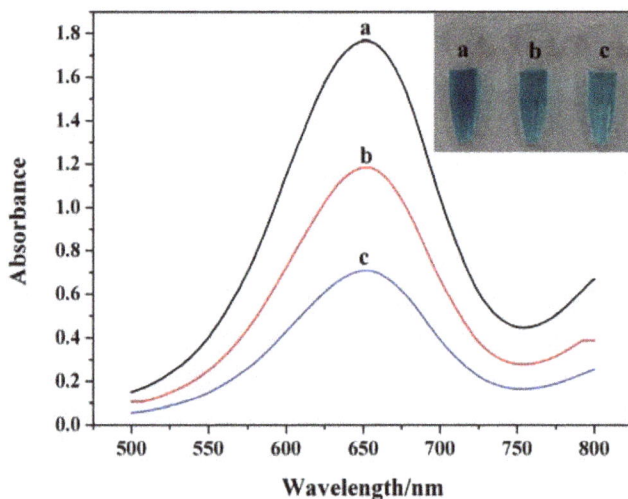

Figure 5. UV-VIS absorption spectrum of TMB at different concentrations of Ag^+ (a) 0; (b) 1.5 nM; and (c) 3.0 nM.

Table 1. Comparison of the kinetic parameter of PtNPs before and after being treated with Ag^+.

Ag^+ (nM)	TMB (K_m/mM)	TMB (V_{max}/M S^{-1})	H_2O_2 (K_m/mM)	H_2O_2 (V_{max}/M S^{-1})
0	0.0995	1.201×10^{-8}	230.8	1.656×10^{-7}
0.5	0.1077	1.045×10^{-8}	255.9	1.372×10^{-7}
2.0	0.1652	0.872×10^{-8}	283.6	1.215×10^{-7}

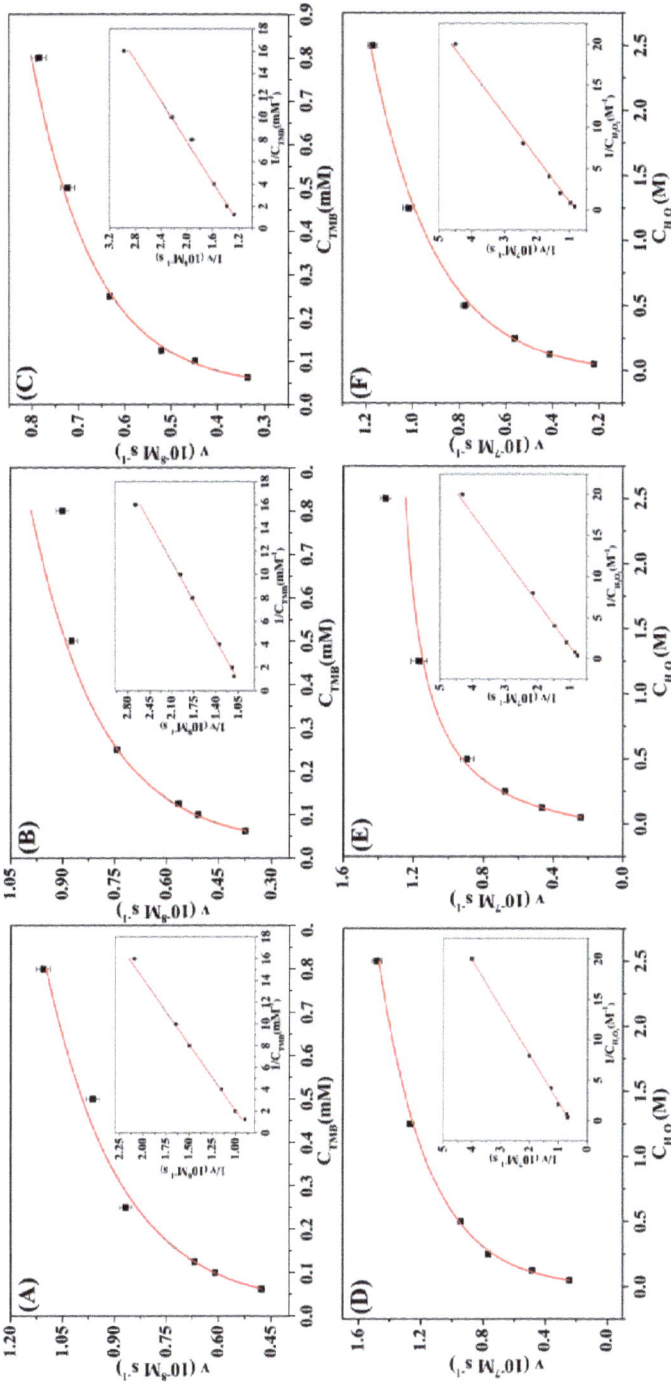

Figure 6. Steady-state kinetic analyses using the Michaelis–Menten model and Lineweaver–Burk model (insets) for PtNPs in the absence (A,D) and presence of 0.5 nM (B,E) and 2.0 nM (C,F) Ag$^+$.

Similar with the interaction between Ag^+ and gold nanoclusters, the possible mechanism of Ag^+ to inhibit the catalytic activity of PtNPs could be related to a Pt-Ag metallic bond [33,34]. The added Ag^+ can first interact with Pt to form a metallic bond, then be reduced by the modified citrate and deposited on the surface of PtNPs. We have performed an XPS spectrum of the citrate-modified PtNPs after being treated with Ag^+ to investigate the inhibition mechanism. After interaction with Ag^+, a new peak of Ag 3d could be observed in the XPS spectrum of PtNPs (Figure 7A). In addition, two well-characterized peaks appeared in the Ag 3d electron spectra of PtNPs after being treated with Ag^+ (Figure 7B). The two signals of Ag $3d_{5/2}$ and Ag $3d_{3/2}$ that arose at binding energies of 367.70 and 373.72 eV corresponded to Ag^0 [35,36]. Theoretically, Pt^0 cannot be oxidized by Ag^+ under conventional conditions due to the inert noble metal properties. The XPS spectra (Figure 7C) also indicated that addition of Ag^+ do not have great effect on the ratio of Pt^0 (58.7%) and Pt^{4+} (41.3%) on the PtNPs surface. Citrate is a thermal reduction reagent (reduction at near-boiling temperature), and the reduction of Ag^+ with citrate is difficult to proceed at room temperature due to the weak reducibility [37,38]. Surprisingly, the citrate adsorbed on the surface of PtNPs could trigger Ag^+ reduction catalyzed by the very reactive Pt surface atoms under mild conditions, which is similar to previous studies that showed the reduction of Hg^{2+} can be catalyzed by citrate-coated gold nanoparticles [39]. These results confirmed that the introduced Ag^+ has been reduced to metallic Ag^0 by the modified citrate, thereby causing changes in the surface chemistry of PtNPs and inhibiting the catalytic activity.

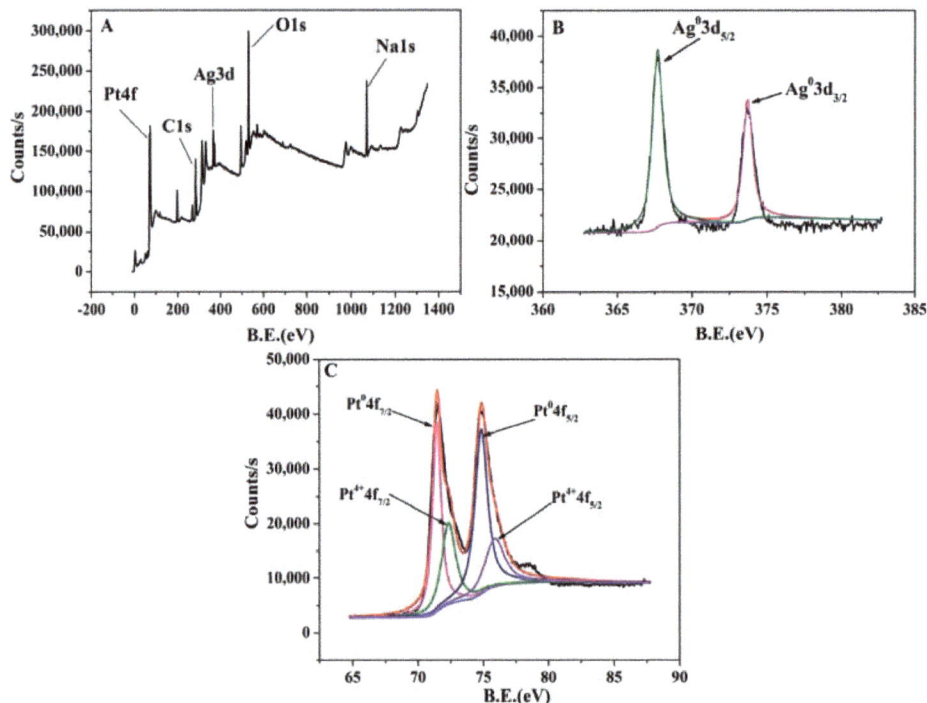

Figure 7. (**A**) The whole XPS spectrum; (**B**) Ag (3d) and (**C**) Pt (4f) XPS spectra of citrate-modified PtNPs after being treated with Ag^+.

3.5. *Optimization of Experimental Conditions*

In order to obtain the best sensing performance, some experimental conditions were optimized. The absorption difference between A_0 and A (recorded as ΔA) was used to evaluate the sensing performance, where A_0 and A represent the absorption signal without and with the addition of 2.0 nM Ag^+, respectively. We found that the amount of PtNPs has a great effect on the absorption signal (Figure 8A). The ΔA value was increased with the increase of PtNP volume up to 8 µL. However, with a further increase in the volume of PtNPs, the ΔA value started to decrease. According to these results, 8 µL PtNPs was used in subsequent experiments. The effect of H_2O_2 concentration on the developed sensor has also been investigated. H_2O_2, which acted as an oxidant, has played an important role in the oxidation of TMB. The ΔA value was remarkably increased with increasing H_2O_2 concentration from 0.05 to 0.5 M, then tends to decrease when the concentration of H_2O_2exceeds 0.5 M (Figure 8B). Therefore, 0.5 M H_2O_2 was used during the sensing process. The reaction time between Ag^+ and PtNPs was also investigated. As shown in Figure 8C, the effect of Ag^+ reaction time on the ΔA value is very small, which reflects that the interaction between Ag^+ and citrate-modified PtNPs is fast. Taking into account the efficiency of the detection and ease of operation, we chose 2 min as the reaction time of Ag^+.

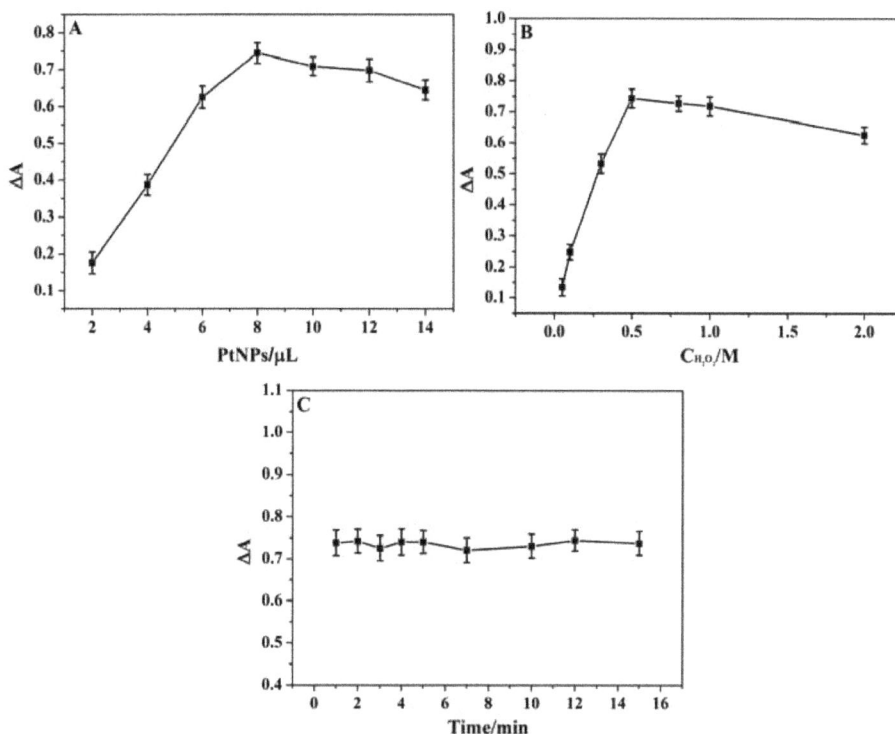

Figure 8. The effect of experimental conditions on the sensing performance (**A**) the volume of citrate-modified PtNPs; (**B**) the concentration of H_2O_2; and (**C**) the reaction time between Ag^+ and PtNPs.

3.6. Sensitivity of the Ag^+ Sensing System

To evaluate the sensitivity and dynamic range of the proposed colorimetric sensor for Ag^+ detection, various concentrations of Ag^+ were tested under the optimal conditions. As shown in Figure 9A, the absorption signal at 652 nm decreased with increasing concentration of Ag^+. The more Ag^+ that was added, the more Ag^0 was formed, which greatly inhibits the catalytic activity of citrate-modified PtNPs. We obtained a good linear response of the absorption signal against the concentrations of Ag^+ in the ranges from 0.01 to 3.0 nM with a correlation coefficient of 0.997 (Figure 9B).

According to triplicate standard deviation over the blank response (3σ), the detection limit of (LOD) Ag^+ was estimated to be 7.8 pM, which was sensitive enough for Ag^+ detection in drinking water.

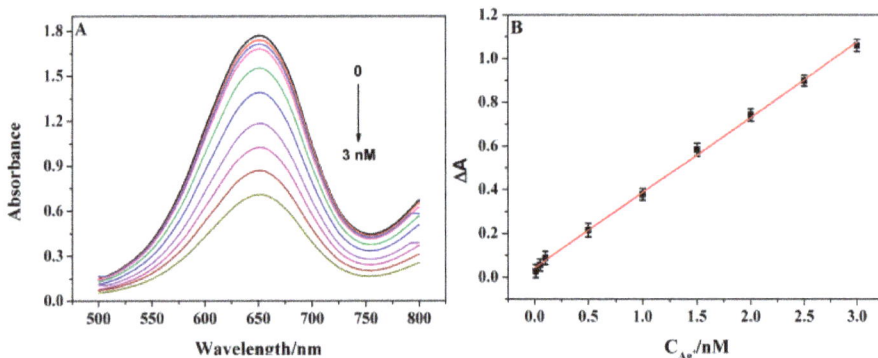

Figure 9. (**A**) Absorbance curves of the sensor for Ag^+ at various concentrations; and (**B**) the corresponding calibration plot of absorbance values against the Ag^+ concentrations (the error bars represent the standard deviation of three measurements).

The analytical performance of the present sensor was compared with other Ag^+ detection methods. As shown in Table 2, the sensitivity of the proposed sensor was higher than that of recently-reported colorimetric, fluorescent and electrochemical methods. Such high sensitivity was attributed to the highly inhibitory effect of Ag^+ on the catalytic activity of PtNPs. The proposed sensor is simple, rapid and economical due to the mild synthesis of PtNPs without the need of special reagents, such as nucleic acid and fluorochrome. The whole sensing process can be finished within twelve minutes.

Table 2. Comparison of our present work with other methods for Ag^+ detection.

Methods	Probes	Linear Range	LOD	References
Colorimetric	Peptide-AuNPs	10~1000 nM	7.4 nM	[40]
Colorimetric	Au@PtNPs	5~100 nM	2.0 nM	[41]
Colorimetric	DNA-AuNPs	1~1000 nM	0.24 nM	[42]
Colorimetric	AuNPs	1~9 μM	0.41 μM	[43]
Colorimetric	BSA-Au clusters	0.5~10 μM	204 nM	[23]
Fluorescence	Proflavine-DNA/MnO$_2$	30~240 nM	9.1 nM	[44]
Fluorescence	DSAI/C-rich DNA	0~4.0 μM	155 nM	[45]
Electrochemistry	DNA/AuNPs	0.1~40 nM	0.05 nM	[46]
Electrochemistry	DNA/Fe$_3$O$_4$-AuNPs	10~150 nM	3.4 nM	[47]
Colorimetric	Citrate-modified PtNPs	0.01~3.0 nM	7.8 pM	This work

3.7. Selectivity and Recovery Performance

In order to investigate the selectivity of the proposed sensor, an interference study was performed with other metal ions that exist in the environment. The absorption intensity was tested under the same conditions, except that other metal ions were used instead of 10 nM Ag^+. As shown in Figure 10, the absorption signal was greatly inhibited by Ag^+. No significant decrease of the absorption signal was observed in the presence of above 100-fold concentration of other ions, except Hg^{2+}. Due to the similar ionic radius and reduction potential between Hg^{2+} and Ag^+, Hg^{2+} could be adsorbed on citrate-capped PtNPs and be reduced by the modified citrate [48,49]. Thus, Hg^{2+} would interfere with the detection. For Ag^+ sensing, EDTA was chosen as a masking agent because it could form more stable complexes with Hg^{2+} than that with Ag^+. After interaction with EDTA, the influence of Hg^{2+} was effectively eliminated. The absorption intensity in the presence of Ag^+ is completely irreversible after the introduction of an excess concentration of EDTA, indicating that Ag^+ interacts with PtNPs through stronger interaction forming an Ag-Pt metallic bond, similar to the Ag-Au metallic bond [33,50]. Therefore, specific detection of Ag^+ can be accomplished by the citrate-modified PtNP-based assay. More importantly, the results of selective experiments are visible to the naked eye, thus no special instruments are required to distinguish the presence or absence of Ag^+.

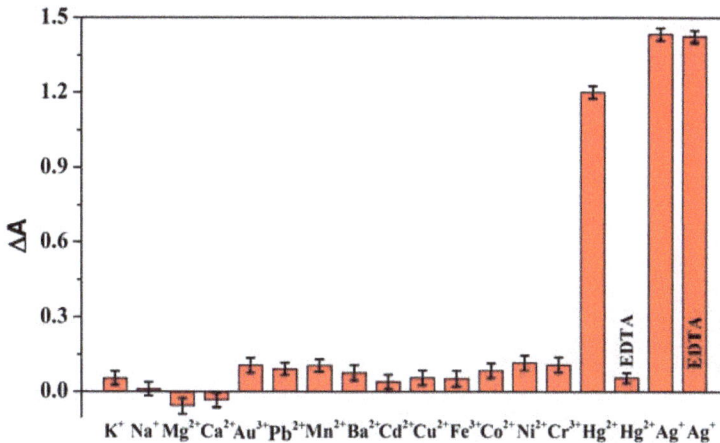

Figure 10. Selectivity investigation of the proposed sensor for Ag^+ detection (the concentration of K^+, Na^+, Mg^{2+}, Ca^{2+} : 0.5 mM, Ag^+ and Hg^{2+}: 10 nM, other ions: 1.0 μM).

The practical application of the designed colorimetric sensor was also tested through determination of Ag^+ in river water samples by the standard addition method. The collected Minjiang River water samples were filtered with a 0.22 μm membrane to remove insoluble matter before Ag^+ detection. Spiked samples were prepared with the further addition of different concentrations of standard Ag^+ to the river water. Each Ag^+ spiked sample was repetitively measured three times. The results are shown in Table 3. The recovery values ranging from 98.0% to 105.0% were obtained, and the relative standard deviation (RSD) was lower than 7%. These results revealed that the developed sensor has acceptable accuracy and reproducibility for the sensing of Ag^+ in real samples.

Table 3. The analysis of Ag^+ in the real water samples.

Sample	Add (nM)	Found (nM)	Recovery (%)	RSD (%)
	0.20	0.21	105.0	6.7
River water	1.00	1.03	103.0	5.4
	2.00	1.96	98.0	5.1

4. Conclusions

In summary, a facile and simple colorimetric sensor was successfully developed for the ultrasensitive detection of Ag^+ with a detection limit down to the pM level. Through efficient and specific inhibition of the peroxidase-mimic activity of citrate-modified PtNPs, highly sensitive and selective detection of Ag^+ in real water samples can be achieved. The whole test can be completed within twelve minutes. There is no need of any expensive regents, complicated separation, or labeling processes during the sensing procedure. Thus, the fabricated sensor is rapid and economical. More importantly, through analysis of Ag^+, the fabricated sensor can provide a new general, high-throughput, and portable sensing platform for indirect detection of various analytes.

Acknowledgments: This work was supported by the National Natural Science Foundation of China (21605019), the Natural Science Foundation of Fujian Province (2015J01083, 2016J01049), and the Fujian High-Level University Project (612014042, 612014015).

Author Contributions: Yi-Wei Wang and Shurong Tang conceived and designed the experiments; Meili Wang and Lixing Wang performed the experiments; Hui Xu and Hongbo Song helped analyze the results of the measured data; Yi-Wei Wang wrote the paper; and Huang-Hao Yang and Lan Zhang have proposed valuable suggestions on the revision of the manuscript. All coauthors reviewed and revised the paper.

Conflicts of Interest: The authors declare no conflict of interest.

References

1. Veitch, N.C. Horseradish peroxidase: A modern view of a classic enzyme. *Phytochemistry* **2004**, *65*, 249–259. [CrossRef] [PubMed]
2. Liu, B.; Liu, J. Surface modification of nanozymes. *Nano Res.* **2017**, *10*, 1125–1148. [CrossRef]
3. Gao, L.; Zhuang, J.; Nie, L.; Zhang, J.; Zhang, Y.; Gu, N.; Wang, T.; Feng, J.; Yang, D.; Perrett, S.; et al. Intrinsic peroxidase-like activity of ferromagnetic nanoparticles. *Nat. Nanotechnol.* **2007**, *2*, 577–583. [CrossRef] [PubMed]
4. Fu, Y.; Zhang, H.; Dai, S.; Zhi, X.; Zhang, J.; Li, W. Glutathione-stabilized palladium nanozyme for colorimetric assay of silver(I) ions. *Analyst* **2015**, *140*, 6676–6683. [CrossRef] [PubMed]
5. Li, W.; Zhang, H.; Zhang, J.; Fu, Y. Synthesis and sensing application of glutathione-capped platinum nanoparticles. *Anal. Methods* **2015**, *7*, 4464–4471. [CrossRef]
6. Zhang, C.; Tang, J.; Huang, L.; Li, Y.; Tang, D. In-situ amplified voltammetric immunoassay for ochratoxin A by coupling a platinum nanocatalyst based enhancement to a redox cycling process promoted by an enzyme mimic. *Microchim. Acta* **2017**, *184*, 2445–2453. [CrossRef]
7. Tong, Y.; Jiao, X.; Yang, H.; Wen, Y.; Su, L.; Zhang, X. Reverse-bumpy-ball-type-nanoreactor-loaded nylon membranes as peroxidase-mimic membrane reactors for a colorimetric assay for H_2O_2. *Sensors* **2016**, *16*, 465. [CrossRef] [PubMed]
8. Wang, Y.M.; Liu, J.W.; Jiang, J.H.; Zhong, W. Cobalt oxyhydroxide nanoflakes with intrinsic peroxidase catalytic activity and their application to serum glucose detection. *Anal. Bioanal. Chem.* **2017**, *409*, 4225–4232. [CrossRef] [PubMed]
9. Wang, Y.W.; Wang, L.; An, F.; Xu, H.; Yin, Z.; Tang, S.; Yang, H.H.; Song, H. Graphitic carbon nitride supported platinum nanocomposites for rapid and sensitive colorimetric detection of mercury ions. *Anal. Chim. Acta* **2017**, *980*, 72–78. [CrossRef] [PubMed]
10. Lin, T.; Zhong, L.; Chen, H.; Li, Z.; Song, Z.; Guo, L.; Fu, F. A sensitive colorimetric assay for cholesterol based on the peroxidase-like activity of MoS_2 nanosheets. *Microchim. Acta* **2017**, *184*, 1233–1237. [CrossRef]
11. Kokura, S.; Handa, O.; Takagi, T.; Ishikawa, T.; Naito, Y.; Yoshikawa, T. Silver nanoparticles as a safe preservative for use in cosmetics. *Nanomed. Nanotechnol.* **2010**, *6*, 570–574. [CrossRef] [PubMed]
12. Sreekumari, K.R.; Nandakumar, K.; Takao, K.; Kikuchi, Y. Silver containing stainless steel as a new outlook to abate bacterial adhesion and microbiologically influenced corrosion. *ISIJ Int.* **2003**, *43*, 1799–1806. [CrossRef]
13. Mijnendonckx, K.; Leys, N.; Mahillon, J.; Silver, S.; Houdt, R.V. Antimicrobial silver: Uses, toxicity and potential for resistance. *Biometals* **2013**, *26*, 609–621. [CrossRef] [PubMed]
14. EPA CASRN. EPA Drinking Water Criteria Document for Silver. *Environ. Prot. Agency* **1989**, *444*, 7440–7444.
15. Ratte, H.T. Bioaccumulation and toxicity of silver compounds: A review. *Environ. Toxicol. Chem.* **1999**, *18*, 89–108. [CrossRef]
16. Krachler, M.; Mohl, C.; Emons, H.; Shotyk, W. Analytical procedures for the determination of selected trace elements in peat and plant samples by inductively coupled plasma mass spectrometry. *Spectrochim. Acta B* **2002**, *57*, 1277–1289. [CrossRef]
17. Musil, S.; Kratzer, J.; Vobecky, M.; Benada, O.; Matousek, T.J. Silver chemical vapor generation for atomic absorption spectrometry: Minimization of transport losses, interferences and application to water analysis. *J. Anal. At. Spectrom.* **2010**, *25*, 1618–1626. [CrossRef]
18. Zorn, M.E.; Wilson, C.G.; Gianchandani, Y.B.; Anderson, M.A. Detection of aqueous metals using a microglow discharge atomic emission sensor. *Sens. Lett.* **2004**, *2*, 179–185. [CrossRef]

19. Mattoussai, H.; Manro, J.M.; Goldman, E.R.; Anderson, G.P.; Sunder, V.C.; Micula, F.V.; Bawendi, M.G. Self-assembly of CdSe-ZnS quantum dot bioconjugates using an engineered recombinant protein. *J. Am. Chem. Soc.* **2000**, *122*, 12142–12150. [CrossRef]

20. Saran, R.; Liu, J. A silver DNAzyme. *Anal. Chem.* **2016**, *88*, 4014–4020. [CrossRef] [PubMed]

21. Li, W.; Chen, B.; Zhang, H.; Sun, Y.; Wang, J.; Zhang, J.; Fu, Y. BSA-stabilized Pt nanozyme for peroxidase mimetics and its application on colorimetric detection of mercury(II) ions. *Biosens. Bioelectron.* **2015**, *66*, 251–258. [CrossRef] [PubMed]

22. Pan, N.; Zhu, Y.; Wu, L.L.; Xie, Z.J.; Xue, F.; Peng, C.F. Highly sensitive colorimetric detection of copper ions based on regulating the peroxidase-like activity of Au@ Pt nanohybrids. *Anal. Methods* **2016**, *8*, 7531–7536. [CrossRef]

23. Chang, Y.; Zhang, Z.; Hao, J.; Yang, W.; Tang, J. BSA-stabilized Au clusters as peroxidase mimetic for colorimetric detection of Ag$^+$. *Sens. Actuators B Chem.* **2016**, *232*, 692–697. [CrossRef]

24. Song, H.; Wang, Y.; Wang, G.; Wei, H.; Luo, S. Ultrathin two-dimensional MnO$_2$ nanosheet as a stable coreactant of 3,3′,5,5′-tetramethylbenzidine chromogenic substrate for visual and colorimetric detection of iron (II) ion. *Microchim. Acta* **2017**, *184*, 3399–3404. [CrossRef]

25. Borodko, Y.; Ercius, P.; Pushkarev, V.; Thompson, C.; Somorjai, G. From single Pt atoms to Pt nanocrystals: Photoreduction of Pt^{2+} inside of a PAMAM dendrimer. *J. Phys. Chem. Lett.* **2012**, *3*, 236–241. [CrossRef]

26. Li, Y.; Tang, L.; Li, J. Preparation and electrochemical performance for methanol oxidation of pt/graphene nanocomposites. *Electrochem. Commun.* **2009**, *11*, 846–849. [CrossRef]

27. Fatih, S.; Gökagaç, G. Different sized platinum nanoparticles supported on carbon: An XPS study on these methanol oxidation catalysts. *J. Phys. Chem. C* **2007**, *111*, 5715–5720.

28. Ishibashi, K.; Fujishima, A.; Watanabe, T.; Hashimoto, K. Quantum yields of active oxidative species formed on TiO$_2$ photocatalyst. *J. Photochem. Photobiol.* **2000**, *134*, 139–142. [CrossRef]

29. Mu, J.; Wang, Y.; Zhao, M.; Zhang, L. Intrinsic peroxidase-like activity and catalase-like activity of Co$_3$O$_4$ nanoparticles. *Chem. Commun.* **2012**, *48*, 2540–2542. [CrossRef] [PubMed]

30. Shi, W.; Wang, Q.; Long, Y.; Cheng, Z.; Chen, S.; Zheng, H.; Huang, Y. Carbon nanodots as peroxidase mimetics and their applications to glucose detection. *Chem. Commun.* **2011**, *47*, 6695–6697. [CrossRef] [PubMed]

31. Liu, X.; Wang, Q.; Zhao, H.; Zhang, L.; Su, Y.; Lv, Y. BSA-templated MnO$_2$ nanoparticles as both peroxidase and oxidase mimics. *Analyst* **2012**, *137*, 4552–4558. [CrossRef] [PubMed]

32. He, W.; Liu, Y.; Yuan, J.; Yin, J.J.; Wu, X.; Hu, X.; Zhang, K.; Liu, J.; Chen, C.; Ji, Y.; et al. Au@ Pt nanostructures as oxidase and peroxidase mimetics for use in immunoassays. *Biomaterials* **2011**, *32*, 1139–1147.

33. Zhang, Y.; Jiang, H.; Wang, X. Cytidine-stabilized gold nanocluster as a fluorescence turn-on and turn-off probe for dual functional detection of Ag$^+$ and Hg^{2+}. *Anal. Chim. Acta* **2015**, *870*, 1–7. [CrossRef] [PubMed]

34. Yue, Y.; Liu, T.Y.; Li, H.W.; Liu, Z.; Wu, Y. Microwave-assisted synthesis of BSA-protected small gold nanoclusters and their fluorescence-enhanced sensing of silver(I) ions. *Nanoscale* **2012**, *4*, 2251–2254. [CrossRef] [PubMed]

35. Chen, Q.; Shi, W.; Xu, Y.; Wu, D.; Sun, Y. Visible-light-responsive Ag–Si codoped anatase TiO$_2$ photocatalyst with enhanced thermal stability. *Mater. Chem. Phys.* **2011**, *125*, 825–832. [CrossRef]

36. Huang, H.; Chen, R.; Ma, J.; Yan, L.; Zhao, Y.; Wang, Y.; Zhang, W.; Fan, J.; Chen, X. Graphitic carbon nitride solid nanofilms for selective and recyclable sensing of Cu^{2+} and Ag$^+$ in water and serum. *Chem. Commun.* **2014**, *50*, 15415–15418. [CrossRef] [PubMed]

37. Xue, C.; Métraux, G.S.; Millstone, J.E.; Mirkin, C.A. Mechanistic study of photomediated triangular silver nanoprism growth. *J. Am. Chem. Soc.* **2008**, *130*, 8337–8344. [CrossRef] [PubMed]

38. Wojtysiak, S.; Kudelski, A. Influence of oxygen on the process of formation of silver nanoparticles during citrate/borohydride synthesis of silver sols. *Colloid. Surf. A* **2012**, *410*, 45–51. [CrossRef]

39. Ojea-Jiménez, I.; López, X.; Arbiol, J.; Puntes, V. Citrate-coated gold nanoparticles as smart scavengers for mercury(II) removal from polluted waters. *ACS Nano* **2012**, *6*, 2253–2260. [CrossRef] [PubMed]

40. Li, X.; Wu, Z.; Zhou, X.; Hu, J. Colorimetric response of peptide modified gold nanoparticles: An original assay for ultrasensitive silver detection. *Biosens. Bioelectron.* **2017**, *92*, 496–501. [CrossRef] [PubMed]

41. Peng, C.F.; Zhang, Y.Y.; Wang, L.Y.; Jin, Z.Y.; Shao, G. Colorimetric assay for the simultaneous detection of Hg^{2+} and Ag$^+$ based on inhibiting the peroxidase-like activity of core-shell Au@ Pt nanoparticles. *Anal. Methods* **2017**, *9*, 4363–4370. [CrossRef]

42. Xi, H.; Cui, M.; Li, W.; Chen, Z. Colorimetric detection of Ag$^+$ based on C-Ag$^+$-C binding as a bridge between gold nanoparticles. *Sens. Actuators B Chem.* **2017**, *250*, 641–646. [CrossRef]
43. Safavi, A.; Ahmadi, R.; Mohammadpour, Z. Colorimetric sensing of silver ion based on anti aggregation of gold nanoparticles. *Sens. Actuators B Chem.* **2017**, *242*, 609–615. [CrossRef]
44. Qi, L.; Yan, Z.; Huo, Y.; Hai, X.M.; Zhang, Z.Q. MnO$_2$ nanosheet-assisted ligand-DNA interaction-based fluorescence polarization biosensor for the detection of Ag$^+$ ions. *Biosens. Bioelectron.* **2017**, *87*, 566–571. [CrossRef] [PubMed]
45. Ma, K.; Wang, H.; Li, X.; Xu, B.; Tian, W. Turn-on sensing for Ag$^+$ based on AIE-active fluorescent probe and cytosine-rich DNA. *Anal. Bioanal. Chem.* **2015**, *407*, 2625–2630. [CrossRef] [PubMed]
46. Wang, J.; Guo, J.; Zhang, J.; Zhang, W.; Zhang, Y. Signal-on electrochemical sensor for the detection of two analytes based on the conformational changes of DNA probes. *Anal. Methods* **2016**, *8*, 8059–8064. [CrossRef]
47. Miao, P.; Tang, Y.; Wang, L. DNA modified Fe$_3$O$_4$@ Au magnetic nanoparticles as selective probes for simultaneous detection of heavy metal ions. *ACS Appl. Mater. Interfaces* **2017**, *9*, 3940–3947. [CrossRef] [PubMed]
48. Lou, T.; Chen, Z.; Wang, Y.; Chen, L. Blue-to-red colorimetric sensing strategy for Hg^{2+} and Ag$^+$ via redox-regulated surface chemistry of gold nanoparticles. *ACS Appl. Mater. Interfaces* **2011**, *3*, 1568–1573. [CrossRef] [PubMed]
49. Tan, G.; Shi, F.; Doak, J.W.; Sun, H.; Zhao, L.; Wang, P.; Uher, C.; Wolverton, C.; Dravid, V.P.; Kanatzidis, M.G. Extraordinary role of Hg in enhancing the thermoelectric performance of p-type SnTe. *Energy Environ. Sci.* **2015**, *8*, 267–277. [CrossRef]
50. Wu, Z.; Wang, M.; Yang, J.; Zheng, X.; Cai, W.; Meng, G.; Qian, H.; Wang, H.; Jin, R. Well-defined nanoclusters as fluorescent nanosensors: A case study on Au$_{25}$(SG)$_{18}$. *Small* **2012**, *8*, 2028–2035. [CrossRef] [PubMed]

![sensors logo] *sensors*

MDPI

Article

Determination of Cadmium in Brown Rice Samples by Fluorescence Spectroscopy Using a Fluoroionophore after Purification of Cadmium by Anion Exchange Resin

Akira Hafuka [1], Akiyoshi Takitani [2], Hiroko Suzuki [3], Takuya Iwabuchi [3], Masahiro Takahashi [2], Satoshi Okabe [2] and Hisashi Satoh [2,*

[1]   Department of Integrated Science and Engineering for Sustainable Society,
       Faculty of Science and Engineering, Chuo University, Tokyo 112-8551, Japan; hafuka.14p@g.chuo-u.ac.jp
[2]   Division of Environmental Engineering, Faculty of Engineering, Hokkaido University,
       Sapporo 060-8628, Japan; aaa.aki.zzz@gmail.com (A.T.); m-takaha@eng.hokudai.ac.jp (M.T.);
       sokabe@eng.hokudai.ac.jp (S.O.)
[3]   Department of Research and Development, Metallogenics Co., Ltd., Chiba 260-0856, Japan;
       hsuzuki@ak-j.com (H.S.); tiwabuchi@ak-j.com (T.I.)
*   Correspondence: qsatoh@eng.hokudai.ac.jp; Tel.: +81-11-706-6277

Received: 22 August 2017; Accepted: 7 October 2017; Published: 9 October 2017

Abstract: Simple analytical methods are needed for determining the cadmium (Cd) content of brown rice samples. In the present study, we developed a new analytical procedure consisting of the digestion of rice using HCl, Cd purification using anion exchange resin, and then determining the Cd content using fluorescence spectroscopy. Digestion with 0.1 M HCl for 10 min at room temperature was sufficient to extract Cd from the ground rice samples. The Cd in the extract was successfully purified in preference to other metals using Dowex 1X8 chloride form resin. Low concentrations of Cd in the eluate could be determined using fluorescence spectroscopy with a fluoroionophore. Overall, the actual limit of quantification value for the Cd content in rice was about 0.1 mg-Cd/kg-rice, which was sufficiently low compared with the regulatory value (0.4 mg-Cd/kg-rice) given by the Codex Alimentarius Commission. We analyzed authentic brown rice samples using our new analytical procedure and the results agreed well with those determined using inductively coupled plasma optical emission spectrometry (ICP-OES). Since the fluoroionophore recognized Zn^{2+} and Hg^{2+} as well as Cd^{2+}, a sample containing high concentration of Zn^{2+} or Hg^{2+} might cause a false positive result.

Keywords: cadmium; rice; simple analytical method; fluorescence spectroscopy; pretreatment

1. Introduction

In recent years, the contamination of agricultural land by heavy metals, such as cadmium (Cd) [1], mercury [2], chromium, copper (Cu), zinc (Zn) and lead [3] has become a major problem worldwide. This has increased the risks to food safety and hence human health because heavy metals can easily be absorbed from the soil by agricultural crops and then transferred into the human body through consumption. Among the heavy metals contaminating agricultural crops, Cd in rice is a great problem, especially in Asian countries [1,4,5]. Compared with other agricultural crops, rice tends to accumulate Cd readily [6] and thus can be a major source of dietary Cd intake for humans living in Asian countries where rice is a staple food [7]. Cd has toxic effects on humans leading to many serious diseases and some cancers [8]. In Japan, Itai-Itai disease occurred from the 1910s to the 1970s because rice, vegetables and drinking water had been contaminated with Cd [9]. Currently, China is also facing a similar situation [1]. In 2006, the Codex Alimentarius Commission set the international standard value of Cd contained in polished rice at 0.4 mg-Cd/kg-rice. Thus, a reliable method for determining the Cd content in rice samples has now become more important.

Currently, the most common analytical procedure for determining the Cd content of rice is a sample pretreatment followed by instrumental analysis using atomic absorption spectrometry (AAS), inductively coupled plasma optical emission spectrometry (ICP-OES), inductively coupled plasma mass spectroscopy (ICP-MS), atomic fluorescence spectrometry (AFS), or electrodes [10–15]. Although their sensitivity is high, these methods are time-consuming and require expensive instruments and complex operations. In contrast, simple analytical methods, based on colorimetry, ultraviolet–visible spectroscopy, and immunoassay have recently been developed to determine Cd in rice samples [16–18]. However, these methods lack a simple pretreatment, have low sensitivity, and are not yet adequate for application to real samples. Fluorescence spectroscopy is an alternative method which has attracted a great deal of attention because of its high sensitivity, simplicity, and versatile instrumentation [19]. Like the other simple methods mentioned above, determining Cd using fluorescence spectroscopy depends largely on the characteristics of the indicator used (e.g., a fluoroionophore)—its sensitivity, selectivity, photo-physical properties, and water solubility. Because no indicator has perfect selectivity towards the target analyte, samples must be purified before determining the Cd content. Zhang et al. have determined the Cd^{2+} content in rice samples using a solid phase extraction (SPE)-assisted fluorometric paper sensor [20]. They purified and preconcentrated the Cd^{2+} using SPE then determined its content using a fluoroionophore immobilized on the test paper. For this method, the limit of detection (LOD) was poor (56 µg-Cd/L-solution) and the digestion method used for pretreating the rice, consisting of microwave irradiation with mixed acid solution, was complicated. Therefore, we have proposed a new analytical procedure consisting of rice digestion, Cd purification, followed by Cd determination using fluorescence spectroscopy as a simple method of analysis for Cd in rice. In the present study, we aim to develop a simple digestion method using 0.1 M HCl, a Cd purification method using anion-exchange resin, then the ratiometric determination of Cd using fluorescence spectroscopy.

2. Materials and Methods

2.1. Standard Methods of Rice Digestion and Metal Determination

Brown rice samples were obtained from rice farmers. The brown rice standard sample (NMIJ CRM 7531-a) was purchased from the National Metrology Institute of Japan (Tsukuba, Japan). Milli-Q water (18.2 MΩ·cm) was used in all experiments described below. Rice digestion and metal determination were conducted based on Japanese standard method (Japanese Ministry of Health). Briefly, a sample of ground rice (10 g) was added to 10.4 M of HNO_3 (50 mL) then the suspension was gently heated. The suspension was cooled to room temperature then concentrated H_2SO_4 (2 mL) was added. The suspension was heated again until its color changed to light yellow or colorless. After cooling to room temperature, the solution was transferred to a volumetric flask (100 mL) and 0.1 M HNO_3 was made up to the mark. The concentrations of Cd, Zn, Cu and Iron (Fe) were determined using an ICP-OES instrument (ICPE-9000, Shimadzu Corporation, Kyoto, Japan).

2.2. Hydrochloric Acid Digestion of Rice Samples

Rice samples (20 g) were ground for 10 s using a laboratory-scale mill (Labo Milser LM-PLUS, Osaka Chemical Co., Ltd., Osaka, Japan). The sample powder was then added to 0.1 M HCl (80 mL) and the mixture stirred at room temperature for 10 min. Following centrifugation at 8000 g for 10 min, the supernatant was filtered through a 1.0-µm-pore-size cellulose ester membrane (A100A047A, Advantec Toyo Kaisha Ltd., Tokyo, Japan) to obtain the extract. The extract solution (9 mL) and concentrated HNO_3 (1 mL) were added in 10 mL-test tubes then ICP-OES measurements were conducted. The Cd concentration of the extract determined using ICP-OES was compared with that determined by the standard method to evaluate the extraction efficiency.

2.3. Purification of Cd by Anion-Exchange Resin

We adopted the method by Kallmann et al. [21] to purify Cd from other metals. It uses the difference in stability between the negatively-charged chloro-complex of Cd and other metals in a HCl solution (Cd chloro-complex is more stable than the chloro-complexes of other metals) and separates Cd from other metals using an anion-exchange resin.

Dowex 1X8 chloride form (44340, Sigma-Aldrich Japan K.K., Tokyo, Japan) was used as the anion-exchange resin. The resin was washed three times with Milli-Q water (10 times the volume of the resin) then soaked in Milli-Q water overnight. A disposable polypropylene column (5 mL; 29922, Thermo Fisher Scientific K.K., Yokohama, Japan) held the resin. The column was washed with Milli-Q water then filled with 10% ethanol to obtain the surface hydrophilicity of the column. A frit was introduced at the bottom of the column then 0.5 mL of the prepared resin was added to give a column volume (CV) of 0.5 mL. Finally, the resin was capped with another frit then washed with Milli-Q water (20 mL) to remove the ethanol completely. The prepared columns were preserved at 4 °C in a refrigerator before use.

For the Cd purification process, water (2 mL) and 0.1 M HCl (5 mL) were passed through the column sequentially before adding the sample. The rice extract solution (40 mL = 80 CV) was passed through the column and the flow-through fraction was obtained. Then, 0.1 M HCl (40 mL = 80 CV) was passed through the column to remove organic compounds and metals (except Cd) to obtain the washout fraction. Finally, water (30 mL = 60 CV) was passed through the column and the Cd was extracted in the elution fraction. The Cd, Zn, Cu, and Fe concentrations of the flow-through, washout and elution fractions were determined using ICP-OES to estimate the elution efficiency.

2.4. Determination of Cd by Fluorescence Spectroscopy

The fluoroionophore, 2,2':6',2''-terpyridine-substituted BODIPY (BDP-TPY), designed in our laboratory, was synthesized according to a previously reported method [22]. The chemical structure of BDP-TPY is shown in Figure 1. Since 2,2':6',2''-terpyridine (TPY) acts as a heavy metal ion receptor, the fluorescence color of BDP-TPY changes upon binding of heavy metals to TPY [22]. A stock solution of BDP-TPY (2 μM) was prepared in acetonitrile. The test solutions were prepared by adding BDP-TPY stock solution (5 mL) to a volumetric flask (10 mL), followed by the elution fraction (4.5 mL) and 80 mM Tris buffer (0.5 mL, pH 10.2). The solutions were then transferred to quartz cells and the fluorescence spectra recorded by a spectrofluorometer (FP-6600, JASCO Corporation, Tokyo, Japan). The Cd concentrations of the elution fraction obtained were compared with those determined using ICP-OES.

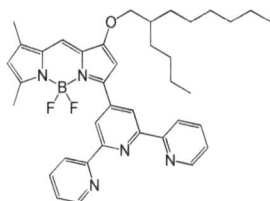

Figure 1. Chemical structure of 2,2':6',2''-terpyridine-substituted BODIPY (BDP-TPY).

3. Results and Discussion

3.1. Acid Digestion of Rice Samples

Rice samples require acid digestion before the metal contents can be determined. In the first step, we optimized the time for grinding the samples. Figure 2 shows the efficiency of extracting Cd from a brown rice sample as a function of grinding time. Standard methods (Section 2.1) were used to digest the rice and determine the metal content. Around 90% of the Cd content was extracted after only 5 s grinding. After 10 s, the extraction efficiencies for Cd reached 100% so this was selected. Therefore, we set grinding time as 10 s for the subsequent experiments. The samples could be ground into a fine powder after 10 s of grinding using a laboratory-scale mill.

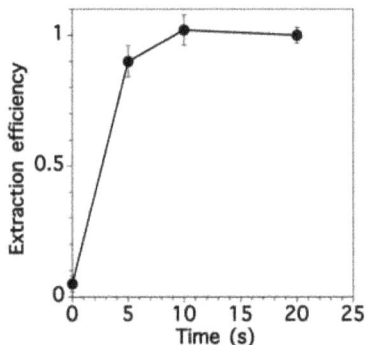

Figure 2. Extraction efficiency of Cd from brown rice samples as a function of grinding time. Error bars indicate the standard deviation, which was determined using three replicates.

The standard method for rice digestion is complicated and time-consuming. Therefore, we developed a simpler method with fluorescence spectroscopy using 0.1 M HCl solution (Section 2.2) which was appropriate for the following purification and fluorescence determination steps [22]. The Cd concentrations in the rice extracts were determined by the standard and our methods (Figure 3). There was an excellent correlation with a determination coefficient of 0.996 between the concentrations determined by the two different methods. This indicated that our simple digestion method was adequate for extracting Cd from the ground rice samples. After 10 min of HCl treatment, the fine rice powder became sticky.

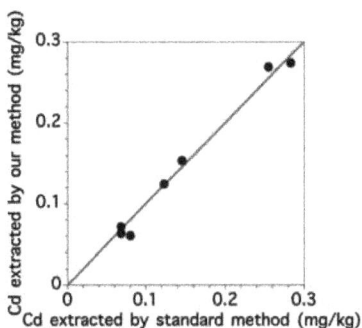

Figure 3. Relationship between Cd concentrations extracted by the standard (ICP-OES) and our developed methods.

3.2. Metals in Column Fractions

The rice extract solutions were run through a polypropylene column packed with anion-exchange resin to purify the Cd. As well as the Cd concentration, we determined the Zn, Cu, and Fe concentrations in each column fraction because these other metals may inhibit the determination of Cd when using fluorescence spectroscopy [23]. Figure 4 shows the fractional amount of these four metals.

In the first step, the negatively-charged chloro-complex of Cd formed during the extraction process was ion exchanged with Cl$^-$ so that the complex was retained on the resin. In contrast, most of the other metals, which were in the free ionic form, passed through the column and collected in the flow-through fraction. During the washout step, 0.1 M HCl solution was passed through the column to remove any residual Zn, Cu, Fe, and organic compounds, especially soluble starch, into the washout fraction. During the elution step, Cd^{2+} was eluted from the column by passing through Milli-Q water. H$_2$O might exchange chloride anion in the chloro-complex of Cd and convert the Cd complex to free Cd ion, which was then washed out by leaving chloride anion on the column. This was collected in the elution fraction because of the decomposition of the chloro-complex form, which was then analyzed during the following fluorescence determination step. It should be noted that the Cd was purified and its concentration was concentrated 1.33 times during these processes.

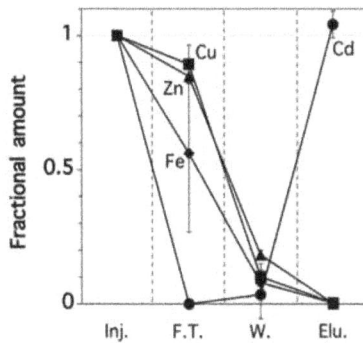

Figure 4. Relative amounts of Cd, Zn, Cu, and Fe from a real rice sample in each column fraction. Inj.: injection; F.T.: flow-through fraction; W.: washout fraction; Elu.: elution fraction. Error bars indicate the standard deviation, which was determined using three replicates.

We determined the Cd and Zn contents in each fraction because Zn can strongly inhibit the determination of Cd using BDP-TPY [23]. The flow-through fraction contained most of the Zn (87%) and the first 20 CV of the washout fraction the remaining Zn content (Figure 5). Cd was rarely detected in the flow-through and washout fractions. Most of the Cd was found in the first 20 CV of the elution fraction and the residual Cd was recovered in the later elution fractions.

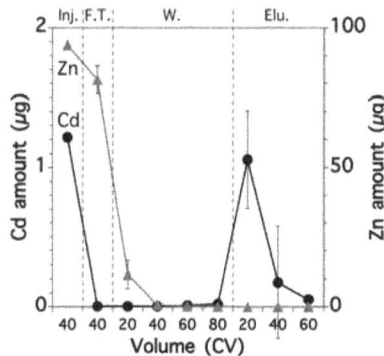

Figure 5. Amounts of Cd and Zn from a real rice sample in each CV of each column fraction. Inj.: injection; F.T.: flow-through fraction; W.: washout fraction; Elu.: elution fraction. Error bars indicate the standard deviation, which was determined using three replicates.

3.3. Determination of Cd by Fluorescence Spectroscopy

We established a calibration curve for Cd by fluorescence spectroscopy with BDP-TPY. Since the fluorescence spectra of BDP-TPY can be affected by pH value due to the protonation reaction [23], fluorescence titration experiments of BDP-TPY with Cd to generate a calibration curve were carried out in Tris buffer. BDP-TPY had a sharp fluorescence peak at 539 nm and its spectra changed as the Cd concentrations of the standard solutions increased. The height of the original fluorescence peak at 539 nm (F539) gradually decreased then a new fluorescence peak at 562 nm (F562) appeared and increased because of the transfer of intramolecular charges (Figure 6A) [22]. This shift in the spectra allowed the ratiometric measurement of Cd by calculating the fluorescence intensity ratio (F539/F562) [24]. The ratiometric measurement has advantages, such as, independency of results on probe concentration, bleaching, optical path length, illumination intensity, etc. F539/F562 increased linearly with Cd concentrations up to 112 µg-Cd/L-solution (1 µM) (Figure 6B), with the regression equation of y = 0.0027x + 0.32 (R^2 = 0.978). The LOD (3s/slope) and limit of quantification (LOQ, 10 s/slope) values for Cd were determined based on the standard deviations of 11 blank solutions. The LOD and LOQ values were estimated to be 3.8 µg-Cd/L-solution (0.03 µM) and 12.7 µg-Cd/L-solution (0.11 µM), respectively. These values were relatively low compared with those of simple analyses previously reported [18,20]. Furthermore, a dynamic range of BDP-TPY was investigated. The linear range of Cd was up to 224 µg-Cd/L-solution (2 µM) (Figure S1 in the Supplementary Material of this paper).

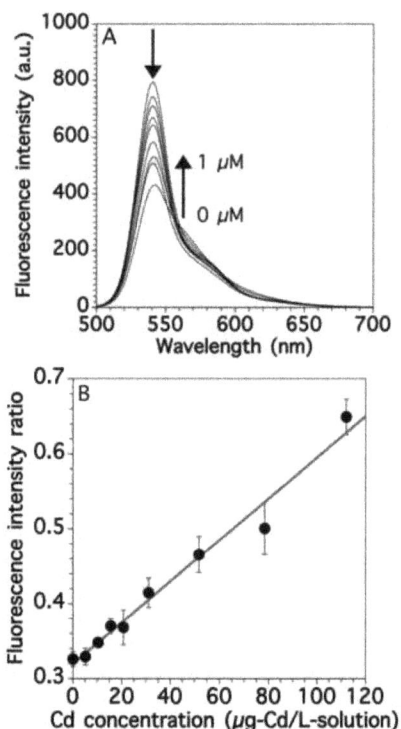

Figure 6. (**A**) Spectra and (**B**) calibration curve for Cd by fluorescence spectroscopy. Error bars indicate the standard deviation, which was determined using three replicates.

Since the elution fraction of the rice extract solutions was acidic, we had to optimize the composition of the buffer solution for the fluorescence determination. Figure S2 in the Supplementary Material of this paper shows the effects of Tris concentrations on the pH of the analytical samples. The pH values of the elution fraction were acidic after the addition of 20 and 60 mM Tris buffer. Above 70 mM, the pH was approximately neutral so we decided to use 80 mM Tris buffer for the fluorescence determination.

We investigated the interfering effect of Zn on Cd determination using fluorescence spectroscopy. Figure 7 shows the relationship between the fluorescence intensity ratio and the Zn concentration while the Cd concentration was kept constant at 33.6 μg-Cd/L-solution (0.3 μM). The ratios gradually increased and an interfering effect was observed at Zn concentrations above 19.5 μg-Zn/L-solution (0.3 μM). Although no Zn was detected in the elution fraction, the Zn concentration should be kept below 19.5 μg-Zn/L-solution when using this method.

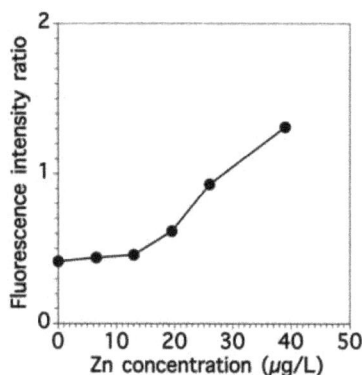

Figure 7. Effect of Zn concentrations on fluorescence intensity ratio (F539/F562) in the sample containing Cd (0.3 μM).

We then investigated the applicability of fluorescence spectroscopy for measuring the Cd content of rice samples both by the standard method using ICP-OES and by our procedure consisting of HCl digestion, column purification and fluorescence spectroscopy. The Cd contents of fifteen brown rice samples were determined using the standard and our methods (Figure 8), revealing an almost linear relationship with a slope of 1.09 and a determination coefficient of 0.964. For most samples, the relative errors in the Cd contents determined by these two methods were within 20%. We thus concluded that Cd contents could be successfully determined in rice samples with complex matrices using our developed procedure. Overall, the actual LOQ value of the Cd content in brown rice was about 0.1 mg-Cd/kg-rice based on the results shown in Figure 8 Since the calibration error exceeded 20% of the calibration span value below 0.1 mg-Cd/kg-rice of Cd contents in brown rice. These values were sufficiently low compared with the regulatory value (0.4 mg-Cd/kg-rice) given by the Codex Alimentarius Commission. Therefore, our method would be suitable for the simple screening of the Cd content in rice samples.

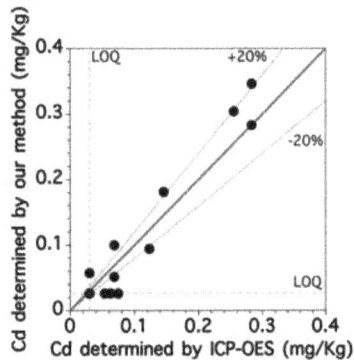

Figure 8. Relationship between concentrations of Cd in brown rice samples determined by fluorescence spectroscopy with the fluoroionophore and by ICP-OES.

4. Conclusions

In the present study, we have developed a simple analytical procedure for determining the Cd content of brown rice samples. We have revealed that digesting a rice sample with 0.1 M HCl for 10 min at room temperature effectively extracted Cd from a ground rice sample. The Cd in the extract was successfully purified in the presence of Zn, Cu, and Fe using a Dowex 1X8 chloride form resin. The Cd content in the elution fraction could be determined using fluorescence spectroscopy with a fluoroionophore with LOD and LOQ values of 3.8 μg-Cd/L-solution and 12.7 μg-Cd/L-solution, respectively. We analyzed authentic brown rice samples using our developed method and the results agreed well with those measured using ICP-OES. Overall, the actual LOQ value for Cd content in brown rice was about 0.1 mg-Cd/kg-rice. Based on these results, we concluded that our developed method would be suitable for the simple screening of the Cd content in rice samples. Nevertheless, as we reported previously, BDP-TPY is a fluoroionophore with high selectivity for Zn, Cd and Hg ions [22]. We also revealed that a chemical structure of an ion receptor of the BODIPY derivative determined the selectivity of it [22] and substitution at the 5-position of an asymmetric BODIPY cation sensor affected the selectivity of it [25]. Based on these results, at present we are trying to develop a novel BODIPY derivative with high selectivity for Cd.

Supplementary Materials: The following are available online at http://www.mdpi.com/1424-8220/17/10/2291/s1. Figure S1: Calibration curve for Cd by fluorescence spectroscopy, Figure S2: Effect of Tris concentration of the buffer solution on the pH of the solution under analysis.

Acknowledgments: This research was partly supported by a Grant-in-Aid for Scientific Research (KAKENHI Grant Nos. 26289178, 17H03328 and 26889054) from the Japan Society for the Promotion of Science and Core Research of Evolutional Science & Technology (CREST) for "Innovative Technology and System for Sustainable Water Use" from Japan Science and Technology Agency (JST).

Author Contributions: A. Hafuka and H. Sato conceived and designed the experiments; A. Takitani and H. Suzuki performed the experiments; T. Iwabuchi, M. Takahashi, and S. Okabe analyzed the data; A. Hafuka wrote the paper; and all authors discussed the results and commented on the manuscript.

Conflicts of Interest: The authors declare no conflict of interest.

References

1. Hu, Y.N.; Cheng, H.F.; Tao, S. The Challenges and Solutions for Cadmium-contaminated Rice in China: A Critical Review. *Environ. Int.* **2016**, *92–93*, 515–532. [CrossRef] [PubMed]
2. Patra, M.; Sharma, A. Mercury toxicity in plants. *Bot. Rev.* **2000**, *66*, 379–422. [CrossRef]
3. Liu, W.H.; Zhao, J.Z.; Ouyang, Z.Y.; Soderlund, L.; Liu, G.H. Impacts of sewage irrigation on heavy metal distribution and contamination in Beijing, China. *Environ. Int.* **2005**, *31*, 805–812. [CrossRef] [PubMed]

4. Horiguchi, H.; Oguma, E.; Sasaki, S.; Okubo, H.; Murakami, K.; Miyamoto, K.; Hosoi, Y.; Murata, K.; Kayama, F. Age-relevant renal effects of cadmium exposure through consumption of home-harvested rice in female Japanese farmers. *Environ. Int.* **2013**, *56*, 1–9. [CrossRef] [PubMed]

5. Kosolsaksakul, P.; Farmer, J.G.; Oliver, I.W.; Graham, M.C. Geochemical associations and availability of cadmium (Cd) in a paddy field system, northwestern Thailand. *Environ. Pollut.* **2014**, *187*, 153–161. [CrossRef] [PubMed]

6. Grant, C.A.; Clarke, J.M.; Duguid, S.; Chaney, R.L. Selection and breeding of plant cultivars to minimize cadmium accumulation. *Sci. Total Environ.* **2008**, *390*, 301–310. [CrossRef] [PubMed]

7. Meharg, A.A.; Norton, G.; Deacon, C.; Williams, P.; Adomako, E.E.; Price, A.; Zhu, Y.G.; Li, G.; Zhao, F.J.; McGrath, S.; et al. Variation in Rice Cadmium Related to Human Exposure. *Environ. Sci. Technol.* **2013**, *47*, 5613–5618. [CrossRef] [PubMed]

8. Nordberg, G.F. Historical perspectives on cadmium toxicology. *Toxicol. Appl. Pharmacol.* **2009**, *238*, 192–200. [CrossRef] [PubMed]

9. Inaba, T.; Kobayashi, E.; Suwazono, Y.; Uetani, M.; Oishi, M.; Nakagawa, H.; Nogawa, K. Estimation of cumulative cadmium intake causing Itai-itai disease. *Toxicol. Lett.* **2005**, *159*, 192–201. [CrossRef] [PubMed]

10. Zheng, F.; Hu, B. Thermo-responsive polymer coated fiber-in-tube capillary microextraction and its application to on-line determination of Co, Ni and Cd by inductively coupled plasma mass spectrometry (ICP-MS). *Talanta* **2011**, *85*, 1166–1173. [CrossRef] [PubMed]

11. Akamatsu, S.; Yoshioka, N.; Mitsuhashi, T. Sensitive determination of cadmium in brown rice and spinach by flame atomic absorption spectrometry with solid-phase extraction. *Food Addit. Contam. Part A* **2012**, *29*, 1696–1700. [CrossRef] [PubMed]

12. Groombridge, A.S.; Inagaki, K.; Fujii, S.; Nagasawa, K.; Okahashi, T.; Takatsu, A.; Chiba, K. Modified high performance concentric nebulizer for inductively coupled plasma optical emission spectrometry. *J. Anal. At. Spectrom.* **2012**, *27*, 1787–1793. [CrossRef]

13. Guo, W.; Zhang, P.; Jin, L.L.; Hu, S.H. Rice cadmium monitoring using heat-extraction electrothermal atomic absorption spectrometry. *J. Anal. At. Spectrom.* **2014**, *29*, 1949–1954. [CrossRef]

14. Wang, Z.Q.; Wang, H.; Zhang, Z.H.; Liu, G. Electrochemical determination of lead and cadmium in rice by a disposable bismuth/electrochemically reduced graphene/ionic liquid composite modified screen-printed electrode. *Sens. Actuator B Chem.* **2014**, *199*, 7–14. [CrossRef]

15. Yu, H.M.; Ai, X.; Xu, K.L.; Zheng, C.B.; Hou, X.D. UV-assisted Fenton digestion of rice for the determination of trace cadmium by hydride generation atomic fluorescence spectrometry. *Analyst* **2016**, *141*, 1512–1518. [CrossRef] [PubMed]

16. Abe, K.; Nakamura, K.; Arao, T.; Sakurai, Y.; Nakano, A.; Suginuma, C.; Tawarada, K.; Sasaki, K. Immunochromatography for the rapid determination of cadmium concentrations in wheat grain and eggplant. *J. Sci. Food Agric.* **2011**, *91*, 1392–1397. [CrossRef] [PubMed]

17. Wen, X.D.; Deng, Q.W.; Guo, J.; Yang, S.C. Ultra-sensitive determination of cadmium in rice and water by UV-vis spectrophotometry after single drop microextraction. *Acta Mol. Biomol. Spectrosc.* **2011**, *79*, 508–512. [CrossRef] [PubMed]

18. Guo, Y.M.; Zhang, Y.; Shao, H.W.; Wang, Z.; Wang, X.F.; Jiang, X.Y. Label-Free Colorimetric Detection of Cadmium Ions in Rice Samples Using Gold Nanoparticles. *Anal. Chem.* **2014**, *86*, 8530–8534. [CrossRef] [PubMed]

19. Lakowicz, J.R. *Principles of Fluorescence Spectroscopy*, 3rd ed.; Springer: New York, NY, USA, 2006.

20. Zhang, Y.; Li, H.; Niu, L.Y.; Yang, Q.Z.; Guan, Y.F.; Feng, L. An SPE-assisted BODIPY fluorometric paper sensor for the highly selective and sensitive determination of Cd^{2+} in complex sample: Rice. *Analyst* **2014**, *139*, 3146–3153. [CrossRef] [PubMed]

21. Kallmann, S.; Steele, C.G.; Chu, N.Y. Determination of Cadmium and Zinc. *Anal. Chem.* **1956**, *28*, 230–233. [CrossRef]

22. Hafuka, A.; Taniyama, H.; Son, S.H.; Yamada, K.; Takahashi, M.; Okabe, S.; Satoh, H. BODIPY-Based Ratiometric Fluoroionophores with Bidirectional Spectral Shifts for the Selective Recognition of Heavy Metal Ions. *Bull. Chem. Soc. Jpn.* **2013**, *86*, 37–44. [CrossRef]

23. Hafuka, A.; Yoshikawa, H.; Yamada, K.; Kato, T.; Takahashi, M.; Okabe, S.; Satoh, H. Application of fluorescence spectroscopy using a novel fluoroionophore for quantification of zinc in urban runoff. *Water Res.* **2014**, *54*, 12–20. [CrossRef] [PubMed]

24. Valeur, B. *Molecular Fluorescence: Principles and Applications*, 2nd ed.; Wiley-VCH: Weinheim, Germany, 2012.
25. Hafuka, A.; Kando, R.; Ohya, K.; Yamada, K.; Okabe, S.; Satoh, H. Substituent Effects at the 5-Position of 3-[Bis(pyridine-2-ylmethyl)amino]-BODIPY Cation Sensor Used for Ratiometric Quantification of Cu^{2+}. *Bull. Chem. Soc. Jpn.* **2015**, *88*, 447–454. [CrossRef]

sensors

MDPI

Article

A Label-Free Fluorescent Array Sensor Utilizing Liposome Encapsulating Calcein for Discriminating Target Proteins by Principal Component Analysis

Ryota Imamura [1], Naoki Murata [1], Toshinori Shimanouchi [2], Kaoru Yamashita [1], Masayuki Fukuzawa [1] and Minoru Noda [1,*]

[1]   Graduate School of Science and Technology, Kyoto Institute of Technology, Matsugasaki, Sakyo-ku, Kyoto 606-8585, Japan; m5621003@edu.kit.ac.jp (R.I.); m5622050@edu.kit.ac.jp (N.M.); yamashita.kaoru@kit.ac.jp (K.Y.); fukuzawa@kit.ac.jp (M.F.)
[2]   Graduate School of Environmental and Life Science, Okayama University, 1-1-1 Tsushima-naka, Kita-ku, Okayama 700-8530, Japan; tshima@cc.okayama-u.ac.jp
*   Correspondence: noda@kit.ac.jp; Tel.: +81-75-724-7443

Received: 16 May 2017; Accepted: 13 July 2017; Published: 15 July 2017

Abstract: A new fluorescent arrayed biosensor has been developed to discriminate species and concentrations of target proteins by using plural different phospholipid liposome species encapsulating fluorescent molecules, utilizing differences in permeation of the fluorescent molecules through the membrane to modulate liposome-target protein interactions. This approach proposes a basically new label-free fluorescent sensor, compared with the common technique of developed fluorescent array sensors with labeling. We have confirmed a high output intensity of fluorescence emission related to characteristics of the fluorescent molecules dependent on their concentrations when they leak from inside the liposomes through the perturbed lipid membrane. After taking an array image of the fluorescence emission from the sensor using a CMOS imager, the output intensities of the fluorescence were analyzed by a principal component analysis (PCA) statistical method. It is found from PCA plots that different protein species with several concentrations were successfully discriminated by using the different lipid membranes with high cumulative contribution ratio. We also confirmed that the accuracy of the discrimination by the array sensor with a single shot is higher than that of a single sensor with multiple shots.

Keywords: biosensor; fluorescence; liposome; protein; array; interaction; cholesterol; principal component analysis (PCA)

1. Introduction

Important biomolecules such as DNA, RNA, proteins and so on are commonly detected by fluorescence techniques based on labeling and staining [1–3]. Both the fluorescent antibody and the immunofluorescent technique are major techniques in the field of biochemistry because they offer high detectability, stability and safety. Especially, immunofluorescence methods such as enzyme-linked immunosorbent assays (ELISA) have been used to detect various proteins [4–6] with high sensitivity. However, ELISA methods involve quite complicated operations, user proficiency, large device size and so on. Therefore, we have attempted to develop new simplified and label-free fluorescent sensing techniques, utilizing phospholipid membrane liposomes, although several immune assays based on liposomes have been reported so far such as a liposome immunosorbent assay and a liposome immune lysis one [7,8].

We have reported preliminary results on an arrayed biosensor utilizing liposomes to encapsulate fluorescent molecules and the time course analysis of the fluorescence [9]. The phospholipid bilayer of the liposome is used as a model cell membrane sensing biomolecule to detect and discriminate external target biomolecules. It is known that the encapsulated molecules leak from the internal aqueous phase after the interactions [10,11]. The phenomenon has been used to characterize the membrane properties and evaluate the interaction against external biomolecules such as proteins [12,13], peptides [14] and the others [15]. In the operation of the fluorescent liposome biosensor, the leakage of fluorescent molecules increases the fluorescence intensity.

Therefore, it is further required: (1) to achieve a higher and more stable output intensity of fluorescence emission; (2) to investigate more different types of liposome phospholipid to improve the sensitivity and discrimination capability between different target biomolecules, and (3) to analyze the arrayed data statistically to increasing the accuracy of the analyzed results.

In this work, we firstly considered a proper concentration of fluorescent calcein molecule for encapsulation in the liposome, as calcein encapsulation is often used to evaluate membrane permeation [16–18]. It is necessary to know the proper calcein concentration in order to emit enough fluorescence when the calcein leaks from the liposome by the liposome-target molecule interaction. Secondly, an important component of cell membrane, cholesterol, was newly incorporated into the liposome membrane because the membrane fluidity changes significantly by incorporation of cholesterol [19,20], which influences on the interaction between the membrane and target molecules. Thirdly, we also evaluated the interaction strengths of different phospholipid liposome species with target molecules to identify the most effective phospholipids. Finally, we proceeded to apply principal component analysis (PCA) for statistical data analyses [21], since most of the microarray techniques need comprehensive analysis because those techniques obtain multi-dimensional data and extract characteristic results. Using PCA, we examined the feasibility and applicability of arrayed sensors with single shots in comparison with single microwell sensors with multiple shots.

2. Materials and Methods

2.1. Materials

The phospholipids we used were 1,2-dipalmitoyl-*sn*-glycero-3-phosphocholine (DPPC, MW = 734.04), 1,2-distearoyl-*sn*-glycero-3-phosphocholine (DSPC, MW = 790.15). Besides, we used DPPC incorporating cholesterol (MW = 386.65) (DPPC:cholesterol = 66 mol%:33 mol%). DPPC, DSPC, and cholesterol were purchased from Avanti Polar Lipids (Alabaster, AL, USA). Target proteins we used were bovine carbonic anhydrase (CAB, MW = 28,400) and lysozyme (MW = 17,307), which were purchased from Sigma Aldrich (St. Louis, MO, USA). The fluorescent molecule encapsulated in liposomes was calcein (MW = 622.53), which was also purchased from Sigma Aldrich. Silpot184 and Silpot184 CAT (hardener) purchased from Toray Dow Corning (Tokyo, Japan) were used for the fabrication of polydimethylsiloxane (PDMS)-based chips. The gel beads, Sepharose 4B, which are used for filtering out the free calcein during liposome preparation, were purchased from GE Healthcare (Uppsala, Sweden). The filtration buffer, phosphate buffered saline (PBS), were purchased from Thermo Fisher Scientific (Yokohama, Japan). All the chemical reagents were of analytical grade.

2.2. Preparation of Liposomes

Three kinds of liposome were prepared using DPPC, DSPC, and DPPC incorporating cholesterol (DPPC:cholesterol = 66 mol%:33 mol%) in the same manner as previously reported [10–12,22]. In brief, three kinds of lipid solutions were prepared in chloroform. They were dried in round-bottom flasks by rotary evaporation under reduced pressure. The obtained homogeneous lipid films were kept under vacuum for more than 3 h and then hydrated with 100 mM calcein solution to form multilamellar vesicles. After five cycles of freezing-thawing treatment, the liposome size was adjusted by the extrusion of liposome suspension through a polycarbonate membrane (pore size: 100 nm). Finally, the free calcein was removed by gel permeation chromatography on Sepharose 4B.

2.3. Fabricated Photometric System

We constructed a photometric system in order to take fluorescent images of array sensor chips. Figure 1 shows a cross-sectional view of the microwells of the sensor array with an illustration of the fluorescence photometric system. Figure 2 shows a fluorescence image from the array sensor. In the measurement, each microwell was filled with a common liposome suspension (10 mM) encapsulating calcein (100 mM) together with a target protein of different species and concentration from each other, thereafter the array was covered with a transparent cover glass. Calcein excitation light (around 495 nm in wavelength) emitted from a blue Light-Emitting Diode (LED) (λ = 495 nm), is reflected by a dichroic beamsplitter and reaches the arrayed microwells. Owing to liposome-protein interaction, calcein molecules leak from the internal aqueous phase of the liposomes, thereafter the fluorescence of the leaked calcein is emitted from the arrayed microwells, passes through the dichroic beamsplitter and a 502 nm longpass filter, successively, and is finally detected by a CMOS imager (WAT-01U2, Watec. Co., Ltd., Tsuruoka, Japan). We evaluated the liposome-protein interaction by investigating the change of intensity of fluorescence emissions using our developed software for analysis [23].

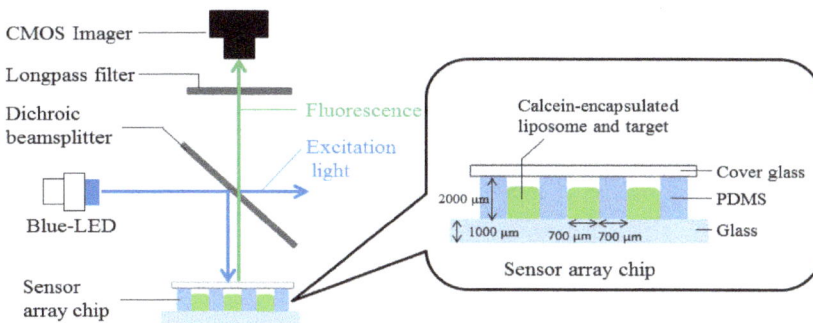

Figure 1. A cross-sectional view of microwells of array sensor with illustration of the fluorescence photometric system. Polydimethylsiloxane (PDMS); Light-Emitting Diode (LED).

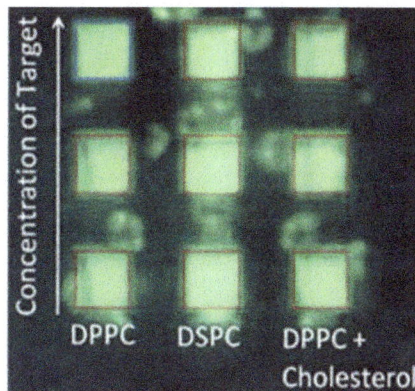

Figure 2. A fluorescent image from the array sensor. Along the horizontal direction, the microwells were filled with different phospholipid liposome species, and along the vertical direction, they were provided with different concentrations of the same kind of target protein. 1,2-dipalmitoyl-sn-glycero-3-phosphocholine (DPPC); 1,2-distearoyl-*sn*-glycero-3-phosphocholine (DSPC).

3. Results and Discussions

3.1. Optimization of Fluorescence of Calcein in Arrayed Microwells

It is known that fluorescence intensity depends on the fluorescent molecule concentration, especially for calcein [24,25]. In our case, the calcein molecules are released depending on how much the liposome membrane permits their permeation. Thus, it is important to prepare calcein with a proper concentration encapsulated in the liposome for increasing and optimizing the fluorescence intensity. To investigate the dependence on the concentration, we prepared calcein solutions diluted by phosphate buffered saline (PBS) with concentration of 0.001, 0.01, 0.1, 0.5, 1, 5, 10, 50, and 100 mM, respectively. After the calcein solutions were introduced into the arrayed microwells, fluorescence measurements were carried out using a fluorescent microscope (BX51, OLYMPUS, Tokyo, Japan) and a halogen lump (U-HGLGPS, OLYMPUS) as an excitation light source filtered at 460–495 nm.

Figure 3 shows the fluorescent intensity measured in the arrayed microwells as a function of calcein concentration. The fluorescent intensity of calcein solution increases rapidly and linearly in low concentration range (as seen in the inset) and is maximum at around 2 mM. For concentrations higher than 2 mM, the fluorescent intensity decreases with increase in concentration. In particular, the fluorescent intensity became almost zero when the concentration increased to 100 mM. This phenomenon where calcein is quenched at high concentrations is called 'self-quenching' [24–26]. Accordingly, we set the calcein concentration encapsulated in liposome as high as 100 mM. In this experiment, based on our our previous results the calcein molecules are estimated to leak by less than a few percent, so the fluorescence intensity of the leaked calcein is much higher than that of the liposomed encapsulating 100 mM calcein.

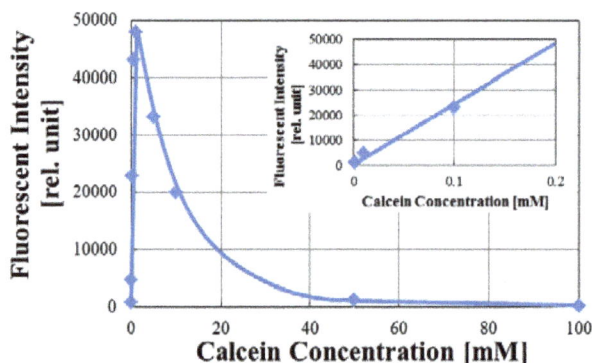

Figure 3. Fluorescent intensity dependence on calcein concentration.

3.2. Effect of Cholesterol on Liposome-Protein Interaction

We measured the liposome-protein interaction by the developed photometric system, and investigated the effect of incorporating cholesterol on the interaction. The suspension of liposome encapsulating 100 mM calcein added with target protein (CAB or lysozyme) was introduced into the arrayed microwell, thereafter the surface of the array chip was sealed by a transparent cover glass. The fluorescence image from the array is shown again in Figure 2. Along the horizontal direction, the microwells were filled with different phospholipid liposome species, and along the vertical direction, they contained different concentrations of the same kind of target protein. Differences in fluorescent intensity among the microwells can be observed dependending on the phospholipid species and concentration of target protein. We investigated the time course of the fluorescence intensity of leaked calcein caused by the interaction between liposomes incorporated with cholesterol and target proteins for 60 min by the photometric system. Moreover, we calculated relative change of fluorescence intensity against the initial intensity (at the start of the measurement). The relative change of fluorescence intensity, defined as RF, is calculated as shown in Equation (1):

$$RF = \frac{\Delta I}{I_0} = \frac{I - I_0}{I_0} \tag{1}$$

Here I is fluorescence intensity after 60 min from the start, and I_0 is the initial fluorescence intensity (at 0 min). Table 1 lists averaged RF's of liposome suspensions of DPPC and DPPC/cholesterol added with 100 µM and 300 µM CAB after 60 min from the start. Regardless of the different liposomes used, RF increases with the increase in CAB concentration. This means that the amount of leaked calcein molecules increases with the concentration of target protein. Also, the RF of DPPC/cholesterol is higher than that of DPPC for the same concentration of CAB. We consider reasonable that incorporating cholesterol into the phospholipid membrane improves the fluidity of liposome. This would lead the observed improvement in sensitivity.

Next, Figure 4 shows the time course of RF's of DPPC/cholesterol added with different target proteins at the same concentration (300 µM), also including the result of a control sample without the presence of those proteins. As mentioned in 3.1, it is noted for the control sample that the calcein molecules are estimated to leak by less than about 2%. From the figure, it is found that RF increased with time and the liposome-CAB interaction is larger than the liposome-lysozyme interaction. We consider that the difference in the behavior of liposome-protein interaction originates from that in the molecular structures of the proteins such as different amount of disulfide (S-S) bonds [27–29]. It is reasonable that the number of S-S bonds influences the intensity of interactions. The lysozyme molecule has four disulfide bonds, thus it has high structural stability and exposing its hydrophobic groups is difficult. However, CAB molecules have no disulfide bond, so the hydrophobic groups of CAB are easily exposed. Therefore, the hydrophobic interaction between the liposome and CAB is stronger than that between the liposome and lysozyme.

Table 1. Averaged RF's of liposome suspension of DPPC and DPPC/Cholesterol added with 100 µM and 300 µM CAB after 60 min from the start.

Target Protein (Conc.)	RF(%) for DPPC	RF(%) for DPPC + Cholesterol
CAB (100 µM)	-1 ± 2	-1 ± 2
CAB (300 µM)	2 ± 1	8 ± 2

1,2-dipalmitoyl-sn-glycero-3-phosphocholine (DPPC); bovine carbonic anhydrase (CAB).

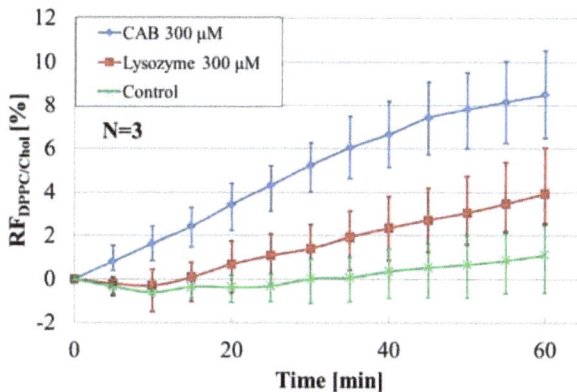

Figure 4. Time course of RF's of DPPC/cholesterol added with different target proteins (300 µM) for 60 min, including the results of control sample without those proteins.

3.3. Effect of Phospholipid Species on Liposome-Protein Interaction

We investigated the time course of RF's of liposome suspensions of DPPC, DPPC/cholesterol, and DSPC added with target proteins for 60 min. In the measurement, we selected 30, 100, 300, and 500 μM as the concentration of CAB, and also 70, 300, 500, and 700 μM as the concentration of lysozyme, respectively. Figure 5 plots the averaged RF's of DSPC vs. that of DPPC measured after 60 min from the initial addition of the target proteins. From Figure 5, it is clear that the RF plots of each different target protein are well separated. Moreover, the plots are also separated and discriminated between the different concentrations. The difference in RF between different phospholipid liposomes is reasonably due to the structure of phospholipid: DPPC and DSPC are composed of saturated fatty acids, which carbon numbers are 16 and 18, respectively. Since the structural stability of lipid molecules with long saturated fatty acid chains is high and the corresponding fluidity becomes low [30], the interaction strength of DPPC is larger than that of DSPC. It is noted from Figure 5 that the RF values of DPPC are relatively higher than those of DSPC, suggesting that the difference of liposome-target protein interaction was caused by that between the species of phospholipid. On the other hand, in Figure 5, it should be noted that the highest RF corresponds to a median concentration, not the highest one. One of the reasons is considered to be that chronological behavior of the calcein leakage would not be always a monotonous increase with time and not proportional to the concentration. It is possible that the behavior would show a specific state of interaction between each different phospholipid and target protein. Another reason may be also due to the fluctuations between different measurements with the same phospholipid species and target protein. Therefore, the above result suggests that it is questionable whether we can obtain a reliable correlation from only simple RF plots. We think that an effective approach is to improve the statistical procedure.

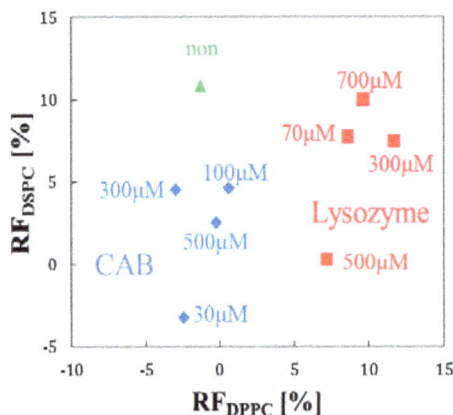

Figure 5. Scatter plot of averaged RF's of DSPC and DPPC after 60 min from the initial addition of target proteins.

The PC group shows neutrality and no electrostatic interaction ability, therefore it is inert to many proteins, suggesting low sensitivity, although we have used the neutral PC membrane because of the simplicity of its experimental treatment and application in sensor devices. As one of effective method, we have proposed the addition of cholesterol into the PC membrane, mentioned above. The exploration and development of different species and surface modification of phospholipid membranes are still preliminary at the present stage.

Here we should consider an osmotic effect due to the high levels of proteins found in the extravesicular environment. Firstly, the possibility of osmotic effect due to the addition of high levels of proteins cannot be ruled out from all our experiments. Osmotic effects depend on the protein concentration according to the van't Hoff equation: $\Pi = C_{pr}RT$ where Π, osmotic pressure; C_{pr}, protein concentration; R, gas constant; T, absolute temperature. If the calcein leakage observed in this work was simply induced by the osmotic effect due to the addition of proteins, the RF value obtained for CAB (300 μM) should be same as that for lysozyme (300 μM). Meanwhile, the time course of RF value for CAB showed was different from that seen for lysozyme, as shown in Figure 4. Thus, the difference in RF value between CAB and lysozyme suggests a difference in the corresponding protein-liposome interactions.

We have previously investigated the protein-liposome interaction by other techniques. We immobilized PC-liposome entrapping calcein into gel beds. CAB and lysozyme were loaded into the PC-liposome-immobilized gel bed. The elution time of CAB was definitely different from lysozyme under strong chemical conditions (such as pHs or denaturants) although small differences in elution time between both was observed under mild conditions (neutral pHs) [31]. Under such mild conditions, we detected the protein-liposome interaction using dielectric dispersion analysis [32]. Protein-liposome interaction (protein 10 μM, lipid 10 mM) could be detected by the reduction of headgroup mobility of PC in the liposomal interface. The dielectric dispersion analysis made it possible to detect the interaction under mild conditions, although this technique requires experimental conditions that include high lipid concentrations and low ion strength conditions. Consequently, we considered that the protein-liposome interaction, that might involve the osmotic effect to a certain extent, could be detected even in the case of high levels of proteins (300 μM).

3.4. Evaluation by Principal Component Analysis

Next, we tried to analyze the obtained data by principal component analysis (PCA). PCA is a dimension reduction approach that has widely used to visualize high dimensional data in metabolomics [33,34]. PCA is a statistical procedure that transforms the data with many variables into a few composition variables without correlation with each other. We select several principal components from the greatest dispersions and analyze the components. The aim of PCA is grouping those correlated variables and replacing the original descriptors by new set called principal components [35]. The principal component scores are described as in Equation (2), where ω_{ij} is weighted proportion with the condition of Equation (3) [10,11]:

$$PC_i = \omega_{i1} RF_{DPPC} + \omega_{i2} RF_{DSPC} + \omega_{i3} RF_{DPPC/Chol.} \tag{2}$$

$$\omega_{i1}{}^2 + \omega_{i2}{}^2 + \omega_{i3}{}^2 = 1 \tag{3}$$

We used a correlation matrix to analyze the measured RF by PCA, thereby we calculated principal component score and finally we created a PCA score plot. Figure 6 shows a PCA score plot, where averaged results (N = 3) are plotted for each protein as a parameter of concentration. It is found that the two target proteins with different concentrations are clearly discriminated. As cumulative contribution ratios larger than 90% are obtained for both PC1 and PC2, we consider that this analyzed result is sufficient to discriminate between the species and concentrations of target proteins.

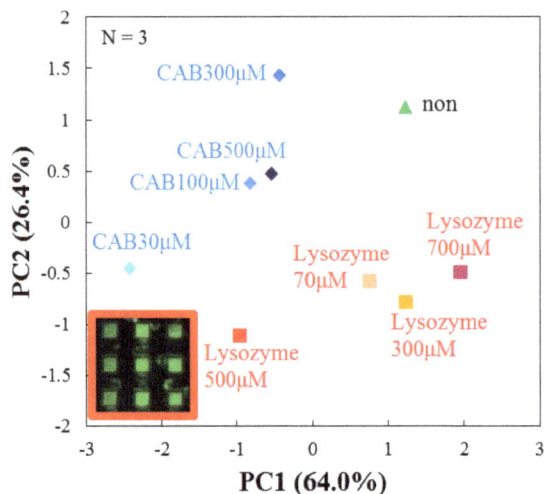

Figure 6. PCA score plot (N = 3) from averaged RF's of DPPC, DSPC and DPPC/cholesterol after 60 min from the initial addition of target proteins by a single-shot of arrayed microwells.

3.5. Performance Analysis of the Array Sensor

Moreover, we examined the performance of arrayed microwells by comparing them with a single microwell. We measured the dispersion across multiple single microwell shots and that across a single-shot of arrayed microwells. Multiple single microwell shots were carried out by sequentially measuring targets in the central microwell of the array chip, focusing on the dispersion that is dependent on different measurements. Single-shots of arrayed microwells were done by simultaneously measuring 9 (=3 × 3) targets in all the microwells of the array, focusing on the dispersion that was dependent on the spatial position of each microwell. Then we compared the dispersions obtained by the two methods. Firstly, we compared the results of the two different measurements analyzed by PCA. Figure 7a shows the result of multiple shots of a single microwell (N = 3). Figure 6 shows again the result of single-shots of arrayed microwells (N = 3). In the case of multiple single microwell shots (Figure 7a), it cannot discriminate the species of target proteins along PC1, because CAB 100 μM cannot be separated from lysozyme 500 μM. The discrimination of the species along PC2 is not good for CAB 100 μM, lysozyme 300 μM, lysozyme 500 μM and no protein. Thus, it is clear that the dispersion dependence on different measurements with single microwells is considerably high. On the other hand, in the case of a single-shot of arrayed microwells in Figure 6, the species are more clearly discriminated along PC1 than the case in Figure 7. Therefore, the results suggest that the dispersion of single-shot arrayed microwells is larger than that of multiple shots of a single microwell in terms of PCA.

Next, we considered the number of measurements for the single-shot of arrayed microwells. Figures 6 and 7b show the results of single-shot of arrayed microwells with N = 3 and 1. For the case of N = 1 (Figure 7b), the protein discrimination is unclear not only along PC1 but also PC2 because some points gather in the center area. However, for N = 3 (Figure 6), the discrimination of target proteins is clearer than that for N = 1, as the plots for CAB maintain a sufficient distance from those for lysozyme. Accordingly, it becomes possible to discriminate the target species and also their concentrations by a single-shot of the array sensor because RFs are well averaged as the number of measurements increases from N = 1 to 3. We again note that the discrimination of protein is more feasible by the arrayed microwells than the single microwell.

Sensors **2017**, *17*, 1630

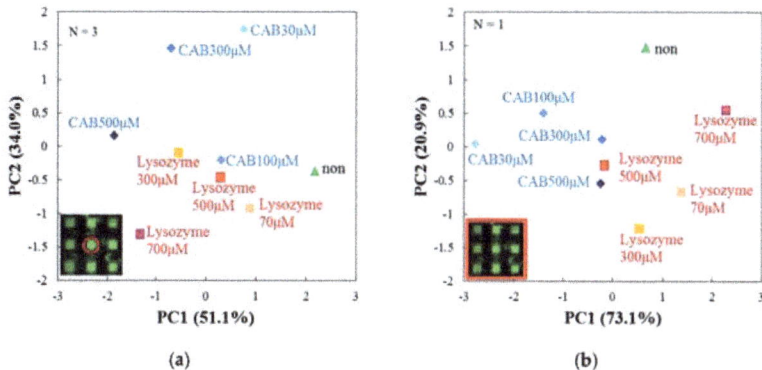

Figure 7. (a) PCA score plot (N = 3) of averaged RF's for multi-shots of single microwell; (b) PCA score plot (N = 1) of RF's for single-shot of arrayed microwells. The RF's are for DPPC, DSPC and DPPC/Cholesterol after 60 min from the initial addition of target proteins.

At the present stage, we have only detected and discriminated high concentrations of target proteins, much different from the detection level possible with ELISA. The sensitivity and discriminability (specific interaction strength) of phospholipids against target molecules is intrinsic and the most important among all the components of this sensor system. Therefore, the work on exploration and development of specific phospholipid and/or surface modification with functional biomolecules such as specific antibodies, sugar chains and so on should be continued. Also, it is inevitably necessary to increase the number of different targets to prove the specificity of this sensor, as we have discriminated only two types of proteins.

4. Conclusions

A fluorescent array sensor utilizing different phospholipid liposomes encapsulating calcein molecules was fabricated to discriminate target proteins and their concentrations. The fluorescence of calcein leaked from inside the liposomes as a result of liposome-protein interactions was detected by a developed photometric system. We examined the proper concentration of calcein encapsulated in the liposome by measuring the fluorescence intensity of calcein as a function of its concentration. Owing to the individual liposome-target protein interactions, the fluorescence intensity time courses obtained from different phospholipid liposomes became different.

We confirmed that: (1) incorporating cholesterol into the phospholipid membrane is an effective way to improve the fluidity of the membrane and the resultant fluorescence, and (2) the fluorescence behaviours of different liposome phospholipids are different. From the measured RF's, PCA successfully discriminated the species and concentration of target proteins (CAB and lysozyme). Also, the protein discrimination by the arrayed sensor with a single-shot was better than that with multiple shots of a single microwell. At the present stage the system capability allows one to approximate the order of concentration of samples, therefore, we hope the next step is to figure out the concentration of unknown tested samples. Finally, we believe that the developed label-free liposome fluorescent arrayed sensor is effective to discriminate different target proteins with different concentrations.

Acknowledgments: This research was supported in part by a Grant-in-Aid for Scientific Research (KAKENHI Grant No. 25249048 and 26630157) from the Japan Society for the Promotion of Science (JSPS).

Author Contributions: R.I., T.S. and M.N. conceived and designed the experiments; R.I. performed the experiments; R.I. and N.M. analyzed the data; T.S. contributed reagents/materials; K.Y. contributed the device process; M.F. contributed the construction of sensing system and analysis tools; R.I. and M.N. wrote the paper.

Conflicts of Interest: The authors declare no conflict of interest.

References

1. Tachi, T.; Kaji, M.; Tokeshi, M.; Baba, Y. Microchip-based immunoassay. *Anal. Sci.* **2007**, *56*, 521–534. [CrossRef]
2. Nyberg, L.; Persson, F.; Åkerman, B.; Westerlund, F. Heterogeneous staining: A tool for studies of how fluorescent dyes affect the physical properties of DNA. *Nucleic Acids Res.* **2013**, *41*, 1–7. [CrossRef] [PubMed]
3. Jung, D.; Min, K.; Jung, J.; Jang, W.; Kwon, Y. Chemical biology-based approaches on fluorescent labeling of proteins in live cells. *Mol. BioSyst.* **2013**, *9*, 862–872. [CrossRef] [PubMed]
4. Sapsford, K.E.; Sun, S.; Francis, J.; Sharma, S.; Kostov, Y.; Rasooly, A. A fluorescence detection platform using spatial electroluminescent excitation for measuring botulinum neurotoxin A activity. *Biosens. Bioelectron.* **2008**, *24*, 618–625. [CrossRef] [PubMed]
5. Van Weemen, B.K.; Schuurs, A.H. Immunoassay using antigen-enzyme conjugates. *FEBS Lett.* **1971**, *15*, 232–236. [CrossRef]
6. Hendrickson, O.D.; Skopinskaya, S.N.; Yarkov, S.P.; Zherdev, A.V.; Dzantiev, B.B. Development of liposome immune lysis assay for the herbicide atrazine. *J. Immunoass. Immunochem.* **2004**, *25*, 279–294. [CrossRef]
7. Tomioka, K.; Nakayama, T.; Kumada, Y.; Katoh, S. Characterization of competitive measurement of antigens by use of antigen-coupled liposomes. *Kagaku Kogaku Ronbunnshu* **2005**, *31*, 346–351. [CrossRef]
8. Kumada, Y.; Tomioka, K.; Katoh, S. Characteristics of liposome immunosorbent assay (LISA) using liposomes encapsulating cpenzyme β–NAD⁺. *J. Chem. Eng. Jpn.* **2001**, *34*, 943–947. [CrossRef]
9. Takada, K.; Fujimoto, T.; Shimanouchi, T.; Fukuzawa, M.; Yamashita, K.; Noda, M. A new discrimination method of target biomolecules with miniaturized sensor array utilizing liposome encapsulating fluorescent molecules with time course analysis. In Proceedings of the 17th International Conference on Miniaturized Systems for Chemistry and Life Sciences, MicroTAS, Freiburg, Germany, 27–31 October 2013; pp. 194–196.
10. Shimanouchi, T.; Oyama, E.; Thi Vu, H.; Ishii, H.; Umakoshi, H.; Kuboi, R. Membranomics research on interactions between liposome membranes with membrane chip analysis. *Membrane* **2009**, *34*, 1–9.
11. Shimanouchi, T.; Oyama, E.; Thi Vu, H.; Ishii, H.; Umakoshi, H.; Kuboi, R. Monitoring of membrane damages by dialysis treatment: Study with membrane chip analysis. *Desalination Water Treat.* **2010**, *17*, 45–51. [CrossRef]
12. Shimanouchi, T.; Ishii, H.; Yoshimoto, N.; Umakoshi, H.; Kuboi, R. Calcein Permeation across phosphatidylcholine bilayer membrane: Effects of membrane fluidity, liposome size, and immobilization. *Colloids Surf. B* **2009**, *73*, 156–160. [CrossRef] [PubMed]
13. Kuboi, R.; Shimanouchi, T.; Yoshimoto, M.; Umakoshi, H. Detection of protein conformation under stress conditions using liposomes as sensor materials. *Sens. Mater.* **2004**, *16*, 241–254.
14. Hertog, A.L.D.; Sang, H.W.F.; Kraayenhof, R.; Bolscher, J.G.M.; Hof, W.V.; Veerman, E.C.I.; Amerongen, A.V.N. Interactions of histatin 5 and histatin 5-derived peptides with liposome membranes: Surface effects, translocation and permeabilization. *Biochem. J.* **2004**, *379*, 556–672. [CrossRef] [PubMed]
15. Davidsen, J.; Jørgensen, K.; Andresen, T.L.; Mouritsen, O.G. Secreted phospholipase A₂ as a new enzymatic trigger mechanism for localised liposomal drug release and absorption in diseased tissue. *Biochim. Biophys. Acta Biomembr.* **2003**, *1609*, 95–101. [CrossRef]
16. Katsu, T.; Imamura, T.; Komagoe, K.; Masuda, K.; Mizushima, T. Simultaneous measurements of K+ and calcein release from liposomes and the determination of pore size formed in a membrane. *Anal. Sci.* **2007**, *23*, 517–522. [CrossRef] [PubMed]
17. Bi, X.; Wang, C.; Ma, L.; Sun, Y.; Shang, D. Investigation of the role of tryptophan residues in cationic antimicrobial peptides to determine the mechanism of antimicrobial action. *J. Appl. Microbiol.* **2013**, *115*, 663–672. [CrossRef] [PubMed]
18. Gopal, R.; Lee, J.K.; Lee, J.H.; Kim, Y.G.; Oh, G.C.; Seo, C.H.; Park, Y. Effect of repetitive lysine-tryptophan motifs on the eukaryotic membrane. *Int. J. Mol. Sci.* **2013**, *14*, 2190–2202. [CrossRef] [PubMed]
19. Komura, S.; Imai, M. Physics of Heterogeneous structures in biomembranes. *J. Phys. Soc. Jpn.* **2013**, *68*, 1–10.
20. Komura, S.; Shirotori, H. Phase transition and phase separation in biomembranes. *J. Phys. Soc. Jpn.* **2005**, *60*, 128–132.
21. Sharma, N.; Litoriya, R. Incorporating Data Mining Techniques on Software Cost Estimation: Validation and Improvement. *IJETAE* **2012**, *2*, 301–309.

22. Noda, M.; Shimanouchi, T.; Okuyama, M.; Kuboi, R. A bio-thermochemical microbolometer with immobilized intact liposome on sensor solid surface. *Sens. Actuators B Chem.* **2008**, *135*, 40–45. [CrossRef]
23. Murata, N.; Imamura, R.; Du, W.; Fukuzawa, M.; Noda, M. An image processing platform for expendable bio-sensing system with arrayed biosensor and embedded imager. In Proceedings of the 3rd International Conference on Applied Computing and Information Technology/2nd International Conference on Computational Science and Intelligence, Okayama, Japan, 12–16 July 2015; pp. 239–242.
24. Hamann, S.; Kiilgaard, J.F.; Litman, T.; Alvarez-Leefmans, F.J.; Winther, B.R.; Zeuthen, T. Measurement of cell volume changes by fluorescence self-quenching. *J. Fluoresc.* **2002**, *12*, 139–145. [CrossRef]
25. Roberts, K.E.; O'Keeffe, A.K.; Lloyd, C.J.; Clarke, D.J. Selective dequenching by photobleaching increases fluorescence probe visibility. *J. Fluoresc.* **2003**, *13*, 513–517. [CrossRef]
26. Wang, T.; Smith, E.A.; Chapman, E.R.; Weisshaar, J.C. Lipid mixing and content release in single-vesicle SNARE-driven fusion assay with 1–5 ms resolution. *Biophys. J.* **2009**, *96*, 4122–4131. [CrossRef] [PubMed]
27. Chen, J.; Liu, Y.; Wang, Y.; Ding, H.; Su, Z. Different effects of L-arginine on protein refolding: Suppressing aggregates of hydrophobic interaction, not covalent binding. *Biotechnol. Prog.* **2008**, *24*, 1365–1372. [CrossRef] [PubMed]
28. Touch, V.; Hayakawa, S.; Fukada, K.; Aratani, Y.; Sun, Y. Preparation of antimicrobial reduced lysozyme compatible in food applications. *J. Agric. Food Chem.* **2003**, *51*, 5154–5161. [CrossRef] [PubMed]
29. Zhang, Z.; Sohgawa, M.; Yamashita, K.; Noda, M. A micromechanical cantilever-based liposome biosensor for characterization of protein-membrane interaction. *Electroanalysis* **2016**, *28*, 620–625. [CrossRef]
30. Oku, N. Novel development of liposome application for development of artificial cell. *Anal. NTS Press.* **2005**, *2005*, 7–8.
31. Yoshimoto, N.; Yoshimoto, M.; Yasuhara, K.; Shimanouchi, T.; Umakoshi, H.; Kuboi, R. Evaluation of temperature and guanidine hydrochloride-induced protein–liposome interactions by using immobilized liposome chromatography. *Biochem. Eng. J.* **2006**, *29*, 174–181.
32. Shimanouchi, T.; Yoshimoto, N.; Hiroiwa, A.; Nishiyama, K.; Hayashi, K.; Umakoshi, H. Relationship between the mobility of phosphocholine headgroup and the protein-liposome interacation: A dielectric sperctoscopic study. *Colloids Surf. B* **2014**, *116*, 343–350. [CrossRef] [PubMed]
33. Yamamoto, H.; Fujimori, T.; Sato, H.; Ishikawa, G.; Kami, K.; Ohashi, Y. Statistical Hypothesis Testing of Factor Loading in Principal Component Analysis and Its Application to Metabolite Set Enrichment Analysis. *BMC Bioinform.* **2014**, *15*, 1–9. [CrossRef] [PubMed]
34. Ma, S.; Dai, Y. Principal component analysis based methods in bioinformatics studies. *Brief. Bioinform.* **2011**, *12*, 714–722. [CrossRef] [PubMed]
35. Ferreira, M.C. Multivariate QSAR. *J. Braz. Chem. Soc.* **2002**, *13*, 742–753. [CrossRef]

sensors

MDPI

Article

Thiolate-Capped CdSe/ZnS Core-Shell Quantum Dots for the Sensitive Detection of Glucose

Samsulida Abd. Rahman [1,2], Nurhayati Ariffin [1], Nor Azah Yusof [2,3,*], Jaafar Abdullah [2,3], Faruq Mohammad [4,*], Zuhana Ahmad Zubir [5] and Nik Mohd Azmi Nik Abd. Aziz [5]

1   Industrial Biotechnology Research Centre (IBRC), SIRIM Berhad, No. 1, Persiaran Dato' Menteri, Section 2, P.O. Box 7035, 40700 Shah Alam, Selangor, Malaysia; sulida@sirim.my (S.A.R.); ahayati@sirim.my (N.A.)
2   Advanced Materials and Nanotechnology Laboratory, Institute of Advanced Technology, Universiti Putra Malaysia, 43400 UPM Serdang, Malaysia; jafar@upm.edu.my
3   Department of Chemistry, Faculty of Science, Universiti Putra Malaysia, 43400 UPM, Serdang, Malaysia
4   Surfactant Research Chair, Department of Chemistry, College of Science, King Saud University, P.O. Box 2455, Riyadh 11451, Saudi Arabia
5   Advance Material Research Centre (AMREC), SIRIM Berhad, Lot 34, Jalan Hi-Tech 2/3, Kulim Hi-Tech Park, 09000 Kulim, Malaysia; zuhana@sirim.my (Z.A.Z.); nikazmi@sirim.my (N.M.A.N.A.A.)
*   Correspondence: azahy@upm.edu.my (N.A.Y.); farooqm1983@gmail.com (F.M.); Tel.: +6-038-946-6782 (N.A.Y.)

Academic Editor: Sheshanath Bhosale
Received: 18 April 2017; Accepted: 30 May 2017; Published: 1 July 2017

Abstract: A semiconducting water-soluble core-shell quantum dots (QDs) system capped with thiolated ligand was used in this study for the sensitive detection of glucose in aqueous samples. The QDs selected are of CdSe-coated ZnS and were prepared in house based on a hot injection technique. The formation of ZnS shell at the outer surface of CdSe core was made via a specific process namely, SILAR (successive ionic layer adsorption and reaction). The distribution, morphology, and optical characteristics of the prepared core-shell QDs were assessed by transmission electron microscopy (TEM) and spectrofluorescence, respectively. From the analysis, the results show that the mean particle size of prepared QDs is in the range of 10–12 nm and that the optimum emission condition was displayed at 620 nm. Further, the prepared CdSe/ZnS core shell QDs were modified by means of a room temperature ligand-exchange method that involves six organic ligands, L-cysteine, L-histidine, thio-glycolic acid (TGA or mercapto-acetic acid, MAA), mercapto-propionic acid (MPA), mercapto-succinic acid (MSA), and mercapto-undecanoic acid (MUA). This process was chosen in order to maintain a very dense water solubilizing environment around the QDs surface. From the analysis, the results show that the CdSe/ZnS capped with TGA (CdSe/ZnS-TGA) exhibited the strongest fluorescence emission as compared to others; hence, it was tested further for the glucose detection after their treatment with glucose oxidase (GOx) and horseradish peroxidase (HRP) enzymes. Here in this study, the glucose detection is based on the fluorescence quenching effect of the QDs, which is correlated to the oxidative reactions occurred between the conjugated enzymes and glucose. From the analysis of results, it can be inferred that the resultant GOx:HRP/CdSe/ZnS-TGA QDs system can be a suitable platform for the fluorescence-based determination of glucose in the real samples.

Keywords: Cdse-ZnS; core-shell quantum dots; semiconducting; glucose sensing; surface modification

1. Introduction

In general, the nanomaterials maintain the special characteristics, which include the substantial surface area, superior reaction surface activity, and higher catalytic efficiency, to mention some [1]. For these reasons, the nanomaterials are considered to be the prospective transducers in enzyme-based bioconjugated sensor-related purposes. The substantial surface area of nanomaterials allows for the efficient adsorption of enzymes to the solid surfaces, in addition to minimizing the enzyme aggregation and protein unfolding, thereby supporting the formation of a more stable enzyme-loaded nanoparticulate system [2,3]. For the majority of biosensor-related applications, the previously reported nanomaterials include gold [4], carbon nanotubes [5], magnetic iron oxide [6], titania [7], silica [8], and quantum dots (QDs) [9–14]. Among all these nanomaterials, QDs have been found to be more favorable for sensor applications as they exhibit broad excitation and narrow emission wavelengths, in addition to allowing their emission wavelengths to be fine-tuned. Additionally, the other attractive features of QDs, i.e., their extremely luminescent and photoresistant properties (due to high surface-to-volume ratio, catalytic efficiency, and reaction activity surface), might be particularly useful in the biosensors sector [1]. The water-soluble QDs have given rise to an increasing passion towards the biosensors and bioimaging because of their biocompatible characteristics in the physiological medium. In addition, the utilization of QDs for the development of enzyme-conjugated systems help in two different ways, i.e., (1) by providing a strong solid support for the immobilization of the enzymes and (2) by acting as a fluorescence sensing probe while taking advantage of the changes in fluorescence intensity [15]. The fluorescence emission wavelengths of QDs are mostly influenced by the changes in particle sizes and surface covered ligands as the charges are strongly inclined by the nature of ligands and all of which finally responsible for the biomolecule interactions [16].

In healthcare and biomedical sector, most of the scientific work reported to date deals with the usage of QDs for diagnostic and biosensory systems such as cancer cell labelling [17], diseased cell imaging [18], drug delivery, and virus detection [19], to mention some. Apart from that, QDs are also the most familiar agents for the sensory detection of glucose levels, and one most popular QD material in that aspect is the CdSe/ZnS core-shell QDs. It was reported that the CdSe/ZnS core-shell QDs are extremely high fluorescence intensity agents in the visible spectrum and that their enhancement in the chemical and photostability is mainly due to the ZnS outer layer [20–22]. The common method for the synthesis of CdSe/ZnS core-shell QDs involves a high temperature organic solvent method, where the particles formed are stabilized in hydrophobic solvents like trioctyl phosphine oxide (TOPO) and oleic acid (OA). Since the QDs formed by this approach result in the generation of hydrophobic particles and in order to avoid this, several methods have been studied to substitute TOPO and/or OA groups with other organic ligands with a hydrophilic nature. However, the exchange of TOPO and/or OA with other organic ligands has been found to generate problems such as a lower quantum yield, a loss of fluorescence intensity, larger-sized particles, and a compromising of biocompatibility [23–26]. Gill et al. (2005) previously reported that the quantum yield of their synthesized QDs was reduced to 50% as the CdSe/ZnS QDs were modified with mercapto-propionic acid (MPA) [24]. In a similar way, Stsiapura et al. (2006) found from their studies that the quantum yield of their synthesized QDs depreciated twofold when mixtures of mercapto-succinic acid (MSA) and thioglycolic acid (TGA; or mercapto-acetic acid (MAA)) were used as ligand exchangers [25]. In addition to quantum yield loss, the QDs were also reported to exhibit less colloidal stability, as they have a greater tendency to form aggregates rather than residing individually in the dispersing medium.

Based on the facts about the synthesis and stability of QDs in the colloidal medium, the present work is aimed to prepare the CdSe/ZnS core-shell QDs with water-soluble behavior and can be developed to serve as a glucose biosensor. For that, we first prepared the CdSe/ZnS core-shell QDs using a high temperature organic solvent method that uses hydrophobic TOPO and OA as capping agents in order to stabilize the particles as soon as they are formed [26]. Further, to enhance their dispersibility in aqueous solutions, the ligand exchange method was applied to modify the surface of QDs, in addition to fine-tuning the surface to match for biosensing applications. For that, the heterobifunctional ligands such as the mercapto-carbonic acid and thiolated ligands were used to replace the TOPO/OA, where the mercapto or thiolated end is bound to the surface of QDs so that the carboxyl moiety remains free and can offer enough water solubility [19,27]. In general, the presence of high molecular weight compounds on the surface restricts the usefulness of the core-shell QDs towards biosensing applications, and this is due to the blockage of electron transfer reaching the core QDs. In our case, the coating of QDs with mercapto-carbonic or thiolated ligands helps to prevent such limitations of electron movement, thereby serving as the best water solubilizing shell to the CdSe/ZnS QDs. Thus, the formed composite was thoroughly characterized by means of transmission electron microscopy (TEM), fluorescence measurements, and further tested to see their effects for the detection of glucose following the QDs loading with that of glucose oxidase enzyme (GOx) and horseradish peroxidase (HRP) enzymes.

2. Materials and Methods

2.1. Chemicals

N-Ethyl-3-(3-dimethylaminopropyl) carbodiimide (EDC), N-hydroxy sulfosuccinimide (sulfo-NHS), glucose oxidase (GOx), and horseradish peroxidase (HRP) were purchased from Sigma (Selangor, Malaysia). The solution of PBS buffer, pH 7, were prepared in house by the addition of disodium hydrogen diphosphate (Na_2HPO_4) and sodium dihydrogen phosphate (NaH_2PO_4) until it reaches pH 7. Both Na_2HPO_4 and NaH_2PO_4 were purchased from Scharlau (Selangor, Malaysia). All the starting materials for enzyme conjugation and glucose detection ordered were of highest grade and were used directly without any further purification.

2.2. Synthesis of CdSe/ZnS-TGA QDs

For the synthesis, we first started with the preparation of 0.1 M Cd and Se precursor solutions separately; and for that about 150 mg of Se powder was dissolved in 20 mL of tri-*n*-octylphosphine (TOP), stirred at 200 °C temperature in nitrogen atmosphere until the solution becomes clear. Similarly, the Cd precursor was prepared by mixing 160 mg of cadmium oxide (CdO), 3.5 mL of OA and 10 mL of 1-octadecene (ODE) for 30 min at 150 °C and further heated up to 280 °C in nitrogen atmosphere. When we see the solution becoming clear, the temperature was reduced to 225 °C and then added about 5 mL of Se precursor solution to the hot flask, stirred for 10 min and stopped the heating process. Now we added about 15 mL of methanol to the reaction mixture so as to stop the size growth and then washing with acetone solution was continued in order to remove any organics, and the washing process was repeated until the solution becomes opaque. The solid precipitate obtained after the centrifugation at 5000 rpm for 15 min was collected by discarding the supernatant, dried in vacuum before using for the shelling process. Similarly, 0.1 M of zinc precursor solution was prepared by adding 0.13 g of ZnO to the mixture of OA and ODE (1.5 mL and 13.5 mL respectively) at 200 °C. Additionally, the sulfur precursor solution (0.1 M) was prepared by dissolving sulfur (0.02 g) in ODE (15 mL) at 200 °C. All the precursor solutions were maintained under nitrogen atmosphere until further use.

For obtaining the CdSe/ZnS (core/shell) QDs, about 0.1 g of CdSe (dissolved in hexane) was added to a solution mixture containing 0.6 mL of ODE and 0.2 mL of OA in a 25 mL reaction vessel, heated up to 100 °C for 30 min. Following the period, nitrogen flow was maintained into the reaction vessel for another 1 h at 100 °C so as to remove the hexane and any other undesired materials of low vapor pressure, and then the solution was heated to 160 °C to allow for shell growth. The amounts of solutions injected into the reaction mixture at each step were as follows: We started with 0.2 mL of Cd and S solutions for the first layer and the temperature was slowly raised to ~180 °C for 5 min. For the second layer, 0.4 mL of each solution and the temperature was slowly raised to ~200 °C for 5 min, and similarly 0.6 mL of each solution for the third layer at ~220 °C temperature for ~5 min, 0.8 mL of each solution for the fourth layer at ~240 °C for 5 min, and 1 mL of each for the fifth layer at 260 °C for 30 min. For each injection, small aliquots were extracted from the mixture after 5 min and mixed with toluene to record the absorption spectrum. Methanol was added to stop the growth, and the mixture was purified using the same method as the core.

In order to convert CdSe/ZnS QDs containing OA ligand (organic soluble) into aqueous dispersible, we first dissolved the CdSe/ZnS powder in toluene, which was then supplemented with ethanol in a 1:1 ratio (i.e., 1 mL of CdSe/ZnS in toluene with 1 mL of ethanol) followed by shaking. After that, the solution mixture was centrifuged at 3000 rpm for 15 min, the supernatant was discarded, and the precipitate was washed with ethanol, and the centrifugation process was repeated 2–3 times. After the required centrifugation, the precipitated product was collected and re-dissolved in toluene; to this, a QD solution, an excess amount of thiol-terminated ligands (*L*-cysteine, *L*-Histidine, TGA, MPA, MSA and MUA), was added and sonicated for 30 min. After the sonication, the mixture was kept at room temperature for one day and then centrifuged at 5000 rpm for 5 min and the obtained precipitate was dried in a vacuum desiccator for 1 h. Finally, the CdSe/ZnS-capped thiolated ligands were dissolved in the PBS pH 7, and the mixture was then centrifuged at 5000 rpm for 5 min, the supernatant was removed, and the collected precipitate was dried in a vacuum desiccator for 1 h. Thus, the obtained CdSe/ZnS capped with respective ligands with an aqueous dispersible nature was re-dissolved in PBS pH 7 until further use.

2.3. Conjugation of CdSe/ZnS-TGA QDs with GOx:HRP Enzymes in an Aqueous System

For the conjugation of enzymes to the QDs, we first added 2.3 μL of EDC and 1 mg of NHS to a beaker containing 5 mL of CdSe/ZnS-TGA. Both solutions were mixed together with a magnetic stirrer for 30 min and then supplemented with 25 μL of GOx and 10 μL of HRP. The stirring was continued for another 2 h before the centrifugation process at 12,000 rpm for 10 min. Following the centrifugation, the supernatant was separated out and the precipitate was re-dissolved in PBS pH 7 and stored at 4 °C for the next usage. All characterizations of the GOx:HRP-conjugated CdSe/ZnS-TGA QDs were performed using the spectrofluorescence technique. The characterization of enzyme conjugation and interaction were also evaluated using different mixing ratios of QDs and enzyme (1:1, 1:2, 1:3, 1:4) in the assay system. Further, the analytical performance of the bioconjugated QDs-enzyme system was evaluated by analyzing the stability, interference, repeatability, and reproducibility.

2.4. Instrumental Analysis

The Ultraviolet–Visible (UV-Vis) absorption spectroscopic measurements were performed on Perkin Elmer LAMBDA 25 UV/VIS system and the samples were prepared by dissolving the CdSe/ZnS particles in an aqueous solution and were run in the wavelength range of 300–800 nm with an absorbance less than 0.1 at a 480 nm wavelength. For the photoluminescence study, the fluorescence spectrums were recorded using the Ocean Optics QE65000 and Flouromax-4 (Horiba Jobin Yvon) spectrophotometers in the wavelength range of 500–700 nm. The samples for the spectrofluorescence study were placed in a quartz cuvette, and the emission spectrum of each sample was measured using the EFOS Novacure spectrophotometer with a mercury light source; all the samples were excited at a wavelength of 375 nm. For the transmission electron microscopy (TEM) analysis, a Philips Technai20 instrument connected with an EDAX analyzer was used (operating at 200 KV, 100X to 1000KX range). The samples for TEM analysis were prepared by placing tiny drops of QD suspension in water on a carbon-coated 400 mesh copper grid.

3. Results and Discussion

The basic principle for the detection of glucose by involving the CdSe/ZnS-capped TGA nanomaterials is shown in Figure 1. Here, the determination of glucose in this research depends on the enzymatic reaction of glucose and the effect of presence of H_2O_2 on the intensity of QDs fluorescence. In the presence of glucose, GOx loaded onto the substrate catalyzes the available glucose into gluconic acid by means of an oxidation process using the oxygen molecule as an electron acceptor and in turn produces H_2O_2 simultaneously. The exchange of the electron happens at the outer surface of the core-shell QDs, whereby H_2O_2 is reduced to oxygen and H_2O, which traps the electron holes at the surface of the QDs. This further results in the generation of non-fluorescent QDs and in some instances causes a reduction in the fluorescence intensity, i.e., the higher glucose concentrations used, the more H_2O_2 produced and thus the greater the quenching effect.

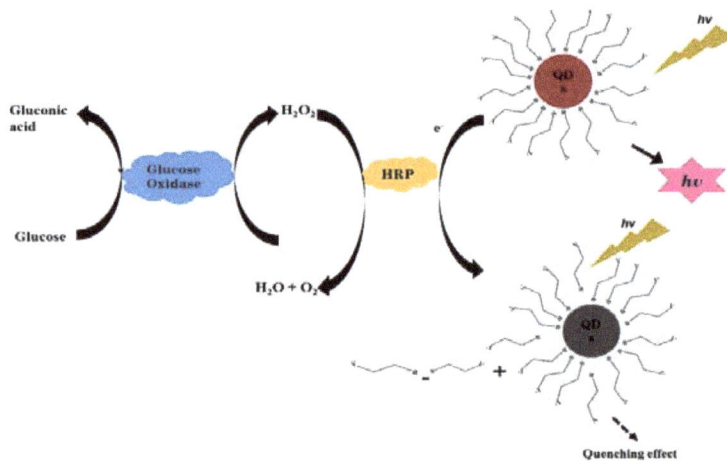

Figure 1. Schematic representation of successive reactions involved in the sensitive detection of glucose by means of GO_X:HRP/CdSe/ZnS-TGA QD system.

TEM images of both CdSe core and CdSe/ZnS core-shell QDs are shown in the Figure 2a,b and from these figures, it can be seen that the sizes of the CdSe and CdSe/ZnS particles are in the range of 3–3.2 nm and 10–12 nm, respectively. It can also be seen from the images that the particles are uniform and monodispersed, thereby providing evidence for the successive coating of the QDs with that of ligands.

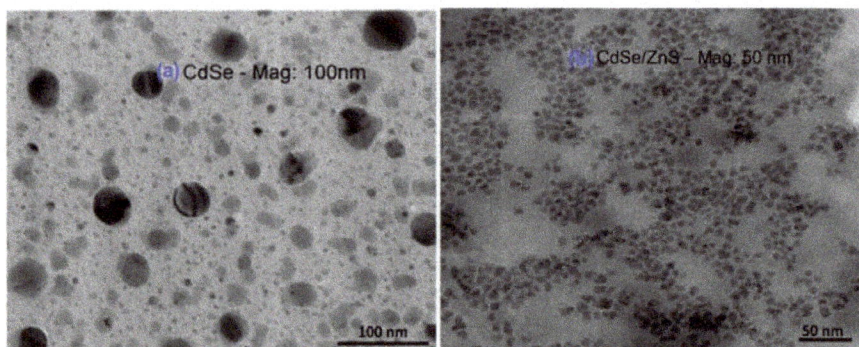

Figure 2. TEM results of (**a**) CdSe core, and (**b**) CdSe/ZnS core-shell QDs with the magnifications of 100 nm and 50 nm, respectively.

For the photoluminescence (PL) measurements in general, the samples are usually dissolved in a non-polar solvent, and for this reason, we have chosen hexane solvent as the dispersing medium for our synthesized QDs. The PL spectrum of the QDs sample before and after its coating with the ZnS shell is shown in Figure 3a. It can be seen from Figure 3a that the PL peak for the CdSe core was observed to be around 532.5 nm, and the full width at half maximum (FWHM) was 28.5 nm, thereby indicating for the formation of the monodispersed nanocrystals. However, the PL peak for the CdSe/Zns core-shell QDs was observed at 572.5 nm, and the PL intensity observed to be quite higher for the encapsulation. Further, the wavelength shifted to about 40 nm between the core and the core/shell type, which may be due to the reheating of core before encapsulating that cause an increase in the diameter of core.

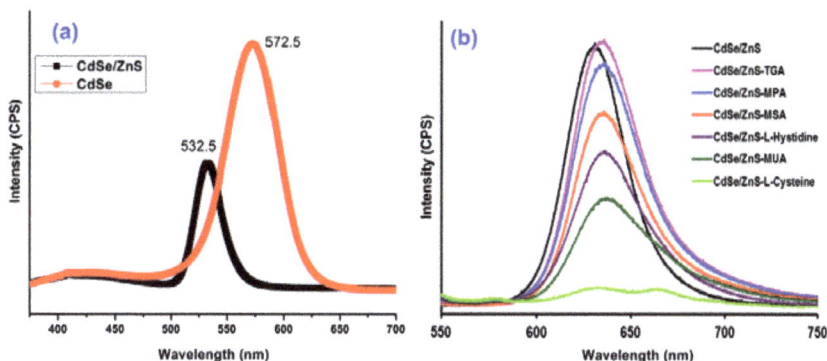

Figure 3. (a) Fluorescence intensity peaks for CdSe and CdSe/ZnS QDs when recorded using PL measurements and (b) fluorescence for the CdSe/ZnS QDs with different surface ligands (TGA, MPA, MSA, *L*-Histidine, MUA, and *L*-Cysteine) using spectrofluorescence measurement.

Further, in order to make the particles water-dispersible, the surface of CdSe/ZnS core-shell QDs (OA organic capping) was replaced with six different hydrophilic ligands (TGA, MPA, MSA, *L*-Histidine, MUA, and *L*-Cysteine). Figure 3b shows the fluorescence intensities of the capped CdSe/ZnS-OA (without modification) as well as those of CdSe/ZnS-TGA, CdSe/ZnS-MPA, CdSe/ZnS-MSA, CdSe/ZnS-*L*-histidine, CdSe/ZnS-MUA, and CdSe/ZnS-*L*-cysteine, respectively. The fluorescent peaks in Figure 3a,b differ in absorption maxima at 572 nm and 620 nm, respectively. This difference in fluorescent peaks is due to the variation in the equipment used and their principles, i.e. we used PL measurements for the fluorescence intensity measurements between CdSe (core) and CdSe/ZnS (core-shell) (Figure 3a), while a spectrofluorometer was employed for the comparison of fluorescence intensity of CdSe/ZnS coated with different ligands. It is clearly indicated here that the fluorescent intensity of CdSe/ZnS capped with *L*-histidine, MUA, and *L*-cysteine seems to decrease compared to pure CdSe/ZnS, and the observation of such results is mostly due to the formation of agglomerated structures and insolubility when they are re-dissolved in PBS buffer, pH 7. However, the CdSe/ZnS capped with TGA, MPA, and MSA showed no signs of agglomeration and are fully dissolved in PBS buffer, pH 7, thereby indicating for their high stability in aqueous dispersions. In general, the fluorescence intensity mostly affected by the presence of surface ligands, i.e. the ligand with the longest chain or complex structure strongly reduces intensity, as compared to the one with simple chain or non-bulky group molecules. Here in our case, TGA maintains only two carbons in its chain and holds a simple structure, which was followed by MPA (three carbon chain), MSA (four carbons), *L*-histidine (four carbons and a substituent having ring structure), MUA (11 carbons), and finally *L*-cystein (3 carbons with spatial arrangement of groups) [28]. Hence, based on the surface ligand and its type, the observed order of fluorescence intensities for the CdSe/ZnS QDs with different ligands can be explained.

The fluorescent intensity of CdSe/ZnS capped with different ligands under different pH conditions can further provide evidence for the functional group changes in accordance with the solution pH. Figure 4 shows the effects of pH buffer ranging from 4–9 on the fluorescent intensity of CdSe/ZnS capped with different organic ligands (as mentioned previously). From the figure, the fluorescent intensity for CdSe/ZnS capped with TGA increased as the pH value increased from 4 to

7, and then became nearly constant when the pH value was 7 or greater. The fluorescent intensity increased 1.5 times for all three solutions when the solution pH increased from 4 to 5, from 5 to 6, and from 6 to 7, respectively. Similar to CdSe/ZnS capped with other ligands (MPA, MSA, L-histidine, MUA and L-cysteine), the fluorescent intensity increased as the solution pH value increased from 4 to 7, while the intensity became nearly constant when the pH value was greater than 7. Additionally, the intensity seems to be increased nearly 1.4 times for the CdSe/ZnS capped with MPA, MSA, and L-histidine, separately, and 1.1 times for the CdSe/ZnS capped with MUA and L-cysteine when the solution pH increased from 4 to 7. When the solution pH value was greater than 7, the mean fluorescent intensity is not affected, meaning that no changes to the surface groups are occurring.

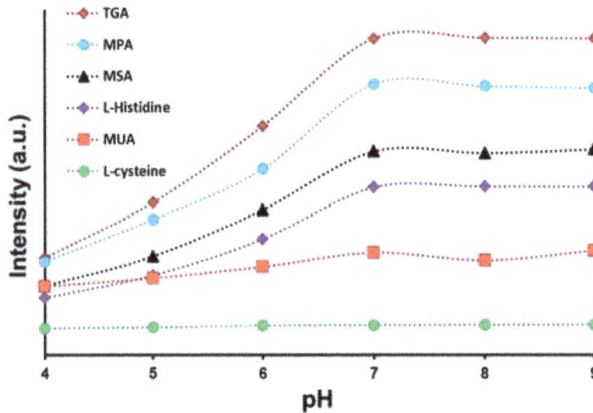

Figure 4. Fluorescence intensity for CdSe/ZnS QDs coated with different surface ligands of TGA, MUA, MPA, MSA, L-cystein, and L-histidine under different solution pH values.

We see from the results that among the three ligands (TGA, MPA, and MSA), the CdSe/ZnS-TGA sample produced the highest fluorescent intensity due to the smaller ligand size as compared to the other two samples (CdSe/ZnS-MPA and CdSe/ZnS-MSA). The chain size in these ligands follow the order of TGA < MPA < MSA, where the decrease in the aggregation levels by means of decreased steric repulsions (more stable particles in solution) can be expected and all of which contributes finally for an enhancement in the mean fluorescence intensity [2,28]. With the same size and ligand chain length principle, the observed highest fluorescence intensity of TGA (comparable to OA-capped CdSe/ZnS QDs) can be explained by relating it with that of MPA, MSA, L-histidine, MUA, and L-cysteine chain length. In addition, it can be mentioned here that the CdSe/ZnS-TGA QDs were observed to be the most efficient probes, as they are water-dispersible, biocompatible, and maintain a fluorescence behavior equal to that of the OA-capped CdSe/ZnS QDs. The biocompatible property is contributed by the OA and TGA groups, as we found from our earlier studies that any ligand which has the functional groups in its structure such as amine, thiol, or unsaturation was observed to maintain some inbuilt anti-oxidative properties. This property is strong enough to protect the cells from the toxic responses generated by the CdSe or ZnS QDs individually or in a combination form [29]. Furthermore, the ligand capping also provides the CdSe/ZnS QDs with a negative surface charge distribution that can provide direct self-assembly with other molecules with a positive charge. Theoretically, the CdSe/ZnS-TGA QDs can be extended for the detection of other analytes if appropriate conditions are established. From the analysis, therefore, we came to a conclusion that the CdSe/ZnS-TGA QDs can serve as the most efficient probes (as compared to MPA, MUA, MSA, L-histidine, and L-cysteine-coated QDs) due to its enhanced fluorescence, water solubility, and biocompatible properties, which can be further exploited for biomedical applications and, in the present case, for glucose sensing through enzyme loading.

The optimization parameters that were studied are of the effect of pH, enzyme ratio, and QD concentration, as these are very important for glucose biosensing and were analyzed using a mixture of GOx/HRP and TGA-QDs, and the results are displayed in Figure 5. Figure 5a shows the optimization of pH buffer on the fluorescence intensity quenching of the CdSe/ZnS-capped TGA. From the graph (Figure 5a), it can be seen that the quenching effect of the CdSe/ZnS-capped TGA increases with an increase in the pH buffer until it reaches an optimum condition of pH 7. This is due to the fact that, at pH 7, the glucose molecule, being in a cyclic hemiacetal form, may exist in the two isomeric forms β-D-glucopyranose (63.6%) and α-D-glucopyranose (36.4%). In association with our QDs probe, some specific binding occurs between the GOx and β-D-glucopyranose forms, while no such binding can occur with the α-D-glucopyranose form. The equilibrium state between the α-β glucopyranose forms seems to be pushed towards the β-side as the amount of consumption is more for the β-form as against the α-form of gluocopyranose. This equilibrium enabled GOx to oxidize all of the glucose in solution. Further, a drop in the intensity of the QDs after pH 7 may be due to the denaturation of the GOx enzyme in the basic environment.

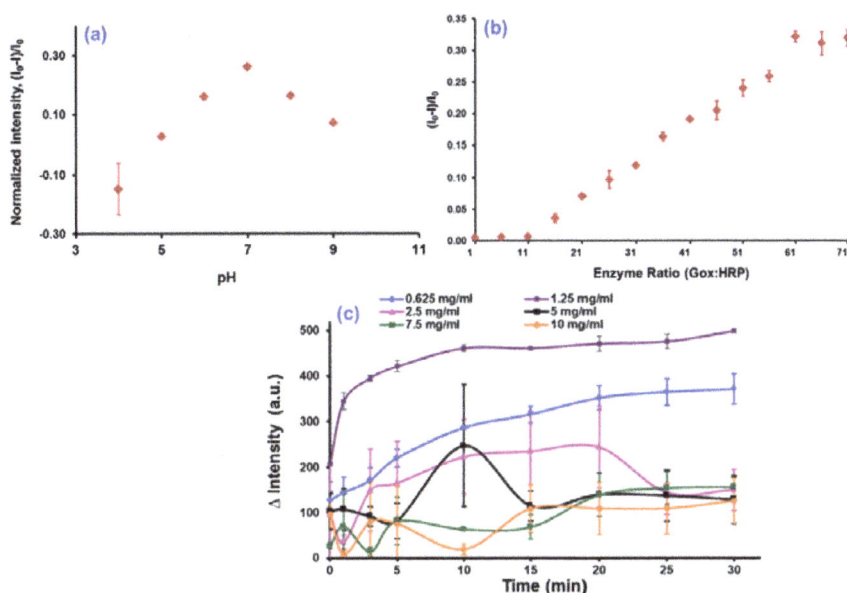

Figure 5. The effect of pH (**a**), enzyme ratio (**b**), and QD concentration (**c**) towards the fluorescence intensity.

The effect of normalized intensity towards the GOx:HRP ratio is shown in Figure 5b, and the graph shows that the optimum ratio of GOx:HRP is 3:2. We observed that the quenching effect of the QDs is greatest when the GOx:HRP enzyme ratio is maintained at 3:2, and the reflection of such an effect may be due to the stability of the system, so the same ratio was repeated for the following experiments. Further, the effect of the QDs concentration studied in the range of 0.625–10.0 mg/mL of CdSe/ZnS and is shown in Figure 5c. The results of the study (Figure 5c) indicate that the highest intensity was obtained for the 1.25 mg/mL concentration of the QDs. However, the intensity of the QDs was decreased when the highest QDs concentration was introduced into the reaction, and this may be due to the quenching effects of CdSe/ZnS at higher concentration, thereby resulting in the enzyme denaturation/degradation. Therefore, based on this study, a 1.25 mg/mL concentration of QDs, a 3:2 ratio of GOx:HRP enzymes ratio, and pH 7 were chosen as the optimal conditions for the subsequent reactions.

The reaction mechanism for glucose detection that depends on the effect of CdSe/ZnS fluorescence intensity quenching was shown to be successful, and the results are shown in Figure 6. From the figure, one can clearly see that the highest intensity was observed for the as-prepared QDs when they are maintained in the absence of GOx:HRP enzymes and glucose. However, in the presence of GOx:HRP and glucose, the reaction occurred and resulted in the highest quenching of QD intensity; thus, we observed the fluorescence intensity to be the lowest, as compared to all other combinations.

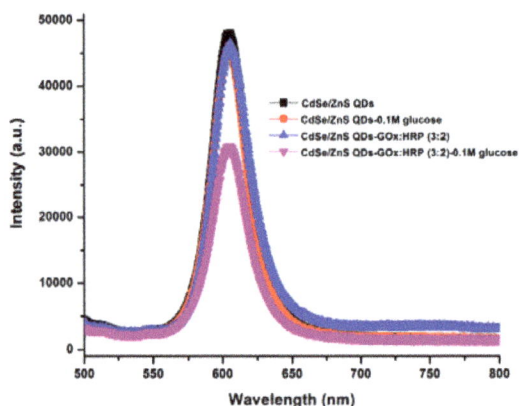

Figure 6. The fluorescence spectrums of CdSe/ZnS QDs along with different combinations of its loading with GOx:HRP enzyme and in the presence of 0.1 M glucose. All spectra were recorded after mixing the components for 30 min.

Further, the calibration studies were performed several times using different concentrations of glucose varying from 0 to 40 mM (0, 0.03, 0.07, 0.15, 0.31, 0.62, 1.25, 2.5, 5, 10, 20, and 40 mM, respectively), and the assay was repeated thrice for each measurement where the results are displayed in Figure 7a–c. In addition, the inset of Figure 7 shows that there is a good linearity between the quenched intensity of the CdSe/ZnS QDs and the glucose levels when recorded in the range of 0–10 mM. From the results, we calculated the corresponding regression coefficient to be about 0.998, while the limit of detection (LOD) obtained was 0.045 mM. The LOD was calculated through $3\sigma/s$, where s is the slope of calibration, while σ is the standard deviation of the corrected blank from the fluorescence signals of the CdSe/ZnS-TGA QDs. The LOD of this study was slightly lower than the other previously reported studies that involve QDs for the glucose detection (Table 1). Further, our proposed method provides a good reproducibility with the relative standard deviation (RSD) of 3.33% for a 10-fold repeated detection of 1.25 mM glucose. Finally, we believe that our method is simple, reliable, and can offer a practical approach for the non-invasive detection of glucose in the real-time samples.

Table 1. Detection limits and ranges for optically based glucose detection using QDs.

QDs	Ligands	Enzymes	Quenching Mechanism	Detection Range	Reference
CdSe/ZnS@ SiO$_2$	Not given	GOD	H$_2$O$_2$	0.5–3 mM	[30]
CdSe/ZnS	MSA	GOD	Acidic change	0.2–10 mM or 2–30 mM	[12]
CdTe	GSH	GOD	H$_2$O$_2$	0.05–1.0 mM	[31]
CdSe/ZnS	TGA	GOD/HRP	H$_2$O$_2$	0.045–10 mM	Our proposed method

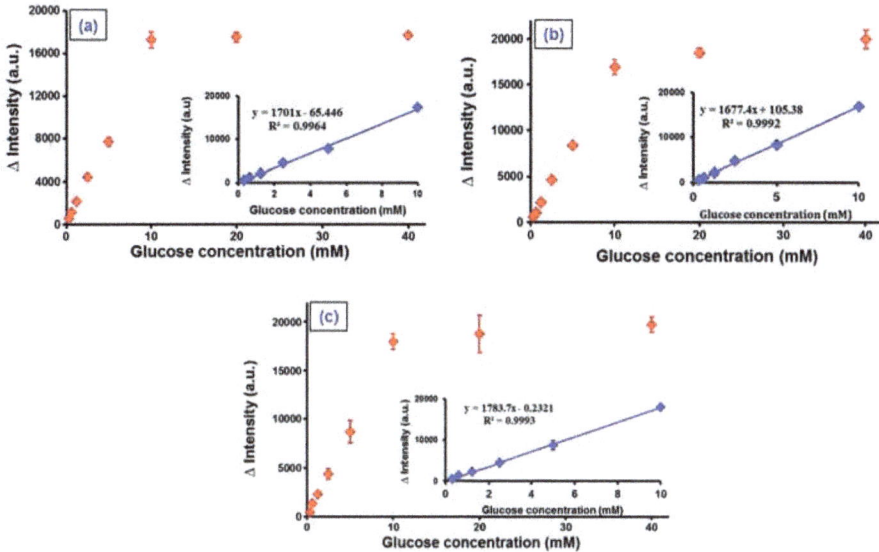

Figure 7. Fluorescence intensity of CdSe/ZnS-capped TGA core shell QDs via various glucose levels from 0 to 40 mM. The inset shows the relationship between intensity and glucose concentration. The calibration curves were produced in triplicate ((**a–c**) labels in the figure corresponds to the repetition thrice) using different batches of CdSe/ZnS.

4. Conclusions

In summary, we have demonstrated the CdSe/ZnS capped TGA water-soluble core-shell QDs for the qualitative and quantitative determination of glucose. The detection principle was based on the fact that the spectrofluorescence signals of the prepared CdSe/ZnS-capped TGA QDs can be quenched by the presence of glucose successively. The parameters for fluorescence quenching reaction includes the solution pH, enzyme ratio, and the QD concentration; further, the optimal conditions for the observation of efficient fluorescence towards glucose detection by the use of QDs have been very well discussed. Since the glucose detection principle is based on the fluorescence quenching effect and we observed that in the presence of 0.1 mM glucose, the fluorescence intensity of the bioconjugated QDs was quenched about 12,000 a.u. Further, the bioconjugated GOx:HRP/QDs-capped TGA was analyzed with known concentrations of glucose and indicated that the quenching of fluorescence intensity is proportionate to the glucose concentration. An extremely good linearity for the glucose determination in the range of 0–10 mM was observed, in addition to obtaining the LOD to be 0.045 mM. Thus, from the analysis of the results in this study, it can be concluded that our synthesized QDs are accurate, sensitive, and can be applied as the fluorescence nanosensor for the detection of glucose in real samples.

Acknowledgments: The authors are grateful for the financial support provided by the Malaysian Government through its Ministry of Science, Technology, and Innovation under Science Fund (03-03-02-SF0330). One of the authors, Dr. Faruq Mohammad acknowledges Deanship of Scientific Research, King Saud University for funding through Vice Deanship of Scientific Research Chairs program.

Author Contributions: S.A.R. was in charge of the experiment and its results. N.A. performed the surface modification, bio-conjugation, and analytical performance of glucose detection. J.A. and N.A.Y. conceived the project and supervised the work. Z.A.Z. and N.M.A.N.A.A. performed the synthesis work. S.A.R. wrote the manuscript and F.M. helped with the writing, necessary corrections, data analysis, and publication process. All authors have contributed to the scientific discussion.

Conflicts of Interest: The authors declare no conflict of interest.

References

1. Narayanan, S.S.; Sarkar, R.; Pal, S.K. Structural and functional characterization of enzyme-quantum dot conjugates: Covalent attachment of CdS nanocrystal to r-chymotrypsin. *J. Phys. Chem. C* **2007**, *111*, 11539–11543. [CrossRef]

2. Gole, A.; Dash, C.; Ramakrishnan, V.; Sainkar, S.R.; Mandale, A.B.; Rao, M.; Sastry, M. Pepsin-gold colloid conjugates: Preparation, characterization, and enzymatic activit. *Langmuir* **2001**, *17*, 1674–1679. [CrossRef]

3. Ohara, T.J.; Rajagopalan, R.; Heller, A. "Wired" enzyme electrodes for amperometric determination of glucose or lactate in the presence of interfering substances. *Anal. Chem.* **1994**, *66*, 2451–2457. [CrossRef] [PubMed]

4. Yu, A.; Liang, Z.; Cho, J.; Caruso, F. Nanostructured electrochemical sensor based on dense gold nanoparticle films. *Nano Lett.* **2003**, *3*, 1203–1207. [CrossRef]

5. Xianbo, Q.L.; Li, L.J.; Yao, X.; Li, J. Direct electrochemistry of glucose oxidase and electrochemical biosensing of glucose on quantum dots/carbon nanotubes electrodes. *Biosens. Bioelectron.* **2007**, *22*, 3203–3209.

6. Huang, S.H.; Liao, M.H.; Chen, D.H. Direct binding and characterization of lipase onto magnetic nanoparticles. *Biotechnol. Prog.* **2003**, *19*, 1095–1100. [CrossRef] [PubMed]

7. Zhang, Y.; He, P.; Hu, N. Horseradish peroxidase immobilized in TiO$_2$ nanoparticle films on pyrolytic graphite electrodes: direct electrochemistry and bioelectrocatalysis. *Electrochim. Acta* **2004**, *49*, 1981–1988. [CrossRef]

8. Hilliard, L.R.; Zhao, X.; Tan, W. Immobilization of oligonucleotides onto silica nanoparticles for DNA hybridization studies. *Anal. Chim. Acta* **2002**, *470*, 51–56. [CrossRef]

9. Cao, L.; Ye, J.; Tong, L.; Tang, B. A new route to the considerable enhancement of glucose oxidase (GOx) activity: The simple assembly of a complex from CdTe quantum dots and GOx, and its glucose sensing. *Chem. Eur. J.* **2008**, *14*, 9633–9640. [CrossRef] [PubMed]

10. Hua, M.; Tiana, J.; Lua, H.T.; Weng, L.X.; Wang, L.H. H$_2$O$_2$-sensitive quantum dots for the label-free detection of glucose. *Talanta* **2010**, *82*, 997–1002. [CrossRef] [PubMed]

11. Massadeh, S.; Nann, T. InP/ZnS nanocrystals as fluorescent probes for the detection of ATP. *Nanomater. Nanotechnol.* **2014**, *4*, 15. [CrossRef]

12. Huang, C.P.; Liu, S.W.; Chen, T.M.; Li, Y.K. A new approach for quantitative determination of glucose by using CdSe/ZnS quantum dots. *Sens. Actuator B Chem.* **2008**, *130*, 338–342. [CrossRef]

13. Wu, W.; Zhou, T.; Shen, J.; Zhou, S. Optical detection of glucose by CdS quantum dots immobilized in smart microgels. *Chem. Commun.* **2009**, *29*, 4390–4392. [CrossRef] [PubMed]

14. Wu, P.; He, Y.; Wang, H.F.; Yan, X.P. Conjugation of glucose oxidase onto Mn doped ZnS quantum dots for phosphorescent sensing of glucose in biological fluids. *Anal. Chem.* **2010**, *82*, 1427–1433. [CrossRef] [PubMed]

15. Saran, A.D.; Sadawana, M.M.; Srivastava, R.; Bellare, J.R. An optimized quantum dot ligand system for biosensing applications: Evaluation as a glucose biosensor. *Colloids Surf. A* **2011**, *384*, 393–400. [CrossRef]

16. Huang, C.P.; Li, Y.K.; Chen, T.M. An investigation of urea detection by using CdSe/ZnS quantum dots. *Biosens. Bioelectron.* **2007**, *22*, 1835–1838. [CrossRef] [PubMed]

17. Xie, R.G.; Kolb, U.; Li, J.X.; Basche, T.; Mews, A. Synthesis and characterization of highly luminescent CdSe-Core CdS/Zn and CdS/ZnS multishell nanocrystals. *J. Am. Chem. Soc.* **2005**, *127*, 7480–7488. [CrossRef] [PubMed]

18. Willard, D.M.; Carillo, L.L.; Jung, J.; Van Orden, A. CdSe-ZnS quantum dots as resonance energy transfer donors in a model protein binding assay. *Nano Lett.* **2001**, *1*, 469–474. [CrossRef]

19. Chan, W.C.W.; Maxwell, D.J.; Gao, X.H.; Bailey, R.E.; Han, M.Y.; Nie, S.M. Luminescent quantum dots for multiplexed biological detection and imaging. *Curr. Opin. Biotechnol.* **2002**, *13*, 40–46. [CrossRef]

20. Sapsford, K.E.; Pons, T.; Medintz, I.L.; Mattoussi, H. Biosensing, with luminescent semiconductor quantum dots. *Sensors* **2006**, *6*, 925–953. [CrossRef]

21. Hines, M.A.; Guyot-Sionnest, P. Synthesis and characterization of strongly luminescing ZnS-Capped CdSe nanocrystals. *J. Phys. Chem.* **1996**, *100*, 468–471. [CrossRef]

22. Dabbousi, B.O.; Rodriguez-Viejo, J.; Mikulec, F.V.; Heine, J.R.; Mattoussi, H.; Ober, R.; Jensen, K.F.; Bawendi, M.G. (CdSe)ZnS Core-Shell Quantum Dots: Synthesis and characterization of a size series of highly luminescent nanocrystallites. *J. Phys. Chem. B* **1997**, *101*, 9463–9475. [CrossRef]

23. Kloepfer, J.A.; Bradforth, S.E.; Nadeau, J.L. Photo-physical properties of biologically compatible CdSe quantum dot structures. *J. Phys. Chem. B* **2005**, *109*, 9996–10003. [CrossRef] [PubMed]

24. Gill, R.; Willner, I.; Shweky, I.; Banin, U. Fluorescence resonance energy transfer in CdSe/ZnS-DNA conjugates: Probing hybridization and DNA cleavage. *J. Phys. Chem. B* **2005**, *109*, 23715–23719. [CrossRef] [PubMed]

25. Stsiapura, V.; Sukhanova, A.; Baranov, A.; Artemyev, M.; Kulakovich, O.; Oleinikov, V.; Pluot, M.; Cohen, J.H.M.; Nabiev, I. DNA-assisted formation of quasi-nanowires from fluorescent CdSe/ZnS nanocrystals. *Nanotechnology* **2006**, *17*, 581–587. [CrossRef]

26. Ariffin, N.; Yusof, N.A.; Abdullah, J.; Zubir, Z.A.; Aziz, N.A.; Azmi, N.M.; Ramli, N.I. Synthesis and surface modification of biocompatible water-soluble core-shell quantum dots. *Adv. Mater. Res.* **2014**, *879*, 184–190.

27. Chan, W.C.W.; Nie, S.M. Quantum dot bioconjugates for ultrasensitive nonisotopic detection. *Science* **1998**, *281*, 2016–2018. [CrossRef] [PubMed]

28. Bwatanglang, I.B.; Mohammad, F.; Yusof, N.A.; Abdullah, J.; Hussein, M.A.; Alitheen, N.B.; Abu, N. Folic acid targeted Mn:ZnS quantum dots for theranostic applications of cancer cell imaging and therapy. *Int. J. Nanomed.* **2016**, *11*, 413–428.

29. Mohammad, F.; Yusof, N.A. Surface ligand influenced free radical protection of superparamagnetic iron oxide nanoparticles (SPIONs) toward H9c2 cardiac cells. *J. Mater. Sci.* **2014**, *49*, 6290–6301. [CrossRef]

30. Cavaliere-Jaricot, S.; Darbandi, M.; Kucur, E.; Nann, T. Silica coated quantum dots: A new tool for electrochemical and optical glucose detection. *Microchim. Acta* **2008**, *160*, 375–383. [CrossRef]

31. Yuan, J.; Guo, W.; Yin, J.; Wang, E. Glutathione-capped CdTe quantum dots for the sensitive detection of glucose. *Talanta* **2009**, *77*, 1858–1863. [CrossRef] [PubMed]

sensors

MDPI

Review

Luminescence-Based Optical Sensors Fabricated by Means of the Layer-by-Layer Nano-Assembly Technique

Nerea De Acha [1,*], Cesar Elosua [1,2], Ignacio Matias [1,2] and Francisco Javier Arregui [1,2]

[1] Department of Electric and Electronic Engineering, Public University of Navarra, E-31006 Pamplona, Spain; cesar.elosua@unavarra.es (C.E.); natxo@unavarra.es (I.M.); parregui@unavarra.es (F.J.A.)

[2] Institute of Smart Cities (ISC), Public University of Navarra, E-31006 Pamplona, Spain

* Correspondence: nerea.deacha@unavarra.es; Tel.: +34-948-166-044

Received: 15 November 2017; Accepted: 4 December 2017; Published: 6 December 2017

Abstract: Luminescence-based sensing applications range from agriculture to biology, including medicine and environmental care, which indicates the importance of this technique as a detection tool. Luminescent optical sensors are required to be highly stable, sensitive, and selective, three crucial features that can be achieved by fabricating them by means of the layer-by-layer nano-assembly technique. This method permits us to tailor the sensors' properties at the nanometer scale, avoiding luminophore aggregation and, hence, self-quenching, promoting the diffusion of the target analytes, and building a barrier against the undesired molecules. These characteristics give rise to the fabrication of custom-made sensors for each particular application.

Keywords: photoluminescence; layer-by-layer nano-assembly technique; nanostructured materials; chemical sensing

1. Introduction

Optical sensing techniques allow for the possibility of making remote [1] and non-invasive measurements [2], as well as working in hazardous environments [3]. These, and many other significant advantages over other detection technologies, have attracted the attention of scientists over the last decades [4,5], and hence this technology has experienced a high development.

Optical sensors can be based on different transduction mechanisms, such as absorption [6,7], resonance [8–10], or photoluminescence [11–13]. The latter consists of emission of light by a material as a consequence of its previous absorption at lower wavelengths (excitation). Depending on the lifetime of this emission (i.e., the average time it takes the intensity to drop by $1/e$), luminescence can be classified as fluorescence (lifetime in the range of ps and ns) or phosphorescence (lifetime greater than ms). The intensity of this emission, as well as its lifetime, can be quenched or enhanced by the variation of different external parameters: pH [14], temperature [15], biomolecules [16], oxygen [17], or metal ion concentration [18]. This modulation of the intensity (and lifetime) by external parameters has been widely employed for the development of luminescence-based sensors, either in solution [19,20], or onto different substrates [21–23].

Among the existing luminescent materials (also known as luminophores), quantum dots (QDs) [24,25], nanoparticles (NPs) [26,27], fluoropolymers [28], dyes [29,30], and porphyrins [31–33] have been the most utilized. For the fabrication of sensors, these luminescent materials are usually entrapped or encapsulated in different matrices [34,35] or shells [36,37], which must be designed to facilitate the interaction between the analyte and the sensing material [38].

There are three key requirements that luminescent sensors must meet: good photostability, and high selectivity and sensitivity [39]. To achieve these characteristics and allow the rapid adsorption/desorption of the target analytes to the sensing films, highly permeable coatings are usually fabricated by means of dip-coating [40], spin-coating [41], sol-gel [42], or xero-gel [43] techniques. However, when utilizing these methods the distribution of the luminophore inside the films cannot be controlled, which gives rise to their aggregation and causes self-quenching, hence significantly reducing the sensors' sensitivity [44]. This can be overcome by fabricating the sensing coatings by means of the layer-by-layer nano-assembly (LbL) technique, which consists of the deposition of oppositely charged materials (typically polyelectrolytes) by electrostatic forces or other attractive forces acting cooperatively, including interactions such as hydrophobic attraction [45]. LbL has been experimentally demonstrated to be a powerful method for the fabrication of luminescence-based sensors, since it is a reproducible technique that allows the utilization of a wide variety of indicators. Furthermore, an accurate selection of materials and assembly parameters not only permits us to modify the permeability of the nanostructure [46], hence promoting the diffusion of target species or forming a barrier against undesirable ones [47], but also allows us to control the layer thickness at the nanometer scale and tailor the space distance between luminescent layers [48]. Taking advantage of this fact, it is possible to tailor the homogeneity of the distribution of the luminophore into the matrix [49] in order to attenuate self-quenching [48].

In recent years, luminescence has become a powerful detection mechanism in a broad range of areas, being the most important sensing tool in different biological applications [50]. This fact, together with the versatility that the LbL technique offers for the fabrication of custom-made sensors [51], has led to the development of a wide variety of luminescent probes built by this technique. Thus, a review in which the principal luminescence-based sensors fabricated by the LbL technique are analyzed is of great interest. This review compiles solution probes as well as multilayer sensing films for different purposes: metal ions detection, dissolved and gaseous oxygen monitoring, and biosensing applications.

2. Luminescent Sensors Based on Encapsulated Indicators

The progress of in vivo measurement systems in recent years has yielded the development of biocompatible optical sensors [52]. A widespread method for the fabrication of this kind of sensor consists of the encapsulation of the sensitive material in multilayered nanostructures (also called shells or capsules). These walls have to meet a key requirement: protecting sensitive molecules from the external environment while allowing fast diffusion of the target analyte [53]. LbL encapsulation of sensors was first described in the early 2000s [54,55] and, since then, many different applications have been reported [56–59].

Particle encapsulation requires a template that is coated with a multilayered nanostructure and then dissolved. The main methods for immobilizing the luminophores inside the shells are diffusion and precipitation [60], which are illustrated in Figure 1.

In the case of encapsulating the sensing molecules by diffusion, the multilayered coatings are adsorbed onto sacrificial templates, usually dissolvable or degradable inorganic polymeric microspheres. Once the capsules are fabricated, the templates are dissolved in order to leave hollow microspheres suspended in solutions of the sensing molecules, which are loaded inside them just by diffusion [61]. Despite this technique offering the possibility of employing almost any dissolvable template, it also exhibits the lowest loading efficiency.

When the luminophores are encapsulated by precipitation, they co-precipitate with the sacrificial templates before the shell is built and, subsequently, templates are diluted [58]. By employing this technique, the highest loading efficiency rates can be obtained.

For an adequate selection of templates, it is important to take into account two factors: firstly, they must be able to keep their structural properties during the coating process and, secondly, they have to be easily dissolved once the encapsulating nanostructure has been attached to their surface [58]. Typical materials used as templates are calcium carbonate ($CaCO_3$) [62], silica [63], latex [64] polystyrene [65], or gold [66] nanoparticles and melamine formaldehyde [67] microspheres.

The dimensions of the shell depend on their shape and size, which can range from nanometers to microns [68].

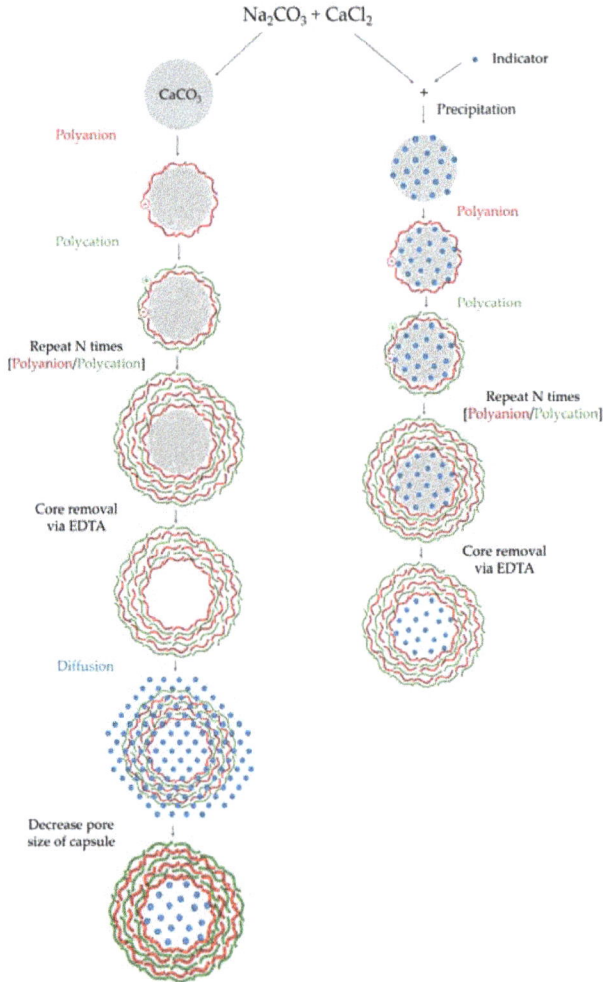

Figure 1. Schematic of the encapsulation techniques.

Owing to the versatility of the LbL method, it has been extensively used for micro- and nanoparticle encapsulation [69,70], usually employing polyelectrolytes. That is because, by varying some properties of polyelectrolyte solutions and controlling the deposition conditions, it is possible to tailor the properties of the capsule walls [71]. For instance, the pH of the solutions not only influences the shell thickness [72], but also its permeability [73] and, hence, the diffusion rate of molecules inward.

As mentioned below, capsules are usually made of different polyelectrolyte combinations. The most used pairs are poly(allylamine hydrochloride) (PAH)/poly(styrene sulfonate) (PSS) [74] and poly(diallyldimethylammonium chloride) (PDDA)/poly(styrene sulfonate) (PSS) [75]. Apart from them, other combinations of materials can be employed, for instance, poly-l-lysine/poly(l-glutamatic acid) [76], chitosan/dextran [63], or poly-l-lysine/heparin [77]. However, their use is not as common as the first mentioned polyelectrolyte pairs.

One of the advantages of this technique is the possibility of developing self-referenced sensors by using a sensitive luminophore and a non-sensitive luminophore that acts as an optical reference [63]. Ideally, both have the same absorption spectrum and a complementary emission spectrum [52], which allows the use of a single excitation light source and the simultaneous monitoring of the emission peaks: one of them will change in the presence of the target analyte, while the other will remain constant. Apart from encapsulating both luminophores [58], it is also common that one of them is assembled as part of the multilayered coating [78].

The main applications of these sensors (metal ion detection, dissolved oxygen sensing, and glucose and lactate monitoring) are analyzed in the following subsections.

2.1. Encapsulated Sensors for Ion Detection

Metal ions are known to be highly toxic materials [79], so considerable efforts have been made towards their detection in aqueous media. Duchesne and co-workers [80] fabricated a potassium sensor by encapsulating the fluorescent indicators by the diffusion technique: using positively charged melamine formaldehyde (MF) particles as templates, multilayer capsules of PSS and PDDA were fabricated. The MF cores were diluted afterwards in HCl, and then the capsules were immersed in solutions of potassium-binding benzofuran isophthalate tetraammonium salt (PBFI), which diffused through the walls to the hollow cores. The PBFI-loaded sensors exhibited a luminescent peak at 500 nm, which moved towards lower wavelengths and increased in intensity upon the addition of K^+ ions in the 0–45 mM range.

Rare-earth (RE) nanocrystals are of great interest for sensing applications because of their optical and chemical properties, such as long lifetime (in the range of μs or ms) [81], high quantum yield, or resistance to photobleaching [82]. A fluorescent sensing system based on RE nanocrystals and CdSe/ZnS quantum dots was fabricated by means of the LbL technique in [83]: PEI-coated $NaYF_4$:Ce,Tb nanorods were separated from the CdSe/ZnS QDs by a PSS/PAH bilayer. Under UV illumination (255 nm), a dual emission was observed, with the luminescent peaks centered at 542 nm (RE nanocrystals) and 650 nm (QDs), with the latter being dominant due to the fluorescence resonance energy transfer between the RE nanorods and the QDs. Under exposure to different concentrations of Cu^{2+} or Ag^+ metal ions in the μM range, the luminescent intensity at 650 nm decreased, while that corresponding to the peak centered at 542 nm remained constant. This can be observed in Figure 2, in which the system was exposed to different metal ions (30 μM): the red color (650 nm) was quenched by Ag^+ ions, while it completely disappeared in the presence of Cu^{2+} ions, demonstrating high selectivity.

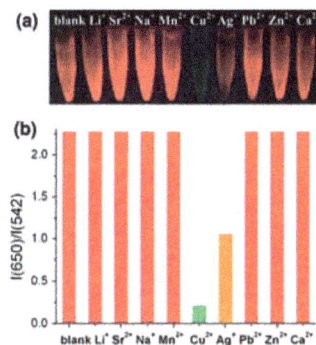

Figure 2. Effect of the different metal ions on the RE–QD composites: (**a**) Cu^{2+} and Ag^+ ions change the color of the RE-QD composites into green and orange, respectively; (**b**) The ratio I(650)/I(542) of the RE-QD composites only decreases in presence of Cu^{2+} and Ag^+. Reprinted from [83] with permission from Springer.

Other encapsulated sensors for the detection of metal ions are summarized in Table 1. Most of them use salts as sensitive indicators (e.g., benzofuran isophthalate tetraammonium salt (PBFI) or sodium-binding benzofuran isophthalate tetraammonium salt (SBFI)), and FluoSpheres as references.

Table 1. Encapsulated sensors for ions detection.

Analyte	Sensitive Indicator	Reference Indicator	Capsule	Detection Range	LOD	Ref.
K^+	PBFI	-	$(PSS/PAH)_5$	0–45 mM		[64]
K^+	PBFI	FluoSpheres	$(PAH/PSS)_4PAH$	0–282 mM		[84]
K^+	PBFI	-	2, 3, and 5 bilayers of {PSS/PDDA}	0–45 mM	-	[80]
K^+	PBFI	Europium FluoSpheres	$(PAH/PSS)_4PAH$	0–120 mM	1 mM	[85]
K^+	PBFI	Europium FluoSpheres	$(PAH/PSS)_4PAH$	0–300 mM	1.2 mM	[86]
Na^+	SBFI	-	2, 3, and 5 bilayers of {PSS/PDDA}	0–100 mM	-	[54]
Cu^{2+}, Ag^+	CdSe/ZnSe QDs	NaYF4:Ce,Tb rare-earth nanocrystals	PSS/PAH	0–35 μM Cu^{2+} 0–90 μM Ag^+	-	[83]
Pb^{2+}	CdSe/CdS QDs	-	chitosan/xylenol orange	0.05–6 μM	20 nM	[59]

2.2. Encapsulated Sensors for Dissolved Oxygen Detection

The most important application of this kind of sensors is the detection of low concentrations of dissolved oxygen for biomedical applications, typically utilizing ruthenium porphyrins as indicators. McShane et al. [52] fabricated a self-referred oxygen sensor by employing tris(2,2'-bipyridyl) dichlororuthenium(II) hexahydrate (Ru(bpy)) as the sensitive material and fluorescein isothiocyanate (FITC) as the reference one, so the use of a single light source at 460 nm was possible, and the simultaneous monitoring of the luminescent peaks at 525 nm (FITC) and 620 nm (Ru(bpy)), allowing ratiometric measurement of the fluorescence (I_{620}/I_{525}). For the capsule fabrication, 2.6 μm melamine formaldehyde microtemplates were coated with {PSS/PAH-FITC}$_n$ or {PSS/PAH-FITC}$_n$PSS shells. After the cores' dilution, the capsules were suspended in solutions of the oxygen-sensitive indicator at different pH values, and Ru(bpy) molecules were loaded by diffusion. The highest loading efficiency was achieved in the case of the {PSS/PAH-FITC}$_5$PSS shells, when the pH value of the Ru(bpy) solution was 10.4. Suspensions of these sensors were bubbled with O_2 and N_2, and it was observed that in the presence of O_2, the intensity at 620 nm (Ru(bpy)) decreased while the intensity at 525 nm (FITC) remained constant. A total decrease of 15% of the I_{620}/I_{525} ratio was measured when only O_2 was bubbled. Moreover, the sensor response was consistent with the dynamic changes of the gas levels over time.

Palladium porphyrin was encapsulated in [58] by the co-precipitation technique: the carboxylate modifies FluoSpheres (FS), used as the reference, and the Pd-meso-tetra(4-carboxyphenyl) porphine (PdTCPP) co-precipitated with the calcium carbonate ($CaCO_3$) nanoparticles. They were first stabilized in poly(vinylsulfonic acid) (PVSA) and then coated with the multilayer $(PDDA/PSS)_{10}$ structure, and the $CaCO_3$ cores were diluted. When the capsules were excited at 405 nm, they emitted luminescence at 515 nm (FS) and 700 nm (PdTCPP). Under exposure to different dissolved oxygen concentrations, the intensity of the emission peak centered at 515 nm remained constant, while that of the peak at 700 nm decreased proportionally to the oxygen concentration. Thus, the ratio I_{700}/I_{515} was used to characterize the sensor, which had a detection limit of 7.62 μM. A schematic illustration of this sensor and its response to oxygen is given in Figure 3.

Other dissolved oxygen probes were developed with different ruthenium porphyrins, such as tris(4,7-diphenyl-1,10-phenanthroline)ruthenium(II) dichloride (Ru(dpp)) or $[Ru(Ph_2phen)_3]^{2+}$. They are summarized in Table 2.

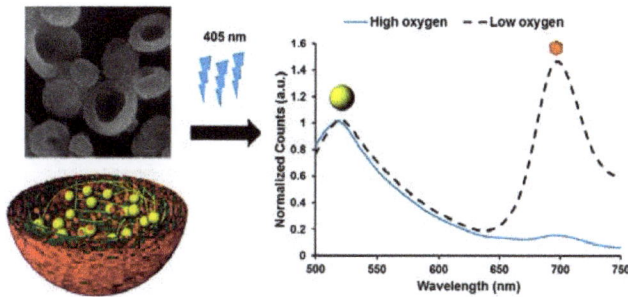

Figure 3. SEM image of the nano-capsules after the dilution of the core (**upper left**), schematic representation of the encapsulated sensors (**lower left**) and luminescence spectra of the sensors when exposed to high and low dissolved oxygen concentrations (**right**). Reprinted with permission from [58]. Copyright: 2014, American Chemical Society.

Table 2. Encapsulated sensors for dissolved oxygen detection.

Sensitive Indicator	Reference Indicator	Capsule	Detection Range	LOD	Ref.
Ru(dpp)	green polystyrene FluoSpheres	{PAH/PSS}$_3$	0–1500 mM	-	[85]
Ru(dpp)	carboxylate-modified nanospheres	(PAH/PSS)$_3$	ON/OFF probe	-	[74]
Ru(dpp)	-	(PSS/PAH)$_4$/PSS	0–0.6 mM	-	[87]
Ru(bpy)	FITC	(PSS/PAH-FITC)$_5$/PSS		-	[52]
Ru(bpy)	FITC	(PSS/PDDA)$_5$/PSS	ON/OFF probe	-	[75]
[Ru(Ph$_2$phen)$_3$]$^{2+}$	carboxylate-modified FluoSpheres	{PAH/PSS}$_3$	0–1.5 mM	-	[88]
PdTCPP	carboxylate-modified FluoSpheres	[PDDA/PSS]$_{10}$	0–250 µM	7.62 µM	[58]

2.3. Encapsulated Sensors for Glucose and Lactate Monitoring

A particular application of dissolved oxygen sensors is glucose monitoring [89]. To this end, glucose binding proteins, usually glucose-oxidase [87] or apo-glucose-oxidase [90], are loaded inside the multilayered shells, as well as the sensitive molecules. For instance, Kazakova et al. [87] fabricated 5 µm-microcapsules made of PAH and PSS loaded with oxygen-sensitive dye (Ru(dpp)) that entrapped glucose-oxidase. This indicator emitted fluorescence between 560 and 700 nm, and its intensity was inversely proportional to the oxygen concentration. Furthermore, glucose concentration was correlated with oxygen reduction during enzymatic degradation by glucose oxidase. An increase in the fluorescent intensity and the calibration curve of this sensor upon the addition of different glucose concentrations is observed in Figure 4.

Kazakova et al. also coated lactate oxidase and peroxide with capsules of PAH and PSS loaded with dihydrorhodamine 123 (DHR123), which was sensitive to hydrogen peroxide. The addition of lactate in the presence of lactate oxidase produced hydrogen peroxide, which oxidized DHR123 in the presence of peroxide, yielding rhodamine123, a molecule that emitted green fluorescence (510–560 nm), as can be observed in Figure 5.

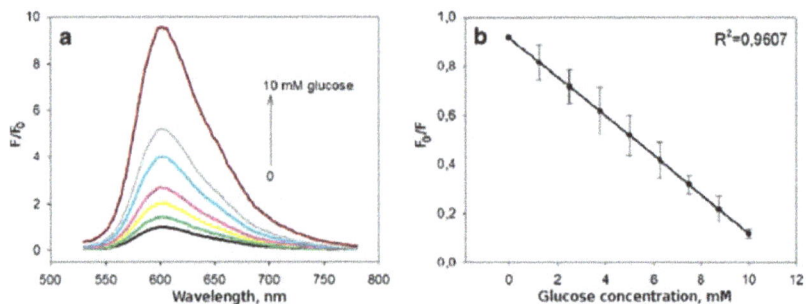

Figure 4. (a) Relative fluorescence intensity upon the addition of different glucose concentrations. F and F_0 represent the fluorescence intensities in the presence (F) and absence (F_0) of glucose; (b) calibration curve of the sensor. Reprinted from [87] with permission from Springer.

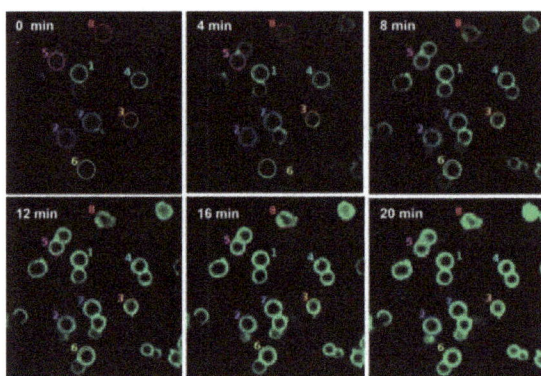

Figure 5. Confocal microscopy images of DHR123-labeled capsules containing lactate oxidase in the presence of 0.23 nM peroxidase and 4 mM lactate. Reprinted from [87] with permission from Springer.

3. Luminophores Immobilized in Multilayer Films for Sensing Applications

Luminescence-based sensors are also fabricated onto different substrates by coating them with films in which the sensitive luminophores are embedded. These luminescent coatings exhibit many different advantages over solution-based sensors, such as the possibility of fabricating them onto almost any kind of substrate [91,92], their easy storage and portability [93], their regeneration and reusability [94], and the good chemical stability of luminophores in solid state [95]. Furthermore, they can be used for vapor/gas detection [96], an application field where the encapsulated probes are hardly used [97].

The luminescent indicators can be entrapped in multilayer structures by different methods, which are displayed in Figure 6. If they are immobilized during the fabrication process, this can be done by direct assembly (in this case they are not neutral materials), or by mixing, covalently linking, or entrapping them inside charged materials. When the immobilization of the indicator occurs after the fabrication of the film, it is carried out by immersing it in a solution of the dye. Examples of these cases are explained in the following subsections.

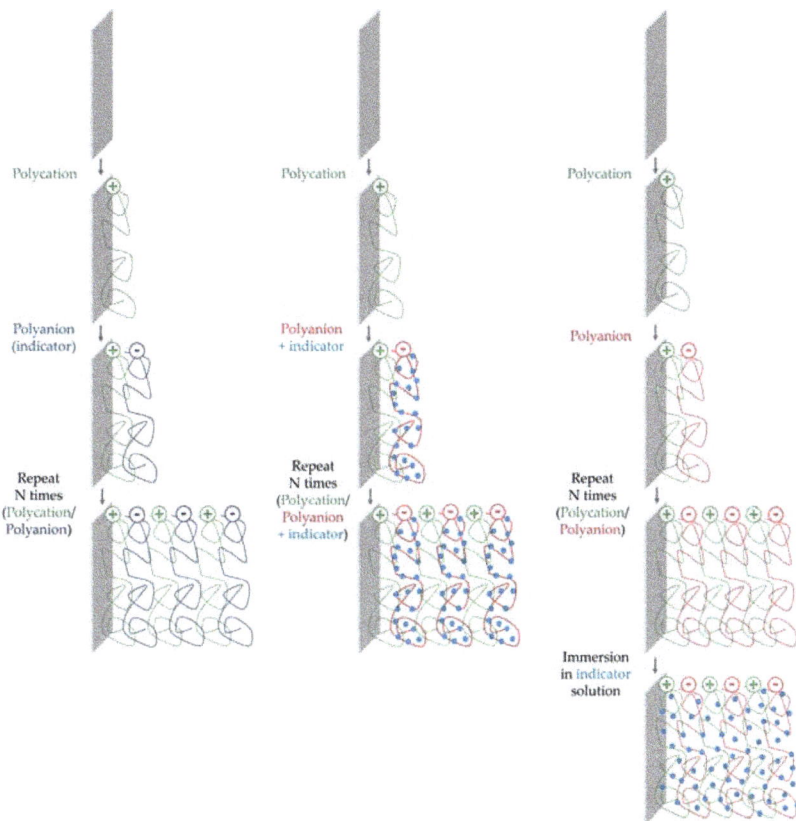

Figure 6. Schematic fabrication luminescent films: the non-neutral indicator is directly assembled into the film (**left** pathway), the neutral indicator is mixed, covalently linked, or entrapped into a charged material and then it is assembled into the coating (**central** pathway), or the fabricated film is immersed into a solution of the dye (**right** pathway).

3.1. LbL Luminescent Coatings for the Detection of Metal Ions

A wide variety of indicators have been employed for the fabrication of luminescent films for the detection of metal ions, ranging from ligand-capped quantum dots to fluorescent conjugated polymers, including porphyrins and water-soluble dyes. To this latter category belongs 1-hydroxypyrene-3,6,8-trisulfonate, HPTS, a luminescent indicator widely used for pH monitoring [98,99]. For its deposition by the LbL technique, Lee and coworkers [100] attached it covalently to the polyanion PAA and, by using PAH as a cationic polyelectrolyte, deposited the multilayer structure [PAH/PAA-HPTS]$_n$ onto glass slides. HPTS emitted luminescence at 485 nm when it was illuminated at 410 nm. The maximum of the luminescent peak decreased linearly, but with different quenching constants, in the presence of electron-deficient metal cations, such as Fe^{3+} and Hg^{2+}, the nitro compound 2,4-dinitrotoluene, DNT, or the dicationic electron acceptor methyl viologen, MV^{2+} [101]. This can be observed in Figure 7, where the Stern–Volmer plots of multilayers of [PAH/PAA-HPTS] for each compound are displayed. Furthermore, this luminescence was not affected by other metal ions, for instance, Ba^{2+}, Ca^{2+}, K^+, Zn^{2+}, Cd^{2+}, and Pb^{2+}.

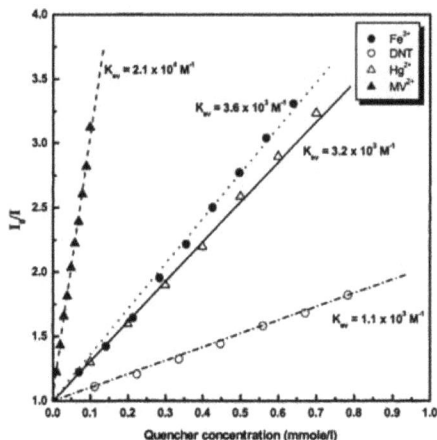

Figure 7. Stern−Volmer plots of multilayer films of PAH/PAA−HPTS as a function of different quencher concentrations. Reprinted with permission from [101]. Copyright © 2000, American Chemical Society.

Different fluorescent probes for mercury(II) ion detection based on the water-soluble porphyrin 5,10,15,20-tetrakis(4-sulfonatophenyl)porphyrin (TPPS) were compared in [102]. PDDA was employed as a cationic polyelectrolyte and TPPS, PSS, or solutions with different ratios of TPPS:PSS were used as anionic counterparts. On one hand, it was observed that, when TPPS and PSS were co-deposited, giving rise to the structure (PDDA/TPPS:PSS)$_n$, higher quantum yields were observed when the PSS:TPPS ratio increased from 1:1 to 1:100. On the other hand, when depositing TPPS alternately (i.e., PDDA/PSS/PDDA/TPPS)$_n$), the amount of adsorbed porphyrin was higher than when as mixed with PSS, and a good quantum yield was also achieved. A detailed analysis of this structure demonstrated that films with one or two tetralayers were most suitable to be used as Hg(II) sensors, since they combined good optical properties with the lowest response time. In the case of (PDDA/PSS/PDDA/TPPS), the sensor was exposed to Hg(II) concentrations in the range between 3.3×10^{-8} and 3.3×10^{-5} M. For higher concentrations, a longer time was required to reach equilibrium as a consequence of the adsorption process of the analyte within the films.

Fluoropolymers have also attracted interest for the development of optical sensors because of their high luminescence quantum yields. For instance, poly(9,9-bis(3'-phosphatepropyl)fluorenealt-1,4-phenylene) sodium salt (PFPNa) was synthesized and deposited with PDDA by means of the LbL [103]. The PFPNa polymer had absorption and luminescence peaks at 364 and 410 nm, respectively, whose intensities were proportional to the pH of the aqueous solutions. The latter was also inversely proportional to Fe^{3+} concentration. Furthermore, the sensor sensitivity was demonstrated to be almost independent of the number of bilayers (Figure 8a), so the 1-bilayer coating was employed as the sensing structure. In this case, fluorescence was quenched 400-fold for 10 μM of Fe^{3+} concentration (Figure 8b), and the detection limit for this metal ion was 10^{-7} M. Another thin-film sensor that employs a fluoropolymer for metal ion detection is reported in [104].

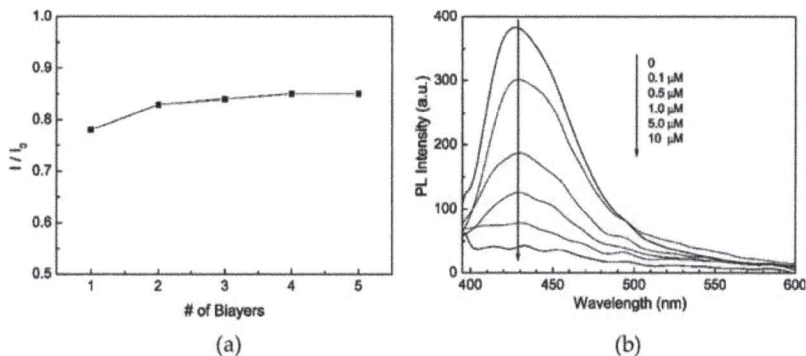

Figure 8. (a) Fluorescence response of (PDDA/PFPNa)$_n$ structures upon addition of 0.1 μM Fe^{3+}; (b) quenching of the fluorescent peak when the sensor (PDDA/PFPNa)$_1$ is exposed to different Fe^{3+} concentrations. Reprinted with permission from [103]. Copyright: 2008, American Chemical Society.

Negatively charged mercaptosuccinic acid (MSA) capped CdTe QDs have been assembled with the cationic polyelectrolyte PDDA onto quartz slides by means of the LbL technique. These QDs exhibited a luminescent peak centered at 589 nm, whose intensity decreased proportionally with the increment of Hg(II) for concentrations ranging from 0.01 μM to 1 μM [105]. Hg(II) removal from the sensing film was possible by adding glutathione (GSH) to the sample solutions, which also led to the recovery of the initial luminescent intensity. In subsequent research, this multilayer structure was employed to monitor Hg^{2+} and Cu^{2+} synchronously [106]: despite both ions having quenched the luminescent intensity, the quenching constant of Hg^{2+} was higher than that of Cu^{2+}. Finally, by cross-linking the outermost layer of PDDA/CdTe QDs multilayers by bovine serum albumin (BSA) [107], a bi-color film was developed: it exhibited two luminescent peaks centered at 553 nm (green) and 657 nm (red), with green being the dominant color. In the presence of Hg(II), the intensity of the first peak decreased (as shown in Figure 9a) and, for Hg(II) concentrations higher than 10^{-6} M, it was totally quenched, which made the sensing film change color from green to red. This color change was detectable by the naked eye, as can be seen in Figure 9b.

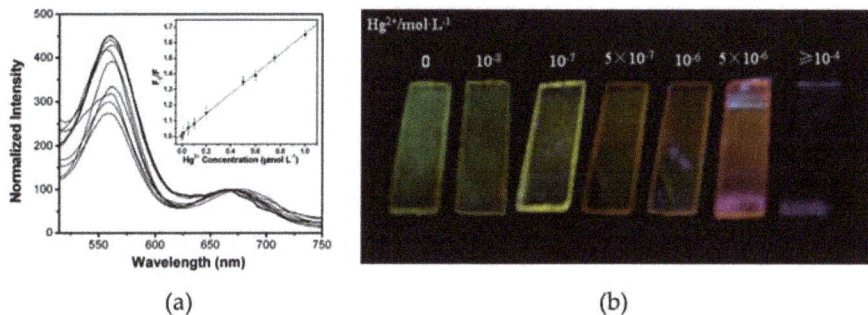

Figure 9. (a) Luminescence spectra of the bi-color film under exposure to different Hg^{2+} concentrations: 0 μM, 0.01 μM, 0.05 μM, 0.1 μM, 0.2 μM, 0.5 μM, 0.6 μM, 0.75 μM, and 1 μM. The inset shows the Stern–Volmer plot of the sensor; (b) Colors of the sensing films under exposure to different Hg^{2+} concentrations: 0 μM, 0.01 μM, 0.1 μM, 0.5 μM, 1 μM, 1.5 μM, and higher than 100 μM. Reprinted from [107] with permission from Elsevier.

Apart from CdTe QDs, carbon nanoparticles have also been employed for the fabrication of Hg(II)-sensitive luminescent films [108]. Their functionalization with PEG200 and N-acetyl-L-cysteine

(NAC) enabled the carbon dots to be assembled with PEI onto the tapered tip of a 600 μm-core optical fiber by means of the LbL technique, as well as the detection of mercury ions. Although the fluorescence quenching mechanism of these sensors was not completely determined, it was likely to be due to the interaction between the –SH groups of NAC and Hg(II) ions. Sensing coatings from one to six bilayers of (PEI/carbon dots) were analyzed: all of them exhibited a reproducible and reversible behavior towards Hg(II) (see Figure 10 for the particular case of the six-layer structure), and it was found that an increase in the number of bilayers led to a decrease in the detection limit (0.1 μM Hg(II) for one bilayer and 0.01 μM Hg(II) for six bilayers) and an increase of the quenching constant. However, cross-sensitivity towards other metal ions was not studied.

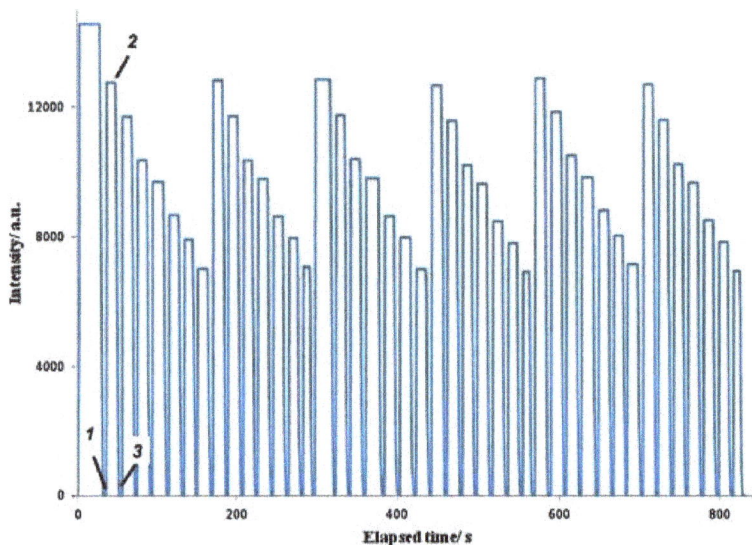

Figure 10. Steady-state fluorescence measurements over time (excitation 380 nm and emission 500 nm) of the dry optical fiber with six layers, followed by three cycles of Hg(II) aqueous solutions with the following concentrations: 0, 0.01, 0.05, 0.1, 0.799, 1.99, and 2.69 μM. (1) The fiber was immersed in water; (2) removed from water; and (3) immersed in Hg(II) 0.01 μM. Reprinted from [108] with permission from Elsevier.

An ultrasensitive Cu(II) sensor was developed by taking advantage of the fluorescence enhancement produced by Ag nanoprisms onto 16-mercaptohexadecanoic acid (16-MHA) capped CdSe quantum dots (QDs) [109]. Si or glass slides were coated with a layer of silver nanoprisms, which was separated from the outer QDs layer by a (PDDA/PSS) spacer of the optimal thickness [110] fabricated by means of LbL. The Ag nanoprisms, together with a UV photobrightening process, enhanced the luminescence, which was selectively quenched by Cu^{2+} ions, with a detection limit as low as 5 nM. The enhancement produced by the photobrightening and the Ag nanoprisms is clearly observable in Figure 11.

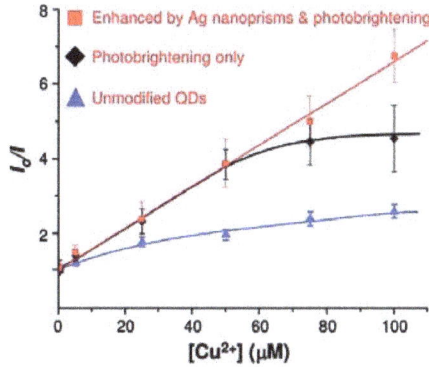

Figure 11. Stern−Volmer plot for the quenching of CdSe PL by Cu^{2+}. The solid red squares (□), black diamonds (◆), and blue triangles (▲) denote the CdSe QDs enhanced by both Ag nanoprisms and photobrightening, the photobrightened QDs, and the unmodified QDs, respectively. Note: the blue and black lines are added as a guide using fits to a third-order polynomial. The red line is a linear fit to the data. Reprinted from [109] with permission from Elsevier.

Table 3 summarizes all the metal ion sensors analyzed in this section.

Table 3. Luminescent films fabricated by LbL for ions detection.

Analyte	Sensitive Indicator	Sensing Film	Detection Range	LOD	Cross-Sensitivity	Ref.
Fe^{3+}, Hg^{2+}	HPTS	[PAH/PAA-HPTS]n	0–0.5 mM Fe^{3+} 0–1 mM Hg^{2+}	1.28 ppm Fe^{3+} 1.79 ppm Hg^{2+}	-	[100, 101]
Hg^{2+}	TPPS	(PDDA/PSS/PDDA/TPPS)	0–3.3×10^{-5} M	$<3.3 \times 10^{-8}$ M	Cd^{2+}, Pb^{2+}, Cu^{2+}	[102]
Hg^{2+}	$PPESO_3$	$(PDDA/PPESO_3)_3$	0–1 mM	10^{-7} M	Fe^{3+}, Al^{3+}	[104]
Fe^{3+}	PFPNa	$(PDAC/PFPNa)_1$	0–10 μM	10^{-7} M	-	[103]
Hg^{2+}	MSA-capped CdTe QDs	$(PDDA/QDs)_{10}$	0–1 μM	$<10^{-8}$ M	Cu^{2+}, Ag1+	[105]
Cu^{2+}, Hg^{2+}	MSA-capped CdTe QDs	$(PDDA/QDs)_5$	0–1 μM Cu^{2+} 0–0.5 μM Hg^{2+}	$<10^{-8}$ M Cu^{2+} $<5 \times 10^{-9}$ M Hg^{2+}	High concentrations of Ni^{2+}, Cr^{3+}, Au^{3+}, Ag^+	[106]
Hg^{2+}	MPA-capped CdTe QDs	(PDDA/QDs)5/PDDA/PSS/ PDDA/(QDs)5/BSA	0.01–1 μM	4.5×10^{-9} M	-	[107]
Hg^{2+}	Carbon dots	$(PEI/Carbon\ dots)_{1-6}$	0.01–2.69 μM for $(PEI/Carbon\ dots)_6$	10^{-8} M for $(PEI/Carbon\ dots)_6$	-	[108]
Cu^{2+}	(16-MHA) capped CdSe QDs	Ag NPs/(PDADMAC/PSS)/QDs	0–100 μM	5×10^{-9} M	-	[109]

3.2. LbL Luminescent Coatings for Dissolved Oxygen Sensing

In neutral solutions singlet oxygen ($^1O_2$) reacts with ascorbate (AscH-) producing H_2O_2, which quenches the fluorescent emission of CdTe QDs [111]. Taking advantage of this reaction, a singlet oxygen sensitive coating was fabricated by means of the LbL technique [112]: first, a glass slide was coated with the base layers $(PDDA/PAA)_3/PDDA$, onto which 10 bilayers of (CdTe QDs/PDDA) were deposited. With the aim of avoiding any interference between the CdTe QDs and the ascorbate, a spacer structure $(PDDA/PAA)_2/PDDA$ was introduced and, finally, two bilayers of PDDA/ascorbate were built. The sensing films were introduced in a phenylalanine solution, which produced singlet oxygen under UV illumination. Then, $^1O_2$ reacted with ascorbate producing H_2O_2, which etched the QDs surface, giving rise to their luminescence quenching, as is shown in Figure 12. This sensing structure detected $^1O_2$ concentration as low as 10^{-15} M. For each concentration, the intensity decreased for 5 min, when a steady stage was reached. This response time was thought to be due to the time required by H_2O_2 to etch the QDs [113].

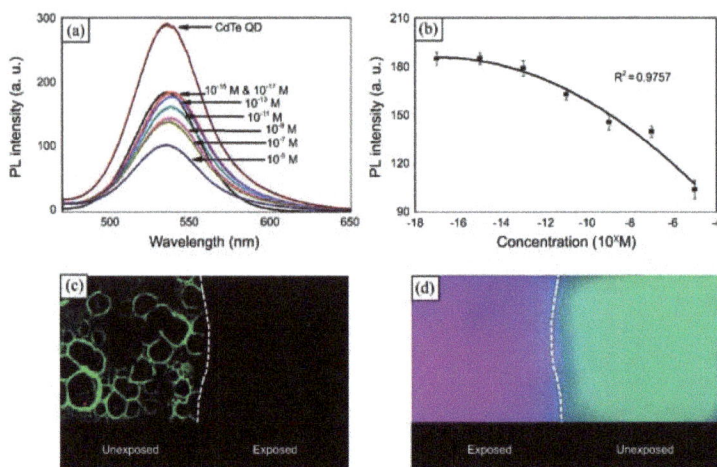

Figure 12. (**a**) Luminescent intensity of the film before and after incubation in solution of different phenylalanine concentrations; and (**b**) the corresponding calibration curve. After incubation and exposition to UV light, singlet oxygen is produced, which reacts with ascorbate, producing H_2O_2, which quenches the luminescence; (**c**) Microscopic image of the interface between a part exposed to phenylalanine (**right**) and a part unexposed (**left**); (**d**) UV image of the phenylalanine-exposed part (**left**), where only the UV light (excitation) is visible, and the unexposed part (**right**), where the luminescence (534 nm, green) is observable. Reprinted from [112] with permission from the Korean Chemical Society.

Other dissolved oxygen sensors were developed by utilizing ruthenium porphyrins as sensitive materials. For instance, Grant and coworkers fabricated a self-referenced optical fiber sensor based on a polymer/polymer-dye multilayer structure, by combining the oxygen sensitive porphyrin bis(2,2'-bipyridine)''-methyl-4-carboxybipyridine-ruthenium-N-succinimidyl-ester bis(hexafluoro-phosphate), Ru(bpy)$_2$(mcbpy), with PAH and the reference dye, FITC, with the same cationic polyelectrolyte [114]. The multilayer architecture (PAH-Ru(bpy)$_2$(mcbpy)/PSS)$_{10}$ + {PAH-FITC/PSS)$_5$ was built on the tip of a 400 μm-core optical fiber, which was connected to a two (200 μm) to one (400 μm) coupler. Under illumination at 450 nm, the sensing film exhibited two luminescent peaks, centered at 524 nm and 630 nm, which corresponded to the two dyes, FITC and Ru(bpy)$_2$(mcbpy) respectively. The fluorescence peak ratio (I_{630nm}/I_{524nm}) changed from 0.82 to 0.75 under dissolved oxygen concentration variations from 0 to 1400 mM. In a subsequent study [115], the number of dye layers was increased up to 15, but no enhancement of the sensor performance was observed.

A study of the adsorption of Ru(bpy) onto planar substrates by LbL was performed in [116]: this porphyrin was attempted to be assembled from a pure dye solution as well as from solutions of different dye-polyion concentrations. In the first case, the multilayer structure PEI/(PSS/PDDA)$_2$/(PSS/Ru(bpy))$_{20}$ was deposited onto glass slides. Despite Ru(bpy) being positively charged, it was observed that it was barely adsorbed to PSS and, what is more, it was desorbed when the substrates were immersed in the anionic solution. When mixing Ru(bpy) with PSS, sensing coatings with the structure PEI/(PSS/PDDA)$_2$/(PSS-Ru(bpy)/PDDA)$_{20}$ and different ratios (1:80, 1:40 and 1:20) of Ru(bpy):PSS were analyzed. As the Ru(bpy):PSS ratio increased, so did the fluorescence intensity. The fluorescence quenching of the sensor fabricated with the 1:20 (Ru(bpy):PSS) ratio exhibited a Stern–Volmer trend, being able to detect changes of less than 3% of the dissolved oxygen concentration in the range between 0 and 12 mg/L. This fact made it suitable for monitoring oxygen concentrations within biological environments. However, a further investigation [117] concluded that the best approach for adsorbing the luminescent dye to the substrate was not polyelectrolyte-dye mixing, but their covalent linkage: this bond prevented any kind of dye desorption when the substrate was immersed into the oppositely charged solution.

3.3. LbL Luminescent Coatings for Gaseous Oxygen Sensing

Another approach to immobilizing these ruthenium porphyrins into the multilayer structures consisted of the fabrication of the multilayer film and its further immersion in the dye solution, with the consequent diffusion of the indicator inside the coating. In [118], three kind of sensors were fabricated onto the tip of a 62.5 μm-core optical fiber. The first type consisted of a hygroscopic polymer membrane made of polyglutamic acid (PGA, anionic material) and poly-Lysine (cationic material). The second coating was a water absorbing polymer membrane composed of PAA (anionic polyelectrolyte) and PAH (cationic polyelectrolyte) and the third one, a porous composite membrane, was a multilayer structure of porous glass beads and PAH built onto a PAA layer. After the deposition of 50 bilayers of each structure onto the optical fiber tips, they were immersed in 80 mM Ru(bpy) solutions. When the sensors were illuminated at 450 nm, no phosphorescence was observed in the case of the water-absorbing polymer membrane, whereas the hygroscopic polymer membrane and the porous composite membrane exhibited a phosphorescent peak centered at 625 nm; the latter was the only one that was quenched when the sensor was exposed to 95–100% oxygen concentrations. Apart from the influence of the multilayer structure on the behavior of the sensors, the effect of the number of bilayers was also studied. It was found that a 125-bilayer structure did not show phosphorescence, which was attributed to the difficulty of introducing the ruthenium porphyrin in such a thick membrane. On the other hand, the five-bilayer sensor had a similar sensitivity to that of the 50-bilayers one, and it also demonstrated high resolution for low oxygen concentrations.

All the ruthenium porphyrins employed up to now are water-soluble, which facilitates their assembly by means of the LbL technique. As is well known, this construction method requires all the materials to be present in water solutions for their deposition onto the substrates, so the water insolubility of certain porphyrins can be an inconvenience. To overcome this fact and facilitate their deposition by the LbL technique, there exists the possibility of entrapping them into micelles [119]. Taking advantage of this method, the water-insoluble platinum(II)-5,10,15,20-tetrakis-(2,3,4,5,6-pentafluorphenyl) porphyrin, Pt-TFPP, was immobilized for the first time, employing the LbL technique in [120]: PAH was used as cationic polyelectrolyte and Sodium Dodecyl Sulfate (SDS) micelles, into which Pt-TFPP was entrapped, were employed as anionic counterpart. The multilayer coating formed by $(PAH/Pt-TFPP_{SDS})_{10}$ was built onto the tip of a 400 μm-core optical fiber, which was connected to a 200 μm-core bifurcated fiber. Under illumination at 390 nm, the intensity of the luminescence peak at 650 nm decreased as the oxygen concentration increased, exhibiting a linear Stern–Volmer plot in the whole range of oxygen concentrations. A comparative study of three different polymeric matrices was carried out in [49], where the structures $(PDDA/Pt-TFPP_{SDS})_{10}$, $(PEI/Pt-TFPP_{SDS})_{10}$, and $(PAH/Pt-TFPP_{SDS})_{10}$ were analyzed in detail. It was shown that the sensitivity was determined by the morphology of the coatings: the rougher the sensing film, the more sensitive the sensor, and the higher the range of oxygen concentrations able to detect the sensor. What is more, the sensors fabricated with PDDA and PEI did not exhibit linear calibration curves; their Stern–Volmer plots had two different quenching constants, indicating that the luminophore was not homogeneously distributed in the matrix. This fact can be seen in Figure 13 and in Table 4, where the calibration curves and the quenching constants of each sensor are displayed.

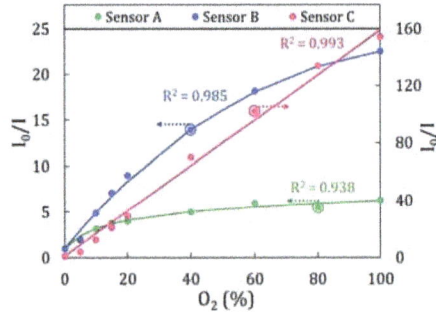

Figure 13. Calibration curves of (PDDA/Pt-TFPP$_{SDS}$)$_{10}$ (Sensor **A**), (PEI/Pt-TFPP$_{SDS}$)$_{10}$ (Sensor **B**), and (PAH/Pt-TFPP$_{SDS}$)$_{10}$ (Sensor **C**). Stern–Volmer plots of Sensors **A** and **B** are adjusted on the left axis, whereas that of Sensor **C** is adjusted on the right axis. Reprinted from [49] with permission from Elsevier.

Table 4. Quenching constants and calibration curves of the three oxygen sensors fabricated employing PDDA, PEI, or PAH as cationic polyelectrolytes. Data obtained from [48].

	Stern–Volmer Constants				Mathematical Model
	f_1	$K_{SV,1}$	f_2	$K_{SV,2}$	
[PDDA/Pt$_{SDS}$]$_5$	0.957	0.0898	0.043	0.0001	$\dfrac{I_0}{I} = \left(\dfrac{0.957}{1+0.0898 \cdot [O_2]} + \dfrac{0.043}{1+0.0001 \cdot [O_2]} \right)^{-1}$
[PEI/Pt$_{SDS}$]$_5$	0.9939	0.1526	0.0061	0.085	$\dfrac{I_0}{I} = \left(\dfrac{0.9939}{1+0.1526 \cdot [O_2]} + \dfrac{0.0061}{1+0.085 \cdot [O_2]} \right)^{-1}$
[PAH/Pt$_{SDS}$]$_5$	1	0.34	0	0	$\dfrac{I_0}{I} = 1 + 0.34 \cdot [O_2]$

With the aim of avoiding self-quenching and enhancing the sensitivity of the sensors, the spacing distance between the luminescent films was increased by introducing PAA layers between the cationic ones [48]. This fact not only affected the sensitivities of the sensors, but also determined the distribution of the luminophores inside the multilayer structure. For instance, in the cases fabricating the sensors with PDDA or PEI, a higher number of spacing layers was needed than in the case of the sensors built with PAH to obtain linear calibration curves and the highest sensitivities. This is shown in Figure 14: for the sensors fabricated with PDDA or PEI, the highest sensitivities are obtained when the luminescent films (Pt$_{SDS}$) are separated by five layers, P(+)/PAA/P(+)/PAA/P(+), where P(+) are PDDA and PEI, respectively. In the case of the sensors fabricated with PAH, only three spacing layers, PAH/PAA/PAH, are enough to reach the maximum sensitivity.

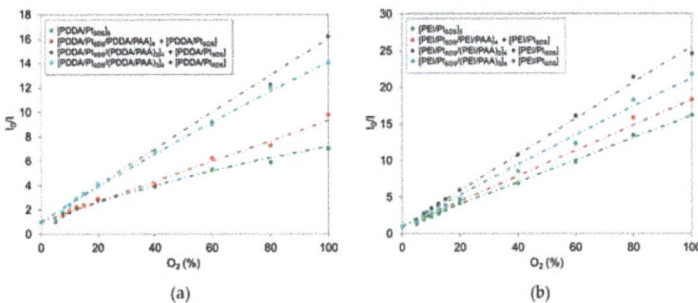

(a)

(b)

Figure 14. *Cont.*

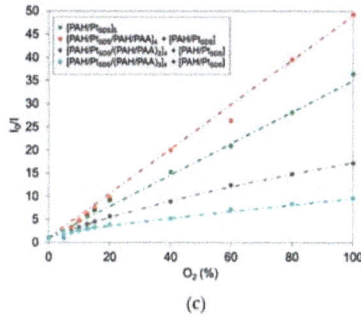

(c)

Figure 14. Stern–Volmer plots of the different sensors fabricated employing (**a**) PDDA, (**b**) PEI, and (**c**) PAH as cationic polyelectrolytes, and PAA as a spacer layer. In the cases of (**a**) PDDA and (**b**) PEI, the maximum of the sensitivity is achieved when luminescent layers are spaced by five layers of polyelectrolytes, whereas in the case of (**c**) PAH, only three layers of polyelectrolytes are necessary to achieve the maximum of the sensitivity. Reprinted from [48] with permission from Elsevier.

3.4. LbL Luminescent Films for Biosensing Applications

Li et al. used the multilayer structure $(PAH/CdTe\ QDs)_x(PAH/PSS)_3(PAH/GOD)_y$ to determine the concentration of blood glucose in real serum samples with good reproducibility and accuracy [113]; the GOD enzyme catalyzed the reaction between oxygen and glucose, producing H_2O_2, which generated defects on the surface of the QDs, quenching their luminescence. Under illumination at 380 m, the initial structure $(PAH/CdTe\ QDs)_{12}(PAH/PSS)_3(PAH/GOD)_3$ showed a luminescence peak centered at 630 nm. The luminescent properties of this coating were analyzed for different temperature and pH values ranging from 28 to 45 °C, and from 6 to 9, respectively. In the case of temperature, the largest quenching rate was obtained for 37 °C, whereas a pH value of 7.4 was chosen as optimal. Figure 15 shows the luminescence quenching of that structure at different temperatures (A), upon different glucose concentrations (B), and the absolute quenching rate of this structure (C).

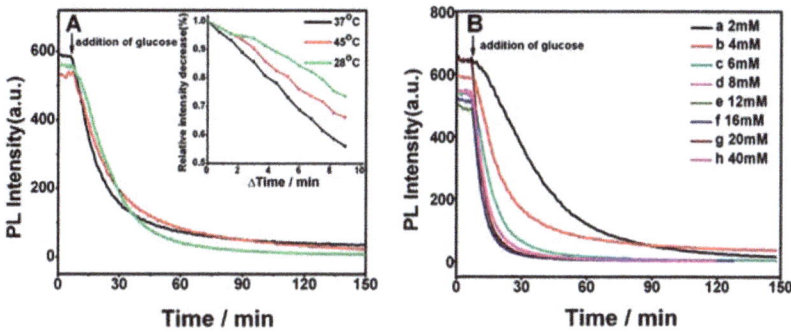

Figure 15. *Cont.*

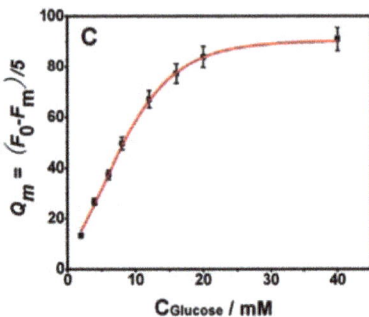

Figure 15. (A) Quenching of the luminescence peak centered at 630 nm of the multilayer structure $(PAH/CdTe)_{12}(PAH/PSS)_3(PAH/GOD)_3$ when it is exposed to a 4 mM glucose solution at different temperatures. The time-dependent luminescence intensity of that peak during the first 9 min of the reaction for each temperature is shown in the inset. **(B)** Luminescence quenching of the same film for different glucose concentrations: **(a)** 2, **(b)** 4, **(c)** 6, **(d)** 8, **(e)** 12, **(f)** 16, **(g)** 20, and **(h)** 40 mM over 150 min; **(C)** quenching rate (Q_m) of the sensor over 5 min as a function of the glucose concentration. F_0 and F_m correspond to the luminescence intensity in the absence (F_0) and presence (F_m) of glucose. All measurements were carried out in a 20 mM phosphate buffer at pH 7.4. Copyright: 2009, American Chemical Society.

Under those conditions (37 °C and pH 7.4), the response upon addition of 4 mM glucose of three different structures (as shown in Figure 16) was studied with the aim of optimizing the number of PAH/CdTeQDs and PAH/GOD bilayers. For a given number of PAH/GOD bilayers (in this case, three), it was observed that the quenching rate of the sensor decreased when the increase in the number of QDs layers was limited to 12. For this number of PAH/CdTe QDs bilayers, the quenching constant increased linearly with the number of enzyme layers as a consequence of the good permeability of the GOD layers towards glucose. The PAH/PSS$_3$ spacer was introduced in order to avoid any kind of influence of the GOD-glucose reaction on the QDs. The structure $(PAH/CdTe\ QDs)_{12}(PAH/PSS)_3(PAH/GOD)_5$ was chosen as optimal for determining the blood glucose concentration in serum samples: it exhibited linear luminescence quenching in the glucose concentration range between 0.5 and 16 mM, with 0.5 mM being the detection limit. Furthermore, no sample pre-treatment was needed.

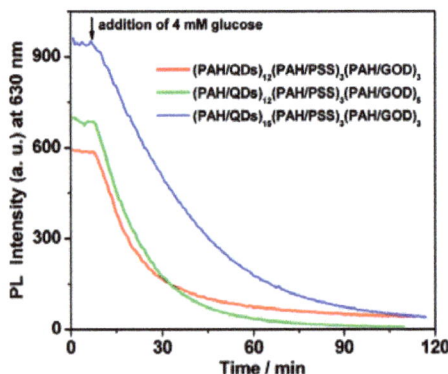

Figure 16. Luminescence quenching at 630 nm (λ_{ex} = 380 nm) when different structures of (PAH/CdTe QDs)$_x$(PAH/PSS)$_3$(PAH/GOD)$_y$ were exposed to 4 mM glucose. All measurements were carried out at 37 °C in a 20 mM phosphate buffer at pH 7.4. Copyright: 2009, American Chemical Society.

Since it was isolated, graphene and some related structures have been employed for diverse sensing applications [121]. Graphene oxide multilayer arrays were assembled by means of the LbL technique by Jung et al. [122] for the fabrication of aptasensor microarrays. These multilayers were prepared by assembling oppositely charged graphene oxide sheets: negatively charged ones (GO^-) were prepared by introducing COOH groups, while positively charged sheets (GO^+) were obtained thanks to the NH_2 groups. An aminated glass slide was coated with 10 bilayers of (GO^-/GO^+), and then a FAM-labeled thrombin aptamer was immobilized on them. The FAM luminescence, centered at 530 nm, was quenched by graphene oxide due to the high energy transfer between the dye and graphene. In the presence of thrombin, fluorescence was recovered owing to the high affinity between aptamers and thrombin. On the other hand, other analytes such as bovine serum albumin (BSA), streptavidin (STA), glucose, and human immunoglobulin (IgG) antibody did not alter the quenched fluorescence of FAM-aptamer-labeled GO multilayers, probing the high specificity of the aptamer-based sensor. For the particular case of 10 bilayers of (GO^-/GO^+) and 2 µM aptamer concentration, the fluorescence intensity from FAM was quenched over 85% and the detection limit for thrombin was 0.001 nM, exhibiting 30-fold higher sensitivity than the solution-based graphene FRET aptasensor [123]. Furthermore, this sensor was reused four times by simply cleaning it with distilled water.

4. Conclusions

It is obvious that luminescence has become a powerful detection mechanism for biological and environmental applications. This sensing method also takes advantage of the wide variety of sensitive luminophores that exists: fluoropolymers, water-soluble and non-soluble porphyrins, or semiconductor quantum dots, among others. These materials can be encapsulated in multilayer shells for their utilization in solution-based probes, or they can be entrapped inside nanostructured films and used as solid-state sensors. In both cases, the LbL nano-assembly technique permits us to tailor the properties of the sensors at the nanometer scale, making feasible the fabrication of custom-made devices that not only exhibit good photostability, but also high selectivity and sensitivity for almost any kind of sensing applications.

The remarkable characteristics of the sensors exposed in this study suggest that the combination of luminescence and the LbL nano-assembly technique is a promising approach for the fabrication of sensing devices for real applications. Environmental and biosensing purposes are probably the most encouraging fields owing to the facility for fabricating arrays of sensor capable of detecting several analytes with a single measurement. Furthermore, a real solution for sensing applications in hazardous environments can be obtained by combining luminescence and the Layer-by-Layer technique with the unique properties of optical fibers.

Acknowledgments: This work was supported by the Spanish State Research Agency (AEI) through the TEC2016-79367-C2-2-R project and the European Regional Development Fund (ERDF-FEDER). Nerea de Acha would also like to acknowledge her pre-doctoral fellowship (reference BES-2014-069692) funded by the Spanish Ministry of Economy and Competitiveness through the TEC2013-43679-R project.

Conflicts of Interest: The authors declare no conflict of interest.

References

1. Sun, C.; Chen, Y.; Zhang, G.; Wang, F.; Liu, G.; Ding, J. Multipoint Remote Methane Measurement System Based on Spectrum Absorption and Reflective TDM. *IEEE Photonics Technol. Lett.* **2016**, *28*, 2487–2490. [CrossRef]
2. Pasinszki, T.; Krebsz, M.; Tung, T.T.; Losic, D. Carbon nanomaterial based biosensors for non-invasive detection of cancer and disease biomarkers for clinical diagnosis. *Sensors* **2017**, *17*, 1919. [CrossRef] [PubMed]
3. Tan, C.H.; Tan, S.T.; Lee, H.B.; Ginting, R.T.; Oleiwi, H.F.; Yap, C.C.; Jumali, M.H.H.; Yahaya, M. Automated room temperature optical absorbance CO sensor based on In-doped ZnO nanorod. *Sens. Actuators B Chem.* **2017**, *248*, 140–152. [CrossRef]

4. Del Villar, I.; Arregui, F.J.; Zamarreño, C.R.; Corres, J.M.; Bariain, C.; Goicoechea, J.; Elosua, C.; Hernaez, M.; Rivero, P.J.; Socorro, A.B.; et al. Optical sensors based on lossy-mode resonances. *Sens. Actuators B Chem.* **2017**, *240*, 174–185. [CrossRef]

5. Qazi, H.H.; Bin Mohammad, A.B.; Akram, M. Recent Progress in Optical Chemical Sensors. *Sensors* **2012**, *12*, 16522–16556. [CrossRef] [PubMed]

6. Baldini, F.; Bacci, M.; Cosi, F.; Del Bianco, A. Absorption-based optical-fibre oxygen sensor. *Sens. Actuators B Chem.* **1992**, *7*, 752–757. [CrossRef]

7. Miki, H.; Matsubara, F.; Nakashima, S.; Ochi, S.; Nakagawa, K.; Matsuguchi, M.; Sadaoka, Y. A fractional exhaled nitric oxide sensor based on optical absorption of cobalt tetraphenylporphyrin derivatives. *Sens. Actuators B Chem.* **2016**, *231*, 458–468. [CrossRef]

8. Zhang, W.; Ye, W.; Wang, C.; Li, W.; Yue, Z.; Liu, G. Silver nanoparticle arrays enhanced spectral surface plasmon resonance optical sensor. *Micro Nano Lett.* **2014**, *9*, 585–587. [CrossRef]

9. Rivero, P.J.; Urrutia, A.; Goicoechea, J.; Matías, I.R.; Arregui, F.J. A Lossy Mode Resonance optical sensor using silver nanoparticles-loaded films for monitoring human breathing. *Sens. Actuators B Chem.* **2013**, *187*, 40–44. [CrossRef]

10. Marques, L.; Hernandez, F.U.; Korposh, S.; Clark, M.; Morgan, S.; James, S.; Tatam, R.P. Sensitive protein detection using an optical fibre long period grating sensor anchored with silica core gold shell nanoparticles. In Proceedings of the SPIE, Santander, Spain, 2–6 June 2014; Volume 9157.

11. Payne, S.J.; Fiore, G.L.; Fraser, C.L.; Demas, J.N. Luminescence oxygen sensor based on a ruthenium(II) star polymer complex. *Anal. Chem.* **2010**, *82*, 917–921. [CrossRef] [PubMed]

12. Bian, W.; Ma, J.; Liu, Q.; Wei, Y.; Li, Y.; Dong, C.; Shuang, S. A novel phosphorescence sensor for Co^{2+} ion based on Mn-doped ZnS quantum dots. *Luminescence* **2014**, *29*, 151–157. [CrossRef] [PubMed]

13. Ferrari, L.; Rovati, L.; Fabbri, P.; Pilati, F. Disposable fluorescence optical pH sensor for near neutral solutions. *Sensors* **2013**, *13*, 484–499. [CrossRef] [PubMed]

14. Lv, H.S.; Huang, S.Y.; Zhao, B.X.; Miao, J.Y. A new rhodamine B-based lysosomal pH fluorescent indicator. *Anal. Chim. Acta* **2013**, *788*, 177–182. [CrossRef] [PubMed]

15. Ross, D.; Gaitan, M.; Locascio, L.E. Temperature measurement in microfluidic systems using a temperature-dependent fluorescent dye. *Anal. Chem.* **2001**, *73*, 4117–4123. [CrossRef] [PubMed]

16. Song, C.; Zhi, A.; Liu, Q.; Yang, J.; Jia, G.; Shervin, J.; Tang, L.; Hu, X.; Deng, R.; Xu, C.; Zhang, G. Rapid and sensitive detection of β-agonists using a portable fluorescence biosensor based on fluorescent nanosilica and a lateral flow test strip. *Biosens. Bioelectron.* **2013**, *50*, 62–65. [CrossRef] [PubMed]

17. Yoshihara, T.; Murayama, S.; Tobita, S. Ratiometric molecular probes based on dual emission of a blue fluorescent coumarin and a red phosphorescent cationic iridium(III) complex for intracellular oxygen sensing. *Sensors* **2015**, *15*, 13503–13521. [CrossRef] [PubMed]

18. Zhang, H.; Zhang, G.; Xu, J.; Wen, Y.; Lu, B.; Zhang, J.; Ding, W. Novel highly selective fluorescent sensor based on electrosynthesized poly(9-fluorenecarboxylic acid) for efficient and practical detection of iron(III) and its agricultural application. *Sens. Actuators B Chem.* **2016**, *230*, 123–129. [CrossRef]

19. Xu, K.X.; Xie, X.M.; Kong, H.J.; Li, P.; Zhang, J.L.; Pang, X.B. Selective fluorescent sensors for malate anion using the complex of phenanthroline-based Eu(III) in aqueous solution. *Sens. Actuators B Chem.* **2014**, *201*, 131–137. [CrossRef]

20. Kuo, S.-Y.; Li, H.-H.; Wu, P.-J.; Chen, C.-P.; Huang, Y.-C.; Chan, Y.-H. Dual colorimetric and fluorescent sensor based on semiconducting polymer dots for ratiometric detection of lead ions in living cells. *Anal. Chem.* **2015**, *87*, 4765–4771. [CrossRef] [PubMed]

21. Chang, J.; Li, H.; Hou, T.; Li, F. Paper-based fluorescent sensor for rapid naked-eye detection of acetylcholinesterase activity and organophosphorus pesticides with high sensitivity and selectivity. *Biosens. Bioelectron.* **2016**, *86*, 971–977. [CrossRef] [PubMed]

22. Chao, M.R.; Hu, C.W.; Chen, J.L. Glass substrates crosslinked with tetracycline-imprinted polymeric silicate and CdTe quantum dots as fluorescent sensors. *Anal. Chim. Acta* **2016**, *925*, 61–69. [CrossRef] [PubMed]

23. Hale, Z.M.; Payne, F.P. Fluorescent sensors based on tapered single-mode optical fibres. *Sens. Actuators B Chem.* **1994**, *17*, 233–240. [CrossRef]

24. Shang, Z.B.; Wang, Y.; Jin, W.J. Triethanolamine-capped CdSe quantum dots as fluorescent sensors for reciprocal recognition of mercury(II) and iodide in aqueous solution. *Talanta* **2009**, *78*, 364–369. [CrossRef] [PubMed]

25. Kim, Y.; Chang, J.Y. Fabrication of a fluorescent sensor by organogelation: CdSe/ZnS quantum dots embedded molecularly imprinted organogel nanofibers. *Sens. Actuators B Chem.* **2016**, *234*, 122–129. [CrossRef]

26. Wang, H.; He, Y. Recent advances in silicon nanomaterial-based fluorescent sensors. *Sensors* **2017**, *17*, 268. [CrossRef] [PubMed]

27. Senkbeil, S.; Lafleur, J.P.; Jensen, T.G.; Kutter, J.P. Gold nanoparticle-based fluorescent sensor for the analysis of dithiocarbamate pesticides in water. In Proceedings of the 16th International Conference on Miniaturized Systems for Chemistry and Life Sciences, MicroTAS 2012, Okinawa, Japan, 28 October–1 November 2012.

28. Gilliard, R.J., Jr.; Iacono, S.T.; Budy, S.M.; Moody, J.D.; Smith, D.W., Jr.; Smith, R.C. Chromophore-derivatized semifluorinated polymers for colorimetric and turn-on fluorescent anion detection. *Sens. Actuators B Chem.* **2009**, *143*, 1–5. [CrossRef]

29. Hashemi, P.; Zarjani, R.A. A wide range pH optical sensor with mixture of Neutral Red and Thionin immobilized on an agarose film coated glass slide. *Sens. Actuators B Chem.* **2008**, *135*, 112–115. [CrossRef]

30. Ensafi, A.A.; Amini, M. A highly selective optical sensor for catalytic determination of ultra-trace amounts of nitrite in water and foods based on brilliant cresyl blue as a sensing reagent. *Sens. Actuators B Chem.* **2010**, *147*, 61–66. [CrossRef]

31. Magna, G.; Catini, A.; Kumar, R.; Palmacci, M.; Martinelli, E.; Paolesse, R.; di Natale, C. Conductive photo-activated porphyrin-ZnO nanostructured gas sensor array. *Sensors* **2017**, *17*, 747. [CrossRef] [PubMed]

32. Chu, C.S.; Chuang, C.Y. Highly sensitive fiber-optic oxygen sensor based on palladium tetrakis (4-carboxyphenyl)porphyrin doped in ormosil. *J. Luminescence* **2014**, *154*, 475–478. [CrossRef]

33. Roales, J.; Pedrosa, J.M.; Guillén, M.G.; Lopes-Costa, T.; Castillero, P.; Barranco, A.; González-Elipe, A.R. Free-base carboxyphenyl porphyrin films using a TiO_2 columnar matrix: Characterization and application as NO_2 sensors. *Sensors* **2015**, *15*, 11118–11132. [CrossRef] [PubMed]

34. Nivens, D.A.; Zhang, Y.; Angel, S.M. A fiber-optic pH sensor prepared using a base-catalyzed organo-silica sol–gel. *Anal. Chim. Acta* **1998**, *376*, 235–245. [CrossRef]

35. Sas, S.; Danko, M.; Bizovská, V.; Lang, K.; Bujdák, J. Highly luminescent hybrid materials based on smectites with polyethylene glycol modified with rhodamine fluorophore. *Appl. Clay Sci.* **2017**, *138*, 25–33. [CrossRef]

36. Asadpour-Zeynali, K.; Mollarasouli, F. A novel and facile synthesis of TGA-capped CdSe@Ag₂Se core-shell quantum dots as a new substrate for high sensitive and selective methyldopa sensor. *Sens. Actuators B Chem.* **2016**, *237*, 387–399. [CrossRef]

37. Lu, D.; Yang, L.; Tian, Z.; Wang, L.; Zhang, J. Core-shell mesoporous silica nanospheres used as Zn^{2+} ratiometric fluorescent sensor and adsorbent. *RSC Adv.* **2012**, *2*, 2783–2789. [CrossRef]

38. Properties and applications of proteins encapsulated within sol–gel derived materials. *Anal. Chim. Acta* **2002**, *461*, 1–36. [CrossRef]

39. Guan, W.; Zhou, W.; Lu, J.; Lu, C. Luminescent films for chemo- and biosensing. *Chem. Soc. Rev.* **2015**, *44*, 6981–7009. [CrossRef] [PubMed]

40. Ertekin, K.; Tepe, M.; Yenigül, B.; Akkaya, E.U.; Henden, H. Fiber optic sodium and potassium sensing by using a newly synthesized squaraine dye in PVC matrix. *Talanta* **2002**, *58*, 719–727. [CrossRef]

41. Brolo, A.G.; Kwok, S.C.; Moffitt, M.G.; Gordon, R.; Riordon, J.; Kavanagh, K.L. Enhanced fluorescence from arrays of nanoholes in a gold film. *J. Am. Chem. Soc.* **2005**, *127*, 14936–14941. [CrossRef] [PubMed]

42. McDonagh, C.; MacCraith, B.D.; McEvoy, A.K. Tailoring of Sol-Gel Films for Optical Sensing of Oxygen in Gas and Aqueous Phase. *Anal. Chem.* **1998**, *70*, 45–50. [CrossRef] [PubMed]

43. Chu, C.-S.; Lo, Y.-L. Fiber-optic carbon dioxide sensor based on fluorinated xerogels doped with HPTS. *Sens. Actuators B Chem.* **2008**, *129*, 120–125. [CrossRef]

44. Zhao, Z.; Lu, P.; Lam, J.W.Y.; Wang, Z.; Chan, C.Y.K.; Sung, H.H.Y.; Williams, I.D.; Ma, Y.; Tang, B.Z. Molecular anchors in the solid state: Restriction of intramolecular rotation boosts emission efficiency of luminogen aggregates to unity. *Chem. Sci.* **2011**, *2*, 672–675. [CrossRef]

45. Decher, G. Fuzzy Nanoassemblies: Toward Layered Polymeric Multicomposites. *Science* **1997**, *277*, 1232–1237. [CrossRef]

46. Elzbieciak, M.; Zapotoczny, S.; Nowak, P.; Krastev, R.; Nowakowska, M.; Warszyński, P. Influence of pH on the structure of multilayer films composed of strong and weak polyelectrolytes. *Langmuir* **2009**, *25*, 3255–3259. [CrossRef] [PubMed]

47. Yang, Y.-H.; Haile, M.; Park, Y.T.; Malek, F.A.; Grunlan, J.C. Super Gas Barrier of All-Polymer Multilayer Thin Films. *Macromolecules* **2011**, *44*, 1450–1459. [CrossRef]
48. De Acha, N.; Elosúa, C.; Matías, I.R.; Arregui, F.J. Enhancement of luminescence-based optical fiber oxygen sensors by tuning the distance between fluorophore layers. *Sens. Actuators B Chem.* **2017**, *248*, 836–847. [CrossRef]
49. De Acha, N.; Elosúa, C.; Martínez, D.; Hernáez, M.; Matías, I.R.; Arregui, F.J. Comparative study of polymeric matrices embedding oxygen-sensitive fluorophores by means of Layer-by-Layer nanoassembly. *Sens. Actuators B Chem.* **2017**, *239*, 1124–1133. [CrossRef]
50. Aslan, K.; Gryczynski, I.; Malicka, J.; Matveeva, E.; Lakowicz, J.R.; Geddes, C.G. Metal-enhanced fluorescence: An emerging tool in biotechnology. *Curr. Opin. Biotechnol.* **2005**, *16*, 55–62. [CrossRef] [PubMed]
51. Ariga, K.; Hill, J.P.; Ji, Q. Layer-by-layer assembly as a versatile bottom-up nanofabrication technique for exploratory research and realistic application. *Phys. Chem. Chem. Phys.* **2007**, *9*, 2319–2340. [CrossRef] [PubMed]
52. McShane, M.J.; Brown, J.Q.; Guice, K.B.; Lvov, Y.M. Polyelectrolyte Microshells as Carriers for Fluorescent Sensors: Loading and Sensing Properties of a Ruthenium-Based Oxygen Indicator. *J. Nanosci. Nanotechnol.* **2002**, *2*, 411–416. [CrossRef] [PubMed]
53. Arregui, F.J. *Sensors Based on Nanostructured Materials*; Springer: New York, NY, USA, 2009; ISBN 978-0-38-777752-8.
54. Duchesne, T.A.; Brown, J.Q.; Guice, K.B.; Lvov, Y.M.; McShane, M.J. Encapsulation and stability properties of nanoengineered polyelectrolyte capsules for use as fluorescent sensors. *Sens. Mater.* **2002**, *14*, 293–308.
55. McShane, M.J. Potential for glucose monitoring with nanoengineered fluorescent biosensors. *Diabetes Technol. Ther.* **2002**, *4*, 533–538. [CrossRef] [PubMed]
56. Bornhoeft, L.R.; Biswas, A.; McShane, M.J. Composite hydrogels with engineered microdomains for optical glucose sensing at low oxygen conditions. *Biosensors* **2017**, *7*, 8. [CrossRef] [PubMed]
57. Rivera-Gil, P.; Nazarenus, M.; Ashraf, S.; Parak, W.J. PH-sensitive capsules as intracellular optical reporters for monitoring lysosomal pH changes upon stimulation. *Small* **2012**, *8*, 943–948. [CrossRef] [PubMed]
58. Biswas, A.; Nagaraja, A.T.; McShane, M.J. Fabrication of nanocapsule carriers from multilayer-coated vaterite calcium carbonate nanoparticles. *ACS Appl. Mater. Interfaces* **2014**, *6*, 21193–21201. [CrossRef] [PubMed]
59. Zhao, Q.; Rong, X.; Chen, L.; Ma, H.; Tao, G. Layer-by-layer self-assembly xylenol orange functionalized CdSe/CdS quantum dots as a turn-on fluorescence lead ion sensor. *Talanta* **2013**, *114*, 110–116. [CrossRef] [PubMed]
60. McShane, M.; Ritter, D. Microcapsules as optical biosensors. *J. Mater. Chem.* **2010**, *20*, 8189–8193. [CrossRef]
61. Antipov, A.A.; Shchukin, D.; Fedutik, Y.; Petrov, A.I.; Sukhorukov, G.B.; Möhwald, H. Carbonate microparticles for hollow polyelectrolyte capsules fabrication. *Coll. Surf. A Physicochem. Eng. Asp.* **2003**, *224*, 175–183. [CrossRef]
62. Biswas, A.; Banerjee, S.; Gart, E.V.; Nagaraja, A.T.; McShane, M.J. Gold Nanocluster Containing Polymeric Microcapsules for Intracellular Ratiometric Fluorescence Biosensing. *ACS Omega* **2017**, *2*, 2499–2506. [CrossRef]
63. Zhang, G.; Shu, F.P.; Robinson, C.J. Design and characterization of a nano-encapsulated self-referenced fluorescent nitric oxide sensor for wide-field optical imaging. In Proceedings of the Annual International Conference of the IEEE Engineering in Medicine and Biology, Lyon, France, 22–26 August 2007.
64. Brown, J.Q.; Lvov, Y.M.; McShane, M.J. Nanoengineered polyelectrolyte microcapsules as fluorescent potassium ion sensors. In Proceedings of the Annual International Conference of the IEEE Engineering in Medicine and Biology, Houston, TX, USA, 23–26 October 2002; Volume 2.
65. Lee, D.; Rubner, M.F.; Cohen, R.E. Formation of nanoparticle-loaded microcapsules based on hydrogen-bonded multilayers. *Chem. Mater.* **2005**, *17*, 1099–1105. [CrossRef]
66. Marinakos, S.M.; Novak, J.P.; Brousseau, L.C., III; House, A.B.; Edeki, E.M.; Feldhaus, J.C.; Feldheim, D.L. Gold particles as templates for the synthesis of hollow polymer capsules. Control of capsule dimensions and guest encapsulation. *J. Am. Chem. Soc.* **1999**, *121*, 8518–8522. [CrossRef]
67. Donath, E.; Sukhorukov, G.B.; Caruso, F.; Davis, S.A.; Möhwald, H. Novel hollow polymer shells by colloid-templated assembly of polyelectrolytes. *Angew. Chem. Int. Ed.* **1998**, *37*, 2201–2205. [CrossRef]

68. Sadovoy, A.; Teh, C. Encapsulated biosensors for advanced tissue diagnostics. In *Woodhead Publishing Series in Biomaterials*; Meglinski, I., Ed.; Elsevier: Amsterdam, The Netherlands, 2015; pp. 321–330, ISBN 978-0-85-709662-3.

69. He, C.; Hu, Y.; Yin, L.; Tang, C.; Yin, C. Effects of particle size and surface charge on cellular uptake and biodistribution of polymeric nanoparticles. *Biomaterials* **2010**, *31*, 3657–3666. [CrossRef] [PubMed]

70. Sukhorukov, G.B.; Donath, E.; Lichtenfeld, H.; Knippel, E.; Knippel, M.; Budde, A.; Möhwald, H. Layer-by-layer self assembly of polyelectrolytes on colloidal particles. *Coll. Surf. A Physicochem. Eng. Asp.* **1998**, *137*, 253–266. [CrossRef]

71. Bertrand, P.; Jonas, A.; Laschewsky, A.; Legras, R. Ultrathin polymer coatings by complexation of polyelectrolytes at interfaces: Suitable materials, structure and properties. *Macromol. Rapid Commun.* **2000**, *21*, 319–348. [CrossRef]

72. Shiratori, S.S.; Rubner, M.F. pH-dependent thickness behavior of sequentially adsorbed layers of weak polyelectrolytes. *Macromolecules* **2000**, *33*, 4213–4219. [CrossRef]

73. Mendelsohn, J.D.; Barrett, C.J.; Chan, V.V.; Pal, A.J.; Mayes, A.M.; Rubner, M.F. Fabrication of microporous thin films from polyelectrolyte multilayers. *Langmuir* **2000**, *16*, 5017–5023. [CrossRef]

74. Zhang, G.; Shitole, P.S.; Pujari, R.A.; Charnani, V.S.; McShane, M.J.; Robinson, C.J. Intrinsic optical signal imaging of a ratiometric fluorescence oxygen nanosensor. In Proceedings of the 2005 3rd IEEE/EMBS Special Topic Conference on Microtechnology in Medicine and Biology, Oahu, HI, USA, 12–15 May 2005; Volume 2005.

75. Guice, K.B.; Lvov, Y.M.; McShane, M.J. Nanoengineered microcapsules for the fluorescent sensing of oxygen. In Proceedings of the 2nd Joint Conference of the IEEE Engineering in Medicine and Biology Society and the Biomedical Engineering Society, Annual International Conference of the IEEE Engineering in Medicine and Biology, Houston, TX, USA, 23–26 October 2002; Volume 2.

76. Zhi, Z.L.; Khan, F.; Pickup, J.C. Multilayer nanoencapsulation: A nanomedicine technology for diabetes research and management. *Diabetes Res. Clin. Pract.* **2013**, *100*, 162–169. [CrossRef] [PubMed]

77. Saxl, T.; Khan, F.; Matthews, D.R.; Zhi, Z.L.; Rolinski, O.; Ameer-Beg, S.; Pickup, J. Fluorescence lifetime spectroscopy and imaging of nano-engineered glucose sensor microcapsules based on glucose/galactose-binding protein. *Biosens. Bioelectron.* **2009**, *24*, 3229–3234. [CrossRef] [PubMed]

78. Acquah, I.; Roh, J.; Ahn, D.J. Dual-fluorophore silica microspheres for ratiometric acidic pH sensing. *Macromol. Res.* **2017**, *25*, 950–955. [CrossRef]

79. Afkhami, A.; Soltani-Felehgari, F.; Madrakian, T.; Ghaedi, H.; Rezaeival, M. Fabrication and application of a new modified electrochemical sensor using nano-silica and a newly synthesized Schiff base for simultaneous determination of Cd^{2+}, Cu^{2+} and Hg^{2+} ions in water and some foodstuff samples. *Anal. Chim. Acta* **2013**, *771*, 21–30. [CrossRef] [PubMed]

80. Duchesne, T.A.; Brown, J.Q.; Guice, K.B.; Nayak, S.R.; Lvov, Y.M.; McShane, M.J. Nanoassembled fluorescent microshells as biochemical sensors. In Proceedings of the SPIE—The International Society for Optical Engineering, San Jose, CA, USA, 23 May 2002; Volume 4624.

81. Stouwdam, J.W.; Van Veggel, F.C.J.M. Near-infrared Emission of Redispersible Er^{3+}, Nd^{3+}, and Ho^{3+} Doped LaF_3 Nanoparticles. *Nano Lett.* **2002**, *2*, 733–737. [CrossRef]

82. Kömpe, K.; Borchert, H.; Storz, J.; Lobo, A.; Adam, S.; Möller, T.; Haase, M. Green-Emitting $CePO_4$:Tb/$LaPO_4$ Core-Shell Nanoparticles with 70 % Photoluminescence Quantum Yield. *Angew. Chem. Int. Ed.* **2003**, *42*, 5513–5516. [CrossRef] [PubMed]

83. Xiang, Y.; Xu, X.-Y.; He, D.-F.; Li, M.; Liang, L.-B.; Yu, X.-F. Fabrication of rare-earth/quantum-dot nanocomposites for color-tunable sensing applications. *J. Nanopart. Res.* **2011**, *13*, 525–531. [CrossRef]

84. Pujari, R.A.; Shitole, P.S.; Charnani, V.S.; McShane, M.J.; Robinson, C.J. Wide-field extrinsic optical signal imaging of fluorescence potassium sensors. In Proceedings of the 2006 3rd IEEE International Symposium on Biomedical Imaging: From Nano to Macro, Arlington, VA, USA, 6–9 April 2006; Volume 2006.

85. Brown, J.Q.; Guice, K.B.; McShane, M.J. Internally-Referenced Chemical Transducers Using Molecular Probes Assembled on Fluorescent Nanoparticles. Proceedings of The IEEE SENSORS, Toronto, ON, Canada, 22–24 October 2003; Volume 2.

86. Brown, J.Q.; McShane, M.J. Core-referenced ratiometric fluorescent potassium ion sensors using self-assembled ultrathin films on europium nanoparticles. *IEEE Sens. J.* **2005**, *5*, 1197–1205. [CrossRef]

87. Kazakova, L.I.; Shabarchina, L.I.; Anastasova, S.; Pavlov, A.M.; Vadgama, P.; Skirtach, A.G.; Sukhorukov, G.B. Chemosensors and biosensors based on polyelectrolyte microcapsules containing fluorescent dyes and enzymes. *Anal. Bioanal. Chem.* **2013**, *405*, 1559–1568. [CrossRef] [PubMed]

88. Guice, K.B.; Caldorera, M.E.; McShane, M.J. Nanoscale internally referenced oxygen sensors produced from self-assembled nanofilms on fluorescent nanoparticles. *J. Biomed. Opt.* **2005**, *10*, 064031. [CrossRef] [PubMed]

89. Pickup, J.C.; Zhi, Z.-L.; Khan, F.; Saxl, T.E. Nanomedicine in diabetes management: Where we are now and where next. *Expert Rev. Endocrinol. Metab.* **2010**, *5*, 791–794. [CrossRef]

90. Chinnayelka, S.; McShane, M.J. Competitive binding assays in microcapsules as "smart tattoo" biosensors. In Proceedings of the IEEE SENSORS, Irvine, CA, USA, 30 October–3 November 2005; Volume 2005.

91. Zamarreño, C.R.; Bravo, J.; Goicoechea, J.; Matias, I.R.; Arregui, F.J. Response time enhancement of pH sensing films by means of hydrophilic nanostructured coatings. *Sens. Actuators B Chem.* **2007**, *128*, 138–144. [CrossRef]

92. Chang-Yen, D.A.; Gale, B.K. An Integrated Optical Glucose Sensor Fabricated Using PDMS Waveguides on a PDMS Substrate. In Proceedings of the SPIE—The International Society for Optical Engineering, San Jose, CA, USA, 23 December 2003; Volume 5345.

93. Chu, C.-S.; Chu, S.-W. Optical oxygen sensor based on time-resolved fluorescence. In Proceedings of the SPIE—The International Society for Optical Engineering, Jeju, Korea, 1 July 2015; Volume 9655.

94. Chan, W.H.; Yang, R.H.; Wang, K.M. Development of a mercury ion-selective optical sensor based on fluorescence quenching of 5,10,15,20-tetraphenylporphyrin. *Anal. Chim. Acta* **2001**, *444*, 261–269. [CrossRef]

95. Lee, D.; Jung, J.; Bilby, D.; Kwon, M.S.; Yun, J.; Kim, J. A novel optical ozone sensor based on purely organic phosphor. *ACS Appl. Mater. Interfaces* **2015**, *7*, 2993–2997. [CrossRef] [PubMed]

96. Yusoff, N.H.; Salleh, M.M.; Yahaya, M. Enhanced the Performance of Fluorescence Gas Sensor of Porphyrin Dye by Using TiO_2 Nanoparticles. *Adv. Mater. Res.* **2008**, *55–57*, 269–272. [CrossRef]

97. Ali, R.; Lang, T.; Saleh, S.M.; Meier, R.J.; Wolfbeis, O.S. Optical sensing scheme for carbon dioxide using a solvatochromic probe. *Anal. Chem.* **2011**, *83*, 2846–2851. [CrossRef] [PubMed]

98. Safavi, A.; Bagheri, M. Novel optical pH sensor for high and low pH values. *Sens. Actuators B Chem.* **2003**, *90*, 143–150. [CrossRef]

99. Wencel, D.; MacCraith, B.D.; McDonagh, C. High performance optical ratiometric sol–gel-based pH sensor. *Sens. Actuators B Chem.* **2009**, *139*, 208–213. [CrossRef]

100. Lee, S.-H.; Kumar, J.; Tripathy, S.K. Fluorescence quenching based thin film sensors employing electrostatic layer-by-layer self-assembly. In Proceedings of the Materials Research Society Symposium, Boston, MA, USA, Fall 2000; pp. 891–896.

101. Lee, S.H.; Kumar, J.; Tripathy, S.K. Thin film optical sensors employing polyelectrolyte assembly. *Langmuir* **2000**, *16*, 10482–10489. [CrossRef]

102. Caselli, M. Porphyrin-based electrostatically self-assembled multilayers as fluorescent probes for mercury(ii) ions: A study of the adsorption kinetics of metal ions on ultrathin films for sensing applications. *RSC Adv.* **2015**, *5*, 1350–1358. [CrossRef]

103. Qin, C.; Cheng, Y.; Wang, L.; Jing, X.; Wang, F. Phosphonate-Functionalized Polyfluorene as a Highly Water-Soluble Iron(III) Chemosensor. *Macromolecules* **2008**, *41*, 7798–7804. [CrossRef]

104. Li, Y.; Huang, H.; Li, Y.; Su, X. Highly sensitive fluorescent sensor for mercury (II) ion based on layer-by-layer self-assembled films fabricated with water-soluble fluorescent conjugated polymer. *Sens. Actuators B Chem.* **2013**, *188*, 772–777. [CrossRef]

105. Li, Y.; Huang, H.; Li, Y.; Su, X. Sensitive Hg (II) ion detection by fluorescent multilayer films fabricated with quantum dots. *Sens. Actuators B Chem.* **2009**, *139*, 476–482. [CrossRef]

106. Ma, Q.; Ha, E.; Yang, F.; Su, X. Synchronous determination of mercury (II) and copper (II) based on quantum dots-multilayer film. *Anal. Chim. Acta* **2011**, *701*, 60–65. [CrossRef] [PubMed]

107. Yang, F.; Ma, Q.; Yu, W.; Su, X. Naked-eye colorimetric analysis of Hg^{2+} with bi-color CdTe quantum dots multilayer films. *Talanta* **2011**, *84*, 411–415. [CrossRef] [PubMed]

108. Gonçalves, H.M.; Duarte, A.J.; Davis, F.; Higson, S.P.; da Silva, J.C.E. Layer-by-layer immobilization of carbon dots fluorescent nanomaterials on single optical fiber. *Anal. Chim. Acta* **2012**, *735*, 90–95. [CrossRef] [PubMed]

109. Chan, Y.-H.; Chen, J.; Liu, Q.; Wark, S.E.; Son, D.H.; Batteas, J.D. Ultrasensitive Copper(II) Detection Using Plasmon-Enhanced and Photo-Brightened Luminescence of CdSe Quantum Dots. *Anal. Chem.* **2010**, *82*, 3671–3678. [CrossRef] [PubMed]

110. Chan, Y.-H.; Chen, J.; Wark, S.E.; Skiles, S.L.; Son, D.H.; Batteas, J.D. Using patterned arrays of metal nanoparticles to probe plasmon enhanced luminescence of CdSe quantum dots. *ACS Nano* **2009**, *3*, 1735–1744. [CrossRef] [PubMed]

111. Kramarenko, G.G.; Hummel, S.G.; Martin, S.M.; Buettner, G.R. Ascorbate reacts with singlet oxygen to produce hydrogen peroxide. *Photochem. Photobiol.* **2006**, *82*, 1634–1637. [CrossRef] [PubMed]

112. Ahmed, S.R.; Koh, K.; Kang, N.L.; Lee, J. Highly Sensitive Fluorescent Probes for the Quantitative Determination of Singlet Oxygen ($^1O_2$). *Bull. Korean Chem. Soc.* **2012**, *33*, 1608–1612. [CrossRef]

113. Li, X.; Zhou, Y.; Zheng, Z.; Yue, X.; Dai, Z.; Liu, S.; Tang, Z. Glucose biosensor based on nanocomposite films of CdTe quantum dots and glucose oxidase. *Langmuir* **2009**, *25*, 6580–6586. [CrossRef] [PubMed]

114. Grant, P.S.; Kaul, S.; Chinnayelka, S.; McShane, M.J. Fiber Optic Biosensors Comprising Nanocomposite Multilayered Polymer and Nanoparticle Ultrathin Films. In Proceedings of the Annual International Conference of the IEEE Engineering in Medicine and Biology, Cancun, Mexico, 17–21 September 2003; Volume 4.

115. Grant, P.; Barnidge, M.; McShane, M. Spectroscopic Fiber Probes for Chemical Sensing Based on LbL Self Assembled Ultra-Thin Films. In Proceedings of the IEEE SENSORS, Toronto, ON, Canada, 22–24 October 2003; Volume 2, pp. 895–898.

116. Chang-Yen, D.A.; Lvov, Y.; McShane, M.J.; Gale, B.K. Electrostatic self-assembly of a ruthenium-based oxygen sensitive dye using polyion–dye interpolyelectrolyte formation. *Sens. Actuators B Chem.* **2002**, *87*, 336–345. [CrossRef]

117. Grant, P.S.; McShane, M.J. Development of multilayer fluorescent thin film chemical sensors using electrostatic self-assembly. *IEEE Sens. J.* **2003**, *3*, 139–146. [CrossRef]

118. Ban, S.; Hosoki, A.; Nishiyama, M.; Seki, A.; Watanabe, K. Optical fiber oxygen sensor using layer-by-layer stacked porous composite membranes. In Proceedings of the SPIE—The International Society for Optical Engineering, San Francisco, CA, USA, 18 April 2016; Volume 9754.

119. Su, F.; Alam, R.; Mei, Q.; Tian, Y.; Youngbull, C.; Johnson, R.H.; Meldrum, D.R. Nanostructured oxygen sensor—Using micelles to incorporate a hydrophobic platinum porphyrin. *PLoS ONE* **2012**, *7*, e33390. [CrossRef] [PubMed]

120. Elosua, C.; De Acha, N.; Hernaez, M.; Matias, I.R.; Arregui, F.J. Layer-by-Layer assembly of a water-insoluble platinum complex for optical fiber oxygen sensors. *Sens. Actuators B Chem.* **2015**, *207*, 683–689. [CrossRef]

121. Tung, T.T.; Nine, M.J.; Krebsz, M.; Pasinszki, T.; Coghlan, C.J.; Tran, D.N.H.; Losic, D. Recent Advances in Sensing Applications of Graphene Assemblies and Their Composites. *Adv. Funct. Mater.* **2017**, in press. [CrossRef]

122. Jung, Y.K.; Lee, T.; Shin, E.; Kim, B.-S. Highly tunable aptasensing microarrays with graphene oxide multilayers. *Sci. Rep.* **2013**, *3*, 3367. [CrossRef] [PubMed]

123. Chang, H.; Tang, L.; Wang, Y.; Jiang, J.; Li, J. Graphene fluorescence resonance energy transfer aptasensor for the thrombin detection. *Anal. Chem.* **2010**, *82*, 2341–2346. [CrossRef] [PubMed]

MDPI

St. Alban-Anlage 66

4052 Basel

Switzerland

Tel. +41 61 683 77 34

Fax +41 61 302 89 18

www.mdpi.com

Sensors Editorial Office

E-mail: sensors@mdpi.com

www.mdpi.com/journal/sensors